Foundation Engineering Analysis and Design

Foundation Engineering
Analysis and Design

An-Bin Huang and Hai-Sui Yu

CRC Press
Taylor & Francis Group
Boca Raton London New York

CRC Press is an imprint of the
Taylor & Francis Group, an **informa** business

CRC Press
Taylor & Francis Group
6000 Broken Sound Parkway NW, Suite 300
Boca Raton, FL 33487-2742

International Standard Book Number-13: 978-1-138-72079-4 (Paperback)
978-1-138-72078-7 (Hardback)

Visit the Taylor & Francis Web site at
http://www.taylorandfrancis.com

and the CRC Press Web site at
http://www.crcpress.com

Printed and bound in the United States of America by
Edwards Brothers Malloy on sustainably sourced paper

This book is dedicated to Chang-Lin Ni (An-Bin Huang's wife) and Xiu-Li Guan (Hai-Sui Yu's wife) for the many sacrifices they made in supporting our academic careers.

Contents

List of figures

List of tables

Preface

This textbook on foundation engineering covers both analysis and design. This book covers in detail all key issues related to the analysis, design, and construction of shallow and deep foundations as well as stability analysis and mitigation of slopes.

On the analytical side, the presentation is unique in the sense that it progressively introduces fundamental critical state soil mechanics and plasticity theories, such as plastic limit analysis and cavity expansion theories, before leading into the theories of foundation, lateral earth pressure, and slope stability analysis. This book prepares students for the more sophisticated nonlinear elastic-plastic analyses in foundation engineering that are commonly used in engineering practice.

On the engineering side, this book introduces up-to-date construction and testing methods used in practice. The connection between theory and practice is emphasized. Through the applications of Excel, design, and analytical methods commonly used in engineering practice can now be taught in classes.

This book is primarily for senior undergraduate students studying civil engineering and graduate students in geotechnical engineering, and can also serve as a reference book for practicing engineers.

The content is divided into nine chapters. Chapter 1 describes basic soil behavior, the concept of critical state soil mechanics, and plasticity theory, which are relevant for foundation analysis. Chapter 2 introduces limit analysis and cavity expansion theory that can be used to determine the bearing capacity of foundations. Chapter 3 details the nature of subsurface exploration and presents a range of techniques for determining the ground conditions and soil properties. Chapter 4 covers key theoretical methods for shallow foundation analysis in terms of bearing capacity and settlement. Chapter 5 introduces the theories of lateral earth pressure, followed by the design of three types of earth-retaining structures: concrete retaining walls, mechanically stabilized earth-retaining walls, and sheet pile walls. Chapter 6 deals with the basic concept and procedures related to the design of braced excavations. Because of the differences in construction procedures between driven piles and bored piles, two chapters are used to describe the issues related to deep foundations. Chapter 7 concentrates on static analysis and design of deep foundations, where most of the design methods can be shared between driven and bored piles. Chapter 8 elaborates on the construction and testing of deep foundations where it is necessary to deal with these issues separately between driven and bored piles because of the significant differences in their construction procedures. The nature and various types of available slope stability analysis and mitigation methods are introduced in Chapter 9.

All chapters include numerous examples to demonstrate the application of various analytical and design methods. Most of the examples are coupled with Excel solutions that can be downloaded by the instructors and students. With Excel, students can easily explore the effects of various parameters on the results of computations. Many lengthy equations such as those related to lateral earth pressure or soil stress increase due to foundation loads can be easily computed with Excel. Some of the charts and tables typically included in foundation engineering textbooks can now be eliminated. With the help of Excel, problems such as those related to methods of slices in slope stability analysis, wave equation analysis, and laterally loaded piles using p–y method can also be solved as explained by many examples in this book. It should be emphasized that these Excel programs are strictly provided as a learning tool and should not be used in engineering practice.

About the Authors

Dr. An-Bin Huang is a retired professor at National Chiao Tung University, Taiwan. He worked as a practicing geotechnical engineer for many years in Chicago before entering academia. He received the 1995 Distinguished Research Award from the Ministry of Science and Technology of Taiwan and was the seventh James K. Mitchell Lecturer of the International Society of Soil Mechanics and Geotechnical Engineering.

Professor Hai-Sui Yu is Pro-Vice-Chancellor at the University of Leeds, UK, and is a Fellow of the Royal Academy of Engineering. Prizes awarded to him include the Chandra Desai Medal of the International Association for Computer Methods and Advances in Geomechanics in 2008, the Telford Medal of the Institution of Civil Engineers London in 2000, the Trollope Medal of the Australian Geomechanics Society in 1998, an Overseas Master awarded by the Chinese Ministry of Education in 2013, and the Outstanding Contributions Medal of the International Association for Computer Methods and Advances in Geomechanics in 2014. He was the first James K. Mitchell Lecturer of the International Society of Soil Mechanics and Geotechnical Engineering.

Chapter 1

Soil behavior and critical state soil mechanics

1.1 INTRODUCTION

The main purpose of a foundation is to ensure that the load from the superstructure is properly transmitted to the ground. Retaining walls are built to hold the soil behind it. Civil engineering projects often involve construction through an uneven terrain, or creation of uneven ground due to excavation where the stability of the natural or man-made slope is of concern. The above subjects, as diverse as they may appear, have one thing in common—they all involve unbalanced loading conditions. The analysis or design is to assure that the ground mass remains stable under the given or expected loading conditions and the built structure performs as required. Traditionally, the analysis has been divided into two broad categories, as follows.

1.1.1 Safety analysis

The safety analysis can also be considered as a limit state analysis. The goal of this analysis is to determine the ratio between the ultimate resistance (capacity) over the given or expected load (demand) that the ground mass or the ground-supporting structure can offer. This ratio is referred to as the factor of safety (FS). Obviously, to maintain stability, an FS larger than 1 is required. This part of the analysis is traditionally done using the limit equilibrium method. In this method, the soil or ground mass around the foundation is divided by a series of slip surfaces into a number of sectors or blocks. The soil is assumed to be rigid outside of the slip surfaces. Ground displacement is allowed only along the slip surfaces as an assembly of rigid blocks. Despite of the fact that the key parameters are in terms of stresses, the analysis is based on the equilibrium of forces acting on the slip surfaces. Shear strength of the involved soil is the key parameter in the safety analysis.

1.1.2 Performance analysis

This part of the analysis deals with predicting the ground displacement under the expected loading conditions. A structure may be stable with FS > 1 but the settlement may be excessive, and that renders the structure not usable. For this type of analysis, we need to consider the stress–strain relationship of the ground material. Compressibility or deformability is the main parameter required in the performance analysis.

Before the advent of computers and numerical methods, a constant compressibility was often used in the performance analysis to facilitate hand calculations. Or, the stress–strain or load–displacement relationship was assumed to be linear. With the help of computers, we can consider non-linear stress–strain relationships, recoverable (elastic) and permanent (plastic) straining of the ground material in our analysis. In this case, the distinction between safety and performance analyses becomes less obvious. Regardless of its details, the analysis is based on our understanding of soil behavior. When pore water is involved, the analysis often needs to consider the "effective stress" and seepage of pore water.

1.1.3 Soil behavior under the framework of critical state soil mechanics

We know from soil mechanics that the soil shear strength generally increases with density. For a clay with the same current overburden stress, the undrained shear strength and stiffness increase with overconsolidation ratio (OCR). There is a trend of pore pressure and volumetric strain developments for a soil element under undrained or drained shearing. In other words, soil strength and compressibility behaviors are interrelated. It would therefore be very helpful if we can integrate the parameters related to strength and compressibility into a common framework. The critical state soil mechanics (CSSM) offers a good effective stress-based framework to serve this purpose. CSSM has also been used as a framework for some of the commonly used constitutive models to analyze non-linear and elasto-plastic soil behaviors. Because of these factors, the authors chose to start this book with a review on soil behavior and introduction to critical state soil mechanics. The critical state concept has validity in relation to the following two separate bodies of engineering experience:

- It gives a simple working model that, as we will see in the remainder of this chapter, provides a rational basis for the discussion of soil strength and compressibility parameters based on simple soil physical property tests; this simple model is valid with the same accuracy as those widely used parameters. The critical state concept is thus well adapted for the traditional "safety + performance" type of engineering analysis and yet offers a rigorous theoretical background.
- The critical state concept forms an integral part of more sophisticated constitutive models such as Cam clay and CASM (described in Section 1.7), and as such it has validity in relation to advanced laboratory tests and complex numerical simulations. From this point of view, the critical state soil mechanics is arguably the best tool that can lead us to the sophisticated non-linear and elasto-plastic analysis in engineering practice.

The objectives of this chapter are as follows:

- Review of the basic theories of effective stress and consolidation of cohesive soils that were covered in the class of soil mechanics (Sections 1.2 and 1.3).
- Introduction to critical state soil mechanics and review of the stress–strain and volumetric/pore pressure behavior of clay and sand under the framework of critical state soil mechanics (Sections 1.4 and 1.5).
- Description of the concept of state parameter and relationships between critical state model and soil engineering properties that include compressibility and shear strength (Section 1.6).
- A brief introduction to the theory of plasticity and soil stress–strain constitutive models that involve the concept of critical state soil mechanics (Section 1.7).

1.2 EFFECTIVE STRESS AND SOIL PROPERTIES

Soils form much of the ground on which foundations are built, and this book is mainly concerned with foundation engineering in soils. Many different types of soils exist but depending on particle sizes and their distribution, soil may be classified into two broad categories: fine-grained (or cohesive) and coarse-grained (or granular) soils. Clay and sand are good examples of these two types of soils.

In general, soils are a three-phase material including solids and voids filled with air and water (i.e., unsaturated soils). In most practical situations, however, good estimates can be made by simply treating soils as a two-phase material (i.e., solids and water or solids and air). The void ratio e, defined as the ratio of the volume of voids to the volume of soil solids, is an important property for describing the density state of a soil mass. The water content w, defined as the ratio of the mass of water to the mass of soil solids, is another widely used soil physical property. For fully saturated soils, the void ratio, e and water content, w can be related by $e = wG_s$ where G_s is the specific gravity of soil particles. As will be demonstrated later, specific volume $\nu = 1 + e$ is also widely used as a density state parameter (Schofield and Wroth, 1968).

One of the key achievements in the early development of soil mechanics was the realization that the mechanical behavior of a fully saturated soil was found to be dependent on the difference between an applied total stress (pressure) and the pore water pressure within the soil (Terzaghi, 1943). This important finding is known as the *effective stress principle*. It is noted that the difference of the applied total pressure and pore water pressure reflects the stress acting upon solid particles in soils and is therefore called the effective stress.

The key mechanical properties of soils that are essential for foundation design include compressibility, strength, and permeability (or hydraulic conductivity). It must be stressed that these soil properties are not generally constant but their values depend on the mechanical state of soils during loading as defined by both void ratios and effective stress levels. This was the key motivation behind the development of CSSM at Cambridge University (Schofield and Wroth, 1968), which has since underpinned many of the advances in soil mechanics and foundation engineering.

It is also very important to note that soil mechanical properties are dependent on the condition of water drainage during loading (i.e., drained or undrained). In particular, soil strength and stiffness properties are significantly different for the same soil tested under fully drained and undrained conditions. Therefore it is essential in foundation engineering that a distinction is made between drained and undrained calculations. In general, for most practical problems involving construction in granular soils such as sand and gravel the loading may usually be assumed to be drained, while for construction involving clays the loading may be assumed to be undrained. However, there are exceptions. For example, when sands or granular soils are subjected to rapid loading such as earthquakes, it may be more appropriate to assume the problem to be of undrained loading.

1.3 ONE-DIMENSIONAL CONSOLIDATION AND TIME EFFECT

In foundation engineering, any design would have to ensure stability (i.e., preventing failure) and serviceability (i.e., acceptable settlement and deformation). While soils will deform under external loading, the response will take time. This is particularly true for saturated clay soils, as their permeability is low and excess pore pressure generated upon loading takes much longer to dissipate when compared with saturated sandy soils where it is much quicker for excess pore pressure to dissipate. Procedures for laboratory consolidation tests, their interpretations for the estimation of the magnitude, and time rate of foundation settlement are well established. The following sections provide a review of these procedures.

1.3.1 The consolidation test

Once an external load is applied to a saturated clay soil, this external load is initially carried by water, leading to an increase of excess pore pressure. With time, the excess pore pressure will reduce and the external load is gradually transmitted to soil skeleton. The time-dependent process during which a soil layer responds to external compression is commonly termed the *process of consolidation*. The most common laboratory tests used to measure such compressibility of a soil are conducted either in one dimension (also known as a consolidometer, where the lateral movement of soil samples is not permitted) or in three dimensions (also known as a triaxial cell). However, the simpler, one-dimensional consolidometer tests were used much earlier in geotechnical engineering to measure soil compressibility for estimating consolidation settlement of foundations over saturated clay soils. Of course, this one-dimensional approach is only an approximation of real foundation engineering problems, but has proved to be relevant for soils under a foundation loading in which the lateral movement of soils may be assumed to be much smaller than the vertical settlement.

Figure 1.1 shows the basic setup of a consolidation test using the consolidometer. A soil sample is placed inside a rigid ring, with one porous stone at the top and one at the bottom of the soil. The vertical stress is increased in steps. After each increment, the load is kept constant until the deformation practically ceases. The deformation at different elapsed times following the load increase is recorded throughout each load increment. The procedure is continued until the desired total vertical stress is reached. An unloading and reloading is often also carried out to measure the consolidation performance of an overconsolidated clay.

The results of consolidation tests are usually plotted in the spaces of $e - \sigma'_v$ or $e - \log \sigma'_v$, where e = void ratio and σ'_v = effective vertical stress. The vertical axis can also be plotted as the specific volume ($v = 1 + e$) as in critical state soil mechanics. An example of a consolidation test result presented in an $e - \log \sigma'_v$ space is given in Figure 1.2. In the linear plot, the consolidation test result has a curved $e - \sigma'_v$ relationship but when plotted in a semi-log space as shown in Figure 1.2, $e - \log \sigma'_v$ relationship can be approximately two segments of straight lines with respective slopes of C_c and C_s, as shown in Figure 1.2.

1.3.2 Preconsolidation stress

From the one-dimensional consolidation test results, we can estimate the preconsolidation stress, σ'_p, which is defined as the maximum past effective overburden stress to which the

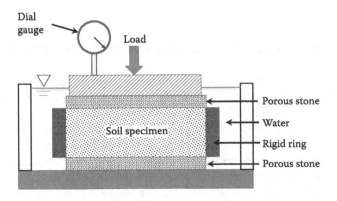

Figure 1.1 Schematic diagram of the consolidation test.

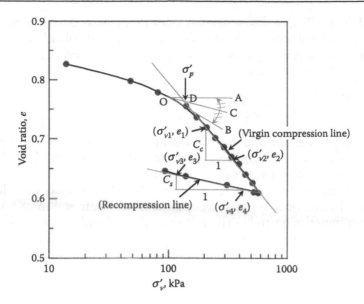

Figure 1.2 Typical results of a consolidation test.

soil sample has been subjected. It can be determined using a graphical procedure proposed by Casagrande (1936), which involves a number of simple steps:

a. Estimate the point O on the consolidation curve that has the sharpest curvature.
b. Draw a horizontal line OA.
c. Draw a line OB that is tangent to the consolidation line.
d. Draw a line OC that bisects the angle AOB.
e. Extend the straight portion of the consolidation line to intersect with OC. The pressure corresponding to the intersect point D will be taken as the preconsolidation pressure σ'_p.

A soil deposit is considered normally consolidated (NC), if its current effective overburden stress, σ'_{vo}, is equal to the preconsolidation stress, σ'_p. It is overconsolidated (OC) if σ'_{vo} is less than σ'_p. For a given soil, the ratio of its σ'_p over σ'_{vo} is called the overconsolidation ratio (OCR = σ'_p/σ'_{vo}). The outermost curve connecting the points when the specimen is in the highest void ratios for any given effective overburden pressure will be referred to as the virgin compression line. The curve connecting the points when the specimen is unloaded from a given preconsolidation pressure will be referred to as the recompression or swelling line. The term "line" is used because in semi-log space, the plots are closer to straight lines than curves. The straight lines will also justify the use of a constant compressibility index, as explained below.

C_c is known as the compression index that represents the compressibility of soils in normally consolidated state. C_c can be determined by arbitrarily choosing two points (σ'_{v1}, e_1) and (σ'_{v2}, e_2) on the virgin compression line (see Figure 1.2) where σ'_{v1} and σ'_{v2} are all greater than σ'_p, and

$$C_c = \frac{(e_1 - e_2)}{(\log \sigma'_{v2} - \log \sigma'_{v1})} = \frac{(e_1 - e_2)}{\log (\sigma'_{v2}/\sigma'_{v1})} \tag{1.1}$$

C_s is called the swelling or recompression index that reflects the compressibility of the soils in overconsolidated state or when the soil is in the stage of unloading or reloading. C_s can be determined by arbitrarily choosing two points (σ'_{v3}, e_3) and (σ'_{v4}, e_4) on the recompression line (see Figure 1.2), and

$$C_s = \frac{(e_3 - e_4)}{(\log \sigma'_{v4} - \log \sigma'_{v3})} = \frac{(e_3 - e_4)}{\log (\sigma'_{v4}/\sigma'_{v3})} \tag{1.2}$$

Often the value of the swelling index is assumed to be 5%–10% of the compression index. Typical values of C_s range from 0.015 to 0.035. Clays with low plasticity and low OCR tend to have lower C_s (Holtz et al., 2010).

If one-dimensional consolidation test results are plotted in the space of $e - \ln p'$ or $v - \ln p'$ where $p' =$ effective mean normal stress, as in critical state soil mechanics, then the slopes of virgin compression and swelling lines would be λ and κ, respectively. As shown in Wood (1990), they may be related to C_c and C_s approximately as follows:

$$C_c = \lambda \ln 10 \cong 2.3\lambda \tag{1.3}$$

$$C_s \approx \kappa \ln 10 \cong 2.3\kappa \tag{1.4}$$

1.3.3 Primary consolidation settlement calculation

Immediately after the stress increase $\Delta\sigma$ in a clay layer, the stress increase is completely taken by the equal increase of excess pore pressure $\bar{u}$. Thus initially $\Delta\sigma = \bar{u}$ and the effective vertical stress σ'_{vo} within the clay layer remains the same. With time, $\bar{u}$ dissipates and eventually $\bar{u} = 0$ as the primary consolidation completes, and at that time $\Delta\sigma$ is completely transferred to effective vertical stress. The process of transferring the applied stress increase from pore pressure to effective stress as the pore pressure dissipates is called primary consolidation. At the end of primary consolidation, the effective vertical stress $\sigma'_v = \sigma'_{vo} + \Delta\sigma$. Coupled with the increase of σ'_v is a change (decrease) of void ratio Δe. Settlement due to primary consolidation of a clay layer in the ground due to an increase in the vertical effective stress $\Delta\sigma$ can be estimated using the results from the one-dimensional consolidation tests of the clay sample. The basic equation of the vertical settlement ΔH of a layer of thickness of H linking to the change of void ratio is derived from the strain definition as follows:

$$\Delta H = \frac{\Delta e}{1 + e_o} H \tag{1.5}$$

where
$e_o =$ initial void ratio of the clay prior to the stress increase

The change (reduction) of void ratio caused by an increase of the effective vertical stress from σ'_{vo} to σ'_v depends on the relative values of the initial effective vertical stress, the final effective vertical stress, and the preconsolidation pressure σ'_p. In other words, we need to establish if the soil layer in the ground is normally consolidated or overconsolidated.

Figure 1.3 shows three possible cases for which the following equations can be used to determine the void ratio reduction, respectively:

a. The case of normal consolidation when $\sigma'_{vo} = \sigma'_p$

$$\Delta e = C_c \log \frac{\sigma'_v}{\sigma'_{vo}} \tag{1.6}$$

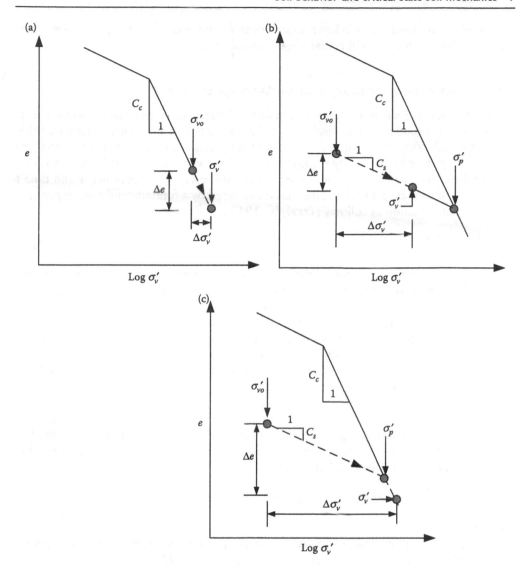

Figure 1.3 Three possible cases for consolidation settlement calculation. (a) Case I: Normally consolidated, $\sigma'_{vo} = \sigma'_p$. (b) Case II: Overconsolidated, $\sigma'_{vo} < \sigma'_v \leq \sigma'_p$. (c) Case III: Over-consolidated, $\sigma'_{vo} < \sigma'_p \leq \sigma'_v$.

b.   The case of overconsolidation when $\sigma'_{vo} < \sigma'_v \leq \sigma'_p$

$$\Delta e = C_s \, \log\frac{\sigma'_v}{\sigma'_{vo}} \tag{1.7}$$

c.   The case of overconsolidation followed by normal consolidation when $\sigma'_{vo} < \sigma'_p \leq \sigma'_v$

$$\Delta e = C_s \log\frac{\sigma'_p}{\sigma'_{vo}} + C_c \log\frac{\sigma'_v}{\sigma'_p} \tag{1.8}$$

If the soil layer is very thick, then it is a normal practice to subdivide it into a number of thinner layers for calculation in order to determine the total settlement. This is because the vertical effective stress for each layer is not a constant but varies with depth, and its average value over each layer will be used in the calculation.

1.3.4 Time rate of primary consolidation settlement

So far, the total settlement or compression of soil layers can be estimated using the one-dimensional consolidation test results from the laboratory. It was noted earlier that this settlement will take time, as it is due to the gradual dissipation of excess pore pressure from the clay layer generated by the application of the vertical stress. It is therefore fundamental to understand the relationship between the excess pore pressure $\bar{u}$ and time t during the consolidation process. This can be described by a fundamental time-dependent consolidation equation as follows (Terzaghi, 1943):

$$\frac{\partial \bar{u}}{\partial t} = c_v \frac{\partial^2 \bar{u}}{\partial z^2} \tag{1.9}$$

where c_v is the coefficient of consolidation that is treated as a material constant for simplicity of solution. It has been shown that typical values of c_v for various clays range approximately from 10^{-4} to 5×10^{-3} cm^2/s (Holtz and Kovacs, 1981).

Once we know the appropriate boundary conditions for the initial excess pore pressure, the above equation can be used to determine the excess pore pressure distribution over a soil layer at any time (Taylor, 1948; Schofield and Wroth, 1968).

For the usual setup in the one-dimensional consolidation test, simulating a clay layer sandwiched between two sand layers as shown in Figure 1.4, we have the following boundary conditions:

a. The excess pore pressure generated due to the external stress increase is assumed to dissipate immediately to zero at both the bottom and top boundaries due to the highly permeable sand at the boundaries. That is:

$$\bar{u} = 0; \quad \frac{\partial \bar{u}}{\partial z} = 0 \quad \text{for } z = 0 \quad \text{and} \quad z = 2H_{dr} \tag{1.10}$$

where

H_{dr} = maximum drainage distance; for a clay layer drained at top and bottom, H_{dr} equals half of the layer thickness H, or $2H_{dr} = H$.

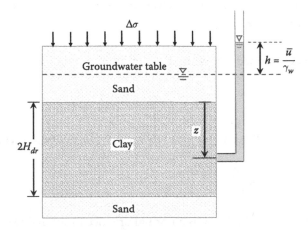

Figure 1.4 Boundary conditions for consolidation solution.

b. The initial excess pore pressure is assumed to be

$$\bar{u} = \bar{u}_0 \quad \text{for } t = 0 \tag{1.11}$$

With the boundary conditions described by Equations 1.10 and 1.11, Equation 1.9 can be solved to arrive at the following analytical solution for normalized excess pore pressure as a function of time t and depth z:

$$\frac{\bar{u}(t)}{\bar{u}_0} = \sum_{n=0}^{n=\infty} \left[\frac{2}{N} \sin\left(\frac{Nz}{H_{dr}} \right) \exp\left(-N^2 T_v \right) \right] \tag{1.12}$$

where
  $\bar{u}(t) =$ excess pore pressure at a given time t and depth z
  $\bar{u}_0 =$ initial pore pressure

$$N = \frac{(2n+1)\pi}{2} \tag{1.13}$$

in which n is an integer, and time factor is defined as

$$T_v = \frac{C_v t}{H_{dr}^2} \tag{1.14}$$

where
  $H_{dr} =$ maximum drainage distance

Following Equation 1.12, at a given depth z, and time represented by T_v, the degree of consolidation, U_z is

$$U_z = 1 - \frac{\bar{u}(t)}{\bar{u}_0} \tag{1.15}$$

Equation 1.12 can be plotted graphically in Figure 1.5, where the degree of consolidation-normalized excess pore pressure varies with time factor and depth.

In foundation engineering design, we are often more interested in how much consolidation has taken place for the entire clay layer at a given time period after the external load application. This can be indicated by an average degree of consolidation U for the entire clay layer that is defined by

$$U = \frac{\Delta H(t)}{\Delta H_{\max}} = \frac{\int_0^{2H_{dr}} \bar{u}_0 \, dz - \int_0^{2H_{dr}} \bar{u}(t) dz}{\int_0^{2H_{dr}} \bar{u}_0 \, dz} = 1 - \sum_{n=0}^{n=\infty} \left[\frac{2}{N^2} \exp\left(-N^2 T_v \right) \right] \tag{1.16}$$

where $U = 0$ indicates no consolidation and $U = 1$ implies full consolidation.

According to Terzaghi (1943), Equation 1.16 may be approximated by the following simpler equations:

a. For $0 \leq T_v \leq 0.217;\ 0 \leq U \leq 0.526$, we have

$$U = \sqrt{\frac{4T_v}{\pi}} \tag{1.17}$$

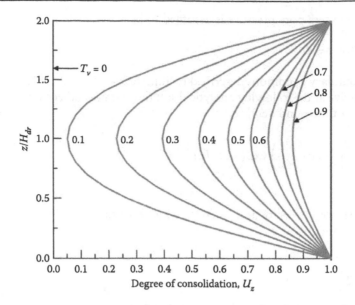

Figure 1.5 One-dimensional consolidation solutions.

b. For $0.217 < T_v \leq 1.781;\ 0.526 < U \leq 1$, we have

$$U = 1 - 10^{[1.0718(1.781 - T_v) - 2]} \tag{1.18}$$

Alternatively, Schofield and Wroth (1968) derived the following approximate solution for the average degree of consolidation:

a. For $0 \leq T_v \leq (1/12)$, we have

$$U = \sqrt{\frac{4T_v}{3}} \tag{1.19}$$

b. For $T_v > (1/12)$, we have

$$U = 1 - \frac{2}{3}\exp\left(\frac{1}{4} - 3T_v\right) \tag{1.20}$$

As shown by Schofield and Wroth (1968), Terzaghi's one-dimensional consolidation theory, represented by Equations 1.16 or 1.19 and 1.20, agrees well with the results obtained in the laboratory on specimens of saturated, remolded Gault clay.

EXAMPLE 1.1

Given

A laboratory consolidation test was carried out on a clay sample. The height of the clay specimen is 25.4 mm with drainage on both sides. Assume that the preconsolidation pressure of the clay sample before consolidation test was less than 150 kPa. When the effective vertical pressure applied reached 150 kPa, the void ratio was 0.96. The void ratio at end of consolidation/compression was 0.88 when the vertical effective pressure increased from 150 to 220 kPa. It was found that the time required for the sample to reach 30% consolidation was 3 min. In a swelling/unloading test following the consolidation test, the vertical effective stress was reduced from 220 to 150 kPa and the void ratio at the end of swelling was 0.9.

Required

a. The compression index of the clay
b. The swelling index of the clay
c. The height of the clay sample at the end of compression
d. The coefficient of consolidation of the clay
e. The final height of the clay sample at the end of swelling

Solution

a. According to Equation 1.1, the compression index can be determined as follows:

$$C_c = \frac{(e_1 - e_2)}{\log(\sigma'_{v2}/\sigma'_{v1})} = \frac{0.96 - 0.88}{\log(220/150)} = 0.48 \tag{1.1}$$

b. According to Equation 1.2, the swelling index can be obtained as follows:

$$C_s = \frac{(e_3 - e_4)}{\log(\sigma'_{v4}/\sigma'_{v3})} = \frac{0.9 - 0.88}{\log(220/150)} = 0.12 \tag{1.2}$$

c. According to Equation 1.5, the height reduction during compression is

$$\Delta H = \frac{\Delta e}{1 + e_0} H = \frac{0.96 - 0.88}{1 + 0.96} \times 25.4\,\text{mm} = 1\,\text{mm} \tag{1.5}$$

So the height of the clay sample at end of compression $= 25.4 - 1 = 24.4$ mm.

d. According to Equation 1.17, the time factor for degree of consolidation of 30% can be calculated as follows:

$$T_v = \frac{\pi U^2}{4} = \frac{3.1415 \times 0.3^2}{4} = 0.07 \tag{1.17}$$

According to Equation 1.14, the coefficient of consolidation is

$$c_v = \frac{T_v (H_{dr})^2}{t} = \frac{0.07 \times (25.4/2)^2}{3} = 3.80\,\text{mm}^2/\text{min} \tag{1.14}$$

e. According to Equation 1.5, the height increase during swelling is

$$\Delta H = \frac{\Delta e}{1 + e_0} H = \frac{0.9 - 0.88}{1 + 0.88} \times 24.4 \approx 0.26\,\text{mm} \tag{1.5}$$

So the height of the clay sample at end of swelling $= 24.4 + 0.26 \approx 24.7$ mm or $= 24.66$ mm.

EXAMPLE 1.2

Given

Consider a normally consolidated clay layer of soil in the field with a thickness of 1.2 m with both upper and lower layers of sands. The clay is the same as that considered in Example 1.1, so that the compression index is $C_c = 0.48$ and the coefficient of consolidation is $c_v = 3.76$ mm^2/min. The initial void ratio of clay was 0.96 before the construction of a foundation on the ground surface when the in-situ effective vertical pressure due to self-weight was 150 kPa. After construction of the foundation, the average vertical effective pressure on the clay layer became 300 kPa.

Required

a. The maximum consolidation settlement of the clay layer
b. The time required for 90% of the consolidation settlement to take place

Solution

a. According to Equations 1.5 and 1.6, the maximum consolidation settlement due to the construction of the foundation is

$$\Delta H = \frac{\Delta e}{1 + e_0} H = \frac{C_c H}{1 + e_0} \log \frac{\sigma_v'}{\sigma_{v0}'} = \frac{0.48 \times 1.2}{1 + 0.96} \times \log \frac{300}{150} = 0.0885 \text{ m} = 88.5 \text{ mm}$$

$$(1.5)(1.6)$$

b. According to Equation 1.18, the time factor for degree of consolidation of 90% can be calculated as follows:

$$T_v = -\frac{1}{1.0718} \log (1 - U) - 0.085 = -0.933 \times \log (1 - 0.9) - 0.085 = 0.848$$

$$(1.18)$$

According to Equation 1.14, the time required for 90% of consolidation to take place is

$$t = \frac{T_v (H_{dr})^2}{C_v} = \frac{0.848 \times (1200 \text{ mm}/2)^2}{3.76 \text{ mm}^2/\text{min}} = 81{,}191 \text{ min} \approx 56.4 \text{ days} \qquad (1.14)$$

1.3.5 Settlement due to secondary consolidation

Figure 1.6 describes a correlation between void ratio and elapsed time in log scale after a stress increase is applied to a clay sample in the one-dimensional consolidation test. It has been observed that the void ratio continues to decrease even after the primary consolidation is completed and excess pore pressure has fully dissipated. This continued consolidation after the excess pore pressure is completely dissipated is called the secondary consolidation. We typically use t_{100} to represent the time when primary consolidation completes. The corresponding void ratio at the end of primary consolidation is e_p. The relationship between e and $\log t$ beyond t_{100}, or in secondary compression, is approximately linear. We define a secondary compression index, C_α according to the slope of the secondary compression line as follows:

$$C_\alpha = \frac{e_1 - e_2}{\log t_2 - \log t_1} = \frac{\Delta e}{\log (t_2/t_1)} \qquad (1.21)$$

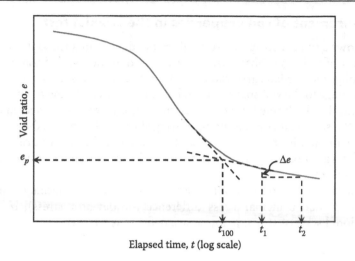

Figure 1.6 Correlation between void ratio and elapsed time.

Table 1.1 Empirical relationships to estimate C_c, C_s, and C_α

Empirical relationship	Reference
$C_c = 0.7(w_{LL} - 0.1)$	Skempton (1944)
$C_s \approx 0.15 \sim 0.25 C_c$	Wroth and Wood (1978)
$\dfrac{C_\alpha}{C_c} \approx 0.04 \pm 0.01$ (inorganic clays and silts)	Mesri and Godlewski (1977)
$\dfrac{C_\alpha}{C_c} \approx 0.05 \pm 0.01$ (organic clays and silts)	Mesri and Godlewski (1977)
$\dfrac{C_\alpha}{C_c} \approx 0.075 \pm 0.01$ (peats)	Mesri and Godlewski (1977)

Note
w_{LL} = liquid limit in decimal.

Table 1.1 shows a series of empirical relationships that can be used to estimate C_c, C_s, and C_α for cohesive soils.

1.4 STRESS–STRAIN BEHAVIOR AND CRITICAL STATE OF CLAYS

It is well known that the stress–strain and strength behavior of soils can be affected by shearing mode (e.g., triaxial, simple shear or plane strain shearing), strain rate and drainage conditions (i.e., drained or undrained), and principal stress direction, among others. A large number of testing devices have been developed and applied to measure the stress–strain and strength behavior of soils. Among these available methods, the triaxial test has been the most widely used to study the mechanical behavior of soils in the laboratory. It can be used to give stress–strain and strength behavior for different initial soil conditions (e.g., current state of stress and stress history) and stress paths (e.g., drained or undrained). The following sections use triaxial test as a basis for discussion. Readers interested in other types of testing methods are referred to Mayne et al. (2009).

1.4.1 Measurement of soil properties in the triaxial test

Figure 1.7 shows a typical setup of a triaxial testing device. In a triaxial test, a cylindrical soil specimen confined in a rubber membrane is placed inside a sealed triaxial cell. The soil specimen is subjected to a lateral or radial stress, σ_r (i.e., cell pressure) and vertical or axial stress, σ_a. Drainage to the soil specimen can be controlled via drainage lines and valves as shown in Figure 1.7. In drained triaxial tests, the drainage valves connected to the burette are open and the specimen volume change during shearing is measured by monitoring the water level fluctuation in the burette. In undrained triaxial tests, all drainage valves are closed and the variation of pore water pressure in the specimen is measured using a pressure transducer.

To facilitate the discussion of triaxial tests, we often use the terms of mean effective normal stress, p' and principal stress difference (or deviator stress), q. In a general three-dimensional state of stress, p' and q are defined as follows:

$$p' = \frac{(\sigma'_1 + \sigma'_2 + \sigma'_3)}{3} \tag{1.22}$$

where

σ'_1, σ'_2, and σ'_3 = major, intermediate, and minor effective principal stress

In axial compression triaxial tests, there are only two independent principal stresses; therefore

$$p' = \frac{(\sigma'_a + 2\sigma'_r)}{3} \tag{1.22a}$$

where

σ'_a = effective axial stress applied to the specimen = σ'_1
σ'_r = effective radial stress applied to the specimen = σ'_3

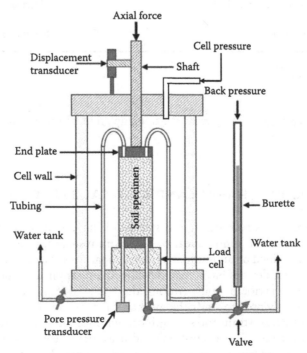

Figure 1.7 A typical triaxial test device setup.

and

$$q = \sigma_1 - \sigma_3 \tag{1.23}$$

Similarly, in the case of triaxial test,

$$q = \sigma_a - \sigma_r \tag{1.23a}$$

where

σ_1 and σ_3 = major and minor principal stress
σ_a = axial stress applied to the specimen
σ_r = radial stress applied to the specimen

Occasionally, we loosely call q shear stress. The corresponding shear strain, ε_q and volumetric strain, ε_p are (Schofield and Wroth, 1968; Wood, 1990):

$$\varepsilon_q = \frac{2(\varepsilon_1 - \varepsilon_3)}{3} = \frac{2(\varepsilon_a - \varepsilon_r)}{3} \tag{1.24}$$

and

$$\varepsilon_p = \varepsilon_1 + \varepsilon_2 + \varepsilon_3 = \varepsilon_a + 2\varepsilon_r \tag{1.25}$$

where

ε_1, ε_2, and ε_3 = strain in major, intermediate, and minor stress direction
ε_a = strain in the axial direction
ε_r = strain in the radial direction

A triaxial test is often divided into two stages. The first stage consolidates the specimen, and the second stage shears the specimen by inducing a principal stress difference to the specimen.

1.4.1.1 The consolidation stage

The specimen can be isotropically consolidated with $\sigma_r/\sigma_a = 1$ during consolidation or anisotropically consolidated where $\sigma_r/\sigma_a \neq 1$ during consolidation. As a special case, the specimen can be K_o consolidated where σ_r is adjusted to maintain zero ε_r while σ_a is increasing. Unloading of σ_r and σ_a can be applied upon reaching a designated consolidation stress to create overconsolidation to the soil specimen. The drainage valves connected to the burette are open (see Figure 1.7) to allow flow of pore water in and out of the specimen during consolidation. For cohesive soil samples retrieved from the field (details of soil sampling are discussed in Chapter 3) or compacted in laboratory, we may also conduct triaxial tests without going through the consolidation stage. In this case, the triaxial test is called unconsolidated.

1.4.1.2 The shearing stage

The soil specimen can be sheared in drained or undrained condition. For a drained test, shearing is conducted slowly enough that the shear-induced excess pore pressure Δu is dissipated completely (i.e., $\Delta u = 0$). In drained test, the soil specimen is allowed to undergo volumetric strain ε_p. On the other hand, in undrained shearing, the soil volume remains constant (i.e., $\varepsilon_p = 0$), but Δu can change. Independent of the choice of drainage conditions, the specimen shearing can be conducted by a combination of variations in σ_a and σ_r from their respective value at the end of consolidation.

A few types of commonly used triaxial tests are described below:

1. Isotropically consolidated, undrained axial compression test (CIUC)—the specimen is consolidated under a stress condition where $\sigma_r/\sigma_a = 1$ and then sheared undrained by keeping $\sigma_r =$ constant while σ_a increases.
2. Isotropically consolidated, drained axial compression test (CIDC)—the specimen is consolidated under $\sigma_r/\sigma_a = 1$ and then sheared in drained condition by keeping $\sigma_r =$ constant while σ_a increases.
3. Anisotropically consolidated, undrained axial compression test (CAUC)—the specimen is consolidated under $\sigma_r/\sigma_a \neq 1$ and then sheared undrained by keeping $\sigma_r =$ constant while σ_a increases.
4. Anisotropically consolidated, drained axial compression test (CADC)—the specimen is consolidated under $\sigma_r/\sigma_a \neq 1$ and then sheared in drained condition by keeping $\sigma_r =$ constant while σ_a increases.
5. The shearing can also be conducted by axial extension where $\sigma_r =$ constant while σ_a decreases. In this case, the above four types of triaxial tests are called CIUE, CIDE, CAUE, and CADE, respectively.
6. The specimen is not consolidated; shearing is conducted by axial compression in undrained condition. The test is then called UU test.

1.4.2 Stress–strain behavior of clays

Figures 1.8 through 1.10 describe the characteristics of typical triaxial axial compression test results on clays. For drained test on a normally consolidated specimen, the stress–strain curve tends to be of strain-hardening (i.e., q increases with axial strain ε_a monotonically), as indicated in Figure 1.8, and there is a contraction in specimen volume ($\varepsilon_p < 0$),

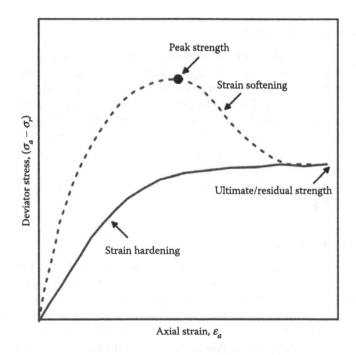

Figure 1.8 Characteristics of $(\sigma_a - \sigma_r) - \varepsilon_a$ relationships from triaxial tests on clays.

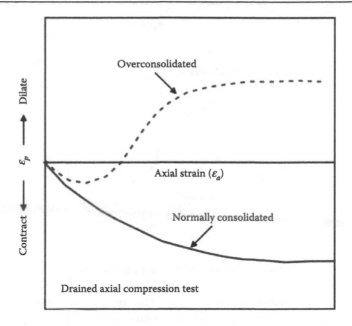

Figure 1.9 Characteristics of $\varepsilon_a - \varepsilon_p$ relationships from drained triaxial tests on clays.

as shown in Figure 1.9. On the other hand, for the drained test of an overconsolidated specimen, the stress–strain curve follows a strain hardening to a peak before a strain softening (i.e., q increases with ε_a and then decreases with ε_a after reaching a peak q, as shown in Figure 1.8). The overconsolidated specimen tends to increase in volume during the triaxial test ($\varepsilon_p > 0$, as shown in Figure 1.9).

For undrained test of a normally consolidated clay specimen, the stress–strain curve tends to be strain-hardening (i.e., q increases with ε_a monotonically, as shown in

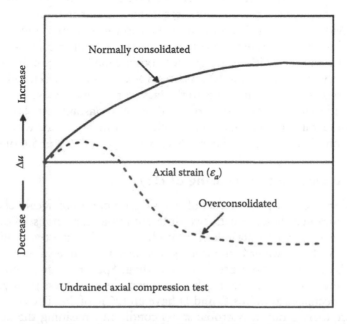

Figure 1.10 Characteristics of $\varepsilon_a - \Delta u$ relationships from undrained triaxial tests on clays.

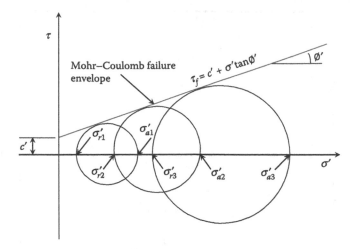

Figure 1.11 Mohr–Coulomb failure envelope and strength parameters.

Figure 1.8) and there is a positive pore pressure buildup ($\Delta u > 0$; see Figure 1.10). On the other hand, for undrained test on an overconsolidated clay specimen, the stress–strain curve also experiences strain hardening but pore pressure buildup is positive initially and then becomes negative as shearing continues ($\Delta u < 0$; see Figure 1.10).

1.4.3 Mohr–Coulomb shear strength parameters of clays

To determine the shear strength parameters, a common procedure is to perform at least three triaxial tests, each with a different σ_r'. For each of the triaxial tests, a Mohr circle is constructed based on the peak strength (i.e., the peak q value in the stress–strain curve shown in Figure 1.8). For tests with strain-hardening behavior, the peak strength is the same as the ultimate strength (see Figure 1.8). Figure 1.11 provides a qualitative description of the Mohr circles from triaxial tests based on effective normal stresses (i.e., from drained tests or undrained tests with pore pressure measurements). A straight line, tangential to all the Mohr circles, is then drawn as the Mohr–Coulomb failure envelope. The slope of this failure envelope is defined as the drained peak friction angle, $\emptyset'$, and the intercept of the failure envelope with the shear stress axis is called drained cohesion, c' (see Figure 1.11). Note that natural soil rarely possesses true cohesion between soil particles, even for clays. The existence of c' can be due to the fact that we use peak strength to construct Mohr circles and that the tests were performed on overconsolidated clays. The use of c' can lead to unsafe foundation designs and should therefore be treated with caution. More discussion on cohesion is presented later in Sections 1.7.7, 4.6.2, and 5.6 of this book.

1.4.4 Critical state in the shearing of clays

As an illustration, we now present a set of stress–strain curves of Weald clay under both drained and undrained shearing with two different initial conditions (Wood, 1990). As shown in Figure 1.12 (with data taken from Henkel, 1959), the two initial conditions are depicted in a plot of water content w against mean effective stress p'. Specimens at points A and B are NC without previous unloading. Specimens at points C and D are OC and have undergone an unloading from a mean effective stress p' of 827 to 34 kPa. Therefore the specimens at points C and D have an OCR of 24. The consolidation and unloading were carried out in isotropic stress conditions, meaning the applied stresses in all directions were equal (i.e., $\sigma_a' = \sigma_r'$).

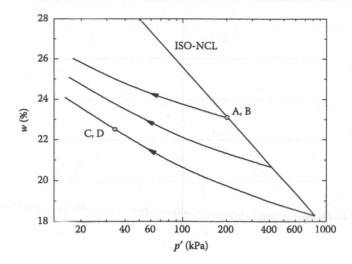

Figure 1.12 Initial conditions or states of Weald clay before shearing.

EXAMPLE 1.3

Given

In Figure 1.12, samples A and B are normally consolidated clays with the initial water content w of 23.2%. Samples C and D are overconsolidated clays with the w of 22.4%. For Weald clay, the specific gravity of the clay particle is known to be $G_s = 2.75$.

Required

Determine the initial void ratios e and specific volume v of samples A, B, C, and D.

Solution

For samples A and B, the initial void ratio is

$$e = wG_s = 0.232 \times 2.75 = 0.638$$

Their initial specific volume is

$$v = 1 + e = 1 + 0.638 = 1.638$$

For samples C and D, the initial void ratio is

$$e = wG_s = 0.224 \times 2.75 = 0.616$$

Their initial specific volume is

$$v = 1 + e = 1 + 0.616 = 1.616$$

The results of the four drained and undrained tests of Weald clay, referred to in Figure 1.12, are given in Figures 1.13 and 1.14 (with data taken from Bishop and Henkel, 1957). The trend of these tests in terms of stress–strain and strain–pore pressure/volume change relationships is consistent with those presented in Figures 1.8 through 1.10.

Figure 1.15 shows the loading paths (or stress paths) in the plots of $q - p'$ and $v - p'$ for those four tests (Wood, 1990). Points 1–4 and 5–8, respectively, represent the loading

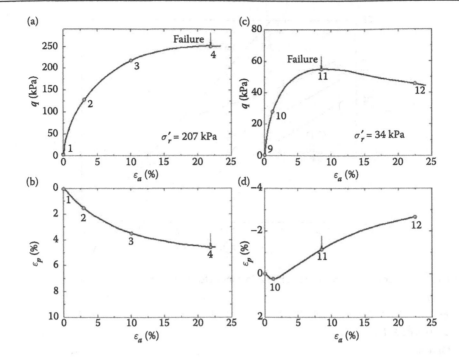

Figure 1.13 Stress–strain curves from drained tests of Weald clay. (a) Stress–strain plot of the NC specimen. (b) Volumetric strain–axial strain plot of the NC specimen. (c) Stress–strain plot of the OC specimen. (d) Volumetric strain–axial strain plot of the OC specimen.

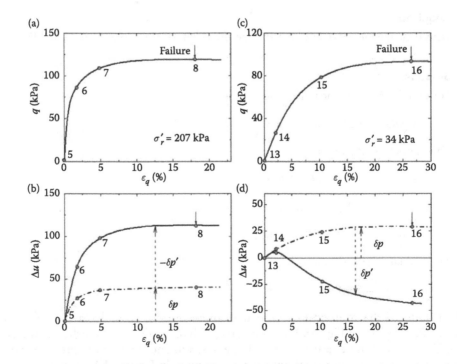

Figure 1.14 Stress–strain curves from undrained tests of Weald clay. (a) Stress–strain plot of the NC specimen. (b) Pore pressure–triaxial shear strain plot of the NC specimen. (c) Stress–strain plot of the OC specimen. (d) Pore pressure–triaxial shear strain plot of the OC specimen.

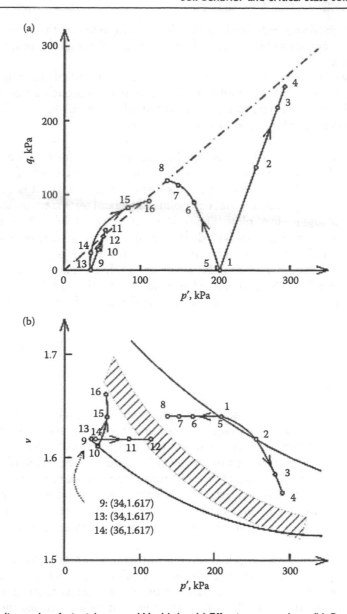

Figure 1.15 Loading paths of triaxial tests on Weald clay. (a) Effective stress plane. (b) Compression plane.

paths of the drained and undrained triaxial tests on the NC specimen [with their initial (w, p') at points A and B in Figure 1.12]. On the other hand, points 9–12 and 13–16 represent the loading paths of the drained and undrained triaxial tests on the OC specimen [with their initial (w, p') at points C and D in Figure 1.12].

There are a number of important observations that we should note from Figure 1.15.

1. For the undrained tests, the loading paths in the plot of $v - p'$ are horizontal because the volume remains constant under undrained conditions.
2. For drained tests, the NC specimen contracts in volume, while the OC specimen dilates during shearing.

3. For both drained and undrained tests of NC or OC specimens, the loading paths in the plot of $q - p'$ appear to be a straight line that passes through the origin (dash line in Figure 1.15a).
4. The end points (stress and strain readings toward the end of shearing where q has reached its ultimate value) of all four tests appear to approach a narrow band of a curve in the plot of $v - p'$, as shown in the shaded area of Figure 1.15b (which is straight in the plot of $v - \ln p'$, as discussed later).

To confirm the above observations based on the results of four tests on Weald clay, Roscoe et al. (1958) collated data on end points of a large number of undrained and drained tests on Weald clay. These data are shown in Figure 1.16. It is clear that these data confirm earlier observations that end points of triaxial tests seem to lie on a straight line in both $q - p'$ and $v - \ln p'$ plots (Figure 16a and c, respectively). The straight line of end points during triaxial tests of Weald clay was termed by Roscoe et al. (1958) as the *critical state line* (CSL). It is also important to note that the CSL is parallel to the isotropical normal consolidation line (ISO-NCL) in the $v - \ln p'$ space, as demonstrated in Figure 1.16b and c. The CSLs are plotted as dash lines in Figure 1.16.

Mathematically, the critical state for a soil is fully determined therefore by the following two equations:

$$q = Mp' \tag{1.26}$$

$$\Gamma = v + \lambda \ln p' \tag{1.27}$$

Equations 1.26 and 1.27 are also marked in Figure 1.16a and c. In other words, when the soil is subjected to shear loading such as the axial compression in triaxial tests, it will ultimately approach a critical state, at which the ratio of deviator stress q and mean effective stress p' become a constant value, M (capital μ) regardless of its initial state (i.e., stress level and void ratio). M is a function of critical state friction angle, $\emptyset'_{cs}$ as explained in Section 1.6. To reach critical state (stress–strain relationship reaches its ultimate state) in a triaxial test, we usually need to perform the test until the axial strain exceeds 20%. Γ and λ are two further critical state constants for soils which are explained in Section 1.6.

The critical state soil mechanics is based on the premise that when a soil sample is subjected to shearing, it will ultimately reach a state in which the soil behaves as a frictional fluid with a constant volume and a constant ratio of shear stress to mean normal stress, regardless of the initial state of the soil. This ultimate state was termed the *critical state* independently by Roscoe et al. (1958) and Parry (1958) and proves to be a powerful reference state for developing a large number of constitutive models to predict the mechanical behavior of soils when subjected to various loading conditions (Schofield and Wroth, 1968; Yu, 2006).

1.5 STRESS–STRAIN BEHAVIOR AND CRITICAL STATE OF SANDS

There are some similarities in the stress–strain behavior between sand and clay. The stress–strain and strain-dilatancy relationships in drained and undrained shearing of a dense sand are similar to those of overconsolidated clay, whereas loose sand is similar to normally consolidated clay in stress–strain and strain-dilatancy relationships (see Figures 1.8 through 1.10).

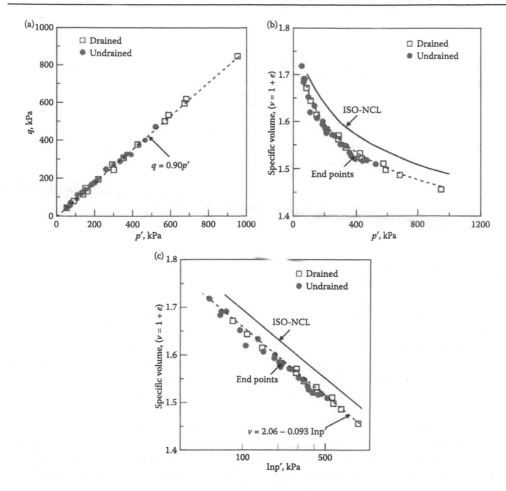

Figure 1.16 End points of undrained tests of Weald clay. (a) Effective stress plane. (b) v–p' compression plane. (c) v–lnp' compression plane. (Adapted from Roscoe, K.H., Schofield, A.N., and Wroth, C.P. 1958. *Geotechnique*, 8, 22–52.)

1.5.1 Mohr–Coulomb shear strength parameters of sands

In most cases, the dilatancy of sand is sensitive to confining stress (Bolton, 1986). The Mohr–Coulomb failure envelope constructed by drawing a tangent to Mohr circles from tests on sand of the same void ratio is usually curved, as shown in Figure 1.17. A straight dash line is added in Figure 1.17 to contrast the curved failure envelope. For sand, $c' = 0$ unless the sand is cemented. For a sand with the same void ratio, the drained peak friction angle $\varnothing'$ decreases with the effective normal stress (Bolton, 1986). It will be demonstrated later from a critical state soil mechanics point of view why $\varnothing'$ (or dilatancy) of sand decreases with the effective normal stress.

1.5.2 Critical state in the shearing of sands

The early development of critical state soil mechanics was largely based on experimental results of clays (Bishop and Henkel, 1957; Roscoe et al., 1958; Schofield and Wroth, 1968). Its extension to sand has been slow, and in fact did not make much progress until the 1980s, partly due to the difficulties of determining critical state lines in the laboratory

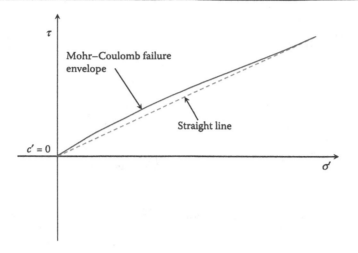

Figure 1.17 Curved Mohr–Coulomb failure envelope.

(Wroth and Bassett, 1965; Vesic and Clough, 1968; Been and Jefferies, 1985; Been et al., 1991).

In addition to the development of critical state concept in the UK, there was an effort in the US devoted to developing a concept of steady state primarily for earthquake liquefaction applications (Castro, 1969; Castro and Poulos, 1977; Poulos, 1981). In fact, the definition of steady state was very similar to that of critical state, and for most practical applications they may be regarded as the same (Been et al., 1991).

It should be noted that the steady state for sands has only been measured after liquefaction in triaxial tests. These tests tended to be carried out on loose sand samples under undrained conditions at which it was possible for liquefaction to occur. On the other hand, critical state workers tend to use drained triaxial tests of dense sand samples to determine the critical state (Been et al., 1991; Klotz and Coop, 2002).

In an important study, Been et al. (1991) reported the results of a large number of triaxial tests for Erksak sand focusing on critical states from drained tests of dense samples and steady states from undrained tests of loose sand samples. They drew a general conclusion that for practical purposes the critical state and steady state may be regarded as identical. This finding has been supported by recent studies.

To illustrate the critical state and stress–strain behavior of sand, we use the results of a large number of drained and undrained triaxial tests on Portaway sand, carried out by Wang (2005). To demonstrate a typical set of stress–strain behavior of sand under drained triaxial tests on Portaway sand, Figure 1.18 shows the stress–strain curves and volumetric strain–axial strain relations for five tests with different initial conditions. It is clear that for the three loose samples (CIDC-1, CIDC-2, and CIDC-3), there was a strain-hardening behavior with volume contraction during the triaxial loading. For the two dense samples (CIDC-4 and CIDC-5), the stress–strain behavior follows a strain-hardening pattern before softening to a critical state in the end. So there tends to be a peak shear strength for dense sand under drained triaxial tests. This peak shear strength is associated with a change of volume contraction to dilation during shearing. At the critical state, the rate of soil dilation approaches zero (i.e., the curves become flat) in all five tests.

Typical stress–strain behavior of Portaway sand in undrained triaxial tests is shown in Figure 1.19. It is clear from these four tests on very loose sand samples that the stress–strain curves under undrained triaxial tests follow a strain-hardening-softening pattern. During the test, the shear strength q drops to a small value due to the buildup of pore

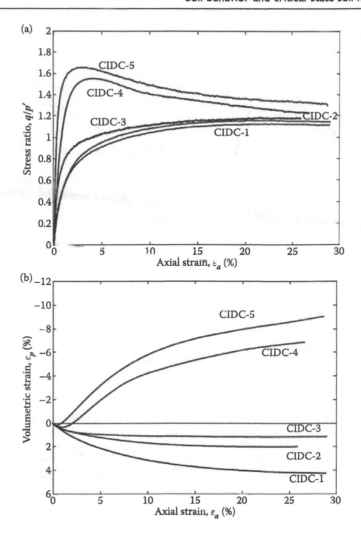

Figure 1.18 Typical stress–strain behavior of Portaway sand under drained triaxial tests. (a) Stress ratio vs. axial strain. (b) Volumetric strain vs. axial strain. (Adapted from Wang, J. 2005. The stress–strain and strength characteristics of Portaway sand. PhD Thesis, University of Nottingham, UK.)

water pressure. In an extreme case (i.e., when the sand is extremely loose), the shear strength q can drop to zero after passing the peak strength. The zero shear strength accompanies $p' \approx 0$ (i.e., $p' - q$ plot of CIUC5 in Figure 1.19), meaning the confining stress is completely offset by pore pressure and the sand behaves like liquid, a phenomenon referred to as soil liquefaction. Liquefaction is therefore often a concern for loose sand under groundwater table (hence saturated). In the field, liquefaction can be triggered by static shearing or a seismic motion, such as an earthquake.

Figure 1.20 shows the loading paths of 24 drained and undrained triaxial tests of Portaway sand in the plot of $v - \ln p'$. The ultimate (or end) point of each test represents a critical state point. The average line of the end points in the $q - p'$ space is the CSL that passes through the origin, as shown in Figure 1.21. Similarly, the ultimate points of the tests in the plot of $v - \ln p'$ also form a straight CSL, as indicated in Figure 1.22. The two CSLs depicted in Figures 1.21 and 1.22 correspond to Equations 1.26 and 1.27, respectively. The equations are also marked on the corresponding figures.

26 Foundation engineering analysis and design

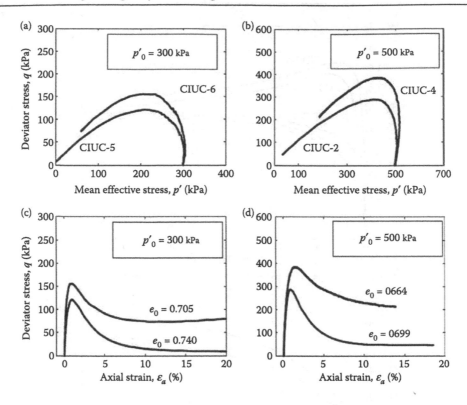

Figure 1.19 Typical stress–strain behavior of Portaway sand under undrained triaxial tests. (a) $q - p'$ plot of CIUC-5 and CIUC-6. (b) $q - p'$ plot of CIUC-2 and CIUC-4. (c) $q - \varepsilon_a$ plot of CIUC-5 ($e_0 = 0.740$) and CIUC-6 ($e_0 = 0.705$). (d) $q - \varepsilon_a$ plot of CIUC-2 ($e_0 = 0.699$) and CIUC-4 ($e_0 = 0.664$). (Adapted from Wang, J. 2005. The stress–strain and strength characteristics of Portaway sand. PhD Thesis, University of Nottingham, UK.)

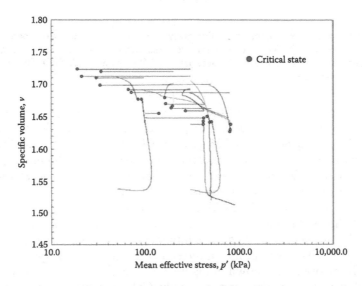

Figure 1.20 Determination of critical states for Portaway sand using both drained and undrained triaxial tests. (Adapted from Wang, J. 2005. The stress–strain and strength characteristics of Portaway sand. PhD Thesis, University of Nottingham, UK.)

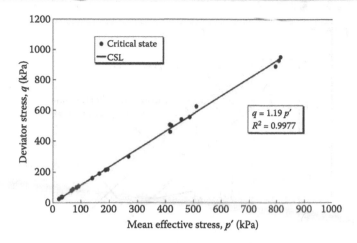

Figure 1.21 The critical state line of Portaway sand in the $q - p'$ space. (Adapted from Wang, J. 2005. The stress–strain and strength characteristics of Portaway sand. PhD Thesis, University of Nottingham, UK.)

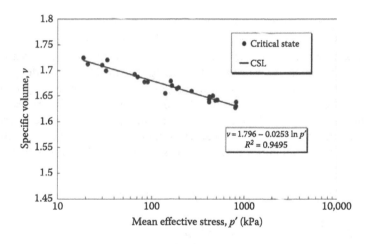

Figure 1.22 The critical state line of Portaway sand in the $v - \ln p'$ space. (Adapted from Wang, J. 2005. The stress–strain and strength characteristics of Portaway sand. PhD Thesis, University of Nottingham, UK.)

1.6 CRITICAL STATE AND SOIL ENGINEERING PROPERTIES

The above observations on particular sets of triaxial test results can be generalized under the framework of critical state soil mechanics so that we can predict the dilatancy (or pore pressure development) of the soil when sheared, based on the state of the soil that is represented by its current values of p' (stress state) and v (density state). To do this, we first plot the $v - p'$ correlations in a semi-natural log, $v - \ln p'$ space. This is slightly different from Figure 1.2, where base 10 log was used in the $e - \log \sigma'_v$ plots. However, the virgin compression and recompression lines continue to be treated as straight lines, as shown in Figure 1.23. In addition, following the observations from Figure 1.16c, we consider the CSL parallel to the virgin compression line. The virgin compression line is also an upper bound of the $v - \ln p'$ relationship. Therefore in Figure 1.23 we see a set of straight lines

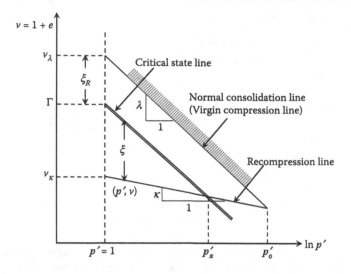

Figure 1.23 The critical state framework in $v - \ln p'$ space. (Adapted from Yu, H.S. 1998. *International Journal for Numerical and Analytical Methods in Geomechanics*, 22, 621–653.)

that include a CSL, a virgin compression line, and there can be numerous recompression lines (depending on where the unloading started). The virgin compression and CSL share the same slope λ, which is equivalent to the Terzaghi's compression index, C_c, described earlier. A recompression line can stem from any point on the virgin compression line and pointing toward the unloading direction. All recompression lines share the same slope of κ that is equivalent to the Terzaghi's recompression index, C_s, described earlier. For convenience, we establish a linear equation to describe each of the three types of straight lines in Figure 1.23 by denoting the intercept of the lines with the unit pressure line $p' = 1$ as follows:

For compression lines $v = v_\lambda - \lambda \ln p'$ (1.28)

For recompression lines $v = v_\kappa - \kappa \ln p'$ (1.29)

And for the CSL, we repeat Equation 1.27,

$$v = \Gamma - \lambda \ln p' \qquad\qquad\qquad (1.27)$$

where
    v_λ, v_κ, and Γ = intercept, in terms of specific volume of the compression, recompression, and CSL with the unit pressure line $p' = 1$.

Note that the intercept values can change, as we use a different unit for p'.

1.6.1 The state parameter

Using the framework of critical state soil mechanics, the soil loading history and current state can be represented by its relative position from the CSL in the $v - \ln p'$ space. A simple measure of this relative position would be the vertical distance in specific volume v from the current state to the CSL, and this quantity has been termed as *the state parameter*,

ξ (Been and Jefferies, 1985; Yu, 1998). Figure 1.23 shows the definition of the state parameter and other critical state indices. Consider a soil sample consolidated along the virgin compression line to an effective mean normal stress of p'_o and unloaded along the recompression line to its current effective mean normal stress, p'. Consider the unloading is significant enough that the unloading line crosses the CSL at an effective mean normal stress, p'_x. The state parameter is defined mathematically in terms of specific volume v and mean effective stress p' as follows:

$$\xi = v + \lambda \ln p' - \Gamma \tag{1.30}$$

where

ξ = state parameter
Γ = intercept in terms of specific volume of the CSL with the unit pressure line $p' = 1$

When a soil sample is sheared, it would ultimately reach the critical state in which it behaves as a frictional fluid with a constant ratio of shear stress to mean effective normal stress (q/p') as described by Equation 1.26, and $v - p'$ correlation satisfies Equation 1.27, regardless of the initial state of the soil. This is the premise of the CSSM established from observations of laboratory tests described above. If the initial state of the soil sample represented by (p', v) sits above the CSL, then its state parameter ξ is >0 before shearing. In this case, the soil sample will contract (i.e., v reduces) if the shearing is drained. Or, p' will reduce (pore pressure will increase) while v remains constant if the shearing is undrained. On the other hand, if ξ is <0 before shearing, the soil sample will dilate (i.e., v increases) if the shearing is drained, or p' will increase (pore pressure will decrease) while v remains constant if the shearing is undrained.

It should be stressed that for clay, the OCR can be equally used to define its state. Following the recompression line in Figure 1.23, the OCR and a spacing ratio r can be defined as follows:

$$\text{OCR} = \frac{p'_0}{p'} \tag{1.31}$$

and

$$r = \frac{p'_0}{p'_x} \tag{1.32}$$

It can be shown from geometry that the state parameter ξ can be expressed as a function of OCR as follows:

$$\xi = (\lambda - \kappa) \ln\left(\frac{r}{\text{OCR}}\right) \tag{1.33}$$

where r is believed to be in a range between 2 and the natural base e (i.e., 2.71828).

It is noted that for soils lying at the normal consolidation or virgin compression line, OCR is 1, and the state parameter reaches a maximum value: $\xi_R = (\lambda - \kappa)\ln r$ (see Figure 1.23), which represents the vertical distance between the virgin compression line and the CSL (Yu, 1998). It is important to note that for soils, especially clays, at a state between the virgin compression and CSL, its OCR > 1 (overconsolidated by definition), but ξ is also >0. In this case, we consider the clay as lightly overconsolidated, and undrained shearing will generate positive pore pressure (p' decreases), a phenomenon that is typical for normally consolidated clays. The typical behavior of an overconsolidated clay is usually not materialized until $\xi < 0$ as correctly predicted by the CSSM, and the corresponding OCR becomes larger than the spacing ratio r.

It is clear that Figure 1.23 has practically combined the soil compressibility and stress–strain behavior. The question then is, how do we relate critical state to other soil engineering properties? This is the main issue in the following sections.

1.6.2 Critical state and engineering properties of clays

Schofield and Wroth (1968) postulated that the standard liquid limit (LL) and plastic limit (PL) tests reflect the remolded shear strength of the clay at the two respective water contents. The PL test causes "crumbling" or tensile failure of the soil sample, similar to a split-cylinder or Brazil test of concrete cylinders. The LL test mimics a miniature slope failure in the banks of the groove created in an LL test. In LL and PL tests, the soil is continually remolded; therefore it must remain at the critical state. We consider q_{LL} and p'_{LL} as the deviator stress (i.e., shear strength) and effective mean normal stress at LL, respectively and q_{PL} and p'_{PL}, as those at PL. Because the soils under these two test conditions are at critical state,

$$\frac{q_{PL}}{p'_{PL}} = \frac{q_{LL}}{p'_{LL}} = M \tag{1.34}$$

in the $q - p'$ space, and

$$\nu_{PL} + \lambda \ln p'_{PL} = \nu_{LL} + \lambda \ln p'_{LL} = \Gamma \tag{1.35}$$

in the $\nu - \ln p'$ space,

where

ν_{PL}, ν_{LL} = specific volume of the clay at PL and LL, respectively
λ = compression index of the clay in $\nu - \ln p'$ space

For a given clay, (ν_{LL}, p'_{LL}) and (ν_{PL}, p'_{PL}) lay on the same CSL. According to test results reported by Schofield and Wroth (1968), it appears that the LL and PL correspond approximately to fixed strengths q, which have a ratio of 1:100 (i.e., $q_{LL}:q_{PL} \cong 1:100$). Following Equation 1.34, p'_{LL} and p'_{PL} should also have a ratio of 1:100. According to Schofield and Wroth (1968), p'_{PL} was consistently around 552 kPa for the family of clays tested as conceptually described in Figure 1.24.

If we extrapolate the CSLs as indicated by the dashed lines in Figure 1.24, they tend to pass through the point Ω with $\nu_\Omega \cong 1.25$ and $p'_\Omega \cong 10,343$ kPa. Substituting the quoted values for p'_{PL}, ν_Ω, and p'_Ω in Equation 1.35, we get

$$\nu_{PL} - 1.25 = \lambda \ln \frac{10,343}{552} = 2.93\lambda \tag{1.36}$$

and

$$\nu_{LL} - 1.25 = \lambda \ln \frac{10,343}{5.52} = 7.54\lambda \tag{1.37}$$

Consider that the clay samples in limit tests are saturated (i.e., degree of saturation $S = 1$) and its particles have an approximate specific gravity, G_s, of 2.7, then $\nu_{PL} = 1 + G_s w_{PL}$, where $w_{PL} = PL$ in decimal, Equation 1.36 can be rearranged as

$$\lambda = 0.341(\nu_{PL} - 1.25) \cong 0.92(w_{PL} - 0.09) \tag{1.38}$$

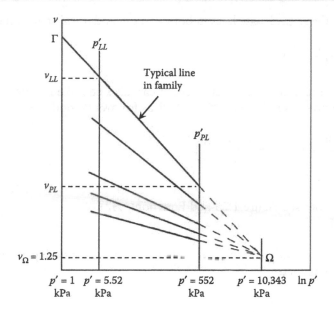

Figure 1.24 Idealized family of critical state lines.

Similarly, if we follow Equation 1.37,

$$\lambda = 0.133(\nu_{LL} - 1.25) \cong 0.36(w_{LL} - 0.09) \tag{1.39}$$

on the basis that $p'_{PL} \cong 100p'_{LL}$. As we learned earlier, C_c (i.e., compression index in 10 based logarithm) $= \lambda \ln 10 = 2.303 \lambda$; hence following Equation 1.39,

$$C_c \cong 0.83(w_{LL} - 0.09) \tag{1.40}$$

which compares well with Skempton's (1944) empirical relationship,

$$C_c = 0.7(w_{LL} - 0.1) \tag{1.41}$$

From the idealized straight lines of Figure 1.23 we can predict that

$$\Gamma = \nu_\Omega + \lambda \ln p'_\Omega = 1.25 + \lambda \ln 10{,}343 = 1.25 + 9.24\lambda \tag{1.42}$$

Through Equation 1.35, Γ can be linked to PL or LL as well. Note again that Γ is the specific volume of the point on the CSL corresponding to unit pressure of 1. It has been demonstrated that we can obtain a reasonably accurate value of λ and an estimate of Γ from simple LL and PL tests.

Derivations that relate undrained shear strength of clays to critical state parameters are beyond the scope of this book. Interested readers are referred to Schofield and Wroth (1968) for details. Instead, only the correlations themselves are described herein. For a clay with a known specific volume ν, its shear strength under undrained loading can be estimated from the critical state lines in both $v - \ln p'$ and $q - p'$ spaces, namely

$$s_u = \frac{M}{2} \exp\left(\frac{\Gamma - v}{\lambda}\right) \tag{1.43}$$

Alternatively, the undrained shear strength can also be linked to its current effective mean stress p' and consolidation history OCR as follows (Wood, 1990):

$$\frac{s_u}{p'} = \frac{M}{2}\left(\frac{OCR}{r}\right)^{(\lambda - \kappa)/\lambda} = \frac{M}{2}\left(\frac{OCR}{r}\right)^{\Lambda} \tag{1.44}$$

where
$\quad r =$ spacing ratio (see Figure 1.23 and Equation 1.32)
$\quad \Lambda = 1 - (\kappa/\lambda)$

For normally consolidated soils (i.e., OCR $= 1$) and considering $r =$ natural base e, Equation 1.44 reduces to

$$\frac{s_u}{p'} = \frac{M}{2} \exp\left(-\frac{\lambda - \kappa}{\lambda}\right) = \frac{M}{2} \exp(-\Lambda) \tag{1.45}$$

and

$$\Lambda = \frac{\lambda - \kappa}{\lambda} \tag{1.45a}$$

Schofield and Wroth (1968) showed that this simple analytical critical state prediction agrees well with experimental observation for Weald clay. This is also consistent with the well-known result that undrained shear strength increases linearly with depth for a normally consolidated deposit.

Using the relationship between the state parameter and OCR expressed in Equation 1.33, the normalized undrained shear strength can also be expressed in terms of the state parameter ξ in a very simple manner: $\xi = (\lambda - \kappa)\ln(r/\text{OCR})$

$$\frac{s_u}{p'} = \frac{M}{2} \exp\left(-\frac{\xi}{\lambda}\right) \tag{1.46}$$

EXAMPLE 1.4

Given

As shown in Figure 1.12, samples A and B are normally consolidated clays with the initial water content of 23.2% and initial mean effective stress of 207 kPa. Samples C and D are overconsolidated clays with the initial water content of 22.4%. The two overconsolidated samples were compressed isotropically to a mean effective stress of 827 kPa and then allowed to swell back isotropically to a mean effective stress of 34 kPa. For Weald clay, the specific gravity of the clay particle is known to be $G_s = 2.75$.

As shown in Figure 1.16a and c, the experimental critical state line for Weald clay may be defined by Equations 1.26 and 1.27 as follows:

$$q = 0.9p' \quad \text{and} \quad v = 2.06 - 0.093 \ln p'$$

Required

a. The OCR of samples C and D
b. The state parameter for samples A and B
c. The state parameter for samples C and D
d. The undrained shear strengths of samples A, B, C and D if they are sheared under the undrained condition

Solution

a. For samples C and D, the current mean effective stress is $p' = 34$ kPa. Their preconsolidation pressure is $p'_0 = 827$ kPa. According to Equation 1.31, the overconsolidation ratio is

$$\text{OCR} = \frac{p'_0}{p'} = \frac{827}{34} \approx 24 \tag{1.31}$$

b. For samples A and B, the initial void ratio is

$$e = wG_s = 0.232 \times 2.75 = 0.638$$

Their initial specific volume is therefore

$$v = 1 + e = 1 + 0.638 = 1.638$$

From the given experimental critical state line, we have $M = 0.9$, $\lambda = 0.093$, $\Gamma = 2.06$. Their current effective mean stress is $p' = 207$ kPa. According to Equation 1.30, the state parameter is

$$\xi = v + \lambda \ln p' - \Gamma = 1.638 + 0.093 \ln(207) - 2.06 = 0.074 \tag{1.30}$$

c. For samples C and D, the current mean effective stress is $p' = 34$ kPa. Their initial void ratio is

$$e = wG_s = 0.224 \times 2.75 = 0.616$$

Their initial specific volume is therefore

$$v = 1 + e = 1 + 0.616 = 1.616$$

According to Equation 1.30, the state parameter is

$$\xi = v + \lambda \ln p' - \Gamma = 1.616 + 0.093 \ln(34) - 2.06 = -0.116 \tag{1.30}$$

d. According to Equation 1.43, we can calculate the undrained shear strength of samples A and B as follows:

$$s_u = \frac{M}{2} \exp\left(\frac{\Gamma - v}{\lambda}\right) = \frac{0.9}{2} \times \exp\left(\frac{2.06 - 1.638}{0.093}\right) = 42 \text{ kPa} \tag{1.43}$$

Alternatively, according to Equation 1.46, we can also determine the undrained shear strength as follows:

$$s_u = \frac{Mp'}{2}\exp\left(-\frac{\xi}{\lambda}\right) = \frac{0.9 \times 207}{2}\exp\left(-\frac{0.074}{0.093}\right) = 42\,\text{kPa} \tag{1.46}$$

For samples C and D, we can determine their undrained shear strength in the same way, using Equation 1.43 as follows:

$$s_u = \frac{M}{2}\exp\left(\frac{\Gamma - v}{\lambda}\right) = \frac{0.9}{2} \times \exp\left(\frac{2.06 - 1.616}{0.093}\right) = 53\,\text{kPa} \tag{1.43}$$

Alternatively, according to Equation 1.46, we can also determine the undrained shear strength as follows:

$$s_u = \frac{Mp'}{2}\exp\left(-\frac{\xi}{\lambda}\right) = \frac{0.9 \times 34}{2}\exp\left(-\frac{-0.116}{0.093}\right) = 53\,\text{kPa} \tag{1.46}$$

1.6.3 Critical state and engineering properties of sands

In practice, OCR has been widely used to describe the state of clay but has been less useful for sand. This is because it is more difficult and often requires very highly applied pressure to define and measure a normal consolidation line for sands. On the other hand, measuring the state parameter ξ does not require knowledge of a normal consolidation line and is therefore more convenient for applications to sands.

In foundation engineering, the shear strength of soils is an important design parameter to ensure foundation stability. For clay, the design needs to be checked against undrained failure. Therefore it is important to measure or estimate undrained shear strength s_u. For sand, the shear strength is often expressed in terms of friction angle, of which the peak friction angle $\emptyset'$ and critical state friction angle $\emptyset'_{cs}$ are the main parameters to be considered.

The critical state friction angle $\emptyset'_{cs}$ reflects the minimum drained friction angle, under which the sand is sheared with no dilatancy. $\emptyset'_{cs}$ is mainly a function of the mineral content of the sand and is independent from density (or relative density) and state of stress. As described in Section 1.5, the end points from triaxial tests on sand specimens with a wide range of initial densities lie on a straight line in the $q-p'$ space passing through the origin that can be defined by Equation 1.26. It can be demonstrated that M relates to $\emptyset'_{cs}$ as follows:

$$\sin\emptyset'_{cs} = \frac{3M}{6 + M} \tag{1.47}$$

which may be written approximately as follows (Wood, 1990):

$$\emptyset'_{cs} \approx 25M \tag{1.48}$$

where
$M =$ slope of the critical state line in the $q-p'$ space

It has been shown that for many sands the critical state friction angle is in a narrow range of 30°–35° (Been and Jefferies, 1985; Bolton, 1986; Yu, 2000).

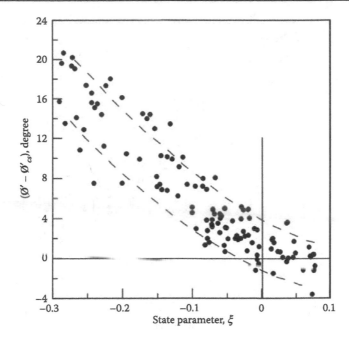

Figure 1.25 The difference of peak and critical state friction angles as a function of state parameter. (Adapted from Been, K. and Jefferies, M.G. 1985. *Geotechnique*, 35, 99–112.)

It is known that the peak friction angle of a sand, $\emptyset'$, measured from triaxial tests depends on its initial density and confining stress, or state parameter, before shearing. Been and Jefferies (1985) collated and analyzed peak friction angle data for a number of sands from the Arctic, Europe, and North and South America. It was found that the peak friction angle correlates very well with the initial state parameter. Figures 1.25 and 1.26 plot, respectively, the difference $(\emptyset' - \emptyset'_{cs})$ and $\emptyset'$ as a function of initial state parameter ξ for various sands.

In engineering design, we usually use peak friction angle as the strength parameter for sands. It is clear from Figures 1.25 and 1.26 that despite the diversity of the sands there is a remarkably good correlation between peak friction angle and state parameter. This is not surprising, because the state parameter rightly combines the effects of both density and stress level that are known to control dilatancy and strength of sand (Bolton, 1986). This experimental evidence gives confidence for using the state parameter in foundation engineering.

Mathematically, the above correlation plotted in Figure 1.25 may be expressed in a simple equation as follows to facilitate the practical applications (Collins et al., 1992; Yu, 1994, 1996, 2000):

$$(\emptyset' - \emptyset'_{cs}) = A[\exp(-\xi) - 1] \tag{1.49}$$

where A is a curve fitting material constant that is in the range of 0.6–0.9 depending on sand types (Yu, 1996). For a given sand and thus a given mineral, $\emptyset'_{cs}$ is close to a constant. According to Equation 1.49, $(\emptyset' - \emptyset'_{cs})$ decreases as ξ becomes larger (more positive). If we further limit the sand to have the same density (same e or v), then larger ξ means higher p'. This is the same as saying that $(\emptyset' - \emptyset'_{cs})$ or $\emptyset'$ decreases with confining stress, and therefore a curved Mohr–Coulomb failure envelope as shown in Figure 1.17.

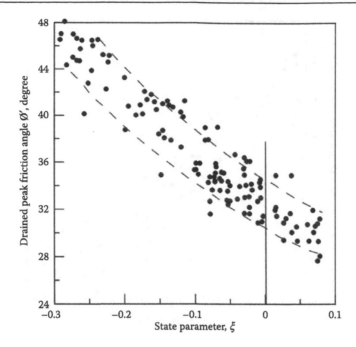

Figure 1.26 Peak friction angle as a function of state parameter. (Adapted from Been, K. and Jefferies, M.G. 1985. *Geotechnique*, 35, 99–112.)

In summary, under the framework of CSSM, we are able to determine the soil mechanical behavior with three soil constants: drained critical state friction angle ($\emptyset'_{cs}$) or M, compression index (λ), and swelling index (κ), and current state of the soil (void ratio e_o, p'_i, q_i, and OCR).

EXAMPLE 1.5

Given

As shown in Figures 1.21 and 1.22, the experimental CSL for Portaway sand can be defined by Equations 1.26 and 1.27 as follows:

$$q = 1.19p' \tag{1.26}$$

and

$$\nu = 1.796 - 0.0253 \ln p' \tag{1.27}$$

The two samples for the drained tests CIDC1 and CIDC4 shown in Figure 1.18 have the initial void ratios of 0.699 and 0.537, respectively. Their initial mean effective stresses are 500 and 300 kPa, respectively. The constant A in Equation 1.49 for Portaway sand is assumed to be 0.8.

Required

a. The critical state friction angle of Portaway sand
b. The state parameters for the samples CIDC1 and CIDC4 before shearing
c. The peak friction angles for the samples CICD1 and CIDC4 during triaxial shear tests

Solution

a. From information on the CSL, we have $\lambda = 0.0253, \Gamma = 1.796, M = 1.19$. According to Equation 1.48, the critical state friction angle is

$$\emptyset'_{cs} \approx 25M = 25 \times 1.19 = 30° \tag{1.48}$$

b. For the sample CIDC1, we have $\nu = 1 + e = 1 + 0.699 = 1.699$ and $p' = 500$ kPa. According to Equation 1.30, the state parameter is

$$\xi = \nu + \lambda \ln p' - \Gamma = 1.699 + 0.0253 \times \ln(500) - 1.796 = 0.06 \tag{1.30}$$

For the sample CIDC4, we have $\nu = 1 + e = 1 + 0.537 = 1.537$ and $p' = 300$ kPa. According to Equation 1.30, the state parameter is

$$\xi = \nu + \lambda \ln p' - \Gamma = 1.537 + 0.0253 \times \ln(300) - 1.796 = -0.115 \tag{1.30}$$

c. According to Equation 1.49, the peak friction angle for the sample CIDC1 is

$$\emptyset' = \emptyset'_{cs} + A[\exp(-\zeta) \quad 1] - \frac{180}{\pi}\left\{30° \times \frac{\pi}{180} + 0.8[\exp(-0.06) - 1]\right\} = 27° \tag{1.49}$$

In the same way, we can calculate the peak friction angle for the sample CIDC4 as follows:

$$\emptyset' = \emptyset'_{cs} + A[\exp(-\xi) - 1] = \frac{180}{\pi}\left\{30 \times \frac{\pi}{180} + 0.8[\exp(0.115) - 1]\right\} \approx 36° \tag{1.49}$$

1.7 THEORY OF PLASTICITY FOR MODELING STRESS–STRAIN BEHAVIOR

For a complete analysis of foundation engineering problems, we need to formulate a stress–strain relationship of soils under a general loading condition. As shown earlier in this chapter, the stress–strain relations are generally complex for real soils, which would involve many aspects such as linear or non-linear elastic behavior, nonlinear elastic-plastic hardening/softening behavior, peak shear strength, and critical (or ultimate) shear strength.

A theoretical framework that can be used to develop and formulate approximate stress–strain relations for metals is known as *the mathematical theory of plasticity* (Hill, 1950). Its further development and application to geotechnical materials such as soil and rock have been covered and treated in a comprehensive manner by Yu (2006).

Given the scope of this textbook, we only review some basic elements of plasticity theory for clay and sand, which will be related to the later chapters. These include key elements of perfect plasticity based on the Tresca, von Mises, and Mohr–Coulomb yield criteria for modeling cohesive and frictional soils. In addition, we present some elements of plasticity theory involving strain-hardening/softening behavior based on the critical state concept.

1.7.1 Yield criterion

Yield criterion (or yield function/surface) is an elastic boundary in the stress space within which material behavior is elastic, and on and beyond which plastic deformation will occur. Figure 1.27 shows a conceptual description of the yield surface in the context of CSSM after Mayne et al. (2009). Consider a group of specimens all with a common preconsolidation stress p'_o, but consolidated to different OCRs and sheared undrained, as shown in Figure 1.27. The normal consolidation line (NCL), recompression line

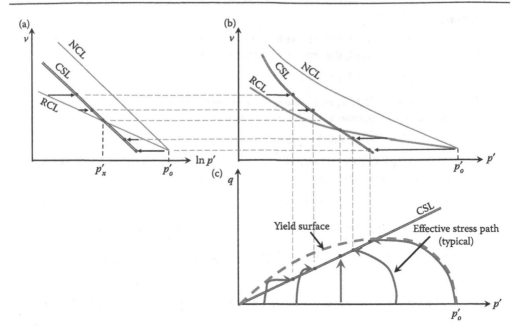

Figure 1.27 Group of effective stress paths forming a yield surface. (a) $v - \ln p'$ compression plane. (b) $v - p'$ compression plane. (c) $q - p'$ stress plane. (Adapted from Mayne, P.W. et al., 2009. *Proceedings, XVII International Conference on Soil Mechanics and Geotechnical Engineering*, Alexandria, Egypt, IOS Press, the Netherlands, pp. 2777–2872.)

(RCL), CSL, and undrained loading paths are presented in $v - \ln p'$ and $v - p'$ space in Figure 1.27a and b, respectively. The end points of all stress paths tend toward the CSL as postulated in CSSM. The convex envelope of the effective stress paths forms a yield surface, as depicted in Figure 1.27c. The top of the yield surface crosses the CSL at about half the magnitude of p'_o. To the right side of this peak, the yield surface lies beneath the CSL, while at the left side, the yield envelope extends above the CSL.

So far the discussion on CSSM and soil constitutive models has been limited to the case of the triaxial test, where there are only two independent stresses (i.e., σ_a and σ_r) and a two-dimensional stress space would suffice for the purpose. In the following discussion, we extend to a general three-dimensional state of stress where there can be three distinct principal stresses, σ_1, σ_2, and σ_3. The above mentioned axial compression triaxial test is a special case, where $\sigma_a = $ major principal stress σ_1, and σ_2 (intermediate principal stress) $= \sigma_3$ (minor principal stress) $= \sigma_r$.

If the yield function/surface is assumed to be fixed and does not depend on previous loading history, then the material is called perfectly plastic. In this case, the yield criterion is only a function of stress tensor σ_{ij}, namely

$$f(\sigma_{ij}) = 0 \tag{1.50}$$

On the other hand, if the yield function/surface changes with plastic deformation history, then the material is termed as strain hardening/softening plastic materials. In this case the yield criterion will be a function of both stress tensor and plastic strain history, namely

$$f(\sigma_{ij}, \alpha) = 0 \tag{1.51}$$

where α denotes a hardening parameter that is a function of plastic strains.

1.7.2 Loading criterion

For solving boundary value problems involving elastic-plastic behavior, we may need to determine what behavior will result from a further stress increment when the stress state is already on the yield surface. Three possible cases are

$$\text{Unloading:} \quad f(\sigma_{ij}, \alpha) = 0 \quad \text{and} \quad df = \frac{\partial f}{\partial \sigma_{ij}} d\sigma_{ij} < 0 \tag{1.52}$$

$$\text{Neutral loading:} \quad f(\sigma_{ij}, \alpha) = 0 \quad \text{and} \quad df = \frac{\partial f}{\partial \sigma_{ij}} d\sigma_{ij} = 0 \tag{1.53}$$

$$\text{Loading:} \quad f(\sigma_{ij}, \alpha) = 0 \quad \text{and} \quad df = \frac{\partial f}{\partial \sigma_{ij}} d\sigma_{ij} > 0 \tag{1.54}$$

Traditionally it is often assumed that for both unloading and neutral loading, material behavior is purely elastic. Plastic deformation will occur only when loading criterion is satisfied.

As detailed by Yu (2006), it must be noted that for perfectly plastic materials, the loading condition as defined by Equation 1.54 is not permissible, as the stress state can only lie on or inside the yield surface. In this case, it is assumed that plastic deformation will occur once the stress state lies on or moves along the yield surface, as defined by Equation 1.53.

1.7.3 Plastic flow rule

Based on the above loading criterion, Hill (1950) shows that the plastic strain rate tensor may be determined by the following equation or flow rule:

$$d\varepsilon_{ij}^p = d\lambda \frac{\partial g}{\partial \sigma_{ij}} \tag{1.55}$$

where g is a plastic potential which may or may not be the same as the yield function, and $d\lambda$ is a positive scalar. The plastic flow rule defined by Equation 1.55 means that the direction of plastic strains is normal to the plastic potential, as shown in Figure 1.28. It can be

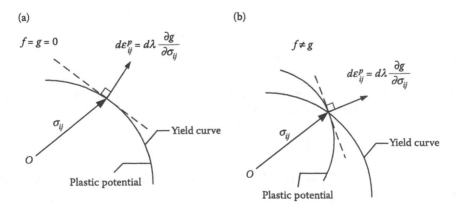

Figure 1.28 Yield surface, plastic potential and plastic flow rule. (a) Associated flow rule. (b) Non-associated flow rule.

used to define the ratios of various components of the plastic strain rate tensor. The flow rule is known as the *associated flow rule* if the plastic potential is identical to the yield function. Otherwise the plastic flow rule is said to be *non-associated*.

1.7.4 Consistency condition

To determine the complete relation between stress and strain for elastic-plastic materials, we would need to make use of the so-called consistency condition proposed by Prager (1949). For perfectly plastic materials, the consistency condition means that the stress state must remain on the yield surface. For strain-hardening materials, consistency means that during plastic flow the stress state remains on the subsequent yield surface. Mathematically the consistency condition is defined as follows:

$$\frac{\partial f}{\partial \sigma_{ij}} d\sigma_{ij} + \frac{\partial f}{\partial \alpha} d\alpha = 0 \tag{1.56}$$

Given the hardening parameter α is a function of plastic strains, the consistency condition can be further reduced to

$$\frac{\partial f}{\partial \sigma_{ij}} d\sigma_{ij} + \frac{\partial f}{\partial \alpha} \frac{\partial \alpha}{\partial \varepsilon_{ij}^p} d\varepsilon_{ij}^p = 0 \tag{1.57}$$

For the special case of perfectly plastic materials, the second term in Equations 1.56 and 1.57 becomes zero.

1.7.5 Elastic-plastic stress–strain relation

Following Yu (2006), we now summarize a general procedure for developing the complete relationship between a stress rate and a strain rate for materials involving elastic-plastic deformation:

a. The total strain rate is divided into elastic and plastic strain rates:

$$d\varepsilon_{ij} = d\varepsilon_{ij}^e + d\varepsilon_{ij}^p \tag{1.58}$$

b. The stress rate is linked to the elastic strain rate through Hooke's law by the elastic stiffness matrix D_{ijkl}:

$$d\sigma_{ij} = D_{ijkl} d\varepsilon_{ij}^e = D_{ijkl}(d\varepsilon_{ij} - d\varepsilon_{ij}^p) \tag{1.59}$$

c. The plastic strain rate is expressed as a function of the plastic potential via the plastic flow rule according to Equation 1.55:

$$d\sigma_{ij} = D_{ijkl}\left(d\varepsilon_{ij} - d\lambda \frac{\partial g}{\partial \sigma_{ij}}\right) \tag{1.60}$$

d. The consistency condition according to Equation 1.56 is then used in conjunction with Equation 1.60 to arrive at

$$d\lambda = \frac{1}{H} \frac{\partial f}{\partial \sigma_{ij}} D_{ijkl} \, d\varepsilon_{kl} \tag{1.61}$$

where H is given by

$$H = \frac{\partial f}{\partial \sigma_{ij}} D_{ijkl} \frac{\partial g}{\partial \sigma_{kl}} - \frac{\partial f}{\partial \alpha} \frac{\partial \alpha}{\partial \varepsilon_{ij}^p} \frac{\partial g}{\partial \sigma_{ij}}$$ (1.62)

e. The elastic-plastic stress–strain relation is derived by substituting Equation 1.61 into Equation 1.60 as follows:

$$d\sigma_{ij} = D_{ijkl}^{ep} d\varepsilon_{kl}$$ (1.63)

where the elastic-plastic stiffness matrix D_{ijkl}^{ep} is given by

$$D_{ijkl}^{ep} = D_{ijkl} - \frac{1}{H} \frac{\partial f}{\partial \sigma_{ij}} D_{ijmn} \frac{\partial g}{\partial \sigma_{mn}} \frac{\partial f}{\partial \sigma_{pq}} D_{pqkl}$$ (1.64)

It is noted that the above procedure for deriving an elastic-plastic stress–strain relation is valid for both strain hardening and perfectly plastic materials. In the case of perfect plasticity, however, the yield function remains unchanged and therefore the expression of H, as defined by Equation 1.62, takes the following form:

$$H = \frac{\partial f}{\partial \sigma_{ij}} D_{ijkl} \frac{\partial g}{\partial \sigma_{kl}}$$ (1.65)

1.7.6 Tresca and von Mises plasticity models for cohesive soils

Using the procedure outlined above, the complete stress–strain relation can be established provided that a yield function and a plastic potential can be specified for a given soil material. For perfectly plastic materials, it is often assumed that a failure criterion is taken as a yield criterion.

For clays under undrained loading conditions, either Tresca's or von Mises' yield criterion can be used. Tresca's yield criterion is defined by

$$f(\sigma_{ij}) = \sigma_1 - \sigma_3 - 2s_u = 0$$ (1.66)

where σ_1 and σ_3 are the major and minor principal stresses, respectively. s_u is the undrained shear strength.

When clays are loaded under undrained conditions, the volume remains constant. It is therefore appropriate to adopt an associated plastic flow rule by treating the yield function as the plastic potential as well:

$$g(\sigma_{ij}) = \sigma_1 - \sigma_3 - 2s_u = 0$$ (1.67)

A slightly better alternative to the Tresca yield criterion is the criterion proposed by von Mises (1913). With von Mises plasticity, both yield function and plastic potential are assumed to be as follows:

$$f(\sigma_{ij}) = g(\sigma_{ij}) = (\sigma_1 - \sigma_2)^2 + (\sigma_2 - \sigma_3)^2 + (\sigma_3 - \sigma_1)^2 - 6k^2 = 0$$ (1.68)

where k is the undrained shear strength of the soil in pure shear. If we make the von Mises yield surface pass through the corners of the Tresca yield surface, then we have the

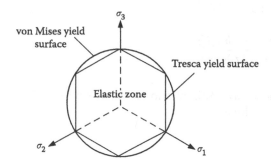

Figure 1.29 Yield surfaces in the principal stress space.

relationship between k and s_u:

$$k = \frac{2}{\sqrt{3}} s_u \qquad (1.69)$$

Figure 1.29 shows the characteristics of Tresca and von Mises yield surface in the principal stress space.

1.7.7 Mohr–Coulomb plasticity model for frictional soils

The yield criterion proposed by Coulomb (1773) was in terms of shear stress τ and normal stress σ_n acting on a plane. It assumes that yielding occurs as long as the shear stress and the normal stress satisfy the following condition:

$$f(\sigma_{ij}) = |\tau| - \sigma_n \tan \varnothing - c = 0 \qquad (1.70)$$

where c and $\varnothing$ are the cohesion and internal angle of friction for the soil.
In terms of the principal stresses, the Mohr–Coulomb yield function can be defined as follows:

$$f(\sigma_{ij}) = \sigma_1 - \sigma_3 + (\sigma_1 + \sigma_3) \sin \varnothing - 2c \cos \varnothing = 0 \qquad (1.71)$$

where σ_1 and σ_3 are the major and minor principal stresses, respectively. The Mohr–Coulomb yield criterion (see Figure 1.30a) allows soil frictional behavior to be considered and thus is widely used in geotechnical engineering.
When $\varnothing = 0$, such as the case of undrained shearing for clays, Equation 1.71 becomes

$$f(\sigma_{ij}) = \sigma_1 - \sigma_3 - 2c = 0 \qquad (1.72)$$

This is identical to Tresca's yield criterion, except that "s_u" was used in Equation 1.66 instead of "c." It should be noted that the undrained shear strength s_u reflects the behavior of cohesive soils (clays) in undrained shearing under $\varnothing = 0$ conditions and s_u has little to do with cohesion between soil particles, which "c" implies. Therefore it is important that we refer to undrained shear strength of clays as s_u instead of c or c_u. Because of the similarity in yield criterion, clays in undrained shearing are also referred to as Tresca soils.
With the Mohr–Coulomb plasticity model, it is widely accepted that the plastic potential can take a similar form as the yield function, but the friction angle is replaced by a

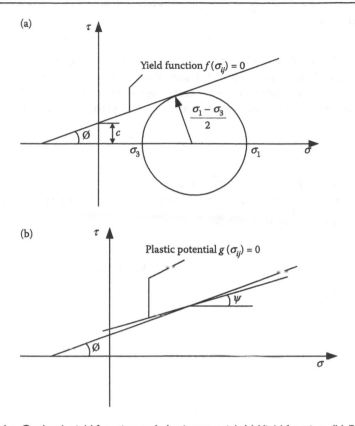

Figure 1.30 Mohr–Coulomb yield function and plastic potential. (a) Yield function. (b) Plastic potential.

dilation angle ψ (Davis, 1968).

$$g(\sigma_{ij}) = |\tau| - \sigma_n \tan \psi = \text{constant} \tag{1.73}$$

which can be shown graphically in Figure 1.30b.

A plastic potential in terms of the principal stresses is given as follows:

$$g(\sigma_{ij}) = \sigma_1 - \sigma_3 + (\sigma_1 + \sigma_3) \sin \emptyset = \text{constant} \tag{1.74}$$

Therefore Mohr–Coulomb's plastic flow rule will be called *associated* if the dilation angle is equal to the friction angle, $\psi = \emptyset$. Otherwise the flow rule will be called *non-associated*, and

$$d\varepsilon_{ij}^p = d\lambda \frac{\partial g}{\partial \sigma_{ij}} \tag{1.75}$$

where g is a plastic potential which may or may not be the same as the yield function and $d\lambda$ is a positive scalar

EXAMPLE 1.6

Given

It is assumed that a Tresca yield criterion following Equation 1.66 and plastic potential of Equation 1.67 can be used to model plastic behavior of clay under undrained loading conditions. A Mohr–Coulomb criterion according to Equation 1.71 and plastic potential of Equation 1.73 are used to model non-associated plastic behavior of sand under drained loading conditions.

Required

a. The plastic volumetric strain rate for clay
b. The ratio of plastic normal strain rate to plastic shear strain rate

Solution

a. According to Equations 1.55 and 1.67, the plastic strain rates are

$$d\varepsilon_{ij}^{p} = d\lambda \frac{\partial g}{\partial \sigma_{ij}} \tag{1.55}$$

$$g(\sigma_{ij}) = \sigma_1 - \sigma_3 - 2s_u = 0 \tag{1.67}$$

$$d\varepsilon_1^{p} = d\lambda \frac{\partial g}{\partial \sigma_1} = d\lambda;$$

$$d\varepsilon_2^{p} = d\lambda \frac{\partial g}{\partial \sigma_2} = 0;$$

$$d\varepsilon_3^{p} = d\lambda \frac{\partial g}{\partial \sigma_3} = -d\lambda$$

Therefore, the plastic volumetric strain rate predicted by the Tresca criterion is

$$d\varepsilon_v^{p} = d\varepsilon_1^{p} + d\varepsilon_2^{p} + d\varepsilon_3^{p} = 0$$

b. According to Equations 1.55 and 1.73, the plastic shear strain rate is

$$d\varepsilon_{ij}^{p} = d\lambda \frac{\partial g}{\partial \sigma_{ij}} \tag{1.55}$$

$$g(\sigma_{ij}) = |\tau| - \sigma_n \tan \psi = \text{constant} \tag{1.73}$$

$$d\varepsilon_t^{p} = d\lambda \frac{\partial g}{\partial \tau} = d\lambda$$

and the plastic normal strain rate is

$$d\varepsilon_n^{p} = d\lambda \frac{\partial g}{\partial n} = -d\lambda \tan \psi$$

Therefore, the ratio of the plastic normal strain rate to the plastic shear strain rate is

$$\frac{d\varepsilon_n^{p}}{d\varepsilon_t^{p}} = -\tan \psi$$

1.7.8 Unified critical state plasticity model for soils

The critical state concept, introduced at Cambridge University, has been used to develop a large number of strain hardening plasticity constitutive soil models (e.g., Roscoe and Burland, 1968; Schofield and Wroth, 1968; Yu, 1998). The models presented by Schofield and Wroth (1968) and Roscoe and Burland (1968) are known as Cam clay and modified Cam clay, respectively. They are only appropriate for modeling clay soils and differ slightly in the shape of the yield surface adopted. On the other hand, the unified critical state plasticity model developed by Yu (1998), known as Clay and Sand Model (CASM), can be applied to model both clay and sand. CASM contains Cam clay as a special case.

The unified critical state yield function of CASM proposed by Yu (1998) can be written in terms of shear stress q and effective mean stress p' as follows:

$$f(\sigma_{ij},\ p'_0) = \left(\frac{q}{Mp'}\right)^n + \frac{\ln p' - \ln p'_0}{\ln r} = 0 \tag{1.76}$$

where n is a new material constant introduced in CASM that is typically in the range of 1.0–5.0 for different soils. The preconsolidation pressure p'_0 is used to control the size of the yield surface during plastic flow (see Figure 1.31) and therefore serves as the hardening parameter (i.e., $\alpha = p'_0$).

By choosing spacing ratio, $r = e$ (natural base e, see Equation 1.32) and $n = 1$, it can be shown that the yield function according to Equation 1.76 of CASM reduces to the yield function of the classic Cam clay model. Details of the derivation are presented in Example 1.8.

If an associated plastic flow rule is used as in most critical state plasticity models, then the plastic potential is taken as identical to the yield function, namely

$$g(\sigma_{ij}) = f(\sigma_{ij}) = \left(\frac{q}{Mp'}\right)^n + \frac{\ln p' - \ln p'_0}{\ln r} = 0 \tag{1.77}$$

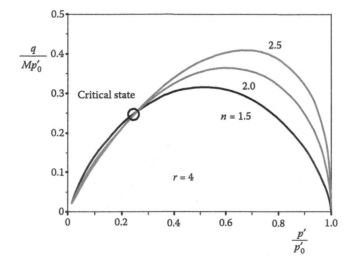

Figure 1.31 Yield surfaces of the unified critical state model CASM normalized with respect to preconsolidation stresses. (Adapted from Yu, H.S. 1998. International Journal for Numerical and Analytical Methods in Geomechanics, 22, 621–653.)

Alternatively, Rowe's stress–dilatancy relation (Rowe, 1962) may be taken as the plastic flow rule, which can then be integrated to give a slightly different plastic potential (Yu, 1998):

$$g(\sigma_{ij}) = 3M \ln\frac{p'}{C} + (3 + 2M) \ln\left(\frac{2q}{p'} + 3\right) - (3 - M) \ln\left(3 - \frac{q}{p'}\right) = 0 \qquad (1.78)$$

where C is a size parameter that can be determined from the above equation with the current stresses (i.e., ensuring the plastic potential passes through the current stress state).

The hardening law is determined by the fact that the preconsolidation pressure increases with the plastic volumetric strain along the normal consolidation line (Wood, 1990), as shown in Figure 1.32.

From the geometry in Figure 1.32b, it can be readily shown (Schofield and Wroth, 1968; Wood, 1990) that the rate of plastic volumetric strain ε_p^p is linked to the rate of preconsolidation pressure as follows:

$$d\varepsilon_p^p = -\frac{dv^p}{v} = (\lambda - \kappa)\frac{dp_0'}{vp_0'} \qquad (1.79)$$

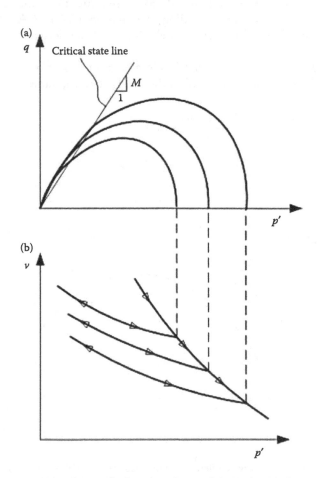

Figure 1.32 Successive yield surfaces with the normal consolidation line and successive swelling lines (a) in the space of $q - p'$; (b) in the space of $v - p'$.

Therefore, the hardening law is given by

$$\frac{\partial \alpha}{\partial \varepsilon_p^p} = \frac{\partial p_0'}{\partial \varepsilon_p^p} = \frac{v p_0'}{\lambda - \kappa} \tag{1.80}$$

In critical state soil mechanics, it is widely assumed that the size of yield surface is not dependent on plastic shear strain ε_q^p, and therefore we have

$$\frac{\partial \alpha}{\partial \varepsilon_q^p} = \frac{\partial p_0'}{\partial \varepsilon_q^p} = 0 \tag{1.81}$$

Elastic volumetric strain rate is linked to the mean effective stress change along the swelling line as follows:

$$d\varepsilon_p^e = -\frac{dv^e}{v} = \kappa \frac{dp'}{v p'} = \frac{dp'}{K} \tag{1.82}$$

where the bulk modulus K is proportional to the mean effective stress defined by

$$K = \frac{v p'}{\kappa} \tag{1.83}$$

On the other hand, elastic shear strain rate is calculated from shear stress change through a constant shear modulus G as in other elasticity models:

$$d\varepsilon_q^e = \frac{dq}{3G} \tag{1.84}$$

As shown in Yu (2006), the plastic flow rule according to Equation 1.55, the consistency of Equation 1.57, and the elastic stress–strain relations according to Equations 1.82 and 1.84 can be used together to give a complete elastic-plastic stress–strain relationship:

$$d\varepsilon_p = \frac{dp'}{K} + \frac{1}{K_p} \frac{\partial g}{\partial p'} \times \left(\frac{\partial f}{\partial p'} dp' + \frac{\partial f}{\partial q} dq \right) \tag{1.85}$$

$$d\varepsilon_q = \frac{dq}{3G} + \frac{1}{K_p} \frac{\partial g}{\partial q} \times \left(\frac{\partial f}{\partial p'} dp' + \frac{\partial f}{\partial q} dq \right) \tag{1.86}$$

which can be conveniently expressed in a matrix form as follows:

$$\begin{bmatrix} d\varepsilon_p \\ d\varepsilon_q \end{bmatrix} = \begin{bmatrix} C_{11} & C_{12} \\ C_{21} & C_{22} \end{bmatrix} \begin{bmatrix} dp' \\ dq \end{bmatrix} = \begin{bmatrix} \dfrac{1}{K} + \dfrac{1}{K_p} \dfrac{\partial g}{\partial p'} \dfrac{\partial f}{\partial p'} & \dfrac{1}{K_p} \dfrac{\partial g}{\partial p'} \dfrac{\partial f}{\partial q} \\ \dfrac{1}{K_p} \dfrac{\partial g}{\partial q} \dfrac{\partial f}{\partial p'} & \dfrac{1}{3G} + \dfrac{1}{K_p} \dfrac{\partial g}{\partial q} \dfrac{\partial f}{\partial q} \end{bmatrix} \begin{bmatrix} dp' \\ dq \end{bmatrix} \tag{1.87}$$

in which K_p is a plastic hardening modulus defined as

$$K_p = -\frac{\partial f}{\partial p_0'} \frac{\partial p_0'}{\partial \varepsilon_p^p} \frac{\partial g}{\partial p'} \tag{1.88}$$

EXAMPLE 1.7

Given

Consider the unified critical state yield criterion CASM defined by Equation 1.76. By setting $n = 1$ and $r = e$, the conventional Cam clay model is recovered from CASM with the following well-known yield function and plastic potential:

$$f(\sigma_{ij}) = g(\sigma_{ij}) = \frac{q}{Mp'} + \ln p' - \ln p'_0 = 0$$

Required

The complete elastic-plastic stress–strain relationship.

Solution

From the Cam clay yield function and plastic potential above, we can obtain

$$\frac{\partial f}{\partial p'} = \frac{\partial g}{\partial p'} = -\frac{q}{M(p')^2} + \frac{1}{p'};$$

$$\frac{\partial f}{\partial q} = \frac{\partial g}{\partial q} = \frac{1}{Mp'};$$

$$\frac{\partial f}{\partial p'_0} = -\frac{1}{p'_0};$$

From Equation 1.79, we have

$$d\varepsilon_p^p = -\frac{dv^p}{v} = (\lambda - \kappa)\frac{dp'_0}{vp'_0} \qquad (1.79)$$

$$\frac{\partial p'_0}{\partial \varepsilon_p^p} = \frac{vp'_0}{\lambda - \kappa};$$

From Equation 1.83, we have

$$K = \frac{vp'}{\kappa} \qquad (1.83)$$

From Equation 1.88, we have

$$K_p = -\frac{\partial f}{\partial p'_0}\frac{\partial p'_0}{\partial \varepsilon_p^p}\frac{\partial g}{\partial p'} \qquad (1.88)$$

$$K_p = -\frac{\partial f}{\partial p'_0}\frac{\partial p'_0}{\partial \varepsilon_p^p}\frac{\partial g}{\partial p'} = \frac{v}{\lambda - \kappa}\left(-\frac{q}{M(p')^2} + \frac{1}{p'}\right)$$

Therefore, following Equation 1.87, we can obtain the following components in the elastic-plastic stress–strain matrix:

$$\begin{bmatrix} d\varepsilon_p \\ d\varepsilon_q \end{bmatrix} = \begin{bmatrix} C_{11} & C_{12} \\ C_{21} & C_{22} \end{bmatrix}\begin{bmatrix} dp' \\ dq \end{bmatrix} = \begin{bmatrix} \dfrac{1}{K} + \dfrac{1}{K_p}\dfrac{\partial g}{\partial p'}\dfrac{\partial f}{\partial p'} & \dfrac{1}{K_p}\dfrac{\partial g}{\partial p'}\dfrac{\partial f}{\partial q} \\ \dfrac{1}{K_p}\dfrac{\partial g}{\partial q}\dfrac{\partial f}{\partial p'} & \dfrac{1}{3G} + \dfrac{1}{K_p}\dfrac{\partial g}{\partial q}\dfrac{\partial f}{\partial q} \end{bmatrix}\begin{bmatrix} dp' \\ dq \end{bmatrix}$$

$$(1.87)$$

$$C_{11} = \frac{1}{K} + \frac{1}{K_p}\frac{\partial g}{\partial p'}\frac{\partial f}{\partial p'} = \frac{\kappa}{\nu p'} + \frac{\lambda - \kappa}{\nu}\left(-\frac{q}{M(p')^2} + \frac{1}{p'}\right);$$

$$C_{12} = \frac{1}{K_p}\frac{\partial g}{\partial p'}\frac{\partial f}{\partial q} = \frac{\lambda - \kappa}{\nu M p'};$$

$$C_{21} = \frac{1}{K_p}\frac{\partial g}{\partial q}\frac{\partial f}{\partial p'} = \frac{\lambda - \kappa}{\nu M p'};$$

$$C_{22} = \frac{1}{3G} + \frac{1}{K_p}\frac{\partial g}{\partial q}\frac{\partial f}{\partial q} = \frac{1}{3G} + \frac{\lambda - \kappa}{\nu M(M p' - q)}$$

EXAMPLE 1.8

Given

Assume that the samples of Weald clay presented in Example 1.4 may be modeled by the special case of CASM as in Example 1.7 (i.e., the conventional Cam clay model) The material parameters of the clay relevant to this special CASM model are given as

$$\kappa = 0.025, \quad G = 1714\,\text{kPa}, \quad \lambda = 0.093, \quad M = 0.9, \quad r = e, \quad n = 1$$

Required

The stress–strain curves of a drained triaxial compression on samples A and C.

Solution

We first consider a sample with a general initial stress state with effective mean stress p'_i and shear stress q_i. During a drained triaxial compression with a constant confining pressure, the mean effective stress p' and shear stress q should satisfy

$$dq = 3dp' \quad \text{or} \quad (q - q_i) = 3(p' - p'_i)$$

Therefore,

$$q = q_i + 3(p' - p'_i)$$

The stress path in the $p' - q$ space always moves from the initial stress state to an initial yield state, then to the critical state. Before the initial yield, the soil response is elastic. From the initial yield to the critical state, the soil response is elastic-plastic. Therefore, we need to determine the stress conditions when the initial yield and critical state are first reached.

Following the yield function of CASM defined as follows:

$$f(\sigma_{ij},\, p'_0) = \left(\frac{q}{Mp'}\right)^n + \frac{\ln p' - \ln p'_0}{\ln r} = 0 \tag{1.76}$$

Cam clay yield function is a special case of $n = 1$, and $r = e$ as given above. The mean effective stress at the initial yield, denoted by p'_y, can thus be derived from the Cam clay yield function:

$$\left(\frac{q}{Mp'}\right)^1 + \frac{\ln p' - \ln p'_0}{\ln e} = \frac{q_i + 3(p'_y - p'_i)}{Mp'_y} + \ln p'_y - \ln p'_{0i} = 0$$

where p'_{0i} is the initial preconsolidation pressure.

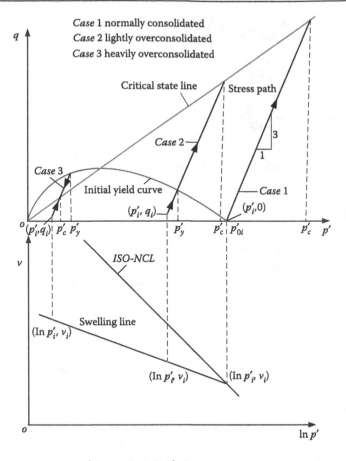

Figure E1.8a Stress paths in the $p' - q$ and $v - \ln p'$ planes.

From Equation 1.26, the mean effective stress at the critical state, denoted by p'_c, can be determined as

$$q = Mp' \tag{1.26}$$

Recall from above that in drained triaxial test, $dq = 3dp'$ (see Figure E1.8a); therefore

$$p'_c = \frac{q_i - 3p'_i}{M - 3}$$

Depending on their initial stress states, soil samples may be classified as normally consolidated, lightly overconsolidated, or heavily overconsolidated, as shown in Figure E1.8a. The corresponding locations of p'_i, p'_y, and p'_c for different initial states are illustrated in Figure E1.8a.

Starting from an overconsolidated state (such as Cases 2 and 3 in Figure E1.8a), response of the first stage of the compression defined by $p'_i \le p' < p'_y$ is elastic

[(p',q) lies within the yield surface]. From the elastic stress–strain relationship, we obtain

$$dv = -v\, d\varepsilon_p = -v\, d\varepsilon_p^e = -\kappa \frac{dp'}{p'};$$

$$d\varepsilon_q = d\varepsilon_q^e = \frac{dq}{3G} = \frac{dp'}{G}$$

Integrating (accumulating dv and $d\varepsilon_q$) along the stress path with respect to p' leads to

$$v = v_i - \kappa \ln\left(\frac{p'}{p'_i}\right);$$

$$\varepsilon_q = \frac{p' - p'_i}{G}$$

It should be noted that for normally consolidated soils the stress state is initially on yield surface; that is, $p'_i = p'_y = p'_{0i}$. Thus, no elastic behavior will be predicted for normally consolidated samples.

During the second stage of the compression after the initial yield, the stress–strain response will be elastic-plastic. For heavily overconsolidated soil $p'_y > p'_c$ (Case 3 in Figure E1.8a), p' starts from p'_i and increases monotonically before reaching the yield surface where $p' = p'_y$. A strain softening then occurs where p' decreases from p'_y to p'_c. In contrast, for lightly overconsolidated and normally consolidated soils, $p'_y < p'_c$, hardening would occur with $p' \geq p'_y$.

Recall from the complete elastic-plastic stress–strain matrix given Equation 1.87 and Example 1.7,

$$\begin{bmatrix} d\varepsilon_p \\ d\varepsilon_q \end{bmatrix} = \begin{bmatrix} C_{11} & C_{12} \\ C_{21} & C_{22} \end{bmatrix} \begin{bmatrix} dp' \\ dq \end{bmatrix} = \begin{bmatrix} \dfrac{1}{K} + \dfrac{1}{K_p}\dfrac{\partial g}{\partial p'}\dfrac{\partial f}{\partial p'} & \dfrac{1}{K_p}\dfrac{\partial g}{\partial p'}\dfrac{\partial f}{\partial q} \\ \dfrac{1}{K_p}\dfrac{\partial g}{\partial q}\dfrac{\partial f}{\partial p'} & \dfrac{1}{3G} + \dfrac{1}{K_p}\dfrac{\partial g}{\partial q}\dfrac{\partial f}{\partial q} \end{bmatrix} \begin{bmatrix} dp' \\ dq \end{bmatrix}$$

$$(1.87)$$

$$dv = -v\, d\varepsilon_p = -v(C_{11}dp' + C_{12}dq)$$

Again, in drained test, $dq = 3\, dp'$; thus, we have

$$dv = -v\, d\varepsilon_p = -v(C_{11} + 3C_{12})dp'$$

$$C_{11} = \frac{\kappa}{vp'} + \frac{\lambda - \kappa}{v}\left(-\frac{q}{M(p')^2} + \frac{1}{p'}\right),$$

and

$$C_{12} = \frac{\lambda - \kappa}{vMp'}$$

Integrating along the stress path with respect to p', from p'_i to p'_y, leads to

$$v = v_y - \lambda \ln\left(\frac{p'}{p'_y}\right) - \frac{(3p'_i - q_i)(\lambda - \kappa)}{M}\left(\frac{1}{p'_y} - \frac{1}{p'}\right)$$

where v_y is the specific volume at the initial yield, which is obtained as

$$v_y = v_i - \kappa \ln\left(\frac{p'_y}{p'_i}\right)$$

Noting from the above result that v is also a function of p' after the initial yield, the relationship between the mean effective stress p' and shear strain ε_q can be established as follows:

$$\varepsilon_q = \int d\varepsilon_q + \varepsilon_{qy} = \int_{p'_y}^{p'} (C_{21} + 3C_{22})dp' + \varepsilon_{qy} = \int_{p'_y}^{p'} C_q dp' + \varepsilon_{qy}$$

where

$$C_q = \frac{\lambda - \kappa}{vM}\left(\frac{1}{p'} + \frac{3}{(M-3)p' + 3p'_i - q_i}\right) + \frac{1}{G}$$

and ε_{qy} is the shear strain at the initial yield which is given as

$$\varepsilon_{qy} = \frac{p'_y - p'_i}{G}$$

Numerical integration techniques may be used to solve ε_q, as it is difficult to obtain an analytical expression of ε_q due to the presence of v.

Noting that before and after the yield, there is the following relationship:

$$dv = -v\,d\varepsilon_p$$

We therefore have

$$\varepsilon_p - \varepsilon_{pi} = -\ln\frac{v}{v_i}$$

where v_i and ε_{pi} are the initial specific volume and volumetric strain, respectively. As $\varepsilon_{pi} = 0$, the volumetric strain is obtained as

$$\varepsilon_p = -\ln\frac{v}{v_i}$$

So far, we have established the relationships between p' and ε_q, ε_q, q; hence all these quantities are interrelated.

For sample A, we have

$$v_i = 1.638; \quad p'_i = 207\,\text{kPa}; \quad q_i = 0; \quad p'_{0i} = 207\,\text{kPa}$$

From the given equations for p'_y (Equation 1.76) and p'_c, $[p'_c = (q_i - 3p'_i)/(M - 3)]$, we obtain

$$p'_y = 207\,\text{kPa}; \quad p'_c = 295.7\,\text{kPa}$$

The specific volume at initial yield is

$$v_y = v_i - \kappa\left(\frac{\ln p'_y}{p'_i}\right) = 1.638$$

Similarly, for sample C, we have

$$v_i = 1.616; \quad p'_i = 34\,\text{kPa}; \quad q_i = 0; \quad p'_{0i} = 827\,\text{kPa}$$

and

$$p'_y = 96\,\text{kPa};$$

$$p'_c = 48.6\,\text{kPa};$$

$$v_y = 1.59$$

It should be noted that a numerical integration technique is used here to calculate the shear strain. As long as the shear strain is obtained, it may be linked to shear stress and volumetric strain. The constitutive responses of samples A and C are presented in terms of stress path in the $p' - q$ plane, shear strain versus shear stress, and shear strain versus volumetric strain in Figure E1.8b–g. It is clearly shown that for the normally consolidated sample A, no peak strength is predicted toward the critical state, while for the heavily overconsolidated sample C, a softening behavior is observed after the peak (reaching yield surface) toward the critical state.

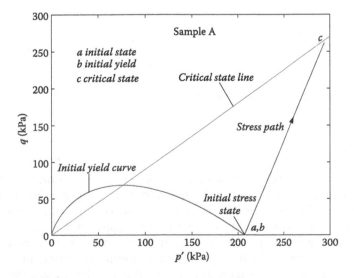

Figure E1.8b Stress path in the $p' - q$ plane.

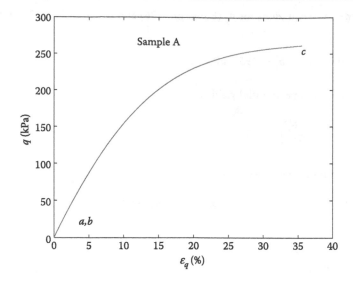

Figure E1.8c Shear strain versus shear stress.

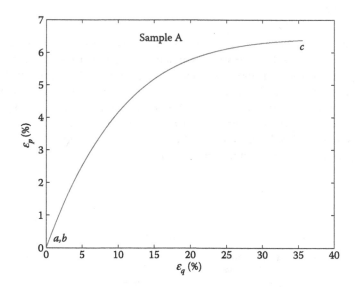

Figure E1.8d Shear strain versus volumetric strain.

1.8 REMARKS

In this chapter, we have reviewed basic principles of consolidation theory and stress–strain behavior of clays and sands in drained and undrained shearing as they relate to foundation engineering analysis. We learned that for certain combinations of density and stress states or stress history, the soil can dilate or contract if sheared in drained condition, or the soil can generate positive or negative pore pressure if sheared in undrained conditions. Regardless of its dilative nature, under continual shearing the soil will eventually reach a state in which it behaves as a frictional fluid with a constant volume and a constant ratio of shear stress q to effective mean stress p'. This ultimate state is called

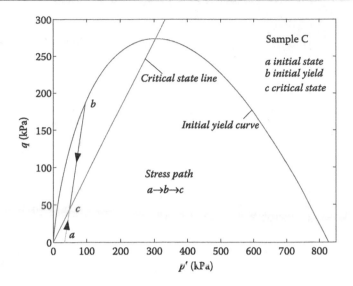

Figure E1.8e Stress path in the $p' - q$ plane.

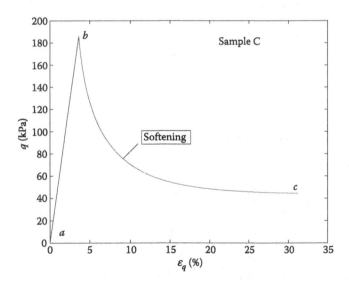

Figure E1.8f Shear strain versus shear stress.

the critical state, a powerful reference state for predicting the mechanical behavior of soils when subjected to various loading conditions. These important observations form the basis for a rather comprehensive framework of critical state soil mechanics that has the following unique features:

- The critical state lines in $q - p'$ and $v - p'$ space serve as the ultimate reference for soil behavior.
- State parameter ξ reflects the difference in void ratio e from the current state to the critical state line, at the same mean effective stress p'.

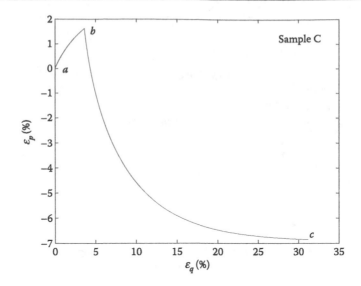

Figure E1.8g Shear strain versus volumetric strain.

- Soil compressibility, shear strength, and dilative nature of the soil during shearing can be predicted with great accuracy under this framework.
- Elastic-plastic constitutive models can be and have been established based on critical state soil mechanics.

HOMEWORK

1.1. According to the definitions of the specific volume v, specific gravity of soil particles G_s and water content w show that

$$v = 1 + G_s w$$

1.2. It is known that the experimental critical state line for a clay may be defined by Equations 1.26 and 1.27 as follows:

$$q = 0.88p' \quad \text{and} \quad v = 2.12 - 0.12 \ln p'$$

The slope of overconsolidation (swelling) line k is measured to be 0.02 and the spacing ratio is $r = 4$. Assume that sample A of the clay has an initial water content of 23.6% and a mean effective stress of 250 kPa, while sample B of the clay has an initial water content of 22.2% and a mean effective stress of 45 kPa. The specific gravity of the clay particle is measured to be $G_s = 2.75$.
  a. Calculate the OCR of samples A and B with reference to Figure 1.23.
  b. Calculate the state parameter for samples A and B.
  c. Calculate the undrained shear strengths of samples A and B when sheared under the undrained condition.

1.3. The friction angle is defined in terms of principal stresses σ_1 and σ_3 as

$$\sin(\emptyset) = \frac{\sigma_1 - \sigma_3}{\sigma_1 + \sigma_3}$$

Assume that the stress ratio q/p' at the critical state for the triaxial compression is M. Derive Equation 1.46.

In Example 1.5, the peak friction angles for Portaway sand during triaxial shear tests were estimated using Equation 1.47. The peak friction angles can also be estimated by using the yield surface. Assume that Portaway sand conforms to the Mohr–Coulomb yield function of Equation 1.70. The material parameters for Portaway sand determined by laboratory tests are given as

$$n = 3.5, \; r = 19.2, \; M = 1.19$$

The slope of overconsolidation (swelling) line k is assumed to be 0.005. Estimate the peak friction angles for samples CICD1 and CIDC4 during triaxial shear tests by using Equation 1.70.

1.4. In Example 1.1, if after the swelling test the clay sample is compressed again with the vertical effective pressure increased from 150 kPa to 400 kPa, calculate the void ratio and final height of the clay at the end of the compression.

1.5. A normally consolidated clay layer of soil in the field with a thickness of 1.4 m is sandwiched with both upper and lower layers of sand. It is found by laboratory tests that for this clay the compression index is $C_c = 0.4$ and the coefficient of consolidation is $c_v = 5 \, \text{mm}^2/\text{min}$. The initial void ratio of clay is 1.0 before the construction of a foundation on the ground surface when the in-situ effective vertical pressure due to soil weight is 170 kPa. After the construction of the foundation, the average vertical effective pressure on the clay layer becomes 400 kPa. Find the maximum consolidation settlement of the clay layer and the degree of consolidation two months (60 days) after construction.

1.6. Assume that von Mises yield criterion (Equation 1.67) and associated flow rule can be used to model the plastic behavior of soil under undrained condition. Derive the plastic volumetric strain rate for clay.

REFERENCES

Been, K. and Jefferies, M.G. 1985. A state parameter for sands. *Geotechnique*, 35, 99–112.

Been, K., Jefferies, M.G., and Hachey, J.E. 1991. The critical state of sands. *Geotechnique*, 37, 285–299.

Bishop, A.W. and Henkel, D.J. 1957. *The Measurement of Soil Properties in the Triaxial Test*. Edward Arnold, London, 228p.

Bolton, M.D. 1986. The strength and dilatancy of sands. *Geotechnique*, 36, 65–78.

Casagrande, A. 1936. Determination of the preconsolidation load and its practical significance. *Proceedings of the 1st International Conference on Soil Mechanics and Foundation Engineering*, Cambridge, MA, Vol. 3, pp. 60–64.

Castro, G. 1969. Liquefaction of sands. PhD thesis, Harvard University.

Castro, G. and Poulos, S.J. 1977. Factors affecting liquefaction and cyclic mobility. *Journal of the Geotechnical Engineering Division, ASCE*, 103, 501–506.

Collins, I.F., Pender, M.J., and Wang, J. 1992. Cavity expansion in sands under drained loading conditions. *International Journal for Numerical and Analytical Methods in Geomechanics*, 16, 3–23.

Coulomb, C.A. 1773. Essai sur une application des regles des maximis et minimis a quelques problemes de statique relatifs a l'architecture. *Mem. pres. Par div. savants*, 7, 343–382.

Davis, E.H. 1968. Theories of plasticity and the failure of soil masses. In: I.K. Lee, ed., *Soil Mechanics—Selected Topics*, Butterworths, London, pp. 341–380.

Henkel, D.J. 1959. The relationships between the strength, pore-water pressure, and volume-change characteristics of saturated clays. *Geotechnique*, 5, 119–135.

Hill, R. 1950. *The Mathematical Theory of Plasticity*. Clarendon Press, Oxford, UK.

Holtz, R.D. and Kovacs, W.D. 1981. *An Introduction to Geotechnical Engineering.* Prentice-Hall, Englewood Cliffs, NJ.

Holtz, R.D., Kovacs, W.D., and Sheahan, T.C. 2010. *An Introduction to Geotechnical Engineering.* Second Edition, Pearson, Upper Saddle River, NJ.

Klotz, E.U. and Coop, M.R. 2002. On the identification of critical state lines for sands. *Geotechnical Testing Journal,* 25(3), 1–14.

Mayne, P.W., Coop, M.R., Springman, S.M., Huang, A.B., and Zornberg, J.G. 2009. Geomaterial behavior and testing. State of The Art Report #1, *Proceedings, XVII International Conference on Soil Mechanics and Geotechnical Engineering,* Alexandria, Egypt, IOS Press, the Netherlands, pp. 2777–2872.

Mesri, G. and Godlewski, P.M. 1977. Time and stress-compressibility interrelationship. *Journal of Geotechnical Engineering Division, American Society of Civil Engineers,* 103(GT5), 417–430.

Parry, R.H.G. 1958. Correspondence on "On the yielding of soils." *Geotechnique,* 8, 183–186.

Poulos, S.J. 1981. The steady state of deformation. *Journal of Geotechnical Engineering, ASCE,* 107, 553–562.

Prager, W. 1949. Recent developments of the mathematical theory of plasticity. *Journal of Applied Physics,* 20, 235.

Roscoe, K.H. and Burland, J.B. 1968. On generalized stress–strain behaviour of wet clay. In: Heyman, J. and Leckie, F.A., eds., *Engineering Plasticity.* Cambridge University Press, London, pp. 535–609.

Roscoe, K.H., Schofield, A.N., and Wroth, C.P. 1958. On the yielding of soils. *Geotechnique,* 8, 22–52.

Rowe, P.W. 1962. The stress-dilatancy relation for static equilibrium of an assembly of particles in contact. *Proceedings of the Royal Society of London, A,* 267, 500–527.

Schofield, A.N. and Wroth, C.P. 1968. *Critical State Soil Mechanics.* McGraw-Hill, London, 310p.

Skempton, A. W. 1944. Notes on the compressibility of clays. *Quarterly Journal of the Geological Society of London,* 100, 119–135.

Taylor, D.W. 1948. *Fundamentals of Soil Mechanics,* Wiley, New York.

Terzaghi, K 1943. *Theoretical Soil Mechanics.* Wiley, New York.

Vesic, A.S. and Clough, G.W. 1968. Behaviour of granular materials under high stresses. *Journal of the Soil Mechanics and Foundations Division, ASCE,* 94(SM3), 661–688.

von Mises, R. 1913. *Mechanik der festen Korper im Korper im plastisch deformation.* Zustand. Nachr. Ges. Wiss, Gottingen, 582p.

Wang, J. 2005. The Stress–strain and strength characteristics of Portaway sand. PhD Thesis, University of Nottingham, UK.

Wood, D.M. 1990. *Soil Behaviour and Critical State Soil Mechanics.* Cambridge University Press, Cambridge, UK.

Wroth, C.P. and Bassett, N. 1965. A stress–strain relationship for the shearing behaviour of a sand. *Geotechnique,* 15, 32–56.

Wroth, C.P. and Wood, D.M. 1978. The correlation of index properties with some basic engineering properties of soils. *Canadian Geotechnical Journal,* 15, 137–145.

Yu, H.S. 1994. State parameter from self-boring pressuremeter tests in sand. *Journal of Geotechnical Engineering, ASCE,* 120, 2118–2135.

Yu, H.S. 1996. Interpretation of pressuremeter unloading tests in sands. *Geotechnique,* 46, 17–31.

Yu, H.S. 1998. CASM: A unified state parameter model for clay and sand. *International Journal for Numerical and Analytical Methods in Geomechanics,* 22, 621–653.

Yu, H.S. 2000. *Cavity Expansion Methods in Geomechanics.* Kluwer Academic, the Netherlands.

Yu, H.S. 2006. *Plasticity and geotechnics.* Springer, New York.

Chapter 2

Limit analysis and cavity expansion methods

2.1 INTRODUCTION

The limit equilibrium method has been widely used to assess the general stability of a soil mass (Terzaghi, 1943), often referred to as the limit state analysis or safety analysis, as described in Chapter 1 (Section 1.1). This includes the calculation of the bearing capacity of foundations and lateral earth pressure behind the retaining walls. The limit equilibrium method assumes the soil as rigid and perfectly plastic, and the calculation is based on rigid mechanics and force equilibrium for an assumed failure mechanism. It must be noted that the limit equilibrium method is an approximate approach, as it does not satisfy all the key stress equilibrium and displacement equations—it only satisfies the force equilibrium of soil–structure interaction in a global sense. Nevertheless, this approach is still widely used in geotechnical practice, mainly due to its simplicity.

The objectives of this chapter are:

- To describe the concept of limit equilibrium method. This method is used extensively throughout this book.
- To introduce a more rigorous limit analysis method that can be used to replace the limit equilibrium method. The limit analysis is founded in the theory of plasticity and can be used to obtain upper- and lower-bound solutions for problems such as foundation bearing capacity and lateral earth pressure (Chen, 1975; Yu, 2006a). The upper- and lower-bound solutions bracket the exact solution in between so that their accuracy is known. In some cases, the upper- and lower-bound solutions derived could be very close, so that the exact solution may be taken as the average of the upper and lower bounds. Limit analysis is particularly suitable for applications to shallow foundations and lateral earth pressure computations.
- To introduce the cavity expansion method. The cavity expansion that includes spherical and cylindrical expansion represents a unique class of boundary value problems that have close similarity to many foundation or geotechnical engineering problems. For example, the state of soil around a penetrating pile or a cone penetrometer is close to that of a spherical cavity expansion. Pressuremeter expansion (described in Chapter 3) is a field replication of cylindrical cavity expansion. This chapter provides a basic introduction to the elastic-plastic cavity expansion method. Unlike limit equilibrium and limit analysis methods, the soil medium can be considered as elastic-plastic in the cavity expansion method. Among other applications, the cavity expansion method has proved to be very useful in determining the bearing capacity of deep foundations (Yu, 2000) and the interpretation of in-situ testing results such as cone penetration and pressuremeter tests (described in Chapter 3).

2.2 LIMIT EQUILIBRIUM METHOD

When a soil mass is subjected to an applied load on its boundary that is gradually increasing, the soil mass initially deforms elastically and then in an elastic-plastic manner until a time when the external load reaches a critical value at which the soil mass collapses (i.e., deformation increases indefinitely under a constant load). Such critical state is called a *limit load state*. The external load at the limit load state is referred to as the *limit load* or *collapse load* for a given soil mass. For the design of a foundation in soil, it is essential that we are able to predict the collapse load, F_c, for the foundation.

The limit equilibrium method is widely used in geotechnical practice (Terzaghi, 1943) for assessing the stability of a soil mass or calculating the collapse load. Most of the subjects covered in this book are related to limit equilibrium analysis. The limit equilibrium method makes use of an assumed failure mechanism that consists of a series of slip surfaces that divide the soil mass into a number of sectors or blocks. All blocks are assumed rigid. Displacements of the blocks are allowed only along the slip surfaces as rigid bodies. The shear strength of the soil mass constitutes the resisting forces along the slip surfaces, and the resisting forces remain constant irrespective of the displacements along the slip surfaces. This analysis method is based on the equilibrium for all the forces applied to the assumed failure mechanism and resisting forces available along the slip surfaces. The maximum load that can be applied to the assumed failure mechanism while maintaining the force equilibrium will be referred to as the collapse load, F_c. According to the plasticity concept described in Chapter 1, the soil mass is assumed to be rigid (i.e., all blocks are rigid) and perfectly plastic (i.e., peak-resisting forces along the slip surfaces remain constant) in limit equilibrium analysis.

To illustrate the concept of limit equilibrium analysis, we now use it to determine the bearing capacity of a strip foundation placed on a cohesive soil in undrained condition ($\emptyset = 0$ and undrained shear strength $= s_u$), as shown in Figure 2.1a. The foundation is infinitely long in the direction that is perpendicular to the paper. The analysis is thus for a unit thickness of the foundation. We consider a surcharge q (force per unit area) applied to the ground surface outside the foundation. The foundation has a width B, the bearing capacity of the foundation q_u equals to the collapse load, F_c divided by the

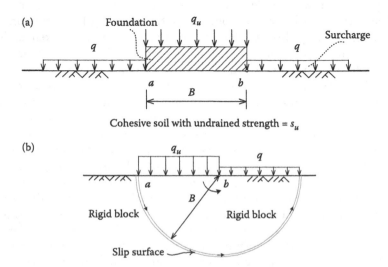

Figure 2.1 Limit equilibrium analysis of a foundation on cohesive soil. (a) The strip foundation and the applied load; (b) the assumed failure mechanism.

area of the foundation (consider a unit thickness), or

$$q_u = \frac{F_c}{(B \times 1)} \tag{2.1}$$

For simplicity, we assume a simple failure mechanism, which consists of a half-circle having a radius B, as shown in Figure 2.1b, and the soil is weightless. The foundation occupies the left half of the half-circle (ab in Figure 2.1b) where the bearing pressure is applied. What we need is to determine the collapse load, F_c ($= q_u \times B$) for the rigid block beneath the foundation that tends to rotate in a counter-clockwise direction about the edge of the foundation b (i.e., the center of the half-circle) due to the applied load, as shown in Figure 2.1b.

At failure, the soil block would rotate about the center b. To ensure overall equilibrium of the soil block, we take and balance the moments about b due to the foundation load, the surcharge, and the shear resistance along the slip surface:

$$(q_u \times B) \times \frac{B}{2} = (q \times B) \times \frac{B}{2} + s_u \times (\pi \times B) \times B \tag{2.2}$$

which gives a limit equilibrium solution for the bearing capacity q_u as

$$q_u = q + 2\pi s_u \tag{2.3}$$

It should be noted that the above problem can be equally solved considering the foundation occupies the right half of the half-circle and rotates in a clockwise direction about point a, and results in the same limit equilibrium solution. The solution is considered correct or exact if the assumed failure mechanism is the same as the failure pattern within the soil mass for the actual loaded foundation. To assure validity of the analysis, we can either verify that the assumed failure mechanism is substantially identical to that in the field, or verify analytically that the current solution is either higher or lower than the correct one for the given loading conditions. To take the former approach, we can perform physical model tests and directly or indirectly measure the actual failure mechanism in soil. Or, we can follow the latter approach and use the limit analysis described below to establish the upper bound and lower bound for the given loading conditions. Loading tests on scaled or full-size foundation systems have been conducted for over half a century. The limit equilibrium analysis-based design methods presented in the following chapters have generally been verified by various physical model tests and/or other types of analytical or numerical analyses. The following sections describe the limit analysis, a more rigorous method, and discuss the correctness of the limit equilibrium method.

2.3 INTRODUCTION TO PLASTIC LIMIT ANALYSIS

Assuming an elastic-perfectly plastic behavior of soils following an associated plastic flow rule (see Section 1.7), two fundamental theorems of plastic collapse for limit analysis have been established (Hill, 1948, 1950; Drucker et al., 1952). They are known as the upper- and lower-bound theorems of limit analysis that can be used to derive an upper and a lower bound of the exact collapse load without having to follow an incremental, step-by-step load-deformation analysis.

Consider a perfectly plastic soil mass loaded by a set of surface tractions F_o on its boundary. When F_o is increased from zero, the soil mass deforms elastically initially and then starts to become partly plastic. When F_o reaches a critical value, called collapse load F_c, a limit state (i.e., plastic collapse) will occur when the deformation continues to increase with the load remaining constant. At plastic collapse, it can be proved (Yu, 2006a) that the stress field within the soil mass remains constant. It then follows that at collapse the elastic strain will be zero and all the strain will be plastic (not recoverable).

The following sections present two important theorems that can be used to obtain a lower bound and an upper bound of the collapse load F_c.

2.3.1 Lower-bound theorem of plastic collapse

To apply the lower-bound theorem, we need to define a statically admissible stress field that is in equilibrium with the surface traction and nowhere violates the yield condition. A load F_c^s is defined as the collapse load that corresponds to a statically admissible stress field.

The lower-bound theorem states:

If all changes in geometry occurring during collapse are neglected, a statically admissible collapse load, F_c^s, is always less than or equal to the exact collapse load F_c (i.e., $F_c^s \leq F_c$). The equality sign is valid only when the statically admissible stress field is the true stress field. In other words, the load derived from a statically admissible stress field, F_c^s, is a lower bound of the true collapse load, F_c.

2.3.2 Upper-bound theorem of plastic collapse

To apply the upper-bound theorem, we need to define a kinematically admissible velocity field that is compatible with the boundary conditions. A load F_c^k is defined as a collapse load that corresponds to a kinematically admissible velocity field.

The upper-bound theorem states:

If all changes in geometry occurring during collapse are neglected, a kinematically admissible collapse load, F_c^k, is always greater than or equal to the exact collapse load F_c ($F_c^k \geq F_c$). The equality sign is valid only when the kinematically admissible velocity field is the true velocity field. In other words, the load derived from a kinematically admissible velocity field, F_c^k, is an upper bound on the true collapse load factor.

The purpose of a limit analysis is to determine F_c^k and F_c^s of a soil mass for the given strength parameters and boundary conditions through computations of the upper- and lower-bound analysis.

2.4 UPPER-BOUND LIMIT ANALYSIS

An upper bound to the collapse load can be derived from an energy equation between the external work and the internal plastic power dissipation with any kinematically admissible failure mechanism. Yu (2006a) suggested the following procedure to carry out an upper-bound limit analysis:

1. Assume a kinematic failure mechanism.
2. Draw a relevant velocity field diagram to give the relationship between various velocity components.

3.   Derive the external work and the internal (plastic) power dissipation and equate them.
4.   Solve the energy equation to obtain an upper-bound solution.

2.4.1 Velocity discontinuity and plastic power dissipation

It is a common practice to assume that the internal power dissipation occurs entirely along the velocity discontinuities (i.e., the slip surface). This means that no plastic deformation takes place in the continuum away from the slip surface. This assumption is used mainly to lend simplicity in the application of the upper-bound theorem.

Refer to Figure 2.2; a slip or failure surface will form in soil mass as long as the normal stress σ_n and shear stress τ acting on it satisfy the Mohr–Coulomb yield criterion:

$$|\tau| = \sigma_n \tan \emptyset + c \tag{2.4}$$

The internal power dissipation rate per unit area, D, of the slip surface can be expressed as follows (Davis and Selvadurai, 2002):

$$D = [\sigma_n d\varepsilon_n^p + |\tau| d\varepsilon_t^p] b = \sigma_n du_n + |\tau| du_t \tag{2.5}$$

where
$b =$ thickness of the slip surface
ε_n^p, $\varepsilon_t^p =$ normal and tangential plastic strain rate, respectively
u_n, $u_t =$ velocity in normal and tangential direction, respectively

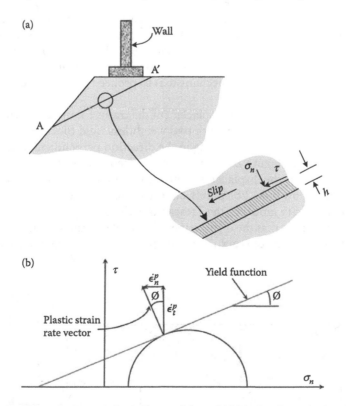

Figure 2.2 Velocity discontinuity and plastic flow at failure. (a) Velocity discontinuity; (b) plastic flow at failure. (Adapted from Davis, R.O. and Selvadurai, A.P.S. 2002. *Plasticity and Geomechanics.* Cambridge University Press, Cambridge, UK.)

The normal and tangential plastic strain rates (or the corresponding normal and tangential velocities) are related by the associated plastic flow rule as shown in Figure 2.2, namely

$$\frac{d\varepsilon_n^p}{d\varepsilon_t^p} = \frac{du_n}{du_t} = -\tan\varnothing \tag{2.6}$$

Substituting Equation 2.6 into Equation 2.5 leads to

$$D = |\tau|du_t - \sigma_n \tan\varnothing\, du_t = (|\tau| - \sigma_n \tan\varnothing)du_t \tag{2.7}$$

By substituting the Mohr–Coulomb yield criterion of Equation 2.4 into Equation 2.7, we obtain a very simple but important expression of plastic power dissipation per unit length of the slip surface (Davis, 1968):

$$D = cdu_t \tag{2.8}$$

The failure of soil mass at collapse may follow a slip surface that represents a surface of discontinuous velocities or displacements. We will now investigate the way by which we can calculate the internal plastic power dissipation along the slip surface.

2.4.2 Velocity fields from displacement diagrams

In order to carry out an upper-bound limit analysis using a failure mechanism, we need to determine the displacements for the external work calculation. We must also determine the velocities across all the slip surfaces for the internal plastic power dissipation calculation. As discussed by Atkinson (1981), the simplest way of determining the velocity or displacement components for a failure mechanism is to proceed graphically by making use of a displacement diagram.

As an illustration for drawing the displacement diagrams for upper-bound analysis, we consider a simple failure mechanism with two weightless rigid blocks under a strip foundation (a foundation that is infinitely long in the direction perpendicular to the paper) on cohesive soil as shown in Figure 2.3 (Davis, 1968; Chen, 1975; Atkinson, 1981).

It should be noted that the displacement direction on a failure surface is controlled by an associated plastic flow rule. For purely cohesive soils (i.e. conditions of $\varnothing = 0$ in the Mohr–Coulomb plasticity model or the Tresca plasticity model), the plastic flow rule predicts no volume change so that the displacement is parallel to the slip surface as the movement normal to the slip surface is zero. On the other hand, for cohesive-frictional soils in

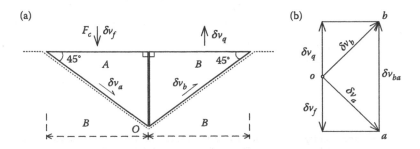

Figure 2.3 A strip foundation on cohesive soils. (a) Failure mechanism; (b) displacement diagram.

the Mohr–Coulomb plasticity model, the displacement direction should make an angle of the internal friction with the slip surface as determined by the associated flow rule.

2.4.3 Upper-bound solution for a foundation on cohesive soil

Now we consider a foundation resting on the surface of a cohesive soil with an undrained shear strength of s_u. There is no surcharge acting on the soil surface outside the foundation, and the soil is assumed as weightless. For simplicity, we assume that at collapse, the soil will fail according to the two-rigid block mechanism with three slip surfaces shown in Figure 2.3a.

Assume that the collapse load of the foundation is denoted by F_c and the foundation has a width of B. From the displacement diagram shown in Figure 2.3b, we can see the relationships between the displacement rate of the foundation and those of the blocks A and B (see Figure 2.3a) are as follows:

$$\delta v_a = \sqrt{2}\delta v_f \tag{2.9}$$

$$\delta v_b = \sqrt{2}\delta v_q = \sqrt{2}\delta v_f \tag{2.10}$$

In addition, the relative displacement rate along the interface between blocks A and B is

$$\delta v_{ba} = 2\delta v_q = 2\delta v_f \tag{2.11}$$

In order to make use of the upper-bound method, we calculate the external work done by the collapse load of the foundation with a vertical displacement increment of δv_f as follows:

$$\delta E = F_c \times \delta v_f = F_c \delta v_f \tag{2.12}$$

The internal plastic power dissipation along the three slip surfaces in the failure mechanism can be determined by

$$\delta W = s_u \times \delta v_a \times \sqrt{2}B + s_u \times \delta v_b \times \sqrt{2}B + s_u \times \delta v_{ba} \times B \tag{2.13}$$

Substituting Equations 2.9 through 2.11 into Equation 2.13 leads to the following simple expression of the internal power dissipation:

$$\delta W = 6Bs_u\delta v_f \tag{2.14}$$

According to the upper-bound theorem, we equate the external work expressed by Equation 2.12 and the internal power dissipation of Equation 2.14, namely

$$F_c\delta v_f = 6Bs_u\delta v_f \tag{2.15}$$

which gives the following collapse load F_c as a function of foundation width B and soil undrained shear strength s_u:

$$F_c = 6Bs_u \tag{2.16}$$

In terms of average pressure, the above upper-bound solution takes the form

$$q_u = \frac{F_c}{B} = 6s_u \tag{2.17}$$

which is an upper bound to the exact solution of $(2 + \pi)s_u$ (Hill, 1950; Bishop, 1953) for this classic problem of bearing capacity of a shallow foundation presented in Chapter 4.

Now we consider the same foundation problem but with an additional surcharge q applied on the soil surface outside the foundation. Again we assume the same failure mechanism as before. In this case, the external work will be done by the collapse load and also by the surcharge, namely

$$\delta E = F_c \times \delta v_f + q \times B \times \delta v_q = (F_c - qB)\delta v_f \tag{2.18}$$

The internal plastic power dissipation will be the same as the case with no surcharge and therefore is expressed by Equation 2.14. Equating Equations 2.18 and 2.14 gives an upper-bound solution:

$$F_c = 6Bs_u + qB \tag{2.19}$$

In terms of average pressure, the above upper-bound collapse load solution takes the following form:

$$q_u = \frac{F_c}{B} = q + 6s_u \tag{2.20}$$

EXAMPLE 2.1

Given

For a shallow strip foundation with a width of B resting on a cohesive soil with undrained shear strength s_u, there is no surcharge acting on the soil surface outside the foundation and assume that soil is weightless. We assume that at collapse, the soil will fail according to the symmetric five-rigid block mechanism shown in Figure E2.1.

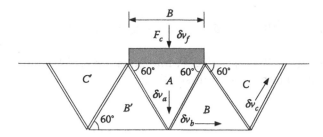

Figure E2.1 A rigid-block failure mechanism for an undrained cohesive soil.

Required

The upper bound to the bearing capacity of the foundation.

Solution

To apply the upper-bound theorem, we need to draw a displacement diagram associated with the assumed failure mechanism. The failure mechanism consists of five rigid blocks—A, B, C, B', and C'. Due to geometric symmetry, block A will move downward together with the foundation. The resulting displacement diagram is shown in Figure E2.1a.

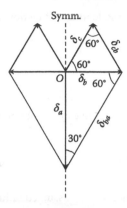

Figure E2.1a The displacement diagram.

It can be shown that the geometric relationships between the displacement rate of the foundation and block A with those of the blocks B (B') and C (C') are as follows:

$$\delta v_a = \delta v_f$$

$$\delta v_b = \tan(30°)\,\delta v_a = \frac{\sqrt{3}}{3}\,\delta v_a$$

$$\delta v_c = \delta v_b = \frac{\sqrt{3}}{3}\,\delta v_a$$

In addition, the relative displacement rates along the interfaces between the soil blocks A, B, and C are

$$\delta v_{ba} = 2\delta v_b = \frac{2\sqrt{3}}{3}\,\delta v_a$$

$$\delta v_{cb} = \delta v_b = \frac{\sqrt{3}}{3}\,\delta v_a$$

In order to make use of the upper-bound method, we calculate the external work done by the collapse load of the foundation with a vertical displacement increment of δv_f as follows:

$$\delta E = F_c \times \delta v_f = F_c \delta v_a$$

Noting the symmetry, the internal plastic power dissipation along the eight slip surfaces in the failure mechanism can be determined by

$$\delta W = 2(s_u \times \delta v_b \times B + s_u \times \delta v_{ba} \times B + s_u \times \delta v_c \times B + s_u \times \delta v_{cb} \times B)$$

Substituting the displacement rate expressions into the above equation leads to the following simple expression of the internal power dissipation:

$$\delta W = \frac{10\sqrt{3}}{3}\,Bs_u\delta v_a$$

According to the upper-bound theorem, we equate the external work and the internal power dissipation, namely

$$F_c\delta v_a = \frac{10\sqrt{3}}{3}\,Bs_u\delta v_a$$

which gives the following collapse load as a function of foundation width and soil undrained shear strength:

$$F_c = \frac{10\sqrt{3}}{3} B s_u$$

In terms of average pressure, the above upper-bound solution takes the form

$$q_u = \frac{F_c}{B} = \frac{10\sqrt{3}}{3} s_u = 5.78 s_u$$

which is the upper-bound solution for the assumed failure mechanism.

We now consider the same foundation problem and failure mechanism described in Figure 2.1 but use the upper-bound method to determine the collapse load for the foundation at failure. The same failure mechanism consists of a half-circle of a rigid, weightless soil block beneath the loaded foundation passing through the edge and rotating about the other edge of the foundation, as shown in Figure 2.4.

At failure, the soil block would rotate about the center of the half-circle. Assume that the soil block has rotated by a small angle of $\delta\omega$. The external work done by the foundation pressure and the surcharge on the soil surface can be determined as follows:

$$\delta E = \int_0^B (x\delta\omega) q_u dx - \int_0^B (x\delta\omega) q dx = \frac{(q_u - q)B^2 \delta\omega}{2} \tag{2.21}$$

On the other hand, the internal plastic power will dissipate along the failure surface during the rotation. The displacement along the failure surface due to a small rotation of $\delta\omega$ will be $B\delta\omega$, and the power dissipation can be calculated by

$$\delta W = s_u \times B\delta\omega \times \pi B = \pi B^2 s_u \delta\omega \tag{2.22}$$

To obtain an upper bound, we equate the external work of Equation 2.21 and the internal plastic power dissipation of Equation 2.22:

$$\frac{(q_u - q)B^2 \delta\omega}{2} = \pi B^2 s_u \delta\omega \tag{2.23}$$

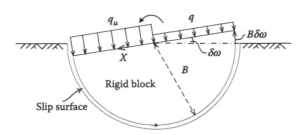

Figure 2.4 Upper bound of a foundation on cohesive soils.

which leads to an upper-bound collapse pressure

$$q_u = q + 2\pi s_u \tag{2.24}$$

The result is the same as Equation 2.3. This shows that with a rigid, translational failure mechanism the collapse load (pressure) derived from an energy equation of the upper-bound method is the same as that calculated from force limit equilibrium analysis.

2.4.4 Upper-bound for a foundation on cohesive-frictional soils

Now we consider a surface foundation resting on a cohesive-frictional, weightless soil with cohesion c and friction angle $\emptyset$, as shown in Figure 2.5a. There is no surcharge acting on the soil surface outside the foundation ($q = 0$ in Figure 2.5a). For simplicity, we assume that at collapse, the soil will fail according to the two-rigid block mechanism with three slip surfaces ac, bc, and cd, as shown in Figure 2.5 (Atkinson, 1981).

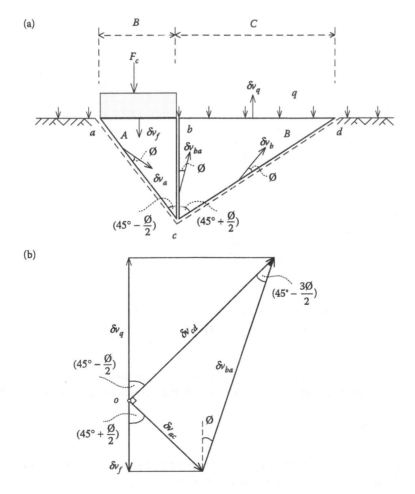

Figure 2.5 Upper bound for a foundation on cohesive-frictional soils. (a) Assumed failure mechanism; (b) displacement diagram.

If the width of the foundation is $ab = B$, the length of sides bd, ac, bc, and cd can be readily derived from geometry as follows:

$$\overline{bd} = B \tan^2\left(45° + \frac{\varnothing}{2}\right) \tag{2.25}$$

$$\overline{ac} = \frac{B}{\cos\left(45° + \dfrac{\varnothing}{2}\right)} \tag{2.26}$$

$$\overline{bc} = B \tan\left(45° + \frac{\varnothing}{2}\right) \tag{2.27}$$

$$\overline{cd} = \frac{B \tan\left(45° + \dfrac{\varnothing}{2}\right)}{\sin\left(45° - \dfrac{\varnothing}{2}\right)} \tag{2.28}$$

With reference to the displacement diagram, the tangential components of displacements along the slip surfaces ac ($= \delta v_a \cos\varnothing$), cd ($= \delta v_b \cos\varnothing$), and bc ($= \delta v_{ba} \cos\varnothing$) can be obtained from geometry as functions of the vertical displacement of the foundation δv_f as follows:

$$\delta v_{ac} = \delta v_a \cos\varnothing = \frac{\cos\varnothing}{\cos\left(45° + \dfrac{\varnothing}{2}\right)} \delta v_f \tag{2.29}$$

$$\delta v_{cd} = \delta v_b \cos\varnothing = \frac{\tan\left(45° + \dfrac{3\varnothing}{2}\right)\cos\varnothing}{\cos\left(45° + \dfrac{\varnothing}{2}\right)} \delta v_f \tag{2.30}$$

$$\delta v_{bc} = \delta v_{ba} \cos\varnothing = \frac{\cos\varnothing}{\cos\left(45° + \dfrac{3\varnothing}{2}\right)\cos\left(45° + \dfrac{\varnothing}{2}\right)} \delta v_f \tag{2.31}$$

To make use of the upper-bound method, we calculate the external work done by the collapse load of the foundation with a vertical displacement increment of δv_f as follows:

$$\delta E = F_c \times \delta v_f \tag{2.32}$$

The internal plastic power dissipation along the three slip surfaces in the failure mechanism assumed in Figure 2.5 can be determined by

$$\delta W = c \times \delta v_{ac} \times \overline{ac} + c \times \delta v_{cd} \times \overline{cd} + c \times \delta v_{bc} \times \overline{bc} \tag{2.33}$$

which can be expressed as a function of the vertical displacement of the foundation:

$$\delta W = cB\delta v_f \left\{ \begin{array}{l} \dfrac{\cos\varnothing}{\cos^2\left(45°+\dfrac{\varnothing}{2}\right)} + \dfrac{\tan\left(45°+\dfrac{\varnothing}{2}\right)\tan\left(45°+\dfrac{3\varnothing}{2}\right)\cos\varnothing}{\cos^2\left(45°+\dfrac{\varnothing}{2}\right)} \\[20pt] + \dfrac{\tan\left(45°+\dfrac{\varnothing}{2}\right)\cos\varnothing}{\cos\left(45°+\dfrac{\varnothing}{2}\right)\cos\left(\dfrac{45°+3\varnothing}{2}\right)} \end{array} \right\} \qquad (2.34)$$

By equating the external work according to Equation 2.32 and internal power dissipation from Equation 2.34, we obtain an upper bound of the collapse load:

$$F_c = cB \left\{ \begin{array}{l} \dfrac{\cos\varnothing}{\cos^2\left(45°+\dfrac{\varnothing}{2}\right)} + \dfrac{\tan\left(45°+\dfrac{\varnothing}{2}\right)\tan\left(45°+\dfrac{3\varnothing}{2}\right)\cos\varnothing}{\cos^2\left(45°+\dfrac{\varnothing}{2}\right)} \\[20pt] + \dfrac{\tan\left(45°+\dfrac{\varnothing}{2}\right)\cos\varnothing}{\cos\left(45°+\dfrac{\varnothing}{2}\right)\cos\left(45°+\dfrac{3\varnothing}{2}\right)} \end{array} \right\} \qquad (2.35)$$

and the average collapse pressure or bearing capacity q_u of the foundation is therefore given by

$$q_u = \dfrac{F_c}{B} = \left\{ \begin{array}{l} \dfrac{\cos\varnothing}{\cos^2\left(45°+\dfrac{\varnothing}{2}\right)} + \dfrac{\tan\left(45°+\dfrac{\varnothing}{2}\right)\tan\left(\dfrac{45°+3\varnothing}{2}\right)\cos\varnothing}{\cos^2\left(45°+\dfrac{\varnothing}{2}\right)} \\[20pt] + \dfrac{\tan\left(45°+\dfrac{\varnothing}{2}\right)\cos\varnothing}{\cos\left(45°+\dfrac{\varnothing}{2}\right)\cos\left(45°+\dfrac{3\varnothing}{2}\right)} \end{array} \right\} c \qquad (2.36)$$

For the special case of zero friction angle, the above solution reduces to the solution of Equation 2.17 for a purely cohesive soil.

We now consider the case when a surcharge, q (see Figure 2.2a), is applied on the soil surface outside the foundation. From the displacement diagram, we can determine the vertical displacement of the soil surface as follows (Atkinson, 1981):

$$\delta v_q = \tan\left(45°+\dfrac{\varnothing}{2}\right)\tan\left(45°+\dfrac{3\varnothing}{2}\right)\delta v_f \qquad (2.37)$$

The external work done by the collapse load and the additional surcharge will be given by

$$\delta E = F_c \times \delta v_f + q \times \overline{bd} \times (-\delta v_q)$$
$$= \left[F_c - qB \tan^3\left(45° + \frac{\varnothing}{2}\right) \tan\left(45° + \frac{3\varnothing}{2}\right) \right] \delta v_f \tag{2.38}$$

The internal power dissipation is the same as that expressed by Equation 2.34. Therefore, the energy equation is used to give the following collapse load expression:

$$F_c = cB \left\{ \begin{array}{c} \dfrac{\cos\varnothing}{\cos^2\left(45° + \dfrac{\varnothing}{2}\right)} + \dfrac{\tan\left(45° + \dfrac{\varnothing}{2}\right) \tan\left(\dfrac{45° + 3\varnothing}{2}\right) \cos\varnothing}{\cos^2\left(45° + \dfrac{\varnothing}{2}\right)} \\[4mm] + \dfrac{\tan\left(45° + \dfrac{\varnothing}{2}\right) \cos\varnothing}{\cos\left(45° + \dfrac{\varnothing}{2}\right) \cos\left(45° + \dfrac{3\varnothing}{2}\right)} \end{array} \right\}$$
$$+ qB \tan^3\left(45° + \frac{\varnothing}{2}\right) \tan\left(45° + \frac{3\varnothing}{2}\right) \tag{2.39}$$

In terms of average pressure q_u, the upper bound takes the following form:

$$q_u = \frac{F_c}{B} = c \left\{ \begin{array}{c} \dfrac{\cos\varnothing}{\cos^2\left(45° + \dfrac{\varnothing}{2}\right)} + \dfrac{\tan\left(45° + \dfrac{\varnothing}{2}\right) \tan\left(\dfrac{45° + 3\varnothing}{2}\right) \cos\varnothing}{\cos^2\left(45° + \dfrac{\varnothing}{2}\right)} \\[4mm] + \dfrac{\tan\left(45° + \dfrac{\varnothing}{2}\right) \cos\varnothing}{\cos\left(45° + \dfrac{\varnothing}{2}\right) \cos\left(45° + \dfrac{3\varnothing}{2}\right)} \end{array} \right\}$$
$$+ q \tan^3\left(45° + \frac{\varnothing}{2}\right) \tan\left(45° + \frac{3\varnothing}{2}\right) \tag{2.40}$$

For comparison, the exact solution for the bearing capacity of a foundation on cohesive-frictional soils with a surcharge, known as the Prandtl solution, is given below (Bishop, 1953; Yu, 2006a):

$$q_u = \frac{F_c}{B} = N_c c + N_q q \tag{2.41}$$

where

$$N_q = \tan^2\left(45° + \frac{\varnothing}{2}\right) e^{\pi \tan\varnothing} \tag{2.42}$$

$$N_c = (N_q - 1)\cot\varnothing \tag{2.43}$$

The Prandtl solution is presented in Chapter 4.

EXAMPLE 2.2

Given

For a shallow strip foundation with a width of B resting on a cohesive-frictional soil with cohesion c, and the internal friction angle $\varnothing$ of $15°$. There is no surcharge pressure acting on the soil surface outside the foundation. We assume that at collapse soils will fail according to the symmetric five-rigid block mechanism shown in Figure E2.2.

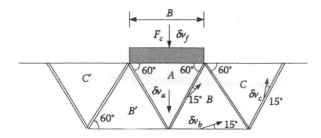

Figure E2.2 A rigid-block failure mechanism for a cohesive-frictional soil.

Required

The upper-bound to the bearing capacity of the foundation.

Solution

To apply the upper-bound theorem, we need to draw a displacement diagram associated with the assumed failure mechanism. The failure mechanism consists of five rigid blocks, A, B, C, B', and C'. Due to geometric symmetry, the block A will move downward together with the foundation. As a result, the displacement diagram can be shown in Figure E2.2a.

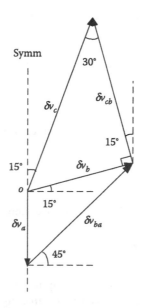

Figure E2.2a The displacement diagram.

It can be shown that the relationships between the velocity of the foundation and block A with those of the blocks B (B') and C (C') are as follows:

$$\delta v_a = \delta v_f$$

$$\delta v_b = \frac{\sin 45°}{\sin 30°}\delta v_a = 1.41\delta v_a$$

$$\delta v_c = \frac{\sin 45° \sin 90°}{\sin 30° \sin 30°}\delta v_a = 2.83\delta v_a$$

In addition, the relative velocities along the interfaces between the soil blocks A, B, and C are

$$\delta v_{ba} = \frac{\sin 105°}{\sin 30°}\delta v_a = 1.93\delta v_a$$

$$\delta v_{cb} = \frac{\sin 45° \sin 60°}{\sin 30° \sin 30°}\delta v_a = 2.45\delta v_a$$

In order to make use of the upper-bound method, we calculate the external work done by the collapse load of the foundation with a vertical displacement increment of δv_f as follows:

$$\delta E = F_c \times \delta v_f = F_c\delta v_f = F_c\delta v_a$$

Noting the symmetry, the internal plastic power dissipation along the eight slip surfaces in the failure mechanism can be determined by

$$\delta W = 2(c \times \delta v_b \times B + c \times \delta v_{ba} \times B + c \times \delta v_c \times B + c \times \delta v_{cb} \times B)$$

Substituting the velocity expressions into the above equation leads to the following simple expression of the internal power dissipation:

$$\delta W = 17.24Bc\delta v_a$$

According to the upper-bound theorem, we equate the external work and the internal power dissipation, namely

$$F_c\delta v_a = 17.24Bc\delta v_a$$

which gives the following collapse load as a function of foundation width and soil cohesion:

$$F_c = 17.24Bc$$

In terms of average pressure, the above upper-bound solution takes the form

$$q_u = \frac{F_c}{B} = 17.24c$$

Upper-bound limit analysis is similar to limit equilibrium analysis in that both approaches make use of a failure mechanism, although upper-bound analysis places more restrictions on the choice of failure mechanism. With an assumed failure mechanism, the upper-bound method is then based on an energy balance equation to derive an upper-collapse load, while the limit equilibrium method is based on force equilibrium to derive an estimated collapse load.

For a class of failure mechanisms with a plane system of rigid blocks separated by thin slip bands or surfaces (also termed *translational failure mechanisms*), it may be proved that the energy balance equation used in upper-bound analysis is equivalent to the force equilibrium used in limit equilibrium analysis (Yu, 2006a). Therefore with a translational failure mechanism, the solution from a limit equilibrium calculation may be regarded as an upper bound.

2.5 LOWER-BOUND LIMIT ANALYSIS

To make use of the lower-bound theorem to determine a lower-bound solution, we need to devise a statically admissible stress field that is in equilibrium with the surface tractions and nowhere violates the yield condition. An appropriate statically admissible stress state may vary smoothly from point to point or there may be sudden jumps or discontinuities, but in both cases equilibrium must be satisfied.

As shown by Prager and Hodge (1951) and Shield (1954), it is useful to employ stress discontinuities in the construction of statically admissible stress fields. This allows us to consider stress fields that may not be physically reasonable under normal circumstances.

2.5.1 Discontinuous stress fields

We consider a mass of solid separated by a stress discontinuity I-I into two regions, 1 and 2, as shown in Figure 2.6.

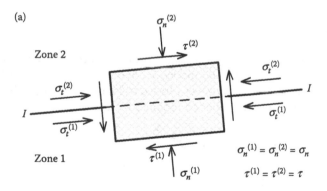

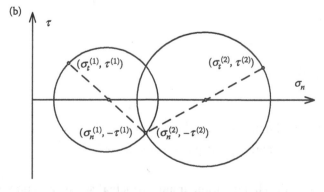

Figure 2.6 (a) Stress states across a discontinuity; (b) their Mohr circles.

Across a discontinuity, the stresses normal to the stress discontinuity on both sides must be the same to satisfy equilibrium. Also, the shear stresses must be continuous. Therefore equilibrium of an element across a stress discontinuity requires that

$$\sigma_n^{(1)} = \sigma_n^{(2)}; \quad \tau^{(1)} = \tau^{(2)} \tag{2.44}$$

However, the normal stresses along the direction of the discontinuity (i.e., $\sigma_t^{(1)}$ and $\sigma_t^{(2)}$) can be different from one side to another. In other words, a stress discontinuity may be defined as a boundary across which the normal stresses along its direction are discontinuous, as shown in Figure 2.6a.

As is well known, a state of stress for any element can be conveniently described by a Mohr circle. The states of stress for an element on both sides of the stress discontinuity can be drawn as two Mohr circles as shown in Figure 2.6b. The coordinates of the intersecting points of these two Mohr circles represent the normal and shear stresses acting on the stress discontinuity.

2.5.2 Discontinuous stress fields in a state of Tresca (undrained cohesive) failure

Now assume that the soil in both sides of a stress discontinuity is in a state of failure according to Tresca yield criterion or $\varnothing = 0$ according to Mohr–Coulomb yield criterion, and this is relevant for cohesive soils with an undrained shear strength of s_u. This means that the two Mohr circles must just touch on the Tresca yield surfaces ($\varnothing = 0$) defined by

$$|\tau| = s_u \tag{2.45}$$

Figure 2.7 shows the directions of the major principal stress on both sides of the stress discontinuity, which are denoted by θ_1 and θ_2. From the geometry of the Mohr circles that are in touch with the Tresca yield surfaces plotted in Figure 2.7b, it can be shown that the jump condition in the mean stress s from region 1 to region 2 is

$$s_2 - s_1 = 2s_u \sin(\theta_2 - \theta_1) = 2s_u \sin \delta\theta \tag{2.46}$$

where

$$s_1 = \frac{\left(\sigma_1^{(1)} + \sigma_2^{(1)}\right)}{2}$$

$$s_2 = \frac{\left(\sigma_1^{(2)} + \sigma_2^{(2)}\right)}{2}$$

$\sigma_1^{(1)}, \sigma_2^{(1)}$ = major and minor principal stress in zone 1
$\sigma_1^{(2)}, \sigma_2^{(2)}$ = major and minor principal stress in zone 2

where $\delta\theta = \theta_2 - \theta_1$ is the change in the major principal stress direction across the stress discontinuity from region 1 to region 2.

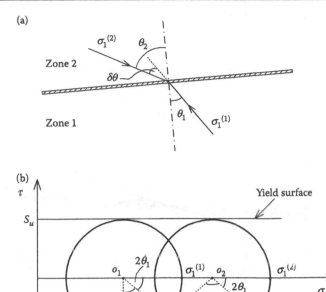

Figure 2.7 (a) Major principal stress across a discontinuity; (b) their Mohr circles at failure with Tresca criterion.

2.5.3 Discontinuous stress fields in a state of Mohr–Coulomb (cohesive-frictional) failure

Now assume that the soil on both sides of a stress discontinuity is in a state of failure according to the Mohr–Coulomb yield criterion, and this is relevant for cohesive-frictional soils. This means that the two Mohr circles must just touch on the Mohr–Coulomb yield surfaces defined by

$$|\tau| = c + \sigma_n \tan \varnothing \tag{2.47}$$

The change of stress across a stress discontinuity in a state of Mohr–Coulomb failure is plotted in Figure 2.8.

Figure 2.8a shows the change in the major principal stress across the stress discontinuity. We now define a mobilized friction angle $\varnothing_d$ along the stress discontinuity in terms of the shear and normal stresses acting on it as follows:

$$\tan \varnothing_d = \frac{|\tau|}{c \cot \varnothing + \sigma_n} \tag{2.48}$$

The mobilized friction angle $\varnothing_d$ on the stress discontinuity is shown in Figure 2.8b. This angle can be linked to the principal stress direction change $\delta\theta = \theta_2 - \theta_1$ by considering the geometry of the Mohr circles (Atkinson, 1981):

$$\sin \varnothing_d = \sin \varnothing \cos \delta\theta \tag{2.49}$$

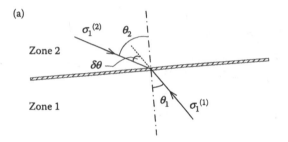

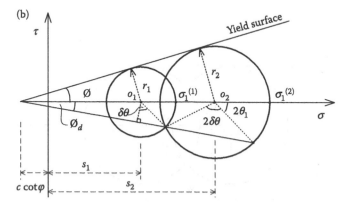

Figure 2.8 (a) Major principal stress across a discontinuity; (b) their Mohr circles at failure with Mohr–Coulomb criterion.

The mean stresses on both sides of the stress discontinuity are further related to the change in the major principal stress direction by the following equation:

$$\frac{s_2 + c\cot\varnothing}{s_1 + c\cot\varnothing} = \frac{\cos(\delta\theta - \varnothing_d)}{\cos(\delta\theta + \varnothing_d)} \tag{2.50}$$

where

$$s_1 = \frac{\left(\sigma_1^{(1)} + \sigma_2^{(1)}\right)}{2}$$

$$s_2 = \frac{\left(\sigma_1^{(2)} + \sigma_2^{(2)}\right)}{2}$$

$\sigma_1^{(1)}$, $\sigma_2^{(1)}$ = major and minor principal stress in zone 1

$\sigma_1^{(2)}$, $\sigma_2^{(2)}$ = major and minor principal stress in zone 2

The mobilized friction angle $\varnothing_d$ on the stress discontinuity is determined from the change of the principal stress direction $\delta\theta = \theta_2 - \theta_1$ by Equation 2.49.

In addition, we can derive a relationship between the angles θ_1 and θ_2 without reference to the mobilized friction angle on the discontinuity. Let's denote the radius of Mohr circle for region 1 by r_1 and that of Mohr circle for region 2 by r_2. Then the normal and shear stresses acting on the stress discontinuity can be expressed, alternatively, from both Mohr circles. First, we consider the shear stress on the stress discontinuity,

$$\tau = r_1 \sin 2\theta_1 = r_2 \sin(180° - 2\theta_2) = r_2 \sin 2\theta_2 \tag{2.51}$$

If we set the following for brevity:

$$\overline{s_1} = s_1 + c \cot \emptyset \qquad (2.52)$$

$$\overline{s_2} = s_2 + c \cot \emptyset \qquad (2.53)$$

we have the following relations from the geometry of the Mohr circles:

$$r_1 = \overline{s_1} \sin \emptyset \qquad (2.54)$$

$$r_2 = \overline{s_2} \sin \emptyset \qquad (2.55)$$

Therefore, we can write the shear stress of Equation 2.51 as follows:

$$\overline{s_1} \sin \emptyset \sin 2\theta_1 = \overline{s_2} \sin \emptyset \sin 2\theta_2 \qquad (2.56)$$

which can also be rewritten as

$$\frac{\overline{s_1}}{\overline{s_2}} = \frac{\sin 2\theta_2}{\sin 2\theta_1} \qquad (2.57)$$

Now we consider the normal stress on the stress discontinuity calculated from both Mohr circles, namely

$$\sigma_n = \overline{s_1} + r_1 \cos 2\theta_1 = \overline{s_2} - r_2 \cos (180° - 2\theta_2) \qquad (2.58)$$

which can be further reduced to

$$\frac{\overline{s_1}}{\overline{s_2}} = \frac{1 + \sin \emptyset \cos 2\theta_2}{1 + \sin \emptyset \cos 2\theta_1} \qquad (2.59)$$

By equating Equations 2.59 and 2.57, we obtain an equation that governs the change in the major principal stress direction across the discontinuity:

$$\frac{\sin 2\theta_2}{1 + \sin \emptyset \cos 2\theta_2} = \frac{\sin 2\theta_1}{1 + \sin \emptyset \cos 2\theta_1} \qquad (2.60)$$

2.5.4 Lower bound for a foundation on cohesive soils

Now we consider a surface foundation resting on a cohesive soil with undrained shear strength s_u. There is a surcharge pressure q acting on the soil surface outside the foundation.

By using the lower-bound method, we now assume the soil is divided into three plastic regions by two vertical stress discontinuities passing through the edges of the foundation, as shown in Figure 2.9a. The stress states for the two regions directly under the surcharge are identical and the regions are denoted as Soil Blocks I and III. The region under the foundation is denoted as Soil Block II. We assume that, at collapse, stress states in soil blocks are uniform and in the state of failure governed by Tresca criterion ($\emptyset = 0$).

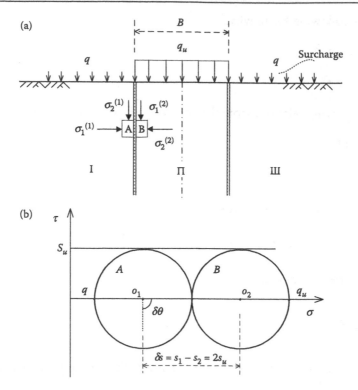

Figure 2.9 Lower bound for a foundation on cohesive soils. (a) Stress state; (b) the Mohr circles at failure.

It is very convenient to use Mohr circles in lower-bound analysis. We now consider the stress states in Blocks I and II.

- For Block I, the minor principal stress is the vertical stress that is equal to the applied surcharge q if self-weight is not considered, namely $\sigma_2^{(1)} = q$. The major principal stress will be in the horizontal direction (i.e., acting on the stress discontinuity), denoted by $\sigma_1^{(1)}$.
- For Block II, the major principal stress is the vertical pressure acting on the foundation at collapse, namely $\sigma_1^{(2)} = q_u$. The minor principal stress will be in the horizontal direction (i.e., acting on the stress discontinuity), which must be the same as the major principal stress on Block I due to stress equilibrium, namely $\sigma_2^{(2)} = \sigma_1^{(1)}$.

Now we construct two Mohr circles for the stress states for both Block I and Block II as shown in Figure 2.9b, having noted that they must touch on the Tresca yield surfaces to be in the state of failure. It is clear from the geometry of the Mohr circles that we have the following relationship:

$$q_u = q + 4s_u \tag{2.61}$$

Note that Equation 2.20 gives an upper bound for this same problem, and taking the mean of these lower and upper bounds we estimate the collapse pressure to be $q + 5s_u$. This estimate is very close to the exact solution of $q + (2 + \pi)s_u$.

Of course, the lower-bound solution of Equation 2.61 can also be obtained mathematically from the following stress jump condition derived earlier:

$$s_2 - s_1 = 2s_u \sin (\theta_2 - \theta_1) = 2s_u \sin \delta\theta \tag{2.62}$$

By noting that $\theta_1 = 0°$ and $\theta_2 = 90°$, we have

$$s_2 - s_1 = (q_u - s_u) - (q + s_u) = 2s_u \sin 90° = 2s_u \tag{2.63}$$

which leads to the same solution of $q_u = q + 4s_u$ as defined by Equation 2.61.

2.5.5 Lower bound for a foundation on cohesive-frictional soils

Now we consider a surface foundation resting on a cohesive-frictional soil with cohesion c and internal friction angle $\varnothing$. There is a surcharge pressure q acting on the soil surface outside the foundation.

Similar to the undrained cohesive case, we assume that the soil is divided into three plastic regions by two vertical stress discontinuities passing through the edges of the foundation, as shown in Figure 2.10a. The stress states for the two regions directly under the surcharge are identical, and they are denoted as Soil Block I. The region under the foundation is denoted as Soil Block II. We assume that at collapse stress states in soil blocks are uniform and in the state of failure governed by the Mohr–Coulomb criterion.

We now construct two Mohr circles for the stress states for both Block I and Block II as shown in Figure 2.10b. Note that they must touch on the Mohr–Coulomb yield surfaces in order to be in the state of failure. It is clear from the geometry of the Mohr circles that we can derive their radii as follows:

$$r_1 = \frac{\sin \varnothing}{1 - \sin \varnothing}(q + c \cot \varnothing) \tag{2.64}$$

$$r_2 = \frac{\sin \varnothing}{1 + \sin \varnothing}(q_u + c \cot \varnothing) \tag{2.65}$$

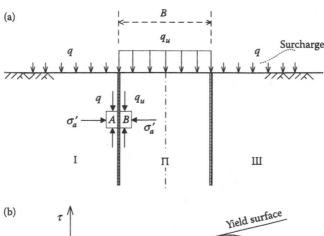

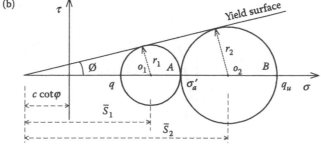

Figure 2.10 Lower bound for a foundation on cohesive-frictional soils. (a) Stress state; (b) the Mohr circles at failure.

In addition, we have the following relationship from the geometry of the Mohr circles:

$$q_u = q + 2r_1 + 2r_2 \tag{2.66}$$

which can be used in conjunction with Equations 2.64 and 2.65 to lead to the following lower-bound solution for the foundation collapse pressure:

$$q_u = \left(\frac{1 + \sin \varnothing}{1 - \sin \varnothing}\right)^2 q + \frac{4c \cos \varnothing}{(1 - \sin \varnothing)^2} = N^2 q + \frac{4Nc}{\cos \varnothing} \tag{2.67}$$

where

$$N = (1 + \sin \varnothing)/(1 - \sin \varnothing) = \tan^2 (45° + \varnothing/2)$$

Note that for a special case of zero friction angle, the above cohesive-frictional solution reduces to the purely cohesive lower-bound solution of Equation 2.61 derived earlier.

It should be noted that the lower-bound solution of Equation 2.67 can also be derived mathematically from the jump condition of Equation 2.59 by noting that $\theta_1 = 0°$ and $\theta_2 = 90°$, namely

$$\frac{\bar{s_1}}{\bar{s_2}} = \frac{q + r_1 + c \cot \varnothing}{q_u - r_2 + c \cot \varnothing} = \frac{1 + \sin \varnothing \cos 2\theta_2}{1 + \sin \varnothing \cos 2\theta_1} = \frac{1 - \sin \varnothing}{1 + \sin \varnothing} \tag{2.68}$$

By using Equations 2.64 and 2.65, the above stress jump condition can be used to lead to the same lower-bound solution defined by Equation 2.67.

EXAMPLE 2.3

Given

For a uniform pressure acting on a purely cohesive soil with a slope next to the loaded area having an angle of 30°, shown in Figure E2.3. The soil undrained shear strength is s_u, and the effect of soil weight is ignored for simplicity (Figure E2.3).

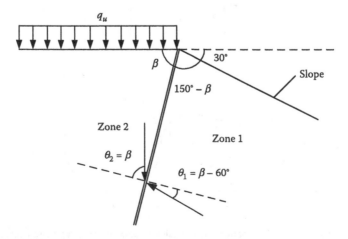

Figure E2.3 A soil slope of 30°.

Required

The lower-bound to the bearing capacity of the foundation.

Solution

To apply the lower-bound theorem, it is best to make use of the Mohr circles. For simplicity, we assume that the soil slope can be divided by a stress discontinuity into two zones, each with a uniform stress state defined by a major and a minor principal stress.

Now assume that the stress discontinuity is inclined from the horizontal by a degree of β. It is readily known that the angle between the stress discontinuity and the soil slope will be $(150° - \beta)$.

We now consider the stress states in Zones 1 and 2:

- For Zone 1, the minor principal stress is zero acting on the soil slope, namely $\sigma_2^{(1)} = 0$. The major principal stress will be acting along the slope direction (i.e., acting on the stress discontinuity), denoted by $\sigma_1^{(1)}$.
- For Zone 2, the major principal stress is the vertical pressure acting on the foundation at collapse, namely $\sigma_1^{(2)} = q_u$. The minor principal stress will be in the horizontal direction (i.e., acting on the stress discontinuity).

Given that the principal stresses from both zones are not acting along or normal to the stress discontinuity, it is clear that there are both normal and shear stress acting on the stress discontinuity.

Now we construct two Mohr circles for the stress states for both Zone 1 and Zone 2 as shown below, having noted that they must touch on the Tresca yield surfaces to be in the state of failure. The coordinates of the interaction point of the two Mohr circles represent the normal and shear stresses acting on the stress discontinuity (Figure E2.3a).

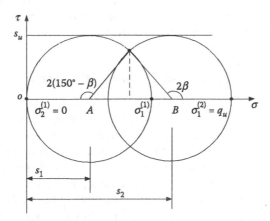

Figure E2.3a Mohr circles in the cohesive soil.

It is clear from the geometry of the Mohr circles that we have the following relationships:

$$2(150° - \beta) = 2\beta \quad \text{and} \quad q_u = 2s_u \cos(180° - 2\beta) + 2s_u$$

The above two equations can be used to give

$$\beta = 75° \quad \text{and} \quad q_u = 3.73s_u$$

Of course, the above lower-bound solution can also be obtained mathematically from the following stress jump condition derived earlier:

$$s_2 - s_1 = 2s_u \sin(\theta_2 - \theta_1) = 2s_u \sin \delta\theta$$

By noting that $\theta_1 = 90° - (150° - \beta) = 15°$ and $\theta_2 = \beta = 75°$, we have

$$s_2 - s_1 = (q_u - s_u) - (s_u) = 2s_u \sin(\theta_2 - \theta_1) = 2s_u \sin 60° = 1.73s_u$$

which leads to the same solution of $q_u = 3.73s_u$ as derived from the Mohr circles.

2.6 CAVITY EXPANSION METHODS

Recall that in limit equilibrium or limit analysis it was assumed that the soil mass was divided by a series of slip surfaces into a number of blocks. All blocks are assumed rigid. Displacements of the blocks are allowed only along the slip surfaces as rigid bodies. While the limit analysis may be suited for estimating the bearing capacity of shallow foundations in soils, it may be less accurate for deep foundations (such as piles) in which the influence of elastic deformation may not be neglected. A better alternative would be to use cavity expansion theory to determine the vertical as well as lateral bearing capacity of pile foundations in soils (Yu, 2000).

Cavity expansion theory is concerned with the stress and displacement fields during the expansion of cavities embedded in an elastic-plastic material (Yu, 2000). In particular, cavity expansion theory can provide a theoretical limit pressure required to expand a cavity from zero radius. It can also give the theoretical limit pressure needed to expand a cavity from a finite radius to a large value. Both cylindrical and spherical cavities can be considered in cavity expansion theory.

For a deep foundation, the bearing capacity includes contributions from both toe bearing and shaft friction. As presented by Yu (2000), it has been established that the spherical cavity limit pressure can be used to estimate the toe bearing capacity, while the shaft friction may be estimated using the cylindrical cavity limit pressure.

2.6.1 The cavity expansion problem in elastic-plastic soils

We consider the problem of a cavity embedded in an elastic-plastic soil which is subjected to an increasing internal cavity pressure. The cavity can be either cylindrical or spherical in shape. Let the initial radius of the cavity be a_o and the soil medium has an outer radius of b_o (which can be assumed as infinite for many cases) as shown in Figure 2.11a. It is assumed that the soil mass is initially subjected to an isotropic stress state of p_o.

The cavity will expand from an initial radius of a_o to a upon increase of the internal pressure applied on the cavity wall from p_o to p as shown in Figure 2.11b. The outer radius increases from b_o to b for a finite outer boundary. For the case of an infinite outer soil boundary, the internal cavity pressure will tend to approach a limit maximum value when the cavity deformation becomes very large. A variety of theoretical solutions for this cavity limit pressure have been derived for soils using various elastic-plastic models. A comprehensive summary of many of these cavity expansion solutions can be found in Yu (2000).

The cavity expansion solution process is generally rather complex, involving both material and geometric non-linearity, and its detailed presentation is beyond the scope of this text. However, we summarize some key solutions for cavity expansion limit pressures that are particularly relevant for deep foundations.

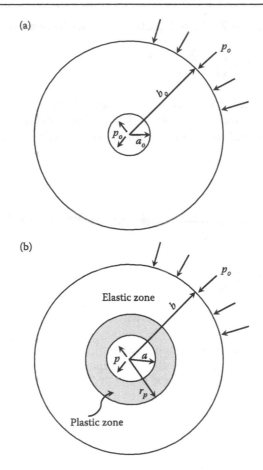

Figure 2.11 Expansion of a cavity in an elastic-plastic soil. (a) Initial conditions before cavity expansion; (b) state of stress during cavity expansion.

2.6.2 Limit pressure for undrained expansion of a spherical cavity in clay

First, we consider the undrained expansion of a spherical cavity in clays (i.e., Tresca soils). When the internal cavity pressure increases, an internal plastic zone in the soil will be formed and its size, denoted by a plastic radius, r_p, will also increase, as shown in Figure 2.11b. By use of stress equilibrium and yield criterion in the plastic zone as well as the radial stress continuity across the elastic-plastic boundary, the internal cavity pressure, p, can be linked to the plastic radius, r_p, theoretically (Yu, 2000). For the case of an infinite outer boundary, this relationship is given below:

$$p = p_o + 4s_u \ln\left(\frac{r_p}{a}\right) + \frac{4s_u}{3} \tag{2.69}$$

where
s_u = undrained shear strength of the clay

At the limit state, the internal cavity pressure approaches a constant limit pressure, and this occurs when the plastic radius reaches the following value:

$$\frac{r_p}{a} = \left\{\frac{E}{6(1-\mu)s_u}\right\}^{\frac{1}{3}} \tag{2.70}$$

By substituting Equation 2.70 into Equation 2.69, we have a cavity limit pressure p_{lim},

$$p_{lim} = p_o + \frac{4s_u}{3}\left\{1 + \ln\left(\frac{E}{6(1-\mu)s_u}\right)\right\} \tag{2.71}$$

where E is Young's modulus and μ is Poisson's ratio. In undrained condition, it is often assumed that the soil is incompressible and therefore Poisson's ratio $\mu = 0.5$. In this case, the plastic radius and limit pressure for spherical cavity expansion can be simply expressed by

$$\frac{r_p}{a} = \left\{\frac{G}{s_u}\right\}^{\frac{1}{3}} \tag{2.72}$$

and

$$p_{lim} = p_o + \frac{4s_u}{3}\left\{1 + \ln\frac{G}{s_u}\right\} \tag{2.73}$$

where G is shear modulus and G/s_u is known as the soil rigidity index and often represented as I_r.

I_r is a very useful parameter that can be used in the interpretation of in-situ test results and design of deep foundations. The following equation reported by Mayne et al. (2002) can be used to estimate the I_r based on OCR and PI (Plasticity Index; $PI = LL - PL$) of the cohesive soil.

$$I_r = \frac{\exp\left(\frac{137-PI}{23}\right)}{\left\{1 + \ln\left[1 + \frac{(OCR-1)^{3.2}}{26}\right]\right\}^{0.8}} \tag{2.74}$$

2.6.3 Limit pressure for undrained expansion of a cylindrical cavity in clay

Now we consider the undrained expansion of a cylindrical cavity in clays (i.e., Tresca soils). Similarly, when the internal cavity pressure is increasing, an internal plastic zone in the soil will be formed and its size denoted by a plastic radius, r_p, will also increase, as shown in Figure 2.11b. As shown in Yu (2000), by use of stress equilibrium and yield criterion in the plastic zone as well as the radial stress continuity across the elastic-plastic boundary, the internal cavity pressure, p, can be linked to the plastic radius, r_p, theoretically. For the case of an infinite outer soil boundary, the cavity pressure-plastic radius relationship is shown to be

$$p = p_o + 2s_u \ln\left(\frac{r_p}{a}\right) + s_u \tag{2.75}$$

At the limit state, the internal cavity pressure approaches a constant limit pressure and this occurs when the plastic radius reaches the following value:

$$\frac{r_p}{a} = \left\{\frac{E}{2(1+\mu)s_u}\right\}^{\frac{1}{2}} = \left\{\frac{G}{s_u}\right\}^{\frac{1}{2}} \tag{2.76}$$

By substituting Equation 2.76 into Equation 2.75, we have the cavity limit pressure

$$p_{\lim} = p_o + s_u \left\{ 1 + \ln \frac{G}{s_u} \right\}$$

or

$$p_{\lim} = p_o + s_u \{ 1 + \ln I_r \} \tag{2.77}$$

2.6.4 Limit pressure for expansion of spherical and cylindrical cavities in cohesive-frictional soils

A large strain solution for cavity expansion in an infinite cohesive-frictional soil using a non-associated plastic flow rule has been derived by Yu (1990) and also presented by Yu and Houlsby (1991). This solution has also been extended by Yu (1992) to the expansion of cavities embedded within a finite soil medium.

The properties of soil are defined by Young's modulus E, shear modulus G, and Poisson's ratio μ, cohesion c, angles of friction and dilation $\varnothing$ and ψ. The initial stress in soil mass is assumed to be isotropic and has a value of p_o. To simplify the presentation, it is possible to combine both cylindrical and spherical cavities. Following Yu (2000), we use the parameter k to indicate cylindrical cavity ($k = 1$) or spherical cavity ($k = 2$) in the solutions.

Several functions of these soil properties recur throughout the analysis, and to abbreviate the mathematics it is convenient to define a number of quantities, all of which are constants in any given analysis:

$$G = \frac{E}{2(1 + \mu)} \tag{2.78}$$

$$M = \frac{E}{1 - \mu^2(2 - k)} \tag{2.79}$$

$$Y = \frac{2c \cos \varnothing}{1 - \sin \varnothing} \tag{2.80}$$

$$\alpha = \frac{1 + \sin \varnothing}{1 - \sin \varnothing} \tag{2.81}$$

$$\beta = \frac{1 + \sin \psi}{1 - \sin \psi} \tag{2.82}$$

$$\gamma = \frac{\alpha(\beta + k)}{k(\alpha - 1)\beta} \tag{2.83}$$

$$\delta = \frac{Y + (\alpha - 1)p_o}{2(k + \alpha)G} \tag{2.84}$$

$$\eta = \frac{(1 + k)\delta[1 - \mu^2(2 - k)]}{(1 + \mu)(\alpha - 1)\beta} \times \left[\alpha\beta + k(1 - 2\mu) + 2\mu - \frac{k\mu(\alpha + \beta)}{1 - \mu(2 - k)} \right] \tag{2.85}$$

$$\chi = \exp \left\{ \frac{(\beta + k)(1 - 2\mu)[1 + (2 - k)\mu][Y + (\alpha - 1)p_o]}{E(\alpha - 1)\beta} \right\} \tag{2.86}$$

When the internal cavity pressure increases from its initial value, the soil is entirely elastic until it reaches a certain value when the cavity wall becomes plastic. It can be shown that this critical cavity pressure p_y is given by

$$p_y = 2kG\left(\frac{Y + (\alpha - 1)p_0}{2(k + \alpha)G}\right) + p_0 = 2kG\delta + p_0 \tag{2.87}$$

When the internal cavity pressure exceeds this value, a plastic zone will form around the cavity. As in the case for undrained expansion, the internal cavity pressure, p, can also be linked to the plastic radius, r_p, theoretically. For the case of an infinite outer soil boundary, the cavity pressure–plastic radius relationship is established as

$$\frac{r_p}{a} = (R)^{\frac{\alpha}{k(\alpha-1)}} \tag{2.88}$$

where
   a = radius of the expanded cavity

and

$$R = \frac{(k + \alpha)[Y + (\alpha - 1)p]}{\alpha(1 + k)[Y + (\alpha - 1)p_0]} \tag{2.89}$$

The cavity pressure–expansion relationship can be expressed in the following explicit form (Yu, 1990, 2000; Yu and Houlsby, 1991):

$$\left(\frac{a}{a_0}\right)^{\beta+\frac{k}{\beta}} = \frac{R^{-\gamma}}{(1 - \delta)^{\beta+\frac{k}{\beta}} - \frac{\gamma}{\chi}\sum_{n=0}^{\infty} A_n(R, \eta)} \tag{2.90}$$

where
   a_0 = initial radius of the cavity before expansion

and

$$A_n(R, \eta) = \begin{cases} \dfrac{\eta^n \ln R}{n!} & \text{if } n = \gamma \\ \dfrac{\eta^n (R^{n-\gamma} - 1)}{n!(n - \gamma)} & \text{otherwise} \end{cases} \tag{2.91}$$

When a cavity is expanded in a plastically deforming material, the cavity pressure does not increase indefinitely, but a limit pressure is approached. Define R_{lim} as that limiting R value defined by Equation (2.89) when $\dfrac{a}{a_0} \to \infty$. By putting $\dfrac{a}{a_0} \to \infty$ in Equation 2.90, the cavity expansion limit pressure can be obtained by finding R_{lim} from the following equation:

$$\sum_{n=0}^{\infty} A_n(R_{lim}, \eta) = \frac{\chi}{\gamma}(1 - \delta)^{\beta+\frac{k}{\beta}} \tag{2.92}$$

Once $R_{\lim}$ is determined from the above equation, the cavity expansion limit pressure $p_{\lim}$ can be readily obtained from the following equation:

$$R_{\lim} = \frac{(k+\alpha)[Y+(\alpha-1)p_{\lim}]}{\alpha(1+k)[Y+(\alpha-1)p_0]}$$

or

$$p_{\lim} = \frac{1}{(\alpha-1)}\left\langle \frac{R_{\lim}\{\alpha(1+k)[Y+(\alpha-1)p_0]\}}{(k+\alpha)} - Y \right\rangle \tag{2.93}$$

While the derivation of $R_{\lim}$ and $p_{\lim}$ using Equations 2.78 through 2.93 is applicable to general cohesive-frictional ($c - \emptyset$) soils, it should be noted that γ, η, and χ become indeterminate for the special case when $\emptyset = 0$. However, it can be confirmed that at very small $\emptyset$ values the solution presented in this section approaches the solution considered earlier for Tresca materials, which is relevant for a purely cohesive soil in undrained condition. Of course when the method is applied to frictional soils in drained condition, appropriate effective angles of friction and dilation should be used.

2.6.5 Application of cavity expansion theory in deep foundations

As shown in Figure 2.12, an axially loaded pile carries the load partly by shear stress generated along the shaft and partly by the normal stress generated at the base of the pile toe.

As summarized by Yu (2000), cavity expansion solutions have been used to predict the behavior of driven piles in both clays (undrained cohesive) and sands (friction only). In particular, cavity expansion theory can be used to model pile behavior in four different ways:

1. The installation of a pile into soil may be modeled as the expansion of a cylindrical cavity from zero radius to the radius of the pile. This will give the stress change in the soil around the pile due to pile installation, which can be used to estimate the shaft friction (Randolph et al., 1979; Houlsby and Withers, 1988; Coop and Wroth, 1989; Collins and Yu, 1996).

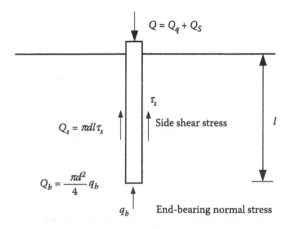

Figure 2.12 An axially loaded driven pile in soil.

Take p_{lim} for cylindrical cavity expansion according to Equation 2.93 as the normal stress on the pile shaft at the end of pile installation, σ_{rr}. The shaft shear stress τ_s may be calculated as follows:

$$\tau_s = \sigma_{rr} \tan \delta_r \tag{2.94}$$

where
δ_r = the residual friction angle at the soil–pile interface.

Experiments suggest that this residual friction angle δ_r ranges from 0.35 to 0.4 times the soil friction angle, Ø (Coop and Wroth, 1989; Bond and Jardine, 1991).

2. Toe bearing capacity of a driven pile can be correlated to the spherical cavity limit pressure in a semi-empirical manner (Gibson, 1950; Ladanyi and Johnston, 1974; Randolph et al., 1994; Yasufuku and Hyde, 1995; Yu, 2000, 2006b).

    As shown in Figure 2.13, the pile toe bearing capacity in clay can be linked to spherical cavity limit pressure by

$$q_b = p_{lim} + \beta_1 s_u \tag{2.95}$$

where the friction coefficient $\beta_1 = 0.0 - 1.0$. The p_{lim} in cohesive soil can be determined using Equation 2.73.

In sand,

$$q_b = (1 + \tan \varnothing \tan \alpha_1) p_{lim} \tag{2.96}$$

where $\alpha_1 = 45° + \varnothing/2$. The p_{lim} in purely frictional soil can be determined using the above procedure for cavity expansion in cohesive-frictional soil but consider spherical cavity expansion ($k = 2$), set $c = 0$.

3. Toe bearing capacity of driven piles in granular soils can also be estimated from a combined cylindrical-spherical cavity expansion method (Yu, 2006b). The basic idea of this relatively new method consists of two steps:

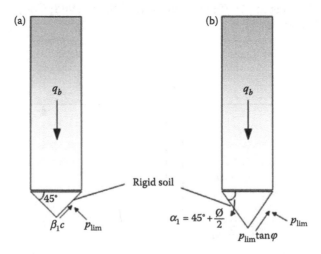

Figure 2.13 Predicting the pile toe bearing capacity from spherical cavity limit pressure for (a) clay and (b) sand. (Adapted from Yu, H.S. 2000. *Cavity Expansion Methods in Geomechanics*. Kluwer Academic Publishers, The Netherlands.)

- Estimate of the size of the plastically deforming zone around the pile using the cylindrical cavity solution.
- Use spherical cavity expansion theory to determine the toe bearing capacity from the above-estimated plastic zone.

This approach was motivated by a large strain finite-element study of cone penetration in sand (Huang et al., 2004), which suggests that the plastic zone behind the pile toe and around the pile shaft is similar to that predicted by the cylindrical cavity expansion theory. Around the pile toe, the elastic-plastic boundary may be assumed to be spherical or elliptical in shape (as shown in Figure 2.14).

By following the above procedure and using the cylindrical cavity expansion solution in cohesive-frictional materials, we first determine $R_{\lim}$ according to Equation 2.92, (for cylindrical cavity expansion, $k = 1$), namely

$$\sum_{n=0}^{\infty} A_n(R_{\lim}, \eta) = \frac{\chi}{\gamma} (1 - \delta)^{\beta + \frac{1}{\beta}} \tag{2.92}$$

The cylindrical cavity limit pressure $p_{\lim}$ is obtained according to Equation 2.93 by setting $k = 1$ and

$$p_{\lim} = \frac{1}{(\alpha - 1)} \left\{ \frac{2\alpha[Y + (\alpha - 1)p_o]R_{\lim}}{(1 + \alpha)} - Y \right\} \tag{2.93}$$

Take $p_{\lim}$ for cylindrical cavity expansion according to Equation 2.93 as σ_{rr} and calculate the shaft shear stress τ_s following Equation 2.94.

The limiting plastic radius r_p can be obtained by using Equation 2.88 for a cylindrical cavity expansion:

$$\frac{r_p}{a} = (R_{\lim})^{\frac{\alpha}{\alpha - 1}} \tag{2.97}$$

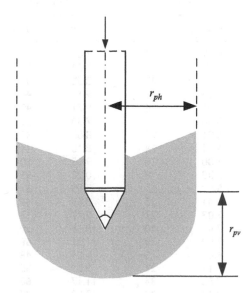

Figure 2.14 Plastic zone around a pile toe or cone tip and shaft from a finite element analysis. (Adapted from Huang, W. et al. 2004. Computers and Geotechnics, Elsevier, 31, 517–528.)

The term r_p/a denotes the relative size of plastic zone generated by the expansion of a cylindrical cavity from zero-radius. This quantity would be the same as the limiting plastic radius when a cylindrical cavity is expanded from a finite radius. In this case, a can be corresponding to the radius of the pile. The toe bearing capacity, q_b, is determined using r_p/a from cylindrical cavity expansion (i.e., Equation 2.97) as follows:

$$\frac{Y + (\alpha - 1)q_b}{Y + (\alpha - 1)p_o} = \frac{3\alpha}{2+\alpha}\left(F\frac{r_p}{a}\right)^{\frac{2(\alpha-1)}{\alpha}} \tag{2.98}$$

where p_o is the initial state of isotropic stress in soil and F is a plastic zone shape factor that accounts for the characteristics of cavity expansion around pile toe. The value of F is taken as 1 if the plastic zone around the pile toe is spherical, otherwise F is less than 1. Pending more numerical and experimental studies, F can be assumed to be between 0.7 and 0.8.

Details of applying the above procedure for a purely frictional soil ($c' = 0$) are demonstrated in Example 2.5. In addition, Table 2.1 shows selected p_{lim}/p_o and q_b/p_o for various $\emptyset$, Ψ, and G values based on cylindrical cavity expansion. To apply Table 2.1, p_o should also be calculated based on effective stress. The value of cavity expansion methods is clearly demonstrated in Table 2.1 where q_b/p_o (details of pile toe bearing capacity q_b are described in Chapter 7) is dependent on sand dilatancy (represented by sand dilation angle ψ, which is a function of confining stress for a given density as described in Chapter 1) and soil rigidity represented by shear modulus G. The conventional bearing capacity equations derived from limit equilibrium analysis use the peak-drained friction angle $\emptyset$ and as in any limit state analysis, the soil is assumed rigid (i.e., the effects of G are not considered). The p_{lim} values shown in Table 2.1 can be used to estimate normal stress, σ_{rr}, on the pile shaft at the end of pile installation

Table 2.1 Selected p_{lim}/p_o and q_b/p_o for various $\emptyset$, Ψ, and G values based on cylindrical cavity expansion ($p_o = 200$ kPa and $\mu = 0.3$)

$\emptyset$, degree	Ψ, degree	G = 10,000 kPa p_{lim}/p_o	q_b/p_o	G = 25,000 kPa p_{lim}/p_o	q_b/p_o	G = 50,000 kPa p_{lim}/p_o	q_b/p_o
25	0	5.37	17	7.02	29	8.61	43
27	0	5.75	19	7.62	33	9.43	50
30	0	6.30	22	8.52	40	10.71	62
32	0	6.68	24	9.12	44	11.55	71
34	0	7.05	26	9.71	49	12.41	80
25	2	5.60	18	7.34	31	9.13	48
27	2	6.01	20	8.03	36	10.02	57
30	2	6.61	24	8.10	44	11.41	71
32	2	7.00	26	9.65	50	12.32	81
34	2	7.39	28	10.28	55	13.26	92
25	5	5.97	21	7.98	37	9.95	57
27	5	6.41	23	8.69	43	10.96	68
30	5	7.07	27	9.77	52	12.51	85
32	5	7.49	30	10.49	59	13.56	98
34	5	7.92	33	11.21	66	14.62	111
25	10	6.62	25	9.04	48	11.48	77
27	10	7.13	29	9.89	55	12.70	91
30	10	7.88	34	11.17	68	14.575	116
32	10	8.37	37	12.02	77	15.85	134
34	10	8.86	41	12.87	91	17.13	153

which is used in Equation 2.94 to estimate the shear stress resistance on the pile shaft. Note that the q_b/p_o and p_{lim}/p_o values shown in Table 2.1 are also p_o dependent. Stress dependency is also observed in bearing capacity of shallow and deep foundations in limit state analysis, as discussed later in Chapters 4 and 7. Equations 2.78 through 2.93 and the associated equations for the determination of q_b and p_{lim} can be easily executed with the help of a spreadsheet program. Readers are encouraged to repeat Table 2.1 and go beyond the variables applied therein.

4. Cavity expansion theory may be used to estimate the limiting pressure of laterally loaded piles in soils (Fleming et al., 1985; Yu, 2000).

   As discussed by Fleming et al. (1985), past experimental research in clay suggests that it is reasonable to assume that pressure exerted by soil in front of a laterally loaded pile approaches the limit pressure according to Equation 2.77.

   For piles in sand, however, it was suggested that at lateral capacity failure, the pressure exerted on the pile shaft, q_l, is close to that required to cause lateral displacement of over 10%–15% of the pile diameter. This pressure can be estimated from cylindrical cavity expansion theory by the following closed-form cavity pressure expansion relation (Yu, 2000):

$$q_l = \frac{2\alpha}{1+\alpha} \left[\frac{1 - \left(\frac{a}{a_0}\right)^{-1-\frac{1}{\beta}}}{1 - (1-\delta)^{1+\frac{1}{\beta}}} \right]^{1/\gamma} p_o \tag{2.99}$$

where p_o is the initial soil lateral pressure and $\alpha, \beta, \gamma, \delta$ are defined by Equations 2.81 through 2.84 by setting $Y = 0$ (purely frictional soil) and $k = 1$ (cylindrical cavity).

Cylindrical cavity expansion limit pressure p_{lim} can be measured in the field using the pressuremeter test (described in Chapter 3).

Examples 2.4 and 2.5 are accompanied by Excel programs. Registered users can download the programs from the publisher's website.

EXAMPLE 2.4

Given

A pile is driven into a clay soil with undrained shear strength s_u of 50 kPa. The in-situ mean stress p_o in the soil is 100 kPa. The shear modulus G of the clay is measured to be 10,000 kPa. Assume that the friction coefficient β_1 of the soil pile end is 0.5.

Required

The toe bearing capacity of the pile using cavity expansion theory through Equation 2.95.

Solution

From Equation 2.95, the toe bearing capacity of the driven pile can be estimated as follows:

$$q_b = p_{lim} + \beta_1 s_u \tag{2.95}$$

where the friction coefficient is given as $\beta_1 = 0.5$ and the spherical cavity limit pressure is given by Equation 2.73 as follows:

$$p_{\text{lim}} = p_o + \frac{4s_u}{3}\left\{1 + \ln\frac{G}{s_u}\right\} \tag{2.73}$$

The toe bearing capacity of the driven pile therefore equals to

$$
\begin{aligned}
q_b &= p_o + \frac{4s_u}{3}\left\{1 + \ln\frac{G}{s_u}\right\} + \beta_1 s_u \\
&= 100 + 4 \times \frac{50}{3} \times \left[1 + \ln\left(\frac{10{,}000}{50}\right)\right] + 0.5 \times 50 = 545\,\text{kPa}
\end{aligned}
$$

EXAMPLE 2.5

Given

A pile with a radius a of 0.4 m is driven into a granular soil with friction angle $\varnothing$ of 30°, and angle of dilation ψ equals 2°. The in-situ mean effective stress p_o is 200 kPa. The shear modulus of the soil G is 10,000 kPa. Poisson's ratio of the soil $\mu = 0.3$.

Required

The toe bearing capacity q_b of the pile using cavity expansion method.

Solution

Determine the relevant intermediate vales:

Consider cylindrical cavity expansion, $k = 1$

$$E = 2(1 + \mu)G = (2)(1 + 0.3)(10{,}000) = 26{,}000\,\text{kPa} \tag{2.78}$$

$$Y = 0 \tag{2.80}$$

$$\alpha = \frac{1 + \sin\varnothing}{1 - \sin\varnothing} = \frac{(1 + \sin 30°)}{(1 - \sin 30°)} = 3.0 \tag{2.81}$$

$$\beta = \frac{1 + \sin\psi}{1 - \sin\psi} = \frac{(1 + \sin 2°)}{(1 - \sin 2°)} = 1.07 \tag{2.82}$$

$$\gamma = \frac{\alpha(\beta + k)}{k(\alpha - 1)\beta} = \frac{(3.0)[(1.07) + 1)]}{(1)[(3.0) - 1](1.07)} = 2.90 \tag{2.83}$$

$$\delta = \frac{Y + (\alpha - 1)p_o}{2(k + \alpha)G} = \frac{(0) + (3.0 - 1)(200)}{2(1 + 3.0)10{,}000} = 0.005 \tag{2.84}$$

$$\eta = \frac{(1 + k)\delta[1 - \mu^2(2 - k)]}{(1 + \mu)(\alpha - 1)\beta} \times \left[\alpha\beta + k(1 - 2\mu) + 2\mu - \frac{k\mu(\alpha + \beta)}{1 - \mu(2 - k)}\right] = 0.008 \tag{2.85}$$

$$\chi = \exp\left\{ \frac{(\beta + k)(1 - 2\mu)[1 + (2 - k)\mu][Y + (\alpha - 1)p_o]}{E(\alpha - 1)\beta} \right\} = 1.0078 \qquad (2.86)$$

Determine $R_{\lim}$ through trial and error using Equation 2.92, where

$$\sum_{n=0}^{\infty} A_n(R_{\lim}, \eta) = \frac{\chi}{\gamma}(1 - \delta)^{\beta + \frac{k}{\beta}} \qquad (2.92)$$

and according to Equation 2.91,

$$A_n(R_{\lim}, \eta) = \begin{cases} \dfrac{\eta^n \ln R_{\lim}}{n!} & \text{if } n = \gamma \\[2ex] \dfrac{\eta^n (R_{\lim}^{n-\gamma} - 1)}{n!(n - \gamma)} & \text{otherwise} \end{cases}$$

$R_{\lim} \approx 4.42$ from trial and error.

$$\frac{r_p}{a} = (R_{\lim})^{\frac{\alpha}{\alpha-1}} = 9.29 \qquad (2.97)$$

and

$$\frac{Y + (\alpha - 1)q_b}{Y + (\alpha - 1)p_o} = \frac{3\alpha}{2 + \alpha}\left(F\frac{r_p}{a}\right)^{\frac{2(\alpha-1)}{\alpha}} \qquad (2.98)$$

Take $F = 0.75$ and $p_o = 200$ kPa. Following Equation (2.98),

$q_b = 4792\,\text{kPa}$

$$p_{\lim} = \frac{1}{(\alpha - 1)}\left\{ \frac{2\alpha[Y + (\alpha - 1)p_o]R_{\lim}}{(1 + \alpha)} - Y \right\} = 1326\,\text{kPa} \qquad (2.93)$$

Or, according to Table 2.1,

$p_{\lim}/p_o = (1326)/(200) = 6.61$
$q_b/p_o = (4792)/(200) = 24$

EXAMPLE 2.6

Given

The same pile and soil conditions in Example 2.5.

Required

The toe bearing capacity q_b of the pile using spherical cavity expansion method and Equation 2.96.

Solution

Determine the relevant intermediate values:

Consider spherical cavity expansion, $k = 2$

$$E = 2(1 + \mu)G = (2)(1 + 0.3)(10{,}000) = 26{,}000 \, \text{kPa} \tag{2.78}$$

$$Y = 0 \tag{2.80}$$

$$\alpha = \frac{1 + \sin \varnothing}{1 - \sin \varnothing} = \frac{(1 + \sin 30°)}{(1 - \sin 30°)} = 3.0 \tag{2.81}$$

$$\beta = \frac{1 + \sin \psi}{1 - \sin \psi} = \frac{(1 + \sin 2°)}{(1 - \sin 2°)} = 1.07 \tag{2.82}$$

$$\gamma = \frac{\alpha(\beta + k)}{k(\alpha - 1)\beta} = \frac{(3.0)[(1.07) + 2]}{(2)[(3.0) - 1](1.07)} = 2.15 \tag{2.83}$$

$$\delta = \frac{Y + (\alpha - 1)p_o}{2(k + \alpha)G} = \frac{(0) + (3.0 - 1)(200)}{2(2 + 3.0)10{,}000} = 0.004 \tag{2.84}$$

$$\eta = \frac{(1 + k)\delta[1 - \mu^2(2 - k)]}{(1 + \mu)(\alpha - 1)\beta} \times \left[\alpha\beta + k(1 - 2\mu) + 2\mu - \frac{k\mu(\alpha + \beta)}{1 - \mu(2 - k)} \right]$$
$$= 0.0094 \tag{2.85}$$

$$\chi = \exp\left\{ \frac{(\beta + k)(1 - 2\mu)[1 + (2 - k)\mu][Y + (\alpha - 1)p_o]}{E(\alpha - 1)\beta} \right\} = 1.0086 \tag{2.86}$$

Determine $R_{\lim}$ through trial and error using Equation 2.92, where

$$\sum_{n=0}^{\infty} A_n(R_{\lim}, \eta) = \frac{\chi}{\gamma} (1 - \delta)^{\frac{\beta + k}{\beta}} \tag{2.92}$$

and according to Equation 2.91,

$$A_n(R_{\lim}, \eta) = \begin{cases} \dfrac{\eta^n \ln R_{\lim}}{n!} & \text{if } n = \gamma \\[2mm] \dfrac{\eta^n (R_{\lim}^{n-\gamma} - 1)}{n!(n - \gamma)} & \text{otherwise} \end{cases}$$

$R_{\lim} \approx 4.542$ from trial and error.

$$p_{\lim} = \frac{1}{(\alpha - 1)} \left\{ \frac{2\alpha[Y + (\alpha - 1)p_o]R_{\lim}}{(1 + \alpha)} - Y \right\} = 1362 \, \text{kPa} \tag{2.93}$$

$$q_b = (1 + \tan \varnothing \tan \alpha_1)p_{\lim} \tag{2.96}$$

$$\alpha_1 = 45° + \varnothing/2$$

Use $\varnothing = 30°$

$$q_b = (1 + \tan \varnothing \tan \alpha_1)p_{\lim} = \left(1 + \tan(30°) \tan\left(45° + \frac{30°}{2}\right)\right)(1362) = 2725 \, \text{kPa}$$

2.7 REMARKS

In this chapter, we introduced the basic concept of limit equilibrium and limit analysis methods. These methods are routinely used for limit state analysis in foundation engineering. In both cases, we assumed soil as rigid and perfectly plastic. It was also mentioned that the limit equilibrium method provides basically the same results as the upper-bound limit analysis. The limit analysis has an important advantage of allowing the lower-bound solution to be obtained, and together with the upper-bound solution they provide a bracket for the exact collapse load.

The cavity expansion has close similarity to the response of soil surrounding the toe of a penetrating pile, a cone penetrometer, or a pressuremeter expansion. For this reason, the cavity expansion limit pressure has been used to estimate the bearing capacity of piles (Chapter 7) and interpretation of cone penetration and pressuremeter tests (Chapter 3). An important advantage of using the cavity expansion method is that the soil rigidity can be considered through the use of rigidity index I_r. This chapter has been largely limited to the description of basic concepts; their applications in various aspects of foundation design are presented in the following chapters. Cavity expansion theories that consider effective stress and pore pressure development have also been developed, but the subject is beyond the scope of this book. Interested readers are referred to Yu (2000).

HOMEWORK

2.1.   For a shallow strip foundation with a width of B resting on a purely cohesive soil with undrained shear strength s_u. The surcharge pressure acting on the soil surface outside the foundation is q. We assume that, at collapse, the soil will fail according to the symmetric five-rigid block mechanism as shown in Example 2.1. Find the upper bound to the bearing capacity of the foundation.

2.2.   For a shallow strip foundation with a width of B resting on a cohesive-frictional soil with cohesion c. The internal friction angle ($\emptyset$) is $20°$. The surcharge pressure acting on the soil surface outside the foundation is q. We assume that at collapse, the soil will fail according to the symmetric five-rigid block mechanism as shown in Example 2.2. Find the upper bound to the bearing capacity of the foundation.

2.3.   For a uniform pressure acting on a cohesive-frictional soil immediately next to a slope having an angle of $30°$, as shown in Example 2.3. The soil cohesion is c and friction angle ($\emptyset$) is $30°$. The effect of soil weight is ignored for simplicity. Calculate the lower bound to the bearing capacity of the slope.

2.4.   A 0.4 m radius pile is driven into a purely cohesive soil with undrained shear strength s_u of 100 kPa. The cohesive soil has a Poisson ratio of 0.5. The in-situ mean stress p_o around the pile toe is 100 kPa. Determine the toe bearing capacity of the pile using cavity expansion theory through Equations 2.73 and 2.95 considering a rigidity index (G/s_u) of 100, 200, and 300. Use $\beta_1 = 0.5$.

2.5.   A pile is driven into a purely frictional soil with an internal friction angle ($\emptyset$) of $30°$. The soil dilation angle (Ψ) is $10°$ and Poisson's ratio is 0.3. The in-situ mean stress in the soil (p_o) is 100 kPa. The shear modulus (G) of the sand is measured to be 50,000 kPa. Predict the toe bearing capacity (q_b) of the pile using cavity expansion theory through Equations 2.97 and 2.98.

REFERENCES

Atkinson, J.H. 1981. *Foundations and Slopes: An Introduction to Applications of Critical State Soil Mechanics*. Wiley, New York, 382p.

Bishop, J.F.W. 1953. On the complete solutions of deformation of a plastic rigid material. *Journal of Mechanics and Physics of Solids*, 2, 43–53.

Bond, A.J. and Jardine, R.J. 1991. Effect of installing displacement piles in a high OCR clay. *Geotechnique*, 41, 341–363.

Chen, W.F. 1975. *Limit Analysis and Soil Plasticity*. Elsevier, New York.

Collins, I.F. and Yu, H.S. 1996. Undrained cavity expansions in critical state soils. *International Journal of Numerical and Analytical Methods in Geomechanics*, 20, 489–516.

Coop, M.R. and Wroth, C.P. 1989. Field studies of an instrumented model pile in clay. *Geotechnique*, 39, 679–696.

Davis, E.H. 1968. Theories of plasticity and the failure of soil masses. In Lee, I.K., ed., *Soil Mechanics—Selected Topics*. Butterworths, London, pp. 341–380.

Davis, R.O. and Selvadurai, A.P.S. 2002. *Plasticity and Geomechanics*. Cambridge University Press, Cambridge, UK.

Drucker, D.C., Prager, W., and Greenberg, H.J. 1952. Extended limit design theorems for continuous media. *Quarterly Applied Mathematics*, 9, 381–389.

Fleming, W.G.K., Weltman, A.J., Randolph, M.F., and Elson, W.K. 1985. *Piling Engineering*. Wiley, London.

Gibson, R.E. 1950. Correspondence. *Journal of Institution of Civil Engineers*, 34, 382–383.

Hill, R. 1948. A variational principle of maximum plastic work in classical plasticity. *Quarterly Journal of Mechanics and Applied Mathematics*, 1, 18–28.

Hill, R. 1950. *The Mathematical Theory of Plasticity*. Clarendon Press, Oxford.

Houlsby, G.T. and Withers, N.J. 1988. Analysis of the cone pressuremeter test in clay. *Geotechnique*, 38, 575–587.

Huang, W., Sheng, D., Sloan, S.W., and Yu, H.S. 2004. Finite element analysis of cone penetration in cohesionless soil. *Computers and Geotechnics, Elsevier*, 31, 517–528.

Ladanyi, B. and Johnston, G.H. 1974. Behaviour of circular footings and plate anchors embedded in permafrost. *Canadian Geotechnical Journal*, 11, 531–553.

Mayne, P.W., Christopher, B.R., and DeJong, J. 2002. Subsurface investigations. Report No. FHWA-NHI-01-031, National Highway Institute Federal Highway Administration U.S. Department of Transportation Washington, DC, 332p.

Prager, W. and Hodge, P.G. 1951. *The Theory of Perfectly Plastic Solids*. Wiley, New York.

Randolph, M.F., Carter, J.P., and Wroth, C.P. 1979. Driven piles in clay—The effects of installation and subsequent consolidation. *Geotechnique*, 29, 361–393.

Randolph, M.F., Dolwin, J., and Beck, R. 1994. Design of driven piles in sand. *Geotechnique*, 44, 427–448.

Shield, R.T. 1954. Stress and velocity fields in soil mechanics. *Quarterly Applied Mathematics*, 12, 144–156.

Terzaghi, K. 1943. *Theoretical Soil Mechanics*. Wiley, New York.

Yasufuku, N. and Hyde, A.F.L. 1995. Pile end-bearing capacity in crushable sands. *Geotechnique*, 45, 663–676.

Yu, H.S. 1990. Cavity expansion theory and its application to the analysis of pressuremeters. DPhil thesis, University of Oxford.

Yu, H.S. 1992. Expansion of a thick cylinder of soils. *Computers and Geotechnics*, 14, 21–41.

Yu, H.S. 2000. *Cavity Expansion Methods in Geomechanics*. Kluwer Academic Publishers, The Netherlands.

Yu, H.S. 2006a. *Plasticity and Geotechnics*. Springer, New York.

Yu, H.S. 2006b. In situ soil testing: From mechanics to interpretation. 1st James K. Mitchell Lecture. *Geomechanics and Geoengineering*, 1, 165–195.

Yu, H.S. and Houlsby, G.T. 1991. Finite cavity expansion in dilatant soils: Loading analysis. *Geotechnique*, 41, 173–183.

Chapter 3

Subsurface exploration for foundation design

3.1 INTRODUCTION

The purpose of subsurface exploration is to provide knowledge of the ground conditions for safe and economical foundation design and potential problems that may be encountered during construction. A successful subsurface exploration should provide the following information:

- Stratigraphy of the ground material, soil/rock properties, and groundwater conditions within the area and depth that will be affected by the proposed structure.
- Geotechnical parameters required for the selection or recommendation of the type and depth of foundations, determination of bearing capacities for the recommended foundation type(s), and estimation of the proposed foundation settlement.
- Design parameters required for related earth or earth-supporting structures, such as embankments, retaining walls, or braced excavations.
- Potential problems to be expected for the construction of the recommended foundation system.

The subsurface exploration should extend beyond the expected depth of the foundation, but to determine the depth of the foundation is one of the purposes for subsurface exploration. This is like trying to solve an equation with unknowns on both sides of the equation. For mathematical problems, we can use an iterative procedure to obtain the solution. To adopt this iterative approach in engineering practice would require repeated site visits and field operations, which can be expensive and impractical. To avoid unnecessary iterative operations, a thorough review of related literature and previous records is imperative to the planning of subsurface exploration. The literature review helps us to decide the methods to be used and scope of the subsurface exploration. A well-planned and executed subsurface exploration assures that the necessary geotechnical design parameters are obtained with reasonable efforts and thus cost. A wide variety of methods and tools are available to fulfill the requirements of subsurface exploration. They include drilling boreholes and taking samples for laboratory testing, field or in-situ testing, and geophysical testing. The objectives of this chapter are to:

- Introduce the general procedure and items to be considered in the planning of a subsurface exploration program
- Describe available techniques and tools for drilling and sampling related to subsurface exploration
- Describe commonly used in-situ testing methods, their operations, and interpretation of the test data as they relate to foundation designs
- Introduce the basic techniques of a few geophysical testing methods commonly used in subsurface exploration

3.2 PLANNING OF SUBSURFACE EXPLORATION

Figure 3.1 shows a flow chart of items to be considered in planning a typical subsurface exploration for foundation designs. The level of detail for each item depends on the scale and importance of the project.

The planning generally should start with collecting the following information:

- Types of structures involved in the project and the expected loading conditions to be imposed on the foundations.
- Geological information, origin of the soil/rock, groundwater, or hydrogeological conditions.
- Terrain information from topographic maps, aerial photographs, or Digital Terrain Model (DTM). Coverage by vegetation or existing building structures. The information is important in assessing the accessibility of the project site and selection of the types of equipment to be used for subsurface exploration.
- Local soil/rock profiles compiled from previous explorations. They may be obtained from a geological survey agency, municipal government, or agricultural department. For an established geotechnical company, it may be possible to find earlier subsurface exploration reports from areas on or close to the project site.

Field reconnaissance is imperative in the planning of subsurface exploration. Objectives of field reconnaissance can include:

- Observation of topography at the project site and conditions of the existing structures surrounding the project site.
- Selection of routes for the drilling/testing vehicles to enter the project site.
- Location of source of water required for borehole drilling.
- Selection of benchmarks or landmarks as references for locating boreholes and determining their surface elevations.

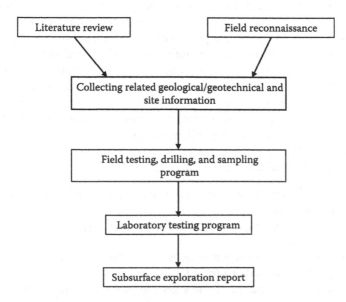

Figure 3.1 Flow chart for planning/execution of a subsurface exploration.

With the knowledge from literature study and field reconnaissance, the engineer is now ready to plan the subsurface exploration. The planning should consider the following issues.

3.2.1 Methods available for the exploration

We have the option to perform borehole drilling, take soil/rock samples (to be referred to as the drilling and sampling method), and then conduct laboratory tests on the samples to obtain the geotechnical parameters, or we can use in-situ testing methods. In this case, a testing device is inserted into the ground to perform experiment in the field. Some of the geophysical testing methods can be conducted from the ground surface and infer profile or cross-sectional information of the ground conditions. The geophysical testing methods can be very useful in complementing the above two methods. It can be advantageous to use a combination of taking samples, performing laboratory tests and in-situ as well as geophysical testing. The exact selection of the exploration methods depends on the project needs, local experience, budget constraints, and availability of the required equipment.

3.2.2 Frequency and depth of exploration

Building codes often stipulate the number of borings per unit area of the building. Usually these codes are adequate for isolated building structures with limited dimensions. However, there are no reliable rules for the number and spacing of explorations that can be applied to all types of projects. The emphasis should be to locate borings to develop cross-sections with sufficient information for the foundation design, with reasonable cost. For isolated structures, it is common to carry explorations to a depth beneath the loaded area of 1.5–2.0 times the least dimension of the foundation. The increase of vertical stress at that depth level should be less than 10% of the bearing pressure imposed by the foundation. The exploration depth can be reduced if hard bearing material such as rock is encountered at shallower depth.

3.2.3 Field testing, drilling/sampling program

In this part of the planning, the type of drill rig and methods to deepen the boreholes and their locations should be specified. Number and frequency of disturbed (representative) and/or undisturbed samples (details to be described later) to be taken in each borehole are defined. The standard penetration test is a field (in-situ) testing method that also takes disturbed soil samples. Other types of field testing (methods to be described later) can also be included in the subsurface exploration. In any case, observation of the groundwater table should always be part of the subsurface exploration.

3.2.4 Laboratory testing program

The types and number of laboratory tests for the samples recovered from field exploration are specified. These tests can include basic physical properties such as grain size distribution, specific gravity, plasticity, soil classification, and unit weight. For cohesive (i.e., clay) soils, shear strength tests such as direct shear, unconfined compression, and triaxial tests on undisturbed soil samples may be required to determine the bearing capacity of foundations. Consolidation tests on undisturbed clay samples are commonly used for foundation settlement analysis. Undisturbed sampling in cohesionless (i.e., sand and gravel) soils is extremely costly and impractical. We usually rely on field testing such as the standard penetration or cone penetration tests to determine the strength and compressibility parameters needed for foundation design in cohesionless soils. Other types of laboratory tests may be necessary depending on the specific local soil conditions and needs for foundation design.

The planning of subsurface exploration needs to reach a balance between cost and risk in foundation design. The cost of subsurface exploration for design should be in the range of 0.5%–1.0% of the construction cost of the project. Cheap or less rigorous subsurface explorations are often associated with higher risk, with potential losses that can far exceed the potential saving.

3.3 DRILLING AND SAMPLING

A typical procedure in the drilling and sampling method generally involves the following steps:

1. Drill a borehole to a given depth.
2. Remove the drilling equipment such as drill rods and drill bit from the borehole.
3. Lower the sampling tool to the bottom of the borehole and take the sample.

The history of drilling a borehole goes back far beyond the beginning of geotechnical engineering. Many drilling methods and tools were developed thousands of years ago (Broms and Flodin, 1988). On the other hand, techniques in taking good quality soil samples were developed mostly in the twentieth century (Hvorslev, 1949). Selection of drilling method is usually independent from that of sampling. The following sections introduce some of the commonly used drilling and sampling techniques.

3.3.1 Drilling methods

This section describes some of the available and commonly used drilling methods for subsurface exploration.

3.3.1.1 Hand auger/shovel

Figure 3.2 shows three possible versions of commonly used hand augers for shallow exploration. The same hand tools have been used for installing the fence post (Figure 3.2a and b). Luo-Yang shovel was used for tomb raiding (Figure 3.2c) in ancient China, but now has become a common drilling tools used by farmers. The depth of boreholes that we can drill with hand augers is usually less than 10 m, and in cohesive soil only.

3.3.1.2 Percussion drilling

Percussion drilling involves repeatedly raising and dropping a heavy drill bit (also called a churn bit) into the borehole as shown in Figure 3.3. The drill bit breaks the rock or soil into small pieces. Water, either from the ground or added by the operator, mixes the crushed or loosened particles into a slurry. When accumulation of slurry becomes excessive, the drill bit is removed from the borehole. A bailer, as shown in Figure 3.4, is lowered into the borehole to remove the slurry from the bottom of the borehole. Frequently used to drill wells or for mineral exploration, this type of drilling has been used for thousands of years, the oldest of all methods for drilling the deep boreholes. Drills can be simple apparatuses consisting of a heavy bit and a rope, and can be operated by machine, hand, or animal. Only a small amount of water is required. Modern versions of percussion drill use steel cables and larger bits, which may weigh over a ton and are powered by gasoline or diesel engines. Whether powered by hand or engine, boreholes in excess of 1000 m can

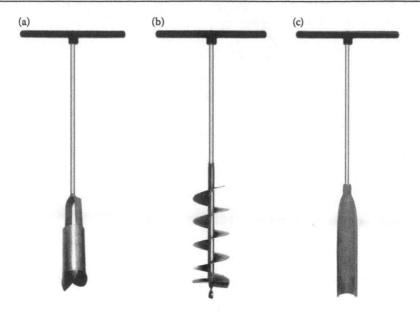

Figure 3.2 Hand augers. (a) Postal auger. (b) Helical auger. (c) Luo-Yang shovel.

bc drilled with this technique. A major drawback for percussion drilling is that it is slow. In ancient times, a deep borehole could take generations to finish using hand-operated equipment. Changes in the character of subsurface materials are determined by the rate of progress, action of the drilling tools, and composition of the slurry. However, the accuracy of this observation technique can be hampered, as the cuttings are removed only intermittently and therefore represent the average material over a considerable depth.

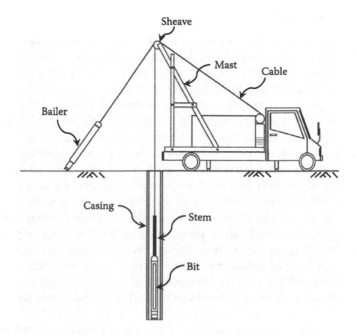

Figure 3.3 Schematic view of the percussion drilling method.

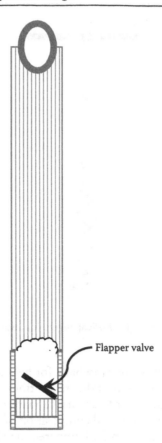

Flapper valve

Figure 3.4 Flat bottom bailer for removing the soil cuttings.

3.3.1.3 Rotary wash boring

In percussion drilling, the drill bit is suspended by a flexible cable which can only lift or drop the bit. In rotary wash boring, the drill bit is attached to the bottom of a hollow metal drill rod. Drilling fluid can be pumped from the ground surface via the hollow drill rod to the drill bit which is located at the bottom of the drill rod. The drill rod is clamped to a rotary drive unit that is powered by a gasoline or diesel engine as shown in Figure 3.5. The drill rod is rotated rapidly while circulating the drilling fluid through the drill bit to remove the cuttings. The drilling fluid can be water or drill mud. Drill mud is made of a mixture of water and expansive clay mineral such as bentonite (montmorillonite) or bio-degradable polymer. Drilling fluid is viscous, with a specific gravity slightly higher than water. The drilling fluid serves as a lubricant between the drill bit and the surrounding soil and helps in stabilizing the borehole and carrying the cutting to the ground surface. Rotation of the drill rod and circulation of drilling fluid or cutting are continuous. Changes in the character of soil can be determined by the resistance to penetration and rotation of the bit and by examination of cuttings brought to the ground surface by drilling fluid. The rotary boring method can be used in clay, sand, and rock. As only the drill bit is making contact with the material encountered at the bottom of the borehole, regardless of its depth, the rotary boring method can be used to extend boreholes to hundreds of meters in depth.

Drill bits are usually made with hardened steel or sometimes engraved with diamond powder. Figure 3.6 shows the picture of a tri-cone drill bit typically used in rotary drilling.

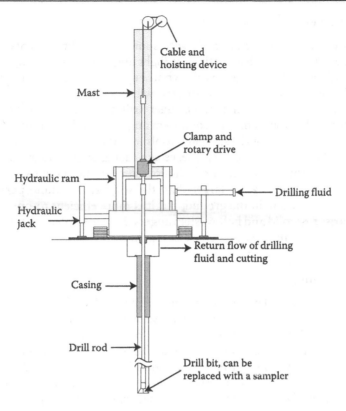

Figure 3.5 Rotary wash boring field equipment setup.

Figure 3.6 A tri-cone drill bit. (Courtesy of An-Bin Huang, Hsinchu, Taiwan.)

3.3.1.4 Power auger

For cohesive soils such as clays or sands with some clay or silt contents, drilling of the borehole can be conducted with a continuous flight auger. Sections of augers are attached to form the continuous flight auger, as shown in Figure 3.7. Rotation of the flight auger brings the soil cuttings to the ground surface without the need for water circulation. When coupled with a powerful engine, the auger drilling can be very efficient. The auger can be solid stem or hollow stem. The solid stem auger has to be removed from the borehole to give room for lowering the sampling tools into the borehole, if soil samples are to be taken. The hollow stem auger can also serve as a casing with enough space to lower the drilling or sampling tools from inside of the hollow stem, without removing the auger. The hollow stem auger is significantly larger than the solid stem counterpart and requires much more torque to rotate in the ground, but it is more efficient to operate. Figure 3.8 shows a comparison of solid and hollow stem augers. Figure 3.9 depicts the field operation of a hollow stem auger along with a truck-mounted drill rig.

3.3.2 Soil sampling

Soil samples can be classified into two categories as either disturbed and undisturbed. The natural structure of a disturbed, but representative soil sample is destroyed in the sampling process. This type of sample can only be used for physical property tests such as specific gravity, liquid limit, plastic limit, grain size distribution, organic content, and soil classification. There is no truly "undisturbed" soil sample in any case. An undisturbed soil sample refers to a sample with its natural structure more or less intact. Laboratory tests such as soil unit weight, consolidation, or triaxial or other types of shearing tests would require undisturbed soil samples.

Disturbed soil samples can be cuttings from the boring operation, samples taken using a shovel, or a sampler that does not meet the requirement of undisturbed sampling.

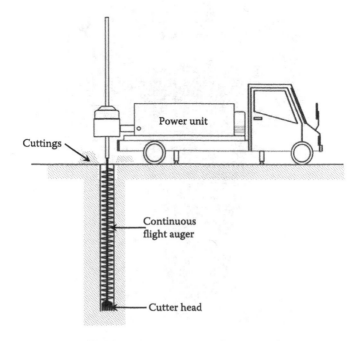

Figure 3.7 Continuous auger boring.

(a)   (b)

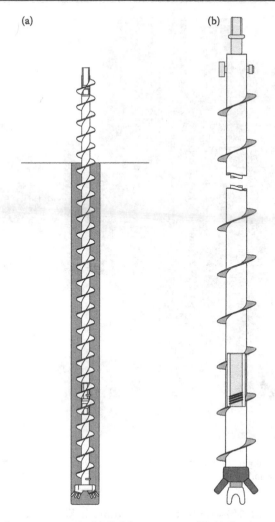

Figure 3.8 Solid and hollow stem auger. (a) Solid stem auger. (b) Hollow stem auger with a drill rod inside and a pilot bit at the tip of the drill rod.

The basic idea of taking a soil sample from a borehole is to push a tube into the soil from the bottom of the borehole, with an intention that sufficient intact soil sample adheres to the inside of the sampling tube when we retrieve and bring this tube to the ground surface. For this purpose, the toe of the sampling tube is sharpened and sometimes bent slightly inward, as shown in Figure 3.10, where D_e (inside diameter at the entry of the sampling tube) is slightly smaller than D_s (inside diameter of the sampling tube above the entry). An inside clearance ratio, C_i, defined as

$$C_i = \frac{D_s - D_e}{D_e} \times 100\%$$

(3.1)

is used to indicate the amount of clearance between the soil sample and inside of the sampling tube. According to American Society of Testing and Materials (ASTM) standard D1587, thin-walled sampling tubes have C_i ranging from 0% to 1.5%. For sampling in soft clays, C_i of 0 or less than 0.5% is used. In stiffer

Figure 3.9 Hollow stem auger in operation, Washington, DC, USA. (Courtesy of An-Bin Huang, Hsinchu, Taiwan.)

formations, larger C_i of 1%–1.5% may be required. With this C_i, soil can enter the sampling tube with minimal friction exerted from the sampling tube during pushing. In the meantime, the clearance is not too excessive. Upon entering the tube, the soil sample expands slightly to develop sufficient adhesion so that it stays in the sampling tube during retrieval.

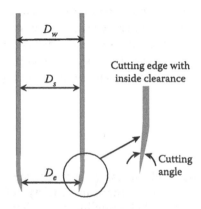

Figure 3.10 Cross-sectional view of a thin-walled sampling tube.

The sampling tube has a finite wall thickness; some soil is inevitably displaced by the insertion of the sampling tube into the soil. An area ratio, C_a, defined as

$$C_a = \frac{D_w^2 - D_e^2}{D_e^2} \times 100\% = \frac{\text{Volume of displaced soil}}{\text{Volume of sample}}$$ (3.2)

is used to indicate the ratio of displaced soil volume over that of the soil sample. Note that D_w = outside diameter of the sampling tube (see Figure 3.10). For undisturbed soil sampling, the area ratio, C_a, should be close to or less than 10%–15%. The cutting edge angle should range from 5° to 15°. Softer soil may require sharper cutting angles of 5°–10°. The tube cutting edge with sharper cutting angles may be easily damaged during sampling.

3.3.3 Taking soil samples with a thin-walled sampling tube

The thin-walled sampling tube is primarily used for taking undisturbed samples in cohesive soils. Recent developments also included the use of a thin-walled tube to take high quality samples in granular soils. The following sections describe a few of these thin-walled tube sampling methods.

3.3.3.1 The thin-walled "Shelby" tube sampler

Figure 3.11 shows a thin-walled tube sampler sometimes referred to as the Shelby tube sampler. In this case, a thin-walled round tube is attached to a head assembly where the sampler can be attached to the bottom of a string of drill rods. The ball check valve in the head assembly helps to develop a vacuum when the sampler is pulled upward during

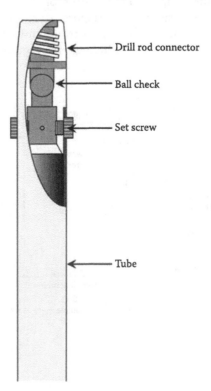

Figure 3.11 The thin-walled "Shelby" tube sampler.

Table 3.1 Dimensions of typical thin-walled sampling tubes

Outside diameter (mm)	Wall thickness (mm)	Length (mm)
50	1.25	1000
75	1.65	1000
125	3.05	1500

retrieval of the sampler. Table 3.1 shows the dimensions of typical thin-walled sampling tubes.

The thin-walled tube sampler is suitable for cohesive soil with sufficient shear strength, so that the combination of adhesion between the soil and sampling tube along with the vacuum created by the ball check valve is sufficient to keep the soil sample inside the tube during retrieval.

3.3.3.2 The Osterberg piston sampler

For soft cohesive soils such as soft clay or silt, the ball check valve and adhesion between the soil and sampling tube are not enough to hold the soil sample inside the sampling tube. By engaging a piston in the sampling tube that makes intimate contact with the top of the soil sample, a much more powerful vacuum can be developed to hold the soil sample inside the tube. Various designs of piston samplers have been proposed in the past, of which the design by Osterberg (1952) is probably the most popular. Figure 3.12 depicts the operation of an Osterberg piston sampler. The piston location in relation to the head assembly is fixed or stationary throughout the sampling procedure. The sampling tube is retracted when lowering the sampler to the bottom of the borehole, making the piston flush with the bottom of the sampling tube (Figure 3.12a) and in full contact with the soil to be sampled. The thin-walled tube is pushed downward by pumping water from ground surface into a chamber above the sampling tube (Figure 3.12b) via the hollow drill

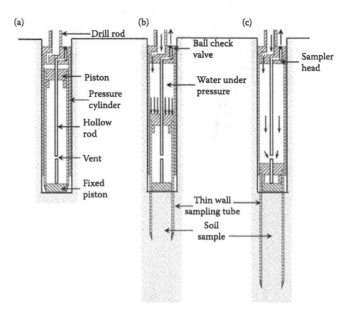

Figure 3.12 Operation of an Osterberg piston sampler. (a) Start of drive. (b) During the drive. (c) End of drive.

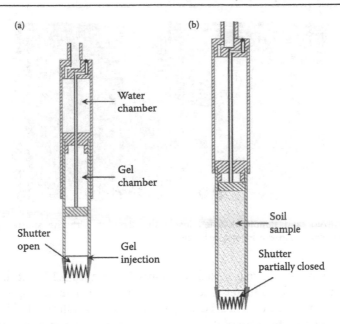

Figure 3.13 The gel-push sampler. (a) During the drive. (b) End of drive. (Adapted from Huang, A.B. et al. 2008. Sampling and field characterization of the silty sand in central and southern Taiwan. *Proceedings, Third International Conference on Site Characterization*, Taipei, pp. 1457–1463.)

rods. The fully extended sampling tube along with the soil sample is then retrieved to the ground surface (Figure 3.12c).

3.3.3.3 The gel-push sampler

For cohesionless soils such as dense sand or gravel, thin-walled sampling is not feasible for taking the soil samples. Excessive friction can develop between soil and surface of the sampling tube, making it very difficult to push the sampling tube into soil. Also, there is a lack of adhesion to hold the soil inside the sampling tube during retrieval. The gel-push sampler as schematically shown in Figure 3.13 was modified from an Osterberg piston sampler (Huang et al., 2008). The sampler injects a polymeric lubricant (the gel) from the sampler shoe to facilitate the penetration of the sampling tube. A shutter or catcher located at the tip of the sampler remains open during pushing (Figure 3.13a). A slight reverse motion by injecting water into the gel chamber triggers the inward bending of the shutter before the sampler is retrieved (Figure 3.13b). The bent or partially closed shutter prevents the sample from falling during retrieving. These unique features make it possible to take undisturbed samples in silty sand or clean sand with the gel-push sampler. Figure 3.14 shows the end of a gel-push sampler after retrieving from borehole with a sand sample retained inside by the shutter in its partially closed position.

3.3.4 The split-spoon sampler

A split-spoon sampler can be used to take disturbed but representative soil samples in a wide variety of soils including sand, gravel, and soft rock in addition to clay and silt. The sampler is split into two halves longitudinally as shown in Figure 3.15a. The two halves are screwed into a drive shoe and head assembly at the bottom and top of the sampler, respectively, to form a full tube. The split spoon is driven into the soil with a drop hammer, as part of a standard penetration test (SPT), an in-situ test method which is

Figure 3.14 Retrieved sand sample retained inside the sampler with the shutter partially closed. (Courtesy of An-Bin Huang, Hsinchu, Taiwan.)

described in Section 3.4. There is a basket retainer (see Figure 3.15a) at the lower end of the split spoon that holds soil after entering the split spoon. The split-spoon soil samples are disturbed with a C_a of 111.5% according to its dimensions shown in Figure 3.15b and Equation 3.2. However, because of its versatility and being part of an in-situ test method, split spoon is the most popular soil sampling method (Mayne et al., 2002), despite the fact that the samples are disturbed.

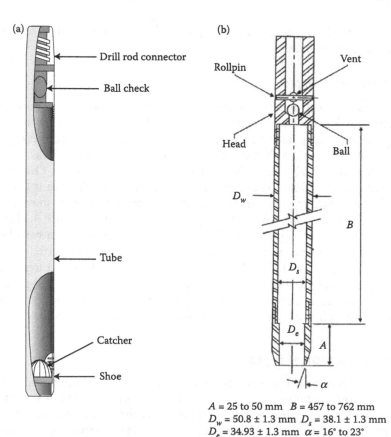

A = 25 to 50 mm B = 457 to 762 mm
D_w = 50.8 ± 1.3 mm D_s = 38.1 ± 1.3 mm
D_e = 34.93 ± 1.3 mm α = 16° to 23°

Figure 3.15 The split-spoon sampler. (a) Assembled sampling tube. (b) Dimensions.

3.3.5 Rock coring

Rock coring is accomplished by rotating a rock sampling tube (core barrel) with a hard cutting bit (coring bit) at the bottom. This cuts an annular hole in the rock mass, thereby creating a cylindrical rock core that is stored in the core barrel. The core barrels are generally operated at speeds from 50 to 1750 rpm (rounds per minute). The harder the rock, the faster the permissible speed. Downward pressure is applied to the bit while rotating. Lower pressure is required for softer rock (Winterkorn and Fang, 1975).

The coring bit is often hardened with diamond impregnation and equipped with discharge channels to facilitate the water circulation for cooling and lubrication during rock coring. Dimensions of typical core sizes are summarized in Table 3.2. Figure 3.16 shows the photograph of a diamond-hardened coring bit. There are single tube and multiple tube core barrel designs (ASTM D2113). Figure 3.17 shows the cross-sectional views of the single tube and double tube core barrels. The use of an inner tube in a double tube core barrel is to provide isolation of the rock core from the erosive action of the drilling fluid and thereby improve the core recovery. A triple tube core barrel is also available with various configurations. The main reason for adding an additional tube is again to offer more protection to the rock core.

A wireline system can be used to take rock cores in a deep borehole. In this case, a thick casing is used to serve the purpose of the drill rod. A wireline (a steel cable) is used to either lower or retrieve the drill bit or core barrel to or from the bottom of the casing/drill rod. The arrangement eliminates the need to remove all the drill rods from the borehole in order to replace drill bit with the core barrel, and thus saves time and effort. The wireline system can also be used when taking the soil samples in deep boreholes.

The rock core recovered from each run is measured for general evaluation of the rock quality. The total length of the recovered rock core is used to determine a recovery ratio:

$$\text{Recovery ratio} = \frac{\text{Length of the recovered core}}{\text{Total length of the rock run}} \qquad (3.3)$$

A recovery ratio close to 1 indicates an intact rock mass. Fractured rocks are associated with recovery ratios close to 0.5 or less. Measurements of the individual length of the rock cores in excess of 100 mm (4 inches) are used to calculate a rock quality designation (RQD) which is defined as follows:

$$\text{RQD} = \frac{\sum \text{length of intact and sound core pieces} > 100\,\text{mm}}{\text{Total length of core run, mm}} \times 100\% \qquad (3.4)$$

Table 3.2 Dimensions of selected core sizes

Size	Diameter of core (mm)	Diameter of borehole (mm)
EX	21.5	37.7
AX	30.1	48.0
AQ wireline	27.1	48.0
BX	42.0	59.9
BQ wireline	36.4	59.9
NX	54.7	75.7
NQ wireline	47.6	75.7
HQ wireline	63.5	96.3
PQ wireline	85.0	122.6

Figure 3.16 Photograph of a diamond coring bit. (Courtesy of An-Bin Huang, Hsinchu, Taiwan.)

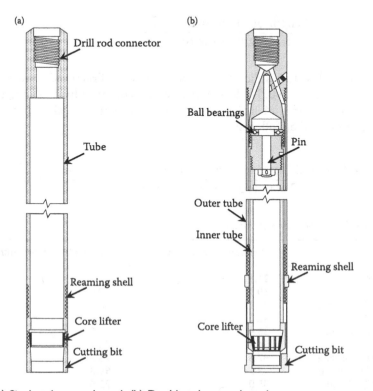

Figure 3.17 (a) Single tube core barrel. (b) Double tube core barrel.

Table 3.3 Relationship between RQD and rock quality

RQD (%)	Classification of rock quality
0–25	Very poor
25–50	Poor
50–75	Fair
75–90	Good
90–100	Excellent

Source: Deere, D.U. 1963. Felsmechanik und Ingieurgeologie, 1(1), 16–22.

Table 3.3 presents a general relationship between RQD and quality of the rock mass (Deere, 1963). Details on recovery ratio and RQD can be found in ASTM standard D 6032.

3.3.6 Sample quality assurance and verification

The most important factor controlling the quality of test results on undisturbed soil sample is maintaining a low level of disturbance. A soil disturbance can occur during drilling/sampling, transportation, and in preparation for laboratory testing (e.g., sample trimming). Possible mechanisms of disturbance can be associated with:

- *Changes in stress*—stress increases when soil enters the sampling tube, stress is relieved after extrusion from the sampling tube
- *Mechanical disturbance*—shear distortions applied to the soil by tube sampling
- *Changes or migration in moisture content/void ratio*—due to absorption/swelling or consolidation of the soil sample caused by stress changes
- *Changes in chemical contents*—due to contact with drilling fluid or sampling tube

Sample disturbance can be managed but not completely avoided. For quality assurance, the following practices and indexes are often applied when taking the undisturbed soil samples.

The length of soil sample and its ratio with the depth of penetration (L_t) are often used as an index of soil sample quality. Refer to Figure 3.18, where L_g = gross length of sample = distance from top of sample to cutting edge of the sampling tube after withdrawal; L_n = net length of sample = distance from top to bottom of the sample after trimming for sealing. The corresponding ratios are gross recovery ratio, R_g and net recovery ratio, R_n, respectively, which are defined as

$$R_g = \frac{L_g}{L_t} \tag{3.5a}$$

and

$$R_n = \frac{L_n}{L_t} \tag{3.5b}$$

In addition to recording the sample length, a visual classification and undrained shear strength testing using a hand tool such as the pocket penetrometer or a torevane is performed from the ends of the sampling tube where the soil sample is exposed. After taking these measurements, the sample is tagged or marked to indicate the location (i.e., borehole

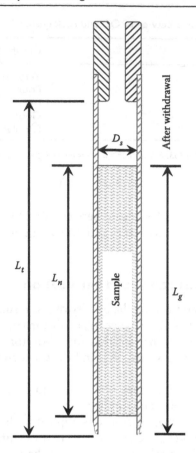

Figure 3.18 Quality index for soil samples.

number) and depth of the sample, and sealed with wax or plugs from both ends of the sampling tube to prevent loss of moisture and to minimize mechanical disturbance of the soil sample.

Soil samples should be stored upright and not be exposed to extreme heat or freezing temperatures, or subject to excessive vibration during shipping (ASTM D4220). Shear wave velocity, V_s, has been used to verify the quality of soil samples. Figure 3.19a shows a comparison between the field and laboratory V_s values for silty sand samples taken with a gel-push sampler presented by Huang and Huang (2007). The field values came from seismic piezo-cone penetration tests (SCPTu), which are discussed in Section 3.4. The laboratory values are taken using a pair of piezo-electric bender elements in a triaxial testing device under confining stress that is compatible with the field conditions. A good quality sample should have shear wave velocity values close to those measured in the field. For cohesive soil samples, a residual suction (negative pore water pressure) should remain in the soil upon extrusion from the sampling tube.

The magnitude of suction remaining in the soil sample has also been used to index the quality of soft clay samples (Tanaka, 2008). When the soil sample is extruded from the sampling tube and exposed to the air, the total mean normal stress experienced by the soil sample is zero. The residual effective stress within the soil sample is thus the remaining suction (negative pore pressure). Figure 3.19b shows the residual effective stress measured on the soft Ariake clay samples and its comparison with the expected effective vertical stress (σ'_{vo}) in the field (Tanaka, 2008). The clay samples were taken

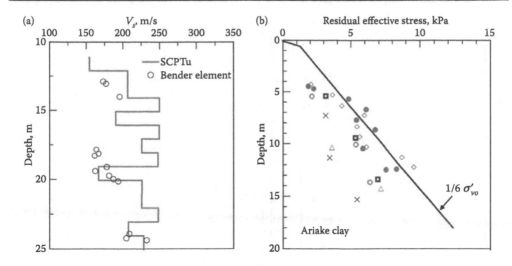

Figure 3.19 Soil sample quality assurance through shear wave velocity and residual effective stress measurements. (a) Shear wave velocity from bender element tests. (Adapted from Huang, A.B. 2008. Sampling and field characterization of the silty sand in central and southern Taiwan. *Proceedings, Third International Conference on Site Characterization,* Taipei, pp. 1457–1463.) (b) Residual effective stress from suction measurements. (Adapted from Tanaka, H. 2008. Sampling and sample quality of soft clays. In Huang and Mayne, eds, *Proceedings, Geotechnical and Geophysical Site Characterization, Taipei,* Taylor & Francis Group, London, pp. 139–157.)

with six types of samplers that include Shelby tube and other types of piston samplers. According to Tanaka (2008), samples with lower residual effective stress usually yield lower unconfined compressive strength. Readers are referred to Tanaka (2008) for details of the six sampling methods and suction measurements on the soil samples.

3.3.7 Observation of groundwater level and field logging

Groundwater level is an important geotechnical parameter to be collected in subsurface exploration. The presence or fluctuation of groundwater can affect the bearing capacity and settlement of foundations. The change of soil samples from moist to saturated conditions or the change of color can all be used as a sign of the presence of groundwater table. For permeable soils, the water level in the borehole can reach an equilibrium with the surrounding groundwater level in minutes. The water level in a borehole can be measured by dropping a tape into the borehole. The plot of water level against time is a good indication if the measurement is stabilized and thus the depth of groundwater table. These measurements can be incorporated in the drilling and sampling operation in the field.

For layered soil conditions with intermittent low-permeable layers, a piezometer may be required for rigorous measurement of the groundwater conditions. Figure 3.20 shows an open-end piezometer installed in a borehole. The open end at the tip of the piezometer is protected by filter material that allows water to enter. The upper part of the open end is blocked by bentonite. Water level in the piezometer in this case reflects the water pressure at the open end.

A field log (Figure 3.21) is used to keep track of the field operation. Details of the project name, borehole location, drill crew, and drilling/sampling methods used at various depths should be recorded. Results from field testing such as blow counts from the SPT, visual description and moisture conditions of the soil samples, and groundwater level based on

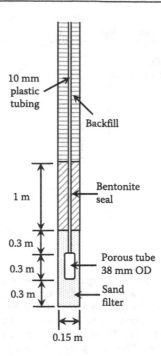

Figure 3.20 The open-end piezometer.

field observations should be recorded and signed by the drill crew. This is an original documentation recorded in the field. In case of any confusion, such as due to missing soil samples or mixture of sample labels, the field log becomes important for clarification.

3.4 IN-SITU TESTING

In-situ testing represents another means of site characterization where the testing device is inserted into the ground and soil is tested as it is in the ground, or in-situ. Advantages of in-situ testing include:

- Larger volume of soil can be tested (in comparison with soil sampling and laboratory testing)
- Continuous soil profile can be produced
- It may be the only practical choice in cohesionless soil
- Soils are tested in their natural state
- Most in-situ test methods are efficient and thus can be less costly

Limitations of in-situ testing are:

- Lack of well-defined boundary and drainage conditions
- The tests often involve complicated strain fields and stress paths
- Soil physical features are not positively identified (no soil sample is taken, except for standard penetration test)

Both in-situ and laboratory testing methods have their inherent merits and disadvantages. Most geotechnical explorations make use of both in-situ and laboratory testing.

Depth, m	Sample type and number	SPT				Foreman: ABC

Project name: xxxx project | Location: No. xx, yy Street, Z city | Drilling method: Rotary boring

Project name: xxxx project		Location: No. xx, yy Street, Z city			Drilling method: Rotary boring	
Borehole number: BH-xxx		Surface elevation: xx m			Casing: 4.2" to 12 m	
Coordinate: N 2644539, E 208303		Datum:	Z city datum		Date started: 12/24/2015	
Borehole depth: 15.5 m		Ground water level: 2.2 m			Date completed: 12/25/2015	

Depth, m	Sample type and number	SPT Depth, m From	To	N	USCS	Foreman: ABC / Helper: CDE / Description
1						0–1.5 m Fill, crushed rock, cobble, and fine sand
2						1.5–4.6 m Brown clayey sand
	S-1	2	2.45	2	SC	
3						
	T-1	3	3.8			
4	S-2	3.55	4	3	ML	
5	S-3	4.55	5	3		
						4.6–10.0 m gray silty sand and clayey sand
6	S-4	5.55	6	3	SM	
7	S-5	6.55	7	6	SC	
8	S-6	7.55	8	4	SM	
9	S-7	8.55	9	7	SM-SC	
10	S-8	9.55	10	15	SM	
						10.0–12.3 m Gray silty sand, with coarse sand and fine gravel
11	S-9	10.55	11	22	SM	
12	S-10	11.55	12	7	SP	
13						12.3–17.5 m gray coarse sand and gravel
	S-11	13.05	13.5	28	SW-SM	
14						
15	S-12	14.55	15	20	SM	
						END

S: Split–spoon sample   T: 76.2 mm OD thin wall tube sample

Page 1 of 1

Figure 3.21 A typical field log.

There are many in-situ testing methods available for a wide variety of soils. The interpretation methods of in-situ test results are largely empirical, and these methods can generally be divided into two groups as follows:

Indirect interpretation: Determine the basic engineering properties such as shear strength, stiffness, compressibility, permeability, and unit weight from the in-situ test and then use the engineering properties for geotechnical design.

Direct interpretation: Relate the in-situ test results directly to design parameters such as the bearing capacity of foundations or liquefaction potential for granular soils.

The following sections introduce five of the commonly used in-situ test methods in the following sequence:

- Standard penetration test (SPT)
- Cone penetration test (CPT)

- Flat dilatometer test (DMT)
- Pressuremeter test (PMT)
- Field vane shear test (FVT)

The respective test equipment, procedures, and the interpretation of the results will be presented in the following sections. Suitability of one approach or another is strictly linked to the aims of the exploration and the method of analysis which is intended to be used in the geotechnical design.

3.4.1 Standard penetration test (SPT)

The development of the SPT can be traced back to the beginning of twentieth century (Massarsch, 2014). SPT (ASTM D1586) is by far the most popular in-situ testing and sampling method in geotechnical site characterization.

3.4.1.1 Equipment and test procedure

In SPT, a thick-walled split-spoon sampler described in Section 3.3 is driven from the bottom of the borehole into the ground with a hammer that weighs 623 N (140 pounds), as shown in Figure 3.22. The hammer is raised by 762 mm (30 inches), creating a potential energy of 474 J before dropping, in every hammer blow. The split-spoon sampler penetrates into the soil as the hammer blow continues. The number of hammer blows for

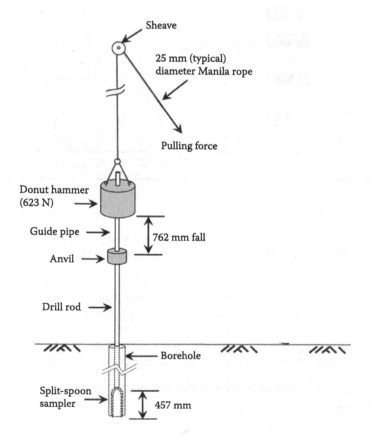

Figure 3.22 Field setup of the standard penetration test.

driving the split spoon by 152.4 mm (6 inches) is recorded. This procedure is repeated for three intervals of 152.4 mm penetration for a total penetration of 457.2 mm (18 inches). The sum of hammer blows for the last two intervals of 152.4 mm or approximately a total of 300 mm (12 inches) penetration is referred to as the standard penetration number, N or simply N value. Upon full penetration, the split-spoon sampler is retrieved to the ground surface. The sampler is split open (hence the name split-spoon sampler) and the soil sample scraped from the sampler and stored in a glass jar. SPT is usually conducted at 1.5 m intervals in the field. The split-spoon sampler combined with SPT can be used in cohesive and cohesionless soils. Since undisturbed sampling is not practical in granular soils, SPT is especially useful for sampling (disturbed sampling) and testing in sand and gravel. SPT is an in-situ penetration test that yields an N value and takes a soil sample in the process. The samples can be used for basic physical property tests that are not affected by sample disturbance. Because of its versatility, SPT and split-spoon sampling is popular worldwide.

The operation of SPT hammer involves the use of a pulling force to lift the hammer, as illustrated in Figure 3.22. Various power mechanisms are now available and are routinely used to raise and release the hammer automatically as required. Human power assisted by a power-driven rotating drum (i.e., a cathead) is also used often. There are different types of SPT hammer designs and control mechanisms in different parts of the world (Schnaid, 2009). The donut and safety hammer as conceptually described in Figure 3.23 are probably the most popular ones used in practice. Characteristics of the drill rod, borehole, design of the hammer, and the inevitable friction existing in the driving system can all affect the energy delivered to the sampler. The delivered energy can vary significantly, usually below the theoretical value of 474 J, making it difficult to interpret N values for a given soil condition. To account for the effects of hammer energy variations, overburden stress, and rod length, it has become a standard practice worldwide to measure the energy delivered to the sampler for a given field SPT setup (Seed et al., 1985). The N values are then corrected to a reference hammer energy that is equivalent to 60% of the potential energy of SPT hammer and an effective overburden stress of 1 atmosphere (100 kPa).

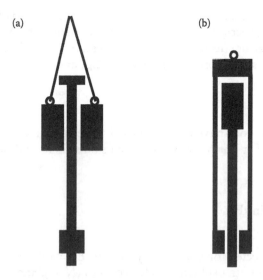

(a)

(b)

Figure 3.23 Donut and safety SPT hammer. (a) Donut hammer with rope. (b) Safety hammer. (Adapted from Schnaid, F. 2009. *In Situ Testing in Geomechanics.* Taylor & Francis, 329p.)

The N values originally collected in the field will be referred to as N_{SPT} in the following discussion.

The hammer energy is measured with a load cell that measures force $F(t)$ and a pair of accelerometers installed immediately below the anvil (see Figure 3.22). Integration of acceleration with time gives velocity, $V(t)$. The hammer energy delivered to the anvil during an SPT hammer blow, E_{SPT}, is

$$E_{SPT} = \int_0^t F(t)V(t)dt \tag{3.6}$$

where t is the time period associated with the stress wave traveling through the measurement location. Details of the hammer energy measurement can be found in ASTM standard D1586.

The ratio of energy, delivered to the anvil over the theoretical or maximum energy of 474 J, is designated as an energy ratio, E_r,

$$E_r = \frac{E_{SPT}}{\text{Maximum energy of 474 J}} \times 100\% \tag{3.7}$$

Measurements from different parts of the world have indicated that E_r can vary from close to 40% to over 80%. The original N_{SPT} is usually converted to N_{60}, or N value corresponds to an energy ratio of 60%, E_{60}, as

$$N_{60} = N_{SPT}C_E \tag{3.8}$$

where C_E is the energy correction factor defined as

$$C_E = \frac{E_{SPT}}{E_{60}} = \frac{E_r}{60} \tag{3.8a}$$

The N_{SPT} value in granular soil with similar relative density tends to increase with overburden stress. For engineering applications in granular soils, the N_{SPT} values are usually corrected to an equivalent effective overburden stress, σ'_{vo}, of 100 kPa, or N_1. The correction is expressed by a stress normalization factor C_N, and

$$N_1 = N_{SPT}C_N \tag{3.9}$$

Many empirical methods have been proposed to determine C_N for SPT under a given σ'_{vo} (Schnaid, 2009). For normally consolidated sands, Liao and Whitman (1985) have suggested that

$$C_N = \sqrt{\frac{100}{\sigma'_{vo}}} \quad (\sigma'_{vo}\text{ in kPa}) \tag{3.9a}$$

An N_{SPT} value corrected for both energy ratio to $E_r = 60\%$ and effective overburden stress σ'_{vo} of 100 kPa (1 atmosphere) is referred to as $(N_1)_{60}$, and

$$(N_1)_{60} = N_{SPT}C_E C_N \tag{3.10}$$

Table 3.4 Estimation of s_u for clays or conditions of residual soils based on N_{60}

Clay	N_{60}	$PI = 50$	$s_u = 4.5N_{60}p_a/100$ (Stroud, 1974)
		$PI = 15$	$s_u = 4.5N_{60}p_a/100$ (Stroud, 1989)
Residual soils (Schnaid, 2009)	N_{60}	0–5	Completely weathered
		5–10	Very weathered (lateritic)
		10–15	Weathered
		>15	Moderately weathered (saprolitic)

Note

s_u and p_a (atmospheric pressure) should use the same unit.

3.4.1.2 Engineering properties of sands according to SPT N values

Note that soil samples can be taken as part of SPT. Many important physical properties such as grain size distribution and soil specific gravity can be determined using laboratory tests on the split-spoon samples. The indirect interpretations of SPT are more related to the estimation of relative density or friction angle in granular soils. Table 3.4 shows available empirical correlations between N_{60} and s_u of clays and conditions of residual soils.

As it is impractical to take undisturbed samples for granular soils, inferring its field relative density, D_r, from N value is an important function of SPT. Terzaghi and Peck (1967) suggested an empirical correlation between D_r and SPT N values. The original correlation has been verified by recent field tests and demonstrated its reasonableness. The relative density (in percent) is related to $(N_1)_{60}$ as follows:

$$D_r\% = 100\sqrt{\left(\frac{(N_1)_{60}}{60}\right)} \tag{3.11}$$

To infer relative density, $(N_1)_{60}$ should be limited to 60; above this value, grain crushing can occur due to high-compressive stress (Mayne et al., 2002).

Hatanaka and Uchida (1996) proposed an empirical equation to estimate peak friction angle, $\varnothing'$, of clean sands using $(N_1)_{60}$ as follows:

$$\varnothing' = [15.4(N_1)_{60}]^{0.5} + 20° \tag{3.12}$$

Note the difference between peak friction angle and critical state friction angle. For details, please refer to Chapter 1.

EXAMPLE 3.1

Given

Table E3.1 shows a set of SPT data that includes the N values for every 1.5 m, and energy efficiencies (in percentage, %) from field measurements in a silty sand. The groundwater table is at 2.6 m below ground surface. The soil unit weight (γ) above groundwater table is 15 kN/m³, saturated soil unit weight (γ_{sat}) below groundwater table is 19 kN/m³. Unit weight of water, γ_w, is 9.81 kN/m³.

Required

Calculate the effective overburden stress (σ'_{vo}) corresponding to each N value and determine the $(N_1)_{60}$ values according to energy efficiency and σ'_{vo}, plot the result of $(N_1)_{60}$ versus depth.

Solution

σ'_{vo} *for a given depth, z:*

Above groundwater table ($z \leq 2.6$ m): $\sigma'_{vo} = z\gamma$

Below groundwater table ($z > 2.6$ m): $\sigma'_{vo} = 2.6\gamma + (z - 2.6)(\gamma_{sat} - \gamma_w)$

Energy correction factor, CE:

$$C_E = \frac{E_{SPT}}{E_{60}} = \frac{E_r(\%)}{60}$$

where E_r is the energy ratio values (in %) from the third column of Table E3.1.

Stress normalization factor, C_N:

$$C_N = \sqrt{\frac{100}{\sigma'_{vo}}}$$

where σ'_{vo} (in kPa) is taken from the 4th column of Table E3.1.

$(N_1)_{60} = N_{SPT}C_E C_N$ where N_{SPT} is the original SPT blow counts in the 2nd column of Table E3.1.

The original N_{SPT} and corrected $(N_1)_{60}$ profiles are shown in Figure E3.1.

Table E3.1 SPT N values and energy efficiency

Depth (m)	N_{SPT} blows/30 cm	E_r (%)	σ'_{vo} (kPa)	C_E	C_N	$(N_1)_{60}$ blows/30 cm
1.5	5	79.0	22.5	1.32	2.11	13.9
3.0	4	78.6	42.6	1.31	1.53	8.0
4.5	5	76.5	56.1	1.27	1.34	8.5
6.0	8	69.1	69.6	1.15	1.20	11.0
7.5	11	80.9	83.1	1.35	1.10	16.3
9.0	9	77.7	96.6	1.29	1.02	11.9
10.5	17	74.6	110.1	1.24	0.95	20.1
12.0	17	79.5	123.6	1.32	0.90	20.2
13.5	10	86.8	137.1	1.45	0.85	12.3
15.0	16	84.4	150.6	1.41	0.81	18.3

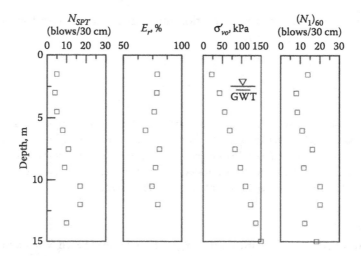

Figure E3.1 Original and corrected N_{SPT} values.

EXAMPLE 3.2

Given

The set of $(N_1)_{60}$ obtained in Example 3.1.

Required

Determine the corresponding relative density using Equation 3.11 and peak friction angles according to Equation 3.12.

Solution

The calculations are conducted by a spreadsheet program, and

$$D_r\% = 100\sqrt{\left(\frac{(N_1)_{60}}{60}\right)} \qquad (3.11)$$

$$\emptyset' = \left[15.4(N_1)_{60}\right]^{0.5} + 20° \qquad (3.12)$$

Results are presented in Table E3.2.

Table E3.2 SPT $(N_1)_{60}$ values and the corresponding D_r and $\emptyset'$

Depth (m)	$(N_1)_{60}$ blows/30 cm	D_r %	$\emptyset'$ degree
1.5	13.9	48	34.6
3.0	8.0	37	31.1
4.5	8.5	38	31.4
6.0	11.0	43	33.0
7.5	16.3	52	35.8
9.0	11.9	44	33.5
10.5	20.1	58	37.6
12.0	20.2	58	37.7
13.5	12.3	45	33.8
15.0	18.3	55	36.8

3.4.1.3 Liquefaction potential assessment based on SPT N values

A classic application of the SPT N values is the assessment of soil liquefaction potential for granular soils. As part of a simplified procedure (Seed and Idriss, 1982; Youd et al., 2001), a correlation between soil cyclic strength and $(N_1)_{60}$ is empirically established based on post-earthquake SPT and field observations that showed signs of liquefaction or no liquefaction. The analysis is usually carried for saturated granular soils below groundwater table, within a depth of 20 m from the ground surface. Through a simplified and semi-empirical procedure, the random ground motion during the given earthquake event is converted into a uniform shear stress wave (i.e., sine wave). A cyclic stress ratio (CSR) is used to reflect the intensity of the earthquake-induced shear stresses in reference to its effective overburden stress from the demand side. CSR is defined as the ratio of $\sigma_d/2\sigma'_{vo}$ where σ_d, σ'_{vo} = amplitude or the maximum cyclic deviator stress and effective overburden stress, respectively, at the depth of SPT. A cyclic resistance ratio (CRR) is used to represent the soil cyclic strength from the supply side. CRR is also defined as the ratio of $\sigma_d/2\sigma'_{vo}$ (Seed and Idriss, 1982; Ishihara, 1993).

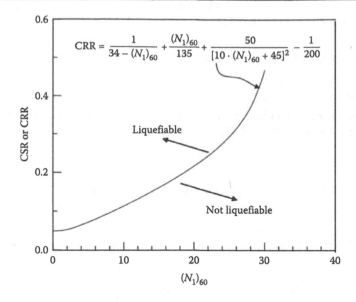

$$CRR = \frac{1}{34 - (N_1)_{60}} + \frac{(N_1)_{60}}{135} + \frac{50}{[10 \cdot (N_1)_{60} + 45]^2} - \frac{1}{200}$$

Figure 3.24 CRR-$(N_1)_{60}$ correlation for clean sands reported by Youd et al. (2001).

The post-earthquake field observation of signs of liquefaction (e.g., sand boil) and SPT forms a pair of CSR-$(N_1)_{60}$, data point. The cyclic strength or CRR-$(N_1)_{60}$ correlation is the boundary curve that divides the CSR-$(N_1)_{60}$ data points of liquefaction and no liquefaction from field observations. To assess if the saturated granular soil at a given site has a potential of liquefaction in a projected future earthquake event, a seismic analysis is conducted to estimate CSR and a series of SPT performed to obtain $(N_1)_{60}$. If the pair of CSR-$(N_1)_{60}$ falls to the upper lefthand side of the CRR-$(N_1)_{60}$ curve, then the tested soil is considered liquefiable. Otherwise, the soil is considered not liquefiable, as shown in Figure 3.24. The CRR-$(N_1)_{60}$ correlation for clean sands (sands with fine particles passing No. 200 sieve less than 5% in weight) shown in Figure 3.24 was reported by Youd et al. (2001). The CRR-$(N_1)_{60}$ correlation can be approximated with an empirical equation as depicted in Figure 3.24. There have been many modifications of the CRR-$(N_1)_{60}$ correlation by various researchers in the past few decades (Idriss and Boulanger, 2006).

3.4.2 Cone penetration test (CPT)

The concept involved in the cone penetration tests can be traced to the Dutch mechanical cone, developed around 1930 (Barentsen, 1936). A rod with a pointed or cone-shaped tip is pushed into the ground while recording the force (therefore the soil resistance) required to push the rod. The use of an outer pipe and inner rod allows the soil resistance reacting on the conical tip to be accurately measured as the cone is pushed into the ground.

3.4.2.1 Equipment and test procedure

The cone, with a cross-sectional area of 10 cm^2 and 60° apex angle, was pushed into the ground manually. The soil resistance experienced by the cone tip was recorded by means of a hydraulic pressure gage on the ground surface. Earlier cone penetrometers measured cone tip resistance only. Cone tip resistance, q_c, is the resistance force measured at the cone

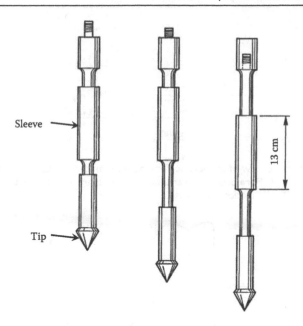

Figure 3.25 Cone penetrometer with a friction sleeve. (Adapted from Begemann, H.K. 1953. Improved method of determining resistance to adhesion by sounding through a loose sleeve placed behind the cone. *Proceedings, 3rd ICSMFE,* Zurich, Switzerland, Vol. 1, pp. 213–217.)

tip divided by the cross-sectional area of the cone (typically at 10 cm^2). The addition of a sleeve behind the cone tip (Begemann, 1953) with a telescope design as shown in Figure 3.25 allows measurement of soil friction imposed on the sleeve, f_s, during penetration (frictional force imposed on the friction sleeve divided by the surface area of the friction sleeve). Charts have been proposed to estimate shaft friction of piles and soil classification using f_s readings.

An electric cone with a friction sleeve was developed in the 1960s. The q_c and f_s measurements were made using electric resistance strain gages. The signals were transmitted to the ground surface via electric cables, and readings were recorded using an automated data logger. This development made CPT a very efficient device in site characterization. ASTM standard (D-5778) and an ISSMFE (International Society for Soil Mechanics and Foundation Engineering) reference test procedure have been established for CPT. The concept of piezocone penetration test (CPTu) was proposed by various researchers in the 1960s and 1970s (Torstensson, 1975; Wissa et al., 1975). The cone penetration-induced pore water pressure is measured by adding an electric piezometer to the cone tip. The porous element or filter that facilitates pore water pressure measurement in CPTu can be located at the tip, face, or behind the cone tip.

Figure 3.26 shows a cross-sectional view of an electric cone penetrometer with a friction sleeve and an electric piezometer (i.e., a piezocone). It should be noted that the location of the porous element can affect the magnitude of the pore water pressure readings and thus the interpretation of the pore pressure readings (Lunne et al., 1997). A separate designation is given for porous element located at different positions as shown in Figure 3.27. Of all the possible porous element locations, u_2 is the most commonly used. Figure 3.28 shows a standard size piezocone (35.6 mm diameter) with a porous (filter) element at u_2 position.

For CPTu with u_2 readings, the pore water pressure infiltrates into the top of the cone tip and results in a lower q_c reading due to unequal cross-sectional area of A_n and

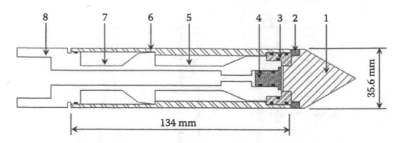

Figure 3.26 Cross-sectional view of an electric cone penetrometer. (1) Cone tip. (2) Porous element. (3) Adapter. (4) Pore pressure transducer. (5) q_c load cell. (6) Friction sleeve. (7) $(q_c + f_s)$ load cell. (8) Connection with rods.

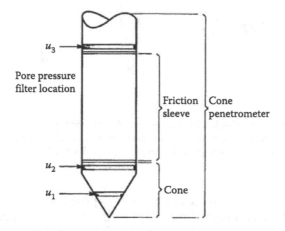

Figure 3.27 Cone penetrometer with porous element at different locations.

Figure 3.28 A piezocone with porous element at u_2 position. (Courtesy of An-Bin Huang, Hsinchu, Taiwan.)

A_c as shown in Figure 3.29. The q_c reading can be too low unless the u_2 effects are accounted for using the following equation:

$$q_t = q_c + u_2(1 - a) \tag{3.13}$$

where

q_t = net cone tip resistance with u_2 correction
a = area ratio of A_n over A_c (typically around 0.8)

The u_2 correction is especially important for CPTu in soft clays where q_c is low and u_2 can be high.

The modern day electric cone penetration test is often conducted from inside of a cone truck as shown in Figure 3.30. Hydraulic jacks are used to level the truck. A hydraulic ram

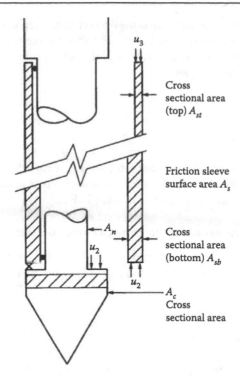

Figure 3.29 Effects of u_2 on cone tip resistance measurement.

is used to push the cone at a rate of 20 mm/sec. The hydraulic ram and data logging system are all housed inside the truck. This arrangement makes CPT mobile, efficient, and productive.

Campanella et al. (1986) reported the concept of seismic cone penetration test where shear wave velocity, V_s, is measured while the cone penetration is suspended in the ground, the cone tip depth is known and stationary. Shear wave is generated by hitting sideways on the hydraulic jack base of the cone truck with a hammer. A seismometer mounted on the hydraulic jack base signals the initiation of the shear wave. Another

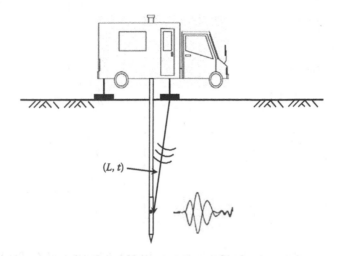

Figure 3.30 Cone truck and field setup for shear wave velocity measurement.

seismometer installed above the cone tip is used to sense the arrival time t (see Figure 3.30) of the shear wave. As the cone tip depth, thus the shear wave travel distance, L (see Figure 3.30), is known, shear wave velocity, $V_s = L/t$. Shear wave velocity relates to the soil maximum shear modulus, G_{max} as

$$G_{max} = \rho V_s^2 \qquad (3.14)$$

where
$\qquad \rho = $ soil mass density

G_{max} is a useful parameter in characterizing the soil properties. The cone penetration tests performed nowadays often involve piezocone with shear wave velocity and typically referred to as seismic piezocone penetration test (SCPTu).

3.4.2.2 Soil classification according to cone penetration tests

Nearly continuous readings of q_t, f_s, and u_2 are obtained at 5–10 mm intervals. The V_s readings are usually taken at 1 m intervals when the penetration is halted to add the 1 m long cone rod. Figure 3.31 shows a representative SCPTu sounding from a silty sand/silty clay deposit.

The term CPT will be used hereinafter to represent the family of cone penetration tests that can include SCPTu or CPTu. The letter "S" and/or "u" will be added only when seismic and/or piezocone penetration is specifically referred to. The nearly continuous readings make the cone penetration test ideally suited to establish the soil stratigraphic information. Many empirical methods have been developed over the last few decades for that purpose, but before we get into the details of CPT-based soil classification it is helpful to notice a general trend that, for a given q_t, CPT in soils with decreasing grain size (or increasing fines content) tends to yield higher f_s. Soil particles passing #200 sieve are referred to as fines. Or, the friction ratio ($R_f = f_s/q_t$ in percent) becomes larger as the soil grain size decreases. It is also important to note that for proper evaluation of soil behavior we should normalize the CPT test results with respect to the vertical effective stress (Wroth 1984, 1988). This normalization is a practical way to eliminate the CPT

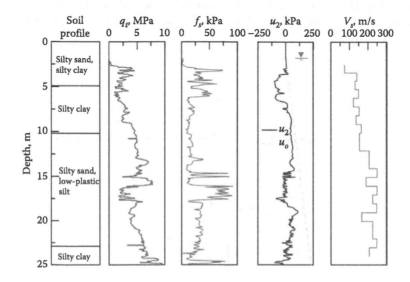

Figure 3.31 SCPTu sounding in a silty sand/silty clay deposit. (Courtesy of An-Bin Huang, Hsinchu, Taiwan.)

depth effect, which is not a function of soil behavior. The following sections introduce two CPT-based soil classification systems based on normalized CPT test results.

3.4.2.3 The Robertson SBTn charts

Of the many available CPT soil classification methods, the series of SBT classification charts proposed and updated by Robertson et al. (1986) and Robertson (1990, 2009, 2016) are likely to be the most commonly used. The SBT index, I_c, proposed by Robertson (2009) can be used to infer the soil behavior, such as OCR, sensitivity, drainage, strength, stiffness, sand-like or clay-like, from CPT sounding. Robertson (2016) emphasized that I_c may not have a dependable relationship with the typical soil classification schemes such as the unified soil classification system, which is based on physical property tests (i.e., sieve analysis, hydrometer, and plasticity tests). The q_t, f_s, and u_2 readings are first normalized with respect to the field vertical (overburden) stress as follows:

$$Q_{tn} = \left(\frac{q_t - \sigma_{vo}}{p_a}\right)\left(\frac{p_a}{\sigma'_{vo}}\right)^n \tag{3.15}$$

$$F_r = \left[\frac{f_s}{q_t - \sigma_{vo}}\right]100(\%) \tag{3.16}$$

$$B_q = \frac{(u_2 - u_o)}{(q_t - \sigma_{vo})} = \frac{\Delta u_2}{(q_t - \sigma_{vo})} \tag{3.17}$$

where
σ_{vo} and σ'_{vo} = total and effective overburden stress, respectively
p_a = atmospheric pressure of the same unit as that of overburden stress and q_t
u_o = the field hydrostatic pressure
Δu_2 = excess pore pressure = $(u_2 - u_o)$

The exponent:

$$n = 0.381 \cdot I_C + 0.05 \cdot (\sigma'_{vo}/p_a) - 0.15 \tag{3.18}$$

where $n \leq 1.0$, and

$$I_c = \sqrt{(3.47 - \log Q_{tn})^2 + (1.22 + \log F_r)^2} \tag{3.19}$$

Figure 3.32 shows a nine-zonal classification of SBTs based on Q_{tn}–F_r correlation and I_c. The determination of I_c requires the value of Q_{tn} which depends on n. But n is a function of I_c according to Equation 3.18. Therefore to use the above procedure for soil classification, particularly for granular soils, it may be advisable to follow an iteration process as follows:

1. Assume a trial n value. A good starting value would be 0.5.
2. Calculate Q_{tn} according to Equation 3.15 and I_c using Equation 3.19.
3. With I_c from step 2, calculate n using Equation 3.18.
4. Repeat steps 1 through 3 with a different n value if necessary until the trial and calculated n values converge.

$n \leq 1.0$, the above iteration procedure is most likely to be applied for CPT in coarse-grained soils (sands) with I_c less than 2.6. For CPT in fine-grained soils, simply

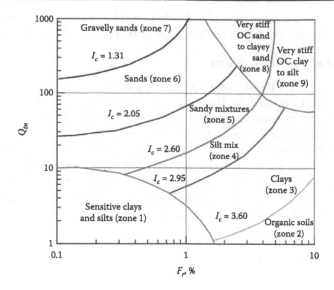

Figure 3.32 Soil behavior type from Q_{tn}–F_r relationship with nine zonal classification. (From Robertson, P.K., 2009. Interpretation of cone penetration tests—a unified approach. *Canadian Geotechnical Journal.* Figure 3.2, p.1342. Reproduced with permission of the NRC Research Press.)

use $n = 1$. For CPT in clean sands, use $n = 0.5$ would suffice in most cases for the calculation of Q_{tn} without going through the iteration process. Equations 3.15 through 3.19 and the zonal boundaries can be easily integrated into a spreadsheet program, making the classification rather efficient. The chart shown in Figure 3.32 is often referred to as the Robertson SBTn chart.

EXAMPLE 3.3

Given

Table E3.3 shows the representative q_t, f_s, σ_{vo}, and σ'_{vo} values for every 2.5 m from the same SCPTu test results shown in Figure 3.31. Use soil total and saturated unit weight, $\gamma = \gamma_{sat} = 19.0 \text{ kN/m}^3$. The groundwater table is at 1.8 m below ground surface.

Table E3.3 Representative values taken from Figure 3.31 and determination of I_c

Depth (m)	2.5	5.0	7.5	10.0	12.5	15.0	17.5	20.0	22.5
q_t, MPa[a]	3.55	2.37	1.97	5.39	4.05	6.73	6.72	6.55	6.56
f_s, kPa[a]	9.36	25.38	11.91	13.20	20.38	15.91	25.07	33.61	26.06
u_2, kPa[a]	8.83	−54.94	43.16	69.65	95.16	82.40	97.12	96.14	47.09
σ_{vo}, MPa[a]	0.048	0.095	0.143	0.190	0.238	0.285	0.333	0.380	0.428
σ'_{vo}, MPa[a]	0.041	0.064	0.087	0.110	0.132	0.156	0.175	0.201	0.224
Q_{tn}[b]	57.0	32.3	20.5	48.5	30.7	45.87	40.9	36.3	33.2
F_r[b]	0.27	1.11	0.65	0.25	0.53	0.25	0.39	0.54	0.43
I_c[b]	1.75	2.34	2.39	1.89	2.20	1.91	2.03	2.14	2.13
n[b]	0.55	0.76	0.80	0.62	0.75	0.65	0.7	0.76	0.77
Zonal No.[b]	6	5	5	6	5	6	6	5	5

Notes

[a] Given values.
[b] Computed values.

Required

Compute the corresponding Q_{tn}, F_r, and I_c values and determine the soil behavior type zonal numbers according to Figure 3.34.

Solution

A spreadsheet computer program provided to registered users was used for the computations and preparation of Table E3.3.

The values of Q_{tn}, F_r, and I_c are calculated using the following equations and iterate the n values according to the procedure described above until the trial and calculated n values converge.

$$Q_{tn} = \left(\frac{q_t - \sigma_{vo}}{p_a}\right)\left(\frac{p_a}{\sigma'_{vo}}\right)^n \tag{3.15}$$

$$F_r = \left[\frac{f_s}{q_t - \sigma_{vo}}\right]100(\%) \tag{3.16}$$

$$n = 0.381 \cdot I_C + 0.05 \cdot (\sigma'_{vo}/p_a) - 0.15 \tag{3.18}$$

and

$$I_c = \sqrt{(3.47 - \log Q_{tn})^2 + (1.22 + \log F_r)^2} \tag{3.19}$$

The computed Q_{tn}, F_r, I_c, and n values along with the SBT zonal numbers selected according to I_c are entered in Table E3.3.

In addition to the Q_{tn}–F_r chart presented in Figure 3.32, Robertson (1990) also proposed a companion Q_{tn}–B_q chart. It is generally deemed that u_2 readings tend to lack repeatability due to possible loss of saturation, especially when CPT is performed onshore above the water table, when the water table is deep, and/or CPT is conducted in very stiff soil. Therefore the Q_{tn}–B_q chart is not used as often.

3.4.2.4 The Schneider et al. Q_t–F_r and Q_t–U_2 charts

Ramsey (2002) showed CPT data in high-OCR clayey soils may be classified as clays, silts, or sands when using the Q_{tn}–F_r and Q_{tn}–B_q charts proposed by Robertson (1990). This bias can be quite severe for CPT in offshore soils when assessing the soil conditions relevant to the design of offshore pipelines or small-scale subsea structures. Schneider et al. (2008, 2012) proposed a pair of charts that used a linear normalization of q_t and u_2 normalization with respect to σ'_{vo} as follows:

$$Q_t = \frac{(q_t - \sigma_{vo})}{\sigma'_{vo}} = \frac{q_{tn}}{\sigma'_{vo}} \tag{3.20}$$

$$U_2 = \frac{(u_2 - u_o)}{\sigma'_{vo}} = \frac{\Delta u_2}{\sigma'_{vo}} \tag{3.21}$$

where

q_{tn} = net corrected cone tip resistance

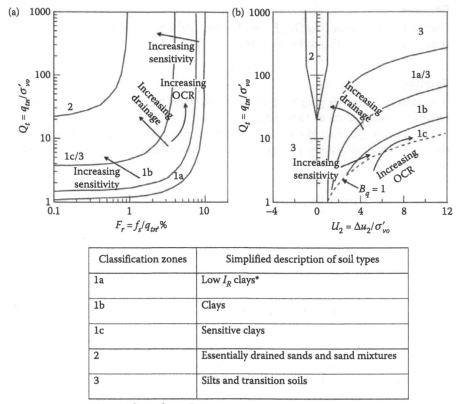

Classification zones	Simplified description of soil types
1a	Low I_R clays*
1b	Clays
1c	Sensitive clays
2	Essentially drained sands and sand mixtures
3	Silts and transition soils

*I_R = rigidity index = G/s_u

Figure 3.33 Q_t–F_r and Q_t–U_2 CPTu soil classification charts. (a) The Q_t–F_r chart. (b) The Q_t–U_2 chart. (From Schneider, J.A. et al. 2012, Comparing CPTU Q–F and Q–$\Delta u_2/\sigma'_{vo}$ soil classification charts. *Géotechnique Letters*. Figure 3.8, p. 214. Reproduced with permission of ICE Publishing, London, UK.)

For tests in transitional soils such as clayey sands, silty sands and silt, clay, and sand mixtures, CPT can be partially drained. The partial drainage also causes partial consolidation. Using the Q_t–F_r alone may not be able to effectively separate the influence of high OCR from partial consolidation. Based on a combination of analytical studies and field data, Schneider et al. (2008, 2012) proposed a pair of Q_t–F_r and Q_t–U_2 charts as shown in Figure 3.33 with a five zonal classification. According to Schneider et al. (2008), Q_t–U_2 space is superior to Q_{tn}–B_q space when evaluating the CPTu for a range of soil data.

EXAMPLE 3.4

Given

For the same CPT data set given in Example 3.3.

Required

Compute the corresponding Q_t, F_r, and U_2, determine the zonal number and plot the Q_t–F_r and Q_t–U_2 data points in the respective Schneider et al. (2012) charts depicted in Figure 3.33.

Solution

The values of Q_t and U_2 are calculated using the spreadsheet computer program and the following equations.

$$Q_t = \frac{(q_t - \sigma_{vo})}{\sigma'_{vo}} \tag{3.21}$$

$$U_2 = \frac{(u_2 - u_o)}{\sigma'_{vo}} \tag{3.22}$$

The calculated friction ratio according to Equation 3.15 and classification using Figure 3.33 are shown in Table E3.4. I_c values obtained from Example 3.3 are included for reference. Data points of Q_t–F_r and Q_t–U_2 are plotted in Figure E3.4.

The data points of Q_t–F_r and Q_t–U_2 are plotted in the respective charts as shown in Figure E3.4.

Table E3.4 Determination of Q_t, F_r, and U_2

Depth (m)	2.5	5.0	7.5	10.0	12.5	15.0	17.5	20.0	22.5
q_t, MPa[a]	3.55	2.37	1.97	5.39	4.05	6.73	6.72	6.55	6.56
f_s, kPa[a]	9.36	25.38	11.91	13.20	20.38	15.91	25.07	33.61	26.06
u_2, kPa[a]	8.83	−54.94	43.16	69.65	95.16	82.40	97.12	96.14	47.09
σ_{vo}, MPa[a]	0.048	0.095	0.143	0.190	0.238	0.285	0.333	0.380	0.428
σ'_{vo}, MPa[a]	0.041	0.064	0.087	0.110	0.132	0.156	0.175	0.201	0.224
F_r[a]	0.27	1.11	0.65	0.25	0.53	0.25	0.39	0.54	0.43
I_c[a]	1.75	2.34	2.39	1.89	2.20	1.91	2.03	2.14	2.13
Q_t[b]	86.13	35.80	21.16	47.47	28.76	41.44	35.77	30.64	27.32
U_2[b]	0.05	−1.36	−0.15	−0.10	−0.07	−0.30	−0.32	−0.41	−0.70
Zonal No. from Q_t–F_r	2	1c/3	1c/3	2	1c/3	2	2	1c/3	1c/3
Zonal No. from Q_t–U_2	2	3	3	2	2	2	2	3	3

Notes
[a] Given values from Example 3.3.
[b] Computed values.

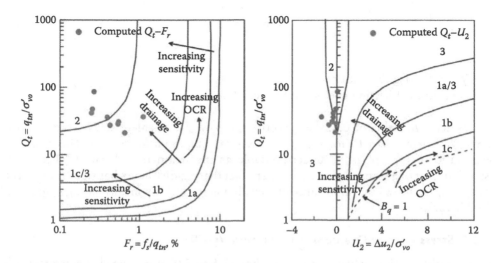

Figure E3.4 Classification according to the Schneider et al. charts.

Eslami and Fellenius (1997) proposed a soil classification system that takes advantage of q_t, f_s, and u_2, along with a procedure to estimate the pile capacity. Details of the Eslami and Fellenius classification method are introduced in Section 7.4.

3.4.3 Engineering properties of clays from CPT

Many methods to estimate shear strength, stress history, and consolidation-related parameters from CPT have been proposed in the past. These methods are mostly empirical, but when applied correctly they can be very effective. Some of the methods are introduced below. CPT is not sensitive to soil stiffness (such as the constrained modulus); correlations between cone tip resistance and stiffness are less reliable and will not be discussed here.

3.4.3.1 Undrained shear strength

At a penetration rate of 20 mm/sec, CPT in clays can be assumed as undrained. In this case, the undrained shear strength of clays, s_u, can be estimated as

$$s_u = \frac{(q_t - \sigma_{vo})}{N_k} \tag{3.22}$$

where

σ_{vo} = total overburden stress
N_k = an empirical factor

The value of N_k depends on the shearing mode the s_u is referring to, plasticity, and stress history of the clay. $N_k = 13.6 \pm 1.9$ has been recommended for s_u in simple shear mode for soft clays (Mayne, 2014). The term $(q_t - \sigma_{vo})$ is often referred to as the net corrected cone tip resistance.

Alternatively, s_u can be related to the total mean normal stress p_o using the spherical cavity expansion theory described in Chapter 2, by treating q_t the same as the toe bearing capacity q_b of a driven pile. By combing Equations 2.73 and 2.110, it can be shown that

$$\frac{(q_t - p_o)}{s_u} \approx \frac{(q_b - p_o)}{s_u} = \left[\beta_1 + \frac{4}{3}\left(1 + \ln\frac{G}{s_u} \right) \right] \tag{3.23}$$

where

$G/s_u = I_r$ = rigidity index
β_1 = empirical friction coefficient between 0 and 1.0

Consider a β_1 value of 0.5, the ratio of $(q_t - p_o)/s_u$ (equivalent to N_k stated above) should vary from 7 to 10 for rigidity indexes ranging from 50 to 500. This relatively lower ratio of $(q_t - p_o)/s_u$ is understandable as the strain rate during CPT is much faster than that of a statically loaded pile. The fact that rigidity index has significant effects on $(q_t - p_o)/s_u$ shows the importance of considering the rigidity in correlating q_t to s_u and why N_k varies.

3.4.3.2 Stress history: Overconsolidation ratio (OCR)

By rough observation, when q_t is greater than $(2.5 \text{ to } 5.0) \times \sigma'_{vo}$ the clay is overconsolidated. A large number of empirical methods have been proposed to estimate OCR from CPT. Ladd and DeGroot (2003) proposed an empirical equation to estimate OCR as follows:

$$\text{OCR} = \left[\frac{(q_t - \sigma_{vo})/\sigma'_{vo}}{S_{CPTu}}\right]^{1/m_{CPTu}} \tag{3.24}$$

where
S_{CPTu} = intercept value of $(q_t - \sigma_{vo})/\sigma'_{vo}$ when OCR = 1 (see Figure 3.34)

The above equation is determined by conducting field CPT and a series of 1-D consolidation tests on undisturbed soil samples at the test site. The correlation between $(q_t - \sigma_{vo})/\sigma'_{vo}$ from CPT and OCR from consolidation tests is approximated with a straight line in a log-log plot as shown in Figure 3.34. The exponent, m_{CPTu}, is the slope of the log (OCR) versus $\log[(q_t - \sigma_{vo})/\sigma'_{vo}]$ plot in Figure 3.34. For Boston blue clay (a sensitive marine clay), the value of m_{CPTu} is around 0.76 (Ladd and DeGroot, 2003). Note that S_{CPTu} equals to N_k times $(s_u/\sigma'_{vo})_{NC}$ (the ratio of reference s_u over effective overburden stress when the soil is normally consolidated or OCR = 1).

Taking advantage of the pore pressure measurements and using the concept of critical state soil mechanics as well as cavity expansion (see Chapters 1 and 2), Mayne (1991) proposed the following equations to estimate OCR for CPTu with u_2 measurements:

$$\text{OCR} = 2\left[\frac{1}{1.95M + 1}\left(\frac{q_t - u_2}{\sigma'_{vo}}\right)\right]^{1.33} \tag{3.25}$$

and

$$M = \frac{6 \sin \varnothing'}{(3 - \sin \varnothing')} \tag{3.26}$$

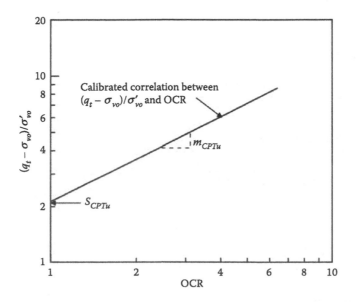

Figure 3.34 A plot of log(OCR) versus $\log\left[(q_t - \sigma_{vo})/\sigma'_{vo}\right]$.

where

M = slope of the critical state line in the q–p' space, where $q = \sigma_1 - \sigma_3$ and $p' = (\sigma'_1 + \sigma'_2 + \sigma'_3)/3$

Note that Equation 3.26 is a rearrangement of Equation 1.46, but in Equation 3.26, $\varnothing'$ refers to the drained peak friction angle, whereas, in Equation 1.46, the $\varnothing'_{cs}$ was the drained critical state friction angle. Please refer to Chapter 1 for the differences between critical state and peak friction angles. Equation 3.25 is effective for a wide range of OCR and compared well with tests in intact, fissured, and cemented clays (Mayne, 1991).

EXAMPLE 3.5

Given

Table E3.5 shows selected q_t, u_2, σ_{vo}, and σ'_{vo} values from field SCPTu in a clay deposit at various depths. Assume $N_k = 14$, $S_{CPTu} = 2.5$ (see Figure 3.34), $m_{CPTu} = 0.76$, and $\varnothing' = 30°$. Use soil total and saturated unit weight, $\gamma = \gamma_{sat} = 19.0 \, \text{kN/m}^3$. The groundwater table is at 1.5 m below ground surface.

Required

Determine s_u and OCR profile using Equations 3.22, 3.24, and 3.25. Plot the profiles of s_u and OCR versus depth.

Solution

Computations are carried out using the spreadsheet program. The values of σ_{vo} and σ'_{vo} are calculated based on depth, soil unit weight and groundwater level. The undrained shear strength s_u:

$$s_u = \frac{(q_t - \sigma_{vo})}{N_k}, \tag{3.22}$$

$N_k = 14$ as given
The OCR:

$$\text{OCR} = \left[\frac{(q_t - \sigma_{vo})/\sigma'_{vo}}{S_{CPTu}} \right]^{1/m_{CPTu}}, \tag{3.24}$$

$S_{CPTu} = 2.5$ and $m_{CPTu} = 0.76$ as given

$$\text{OCR} = 2 \left[\frac{1}{1.95M + 1} \left(\frac{q_t - u_2}{\sigma'_{vo}} \right) \right]^{1.33} \tag{3.25}$$

and

$$M = \frac{6 \sin \varnothing'}{(3 - \sin \varnothing')} \tag{3.26}$$

$\varnothing' = 30°$ as given.
The results are tabulated in Table E3.5 and profiles are shown in Figure E3.5.

Table E3.5 Determination of s_u and OCR

Depth (m)	q_t (MPa)[a]	u_2 (MPa)[a]	σ_{vo} (MPa)[b]	σ'_{vo} (MPa)[b]	s_u (kPa)[b]	OCR[b] equation 3.24	OCR[b] equation 3.25
1.50	0.250	0.038	0.029	0.029	15.8	4.4	5.8
1.75	0.253	0.051	0.033	0.031	15.7	4.0	4.9
2.00	0.259	0.062	0.038	0.033	15.8	3.6	4.3
2.25	0.262	0.074	0.043	0.035	15.7	3.3	3.7
2.50	0.265	0.085	0.048	0.038	15.5	3.0	3.2
2.75	0.268	0.096	0.052	0.040	15.4	2.7	2.8
3.00	0.272	0.105	0.057	0.042	15.4	2.5	2.5
3.25	0.279	0.114	0.062	0.045	15.6	2.4	2.3
3.50	0.285	0.123	0.067	0.047	15.6	2.3	2.1
3.75	0.295	0.131	0.071	0.049	16.0	2.2	2.0
4.00	0.310	0.138	0.076	0.051	16.7	2.2	2.0
4.50	0.341	0.154	0.086	0.056	18.2	2.2	2.0
5.00	0.371	0.169	0.095	0.061	19.7	2.2	2.0
5.50	0.402	0.184	0.105	0.065	21.2	2.2	2.0
6.00	0.432	0.199	0.114	0.070	22.7	2.2	2.0
8.00	0.554	0.259	0.152	0.088	28.7	2.2	2.0
10.0	0.675	0.319	0.190	0.107	34.7	2.2	2.0

Notes

[a] Given values.
[b] Computed values.

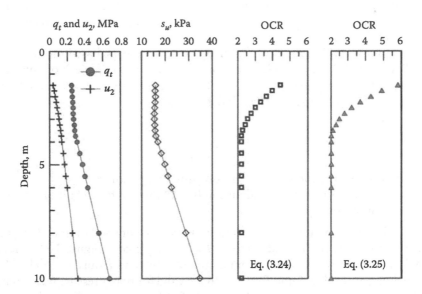

Figure E3.5 Profiles of the computed s_u and OCR.

3.4.3.3 Coefficient of consolidation

With the piezometer readings, a pore pressure dissipation test can be conducted by suspending the cone penetration in ground and keeping track of the decay of pore pressure measurements with time. The procedure provides a convenient measurement of the horizontal coefficient of consolidation, c_h, an important design parameter for ground improvement of cohesive soils by preloading and vertical drains.

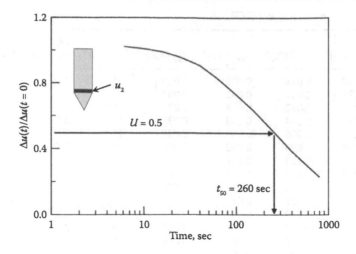

Figure 3.35 Variation of $\Delta u(t)/\Delta u(t=0)$ versus time according to a CPTu dissipation test. (Campanella and Robertson, 1988.)

For CPT in saturated, normally, or lightly consolidated clays, the disturbance of cone penetration generates positive pore pressure radially away from the cone tip as the surrounding soil consolidates. We record the decay of excess pore pressure, $\Delta u(t)$, as time progresses. $\Delta u(t) = 0$ when the consolidation is completed. Figure 3.35 shows the record of a CPTu dissipation test results. In this particular case, pore pressure was measured at u_2 location. The ratio of $\Delta u(t)/\Delta u(t=0)$ also reflects the average degree of consolidation (U) for the surrounding soil.

Teh and Houlsby (1991) developed a theoretical framework based on the strain path method (Baligh, 1985) for the determination of horizontal coefficient of consolidation, c_h, from pore pressure dissipation test. A theoretical correlation between $\Delta u(t)/\Delta u(t=0)$ and time factor T^* is proposed, and T^* is defined as

$$T^* = \frac{c_h t}{r^2 \sqrt{I_r}} \tag{3.27}$$

where

   $t =$ elapsed time after the start of pore pressure dissipation test
   $I_r =$ rigidity index of the surrounding soil (see Chapter 2 for definition of I_r)
   $r =$ radius of the cone penetrometer

Table 3.5 shows the values of T^* for various degrees of consolidation U. A curve showing the correlation between T^* and $\Delta u(t)/\Delta u(t=0)$ is depicted in Figure 3.36. The curve in Figure 3.36 is valid for clays with I_r from 25 to 500 (Teh and Houlsby, 1991). Note that the data included in Table 3.5 and Figure 3.36 are valid for pore pressure measurements at u_2 position. Please refer to Teh and Houlsby (1991) for corresponding values related to u_1 measurements.

Equation 2.74 described in Chapter 2, reported by Mayne et al. (2002), can be used to estimate I_r based on OCR and PI of the cohesive soil around the cone tip. That equation is repeated below:

$$I_r = \frac{\exp\left(\dfrac{137 - PI}{23}\right)}{\left\{1 + \ln\left[1 + \dfrac{(OCR - 1)^{3.2}}{26}\right]\right\}^{0.8}} \tag{3.28}$$

Table 3.5 Theoretical time factors for u_2 pore pressure measurements

Degree of consolidation (%)	T*
20	0.038
30	0.078
40	0.142
50	0.245
60	0.439
70	0.804
80	1.60

Source: Teh, C.I. and Houlsby, G.T. 1991. An analytical study of the cone penetration test in clay, *Geotechnique*, Table 2, p. 31, reproduced with permission of the ICE Publishing, London, UK.

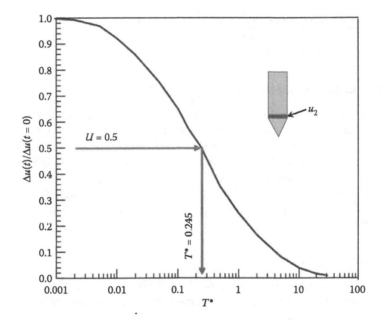

Figure 3.36 Correlation between time factor and degree of consolidation. (From Teh, C.I. and Houlsby, G.T. 1991. An analytical study of the cone penetration test in clay. *Geotechnique*. Figure 3.19, p.31. Reproduced with permission of ICE Publishing, London, UK.)

The procedure for the determination of c_h from pore pressure dissipation test is explained in Example 3.6. Note that due to soil anisotropy, c_h is often higher than c_v (coefficient of consolidation in vertical direction) for the same soil.

EXAMPLE 3.6

Given

For the pore pressure dissipation test result depicted in Figure 3.35, consider the surrounding cohesive soil has an OCR of 3, *PI* of 30. A standard cone penetrometer with a radius of 17.8 mm was used in the CPTu.

Required

Estimate the I_r using Equation 3.28, and determine the c_h.

Solution

$PI = 30$ and $OCR = 3$,

$$I_r = \frac{\exp\left(\dfrac{137 - PI}{23}\right)}{\left\{1 + \ln\left[1 + \dfrac{(OCR-1)^{3.2}}{26}\right]\right\}^{0.8}} = \frac{\exp\left(\dfrac{137 - (30)}{23}\right)}{\left\{1 + \ln\left[1 + \dfrac{(3-1)^{3.2}}{26}\right]\right\}^{0.8}} = 85 \quad (3.28)$$

Refer to Figure 3.35, the time when degree of consolidation reaches 50%, or $U = \Delta u$ $(t)/\Delta u(t=0)0.5$, $t_{50} = 260$ sec. The time factor corresponds to $U = 0.5$ is 0.245 according to Table 3.5. Following Equation 3.27,

$$c_h = \frac{(T^* r^2 \sqrt{I_r})}{t} = \frac{(0.245)(17.8^2)\sqrt{85}}{(260)} = 2.8 \text{ mm}^2/\text{sec}$$
$$= 2.8 \times 10^{-2} \text{ cm}^2/\text{sec} \quad (3.27)$$

3.4.4 Engineering properties of sands from CPT

For CPT in sands, q_t can be affected by relative density (D_r), effective vertical (σ'_{vo}) and horizontal stresses (σ'_{vo}) around the cone tip, compressibility, and age of sands. As it is not practical to take undisturbed samples in sand, procedures to estimate engineering parameters for sands from CPT have been developed mainly based on laboratory calibration chamber tests (Huang and Hsu, 2004, 2005). In this case, soil specimens with known density and stress states are prepared in a calibration chamber where CPT is performed. The following section introduces a few empirical equations developed for the estimation of D_r from CPT.

3.4.4.1 Estimation of D_r

Figure 3.37 shows a correlation among q_t, D_r, and effective vertical stress, σ'_{vo} proposed by Schmertmann (1977) for normally consolidated sand, according to their CPT calibration tests.

It should be noted that the cone tip resistance is also sensitive to the in-situ horizontal stress. The application of Figure 3.37 should be limited to CPT in normally consolidated sands with $K(= \sigma'_{vo}/\sigma'_{ho})$ values around 0.5. For this reason, the use of effective mean normal stress $\sigma'_{oo} [= (\sigma'_{vo} + 2\sigma'_{ho})/3]$ may be more desirable. Two empirical equations that relate q_t to D_r and σ'_{oo} for normally and overconsolidated sands have been proposed by Jamiolkowski et al. (1988) and Huang and Hsu (2005) respectively, based on calibration chamber tests, as follows:

$$q_t = 0.205(\sigma'_{oo})^{0.51}\exp\left(\frac{2.93D_r}{100}\right) \quad (3.28a)$$

and

$$q_t = 0.369(\sigma'_{oo})^{0.5}\exp\left(\frac{2.34D_r}{100}\right) \quad (3.28b)$$

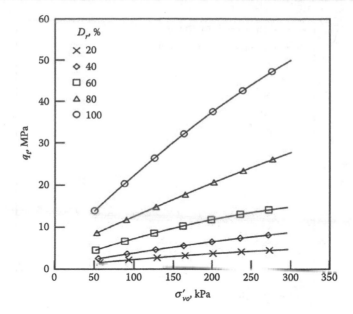

Figure 3.37 q_t as a function of D_r and σ'_{vo}. (Adapted from Schmertmann, J. 1977. Guidelines for cone penetration test—Performance and design. U.S. Department of Transportation, FHWA-TS-78-209, Federal Highway Administration, Washington, DC, 145p.)

where

q_t = cone tip resistance in MPa
σ'_{oo} = effective mean normal stress in kPa
D_r = relative density in %

Comparisons between the above two empirical equations with CPT calibration tests in Da Nang sand reported by Huang and Hsu (2005) are presented in Figure 3.38. In the CPT calibration tests, the K values ranged from 0.4 to 1.8.

Figure 3.39 shows a correlation between normalized cone tip resistance, q_{t1N} and D_r for silica and quartz sands with various degrees of compressibility (Jamiolkowski et al., 2001) based on 456 CPT calibration chamber test data. For sand with medium compressibility,

$$D_r = 100[0.268 \ln(q_{t1N}) - 0.675] \tag{3.29}$$

and q_{t1N} is defined as

$$q_{t1N} = \frac{q_t}{(\sigma'_{vo} \cdot p_a)^{0.5}} \tag{3.30}$$

where

p_a = atmospheric pressure in the same unit as q_t and σ'_{vo}

Robertson and Campanella (1983) proposed the following equation to estimate drained peak friction angle, $\varnothing'$ from q_t for CPT in normally consolidated clean quartz sand based on their compilation of calibration chamber test data:

$$\varnothing'(\text{radian}) = \arctan[0.1 + 0.38 \log(q_t/\sigma'_{vo})] \tag{3.31}$$

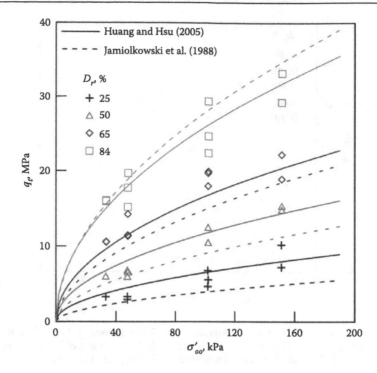

Figure 3.38 q_t as a function of D_r and σ'_{oo}. (From Huang, A.B. and Hsu, H.H. 2005. Cone penetration tests under simulated field conditions. *Geotechnique*. Figure 14, p. 353. Reproduced with permission of ICE Publishing, London, UK.)

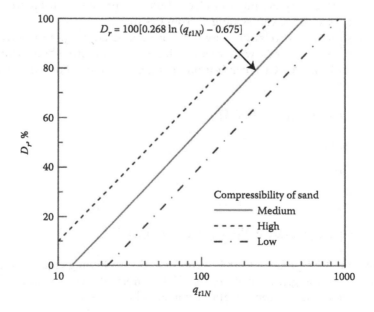

Figure 3.39 Effects of sand compressibility on D_r-q_{t1N} correlation for normally consolidated quartz and silica sands reported by Jamiolkowski et al. (2001).

EXAMPLE 3.7

Given

Table E3.7 shows a set of q_t, f_s, σ_{vo}, and σ'_{vo} values from the same SCPTu test results shown in Figure 3.31. Use soil total and saturated unit weight, $\gamma = \gamma_{sat} = 190 \text{ kN/m}^3$. The groundwater table is at 1.8 m below ground surface. The given data along with the calculated I_c values are included in Table E3.7. Consider $K(= \sigma'_{ho}/\sigma'_{vo}) = 0.5$ and 0.8.

Required

Determine the relative density using Equations 3.28a, b and 3.29. Determine the peak friction angle; consider the sand as normally consolidated using Equation 3.31.

Solution

Compute σ'_{ho} according to the given σ'_{vo} and K values of 0.5 and 0.8. Enter the respective σ'_{ho} values into Table E3.7.

Compute σ'_{oo} given σ'_{vo} and σ'_{ho} from above calculation, and

$$\sigma'_{oo} = \frac{(\sigma'_{vo} + 2\sigma'_{ho})}{3}$$

Rearranging Equation 3.28a, we have

$$D_r = \frac{100}{2.93} \ln\left[\frac{q_t}{0.205(\sigma'_{oo})^{0.51}}\right] \tag{3.28a}$$

Rearranging Equation 3.28b results in

$$D_r = \frac{100}{2.34} \ln\left[\frac{q_t}{0.369(\sigma'_{oo})^{0.5}}\right] \tag{3.28b}$$

Table E3.7 Determination of D_r and $\varnothing'$

Depth (m)[a]	2.5	10.0	15.0	17.5
q_t, MPa[+]	3.55	5.39	6.73	6.72
I_c[a]	1.75	1.89	1.91	2.03
σ_{vo}, MPa[a]	0.048	0.190	0.285	0.333
σ'_{vo}, MPa[a]	0.041	0.110	0.156	0.175
$\sigma'_{ho} = 0.5\sigma'_{vo}$ MPa[b]	0.021	0.055	0.078	0.0875
σ'_{oo} (K = 0.5), kPa[b]	27.3	73.3	104.0	116.7
D_r (Equation 3.28a), %*	43	39	41	39
D_r (Equation 3.28b), %*	26	23	25	22
$\sigma'_{ho} = 0.8\sigma'_{vo}$ MPa*	0.033	0.088	0.125	0.14
σ'_{oo} (K = 0.8), MPa[b]	35.5	95.3	135.2	151.7
D_r (Equation 3.28a), %[b]	35	32	34	32
D_r (Equation 3.28b), %[b]	20	17	19	17
q_{t1N} (Equation 3.30)[b]	55.4	51.4	53.9	50.8
D_r (Equation 3.29), %[b]	40	38	39	38
q_t/σ'_{vo}[b]	86.6	49.0	43.1	38.4
$\varnothing'$ (Equation 3.31), deg.[b]	39.9	36.6	35.8	35.1

Notes
[a] Given values.
[b] Computed values.

Compute the respective D_R (%) values from the above two equations.

$$q_{t1N} = \frac{q_t}{\left(\sigma'_{vo} \cdot p_a\right)^{0.5}} \tag{3.30}$$

$$D_r = 100[0.268 \ln (q_{t1N}) - 0.675] \tag{3.29}$$

$$\emptyset' = \arctan[0.1 + 0.38 \log(q_t/\sigma'_{vo})] \tag{3.31}$$

Jefferies and Been (2006) reported that a linear correlation between the normalized cone tip resistance $(q_t - \sigma_{00})/\sigma'_{oo}$ and state parameter ξ can be established in a semi-log plot. Figure 3.40 shows such a plot based on CPT calibration chamber tests in Da Nang sand (Huang and Chang, 2011). This linear correlation offers a great potential for inferring the state parameter from CPT in sand.

3.4.4.2 Liquefaction potential assessment based on CPT

As in the case of SPT, CPT has also been used to assess the liquefaction potential for granular soils, under the framework of simplified procedure. A correlation between soil CSR or CRR with normalized cone tip resistance, q_{t1N}, calculated using Equation 3.30 is empirically established based on post-earthquake CPT and field observations that showed signs of liquefaction or no liquefaction. The CRR-q_{t1N} correlation is again the boundary curve that divides the data points of liquefaction and no liquefaction. The CRR-q_{t1N} correlation shown in Figure 3.41 was proposed by Robertson and Wride (1998). The CRR-q_{t1N} correlation can be approximated by a pair of empirical equations reported by Youd et al. (2001) as depicted in Figure 3.41. As in the case of SPT-based simplified procedure, there have been many modifications of the CRR-q_{t1N} correlation by various researchers in the past few decades (Juang et al., 2006).

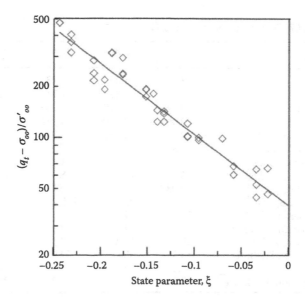

Figure 3.40 A semi-log plot of normalized q_t versus state parameter. (Adapted from Huang, A.B. and Chang, W.J. 2011. *Fifth International Conference on Earthquake Geotechnical Engineering*, Chilean Geotechnical Society, Santiago, Chile, pp. 63–104.)

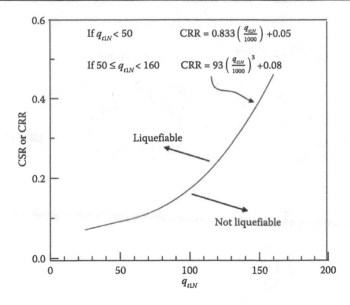

$$\text{If } q_{t1N} < 50 \qquad CRR = 0.833\left(\tfrac{q_{t1N}}{1000}\right) + 0.05$$

$$\text{If } 50 \leq q_{t1N} < 160 \qquad CRR = 93\left(\tfrac{q_{t1N}}{1000}\right)^3 + 0.08$$

Liquefiable

Not liquefiable

Figure 3.41 CRR-q_{t1N} correlation for clean sands proposed by Robertson and Wride (1998). (From Robertson, P.K. and Wride, C.E. 1998. Evaluating cyclic liquefaction potential using the cone penetration test. *Canadian Geotechnical Journal.* Figure 3.11, p. 452. Reproduced with permission of the NRC Research Press, NRC, Ottawa, Canada.)

CPT has also been used to estimate the foundation bearing capacities mainly based on q_t and f_s values. Details of these empirical methods are described in the design of shallow and deep foundations.

3.4.4.2.1 Field vane shear tests (FVT)

The field vane test was developed in Sweden in the early twentieth century. The test is primarily used for the determination of undrained shear strength of saturated clays with s_u less than 200 kPa. The vane consists of four rectangular blades fixed at 90° angles to each other. In performing the field test, the vane is pushed to the designated depth and followed by measuring the torque required to rotate the blades. Standards (i.e., ASTM D2573) have been established to specify the vane shear test equipment and procedure. A variety of vane sizes and shapes are available. Most of the vanes have a rectangular shape with a height-to-diameter ratio (H/D) of 2. The height of the blades generally ranges from 70 to 200 mm. Larger vanes are used for softer soils. The vane blades can be made of different alloys of steel and subject to hardening process. The blade should have a thickness not more than 3 mm. A rod is used to connect the vane to a measuring unit on the ground surface as shown in Figure 3.42. The measuring unit facilitates rotation of the vane via the rod at a constant rate, while measuring the torque required to maintain the rotation. The rod is protected by a casing to prevent friction from the surrounding soil and affecting the torque readings. The vane is retracted inside the protecting shoe during insertion to prevent damage to the field vane.

Upon insertion to the desired depth, the vane shear test should start within 5 minutes. Prolonged waiting can yield higher shear strength due to consolidation of the surrounding soil. Rotation of the vane should proceed at a rate of 6°/minute (0.1°/sec), and the relationship between torque and angular rotation is recorded. The maximum torque is used to

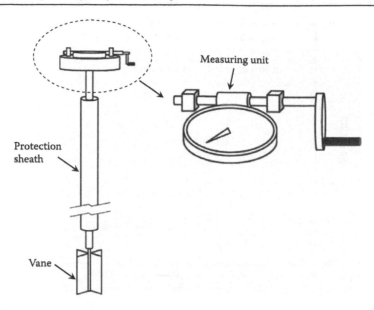

Figure 3.42 Field vane shear device.

calculate the peak undrained shear strength of the soil. The vane is then extensively rotated to completely remold the soil on the shear surface. The residual undrained shear strength and soil sensitivity are then computed based on the torque measurement in this stage.

Rotation of the vane generates a drum-shaped shear surface, as shown in Figure 3.43, with shear stress acting on the surface of the cylindrical drum. Studies have indicated that the mobilized shear stress, τ_V, along the vertical edges of the vane is more or

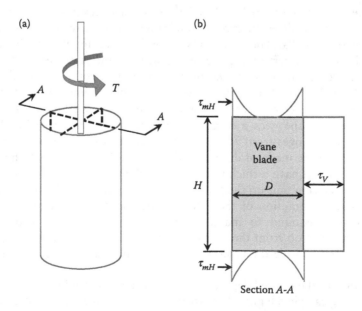

Figure 3.43 Shear stress distribution on a vane. (a) Drum-shaped shear surface. (b) Assumed shear stress distribution.

less uniform except for localized area near the corner of the blade. Assuming uniform shear stress distribution on the vertical surface, the vertical torque component, T_V, is

$$T_V = \pi D H \tau_V \frac{D}{2} = \frac{\pi D^2 H \tau_V}{2} \qquad (3.32)$$

The distribution of τ_H on the horizontal top and bottom surfaces is highly nonlinear, with τ_H increases from center of the vane toward the edge. Wroth (1984) proposed to express τ_H as a polynomial equation,

$$\frac{\tau_H}{\tau_{mH}} = \left(\frac{r}{D/2}\right)^n \qquad (3.33)$$

where
  r = radial distance from the center of the blade
  τ_{mH} = maximum τ_H at the edge of the vane ($r = D/2$)

If τ_H is uniformly distributed, $n = 0$. Wroth (1984) suggested that n should be around 5. Integrating τ_H along the horizontal surfaces from $r = 0$ to $D/2$ yields the horizontal torque component, T_H, as

$$T_H = \frac{\pi D^3 \tau_{mH}}{2(n+3)} \qquad (3.34)$$

The total torque associated with rotation of the vane, T, is thus

$$T = T_V + T_H \qquad (3.35)$$

Schnaid (2009) reported a general formula to correlate T with soil undrained shear strength that considers strength anisotropy and possible nonlinear τ_H distribution. Assume the undrained shear strength on the vertical surface, $S_{uV} = \tau_V$, and undrained shear strength on the horizontal surfaces $s_{uH} = \tau_{mH}$. Soil anisotropy, b, is expressed as

$$b = s_{uV}/s_{uH} \qquad (3.36)$$

Incorporating Equations 3.32, 3.34, 3.35, and 3.36 results in a general formula for deriving s_{uH} from T as

$$s_{uH} = \frac{n+3}{D + Hb(n+3)} \frac{2T}{\pi D^2} \qquad (3.37)$$

By substituting $(n + 3)$ with "a" and assuming isotropic undrained shear strength ($b = 1$ and $s_{uH} = s_{uV} = s_u$), Equation 3.37 becomes

$$s_u = \frac{2T}{\pi D^2 \left(H + \frac{D}{a}\right)} \qquad (3.38)$$

where
  $a = 3.0$ for uniform stress distribution on the horizontal shear surfaces
  $a = 3.5$ for parabolic stress distribution on the horizontal shear surfaces
  $a = 4.0$ for triangular stress distribution on the horizontal shear surfaces

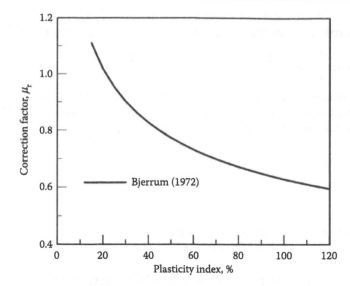

Figure 3.44 Correlations between μ_r and soil plasticity index.

For practical purposes, the soil is generally assumed to be isotropic ($b = 1$), and shear stresses along the vertical and horizontal shear surfaces are all equal to an undrained shear strength s_u ($a = 3$) and $H/D = 2$ for a standard vane, then

$$s_u = \frac{6T}{7\pi D^3} \tag{3.39}$$

It is well recognized that the undrained shear strength obtained from vane shear test can be affected by soil disturbance, strain rate, strength anisotropy, and partial consolidation. Some of these factors may compensate each other. Instead of dealing with these factors individually, Bjerrum (1972) proposed a correction factor μ_r based on back calculated undrained shear strength from field embankment failures, and

$$s_{u(field)} = \mu_r s_{u(vane)} \tag{3.40}$$

and

$$\mu_r = 2.5(PI)^{-0.3} \leq 1.1 \tag{3.40a}$$

where

$s_{u(vane)} =$ undrained shear strength from vane shear test
$s_{u(field)} =$ projected undrained shear strength in the field
$PI =$ plasticity index of the clay in percentage

Figure 3.44 shows the correlation between μ_r and soil plasticity index according to Bjerrum (1972).

EXAMPLE 3.8

Given

A vane with a height, H, of 150 mm, and a diameter, D, of 75 mm was used to perform a field vane shear test in a soft clay with a plasticity index, PI of 50%. The recorded peak torque, T, was 0.1 kN-m.

Required

Determine the undrained shear strength of the clay s_u using Equation 3.38, consider $a =$ 3, 3.5, and 4. Determine the correction factor μ_r following Equation 3.40a by Bjerrum (1972).

Solution

According to Equation 3.38,

$$s_u = \frac{2T}{\pi D^2 \left(H + \frac{D}{a}\right)} = (2)(0.1)(10^9)/[(3.1416)(75^2)((150) + (75)/(a))] \qquad (3.38)$$

when

$a = 3.0$, $s_u = 64.7\,\text{kPa}$
$a = 3.5$, $s_u = 66.0\,\text{kPa}$
and when $a = 4.0$, $s_u = 67.1\,\text{kPa}$
According to Equation 3.40a or Figure 3.44,

$$\mu_r = 2.5(\text{PI})^{-0.3} = (2.5)(50)^{-0.3} = 0.77$$

3.4.5 Flat dilatometer tests (DMT)

Flat dilatometer, DMT (see Figure 3.45), was developed in Italy by Prof. Silvano Marchetti in the 1970s (Marchetti, 1980). The flat dilatometer is a stainless steel blade having a thin, inflatable circular steel membrane mounted flush on one side. The blade is connected to a control unit on the ground surface by a pneumatic-electric tubing (that transmits air pressure and electric signal). The control unit is equipped with

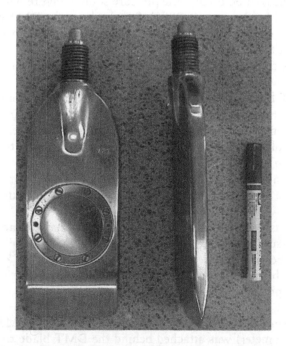

Figure 3.45 The DMT blade. (Courtesy of An-Bin Huang, Hsinchu, Taiwan.)

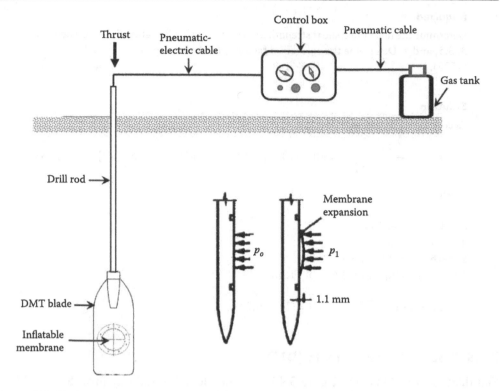

Figure 3.46 DMT system setup. (Adapted from Marchetti, S. et al. 2001. The Flat Dilatometer Test (DMT) in soil investigations. A Report by the ISSMGE Committee TC16. May 2001, 41p.)

pressure regulator, pressure gauges, an audio-visual signal, and a vent valve. High-pressure nitrogen gas is often used as the pressure source. Figure 3.46 shows the typical DMT system setup (Marchetti et al., 2001). Standards on the DMT equipment and test procedure have been established (i.e., TC16 DMT report by Marchetti et al. [2001] and ASTM standard D-6635).

A typical DMT is conducted according to the following procedure:

1. Insert the DMT blade into the soil at a rate of 20 mm/sec to reach the designated depth; perform the test, take A, B, and optional C pressure readings.
2. A reading is taken as the pneumatic pressure required for the membrane on the DMT blade to expand laterally toward soil at its center, just enough to separate the membrane from the blade body.
3. B readings is taken as the pneumatic pressure required for the membrane on the DMT blade to expand laterally toward soil by approximately 1.10 mm at its center.
4. The membrane is deflated after B reading. The optional C reading is taken as the pneumatic pressure when the membrane on the DMT blade regains its contact with the DMT body.
5. Push the DMT blade to the next designated depth, usually at 20 to 30 cm deeper. Repeat the test procedures 2–4.

Marchetti (2014) introduced the concept of a seismic DMT or SDMT. A seismic sensing module (i.e., accelerometer) was attached behind the DMT blade to measure the shear wave velocity in a manner similar to that in SCPTu.

The A, B, and C readings are corrected for membrane stiffness and pressure gage zero offset to obtain p_o, p_1, and p_2 respectively as follows:

$$p_o(\text{corrected } A - \text{pressure}) = 1.05(A - Z_M + \Delta A) - 0.05(B - Z_M - \Delta B) \qquad (3.41)$$

$$p_1(\text{corrected } B - \text{pressure}) = (B - Z_M - \Delta B) \qquad (3.42)$$

$$p_2(\text{corrected } C - \text{pressure}) = C - Z_M + \Delta A \qquad (3.43)$$

where

ΔA = internal suction required in free air to hold the membrane in contact with the DMT blade (i.e., A position)

ΔB = internal pressure required in free air to expand the membrane by 1.1 mm at its center (i.e., B position)

Z_M = offset of the pressure gage when vented to atmospheric pressure

Three intermediate, index parameters are derived from p_o and p_1 readings:

$$\text{DMT modulus,} \quad E_D = 34.7(p_1 - p_o) \qquad (3.44)$$

$$\text{Material index,} \quad I_D = \frac{(p_1 - p_o)}{(p_o - u_o)} \qquad (3.45)$$

$$\text{Horizontal stress index,} \quad K_D = \frac{p_o - u_o}{\sigma'_{vo}} \qquad (3.46)$$

where

u_o = hydrostatic pressure prior to DMT penetration

σ'_{vo} = effective vertical (overburden) stress

Interpretation of DMT results is predominantly based on empirical correlations using these intermediate parameters. Figure 3.47 shows a profile from DMT sounding in Jiayi, Taiwan.

I_D has a reasonable correlation with soil types, thus is a useful parameter to establish soil stratigraphy. A combination of I_D and E_D can be used to estimate soil unit weight, γ. Figure 3.48 shows the correlations among I_D, E_D soil classification and γ/γ_w, where γ_w is the unit weight of water. Additional correlation for γ/γ_w less than 1.6 was added by Lee et al. (2013).

The value of p_o is related to the in-situ horizontal stress. K_D, as a normalized $(p_o - u_o)$ with respect to σ'_{vo}, according to Equation 3.46, can thus be viewed as the at-rest horizontal earth pressure coefficient (K_o) amplified by the penetration of DMT blade. The difference between p_o and p_1 is the basis for evaluating the soil compressibility.

C reading is believed to be strongly influenced by the pore water pressure adjacent to the DMT blade during the test. For free-draining material such as sand, the excess pore water pressure induced by DMT penetration is quickly dissipated, thus C reading is close to the hydrostatic pressure. In soft clay, the DMT penetration-induced pore water pressure is significantly higher than hydrostatic pressure and its permeability is low. In this case, C reading can be significantly higher than hydrostatic pressure.

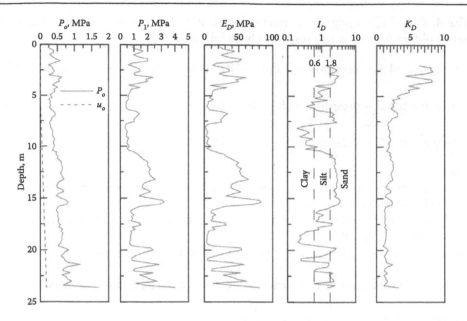

Figure 3.47 Results from DMT sounding in Jiayi, Taiwan. (Courtesy of An-Bin Huang, Hsinchu, Taiwan.)

3.4.6 Geotechnical parameters of clays from DMT

For fine grain soils, engineering parameters such as overconsolidation ratio, K_o, shear strength, constrained modulus, and coefficient of consolidation can be estimated from DMT.

3.4.6.1 Overconsolidation ratio

Marchetti (1980) proposed a correlation between K_D and over consolidation ratio (OCR) for clays as follows:

$$OCR = (0.5K_D)^{1.56} \tag{3.47}$$

When OCR = 1, K_D is consistently close to 2. The above correlation compares favorably with the numerical study by Yu (2004) considering DMT as a flat cavity expansion. The above K_D-OCR correlation should not be extended to granular soils, however.

3.4.6.2 Undrained shear strength

In soil mechanics, we have learned that for clays there is a good correlation between the normalized undrained shear strength, s_u/σ'_{vo} and OCR. It is thus reasonable to extend the K_D-OCR correlation to K_D-s_u/σ'_{vo}. Marchetti (1980) proposed an empirical equation to estimate undrained shear strength, s_u, from K_D as

$$s_u = 0.22\sigma'_{vo}(0.5K_D)^{1.25} \tag{3.48}$$

3.4.6.3 Constrained modulus

The empirical procedure proposed by Marchetti (1980) relates constrained modulus (M_{DMT}) to E_D through a reduction factor, R_M, as

$$M_{DMT} = R_M E_D \tag{3.49}$$

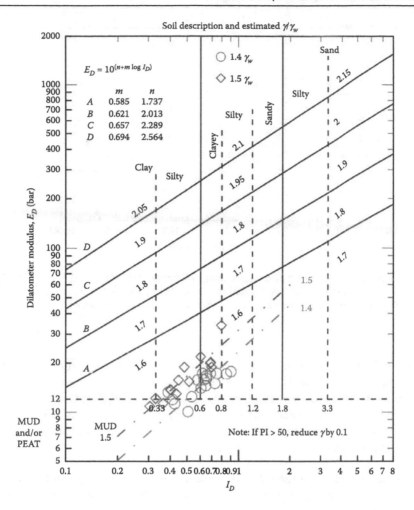

Figure 3.48 Inferring the soil type and unit weight through I_D and E_D. (Adapted from Marchetti, S. and Crapps, D.K. 1981. Flat dilatometer manual, Internal Report of G.P.E. Inc. With additions by Lee et al., 2013.)

The value of R_M is related to I_D, with the following equations:

if $I_D \leq 0.6$	$R_M = 0.14 + 2.36 \log K_D$
if $I_D \geq 3.0$	$R_M = 0.5 + 2 \log K_D$
if $0.6 < I_D < 3.0$	$R_M = R_{Mo} + (2.5 - R_{Mo}) \log K_D$
	$R_{Mo} = 0.14 + 0.15(I_D - 0.6)$
if $I_D > 10$	$R_M = 0.32 + 2.18 \log K_D$

R_M is always ≥ 0.85 (3.50)

Experience has indicated that Equations 3.49 and 3.50 can generally produce reasonable results for clays, since these equations cover a wide range of I_D values, their applications can extend beyond clays.

3.4.6.4 Coefficient of consolidation

In fine-grained soils, the dissipation of excess pore water pressure induced by DMT penetration is slow. By monitoring the decay of this excess pore water pressure, it is possible to

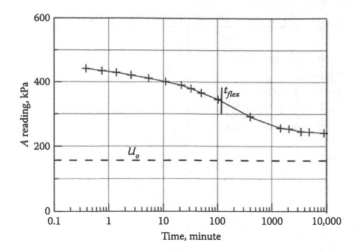

Figure 3.49 Dissipation test using *A* pressure readings. (Adapted from Marchetti, S. and Totani, G. 1989. c_h evaluations from DMTA dissipation curves. *Proceedings, XII International Conference on Soil Mechanics and Foundation Engineering*, Rio de Janeiro, A.A. Balkema, Vol. 1, pp. 281–286.)

deduce coefficient of consolidation in the horizontal direction, c_h. This is the DMT version of the pore pressure dissipation test. As no pore pressure is directly measured in DMT, the dissipation test is conducted via *A* or *C* readings. In either case, *A* or *C* readings are repeatedly recorded along with the elapsed time, following an interruption of DMT penetration. The interpretation of dissipation test result is similar. For the case of using *A* readings, use the following procedure:

1. Plot *A*–log*t* curve using the DMT dissipation test data as shown in Figure 3.49
2. Identify the contraflexure point on the curve and the associated time value, t_{flex}
3. Determine c_h as

$$c_h(OC) = \frac{7\,cm^2}{t_{flex}} \tag{3.51}$$

Equation 3.51 should only be used in overconsolidated (OC) clays. For normally consolidated (NC) clays, c_h may be significantly overestimated using Equation 3.51.

3.4.7 Geotechnical parameters of sands from DMT

For coarse-grained soils, friction angle and stiffness can be estimated from DMT. A chart was proposed earlier to estimate drained friction angle of sand, $\emptyset'$ based on K_o and q_t (or K_D). K_o is not easy to determine and has to be estimated. Taking into account the relationship between q_t and K_D, Marchetti (1997) proposed the following equation to estimate a lower-bound drained friction angle, $\emptyset'_{DMT}$, value as

$$\emptyset'_{DMT} = 28° + 14.6°\log K_D - 2.1°\log^2 K_D \tag{3.52}$$

As described in Chapter 1, state parameter ξ reflects combined effects of soil density and stress states for granular soils. Thus, dilatancy and frictional strength are strongly influenced by state parameter. Yu (2004) proposed a correlation between normalized horizontal stress index, K_D/K_o, and state parameter, ξ, as

$$\xi = -0.002\left(\frac{K_D}{K_o}\right)^2 + 0.015\left(\frac{K_D}{K_o}\right) + 0.0026 \qquad (3.53)$$

where the ratio of K_D/K_o can vary between 4 and 14 (Yu, 2004). Methods of using K_D and E_D have also been explored to assess the liquefaction potential for granular soils under the framework of simplified procedure (Tsai et al., 2009). These developments, although still in their early stages, have demonstrated great potential.

EXAMPLE 3.9

Given

Table E3.9 shows a set of readings and intermediate index parameters selected from the same DMT test results shown in Figure 3.47. The I_D, E_D, and K_D values are given. Use soil total and saturated unit weight, $\gamma = \gamma_{sat} = 19.0$ kN/m^3. The groundwater table is at 1.8 m below ground surface. Assume $K_o\ (= \sigma'_{ho}/\sigma'_{vo}) = 0.5$.

Required

Classify the soil using I_D, determine the unit weight γ/γ_w (Figure 3.48) and constrained modulus, M_{DMT} (Equation 3.49). For soil layers classified as clay, determine their OCR (Equation 3.47) and s_u (Equation 3.48). For soil layers classified as sand, determine their $\varnothing'_{DMT}$ (Equation 3.52) and ξ (Equation 3.53).

Solution

Computations are conducted using the spreadsheet program.
   Soil classification was done based on I_D and Figure 3.48.

$$\text{OCR} = (0.5K_D)^{1.56} \qquad (3.47)$$

$$s_u = 0.22\sigma'_{vo}(0.5K_D)^{1.25} \qquad (3.48)$$

$$M_{DMT} = R_M E_D \qquad (3.49)$$

$$\varnothing'_{DMT} = 28° + 14.6° \log K_D - 2.1° \log^2 K_D \qquad (3.52)$$

$$\xi = -0.002\left(\frac{K_D}{K_o}\right)^2 + 0.015\left(\frac{K_D}{K_o}\right) + 0.0026 \qquad (3.53)$$

The results are tabulated in Table E3.9.

Table E3.9 Interpretation of DMT data shown in Figure 3.47

Depth[a] (m)	σ'_{vo} (kPa)[a]	I_D[a]	K_D[a]	E_D (MPa)[a]	E_D[a] bar	Soil[b] type	γ/γ_w[b]	R_M[b]	M_{DMT} (MPa)[b]	OCR[b]	s_u (kPa)[b]	$\varnothing'_{DMT}$[b] deg.	ξ[b]
2.6	0.05	2.5	8.25	36.07	360	Silty sand	2.05	2.33	83.92	–	–	39.6	−0.29
8.2	0.16	0.2	2.00	1.82	18.2	Clay	1.65	0.85	1.55	I	35.4	–	–
15.0	0.29	3.2	2.00	55.75	557	Silty sand	2.10	1.10	61.43	–	–	32.2	0.03
18.6	0.36	0.2	2.00	4.01	40.1	Clay	1.80	0.85	3.41	I	80.2	–	–

Notes

[a] Given values.
[b] Computed values.
–: not applicable.

3.4.8 Pressuremeter tests (PMT)

A pressuremeter, according to Amar et al. (1991), is "A cylindrical probe that has an expandable flexible membrane designed to apply a uniform pressure to the walls of a borehole." There have been reports on earlier attempts of inserting a cylindrical probe into the ground and performing an expansion test. The term *pressuremeter* was first used by Ménard to describe the equipment he developed in 1955 (Ménard, 1957; Clarke, 1995). To perform the test, a borehole slightly larger than the pressuremeter probe is drilled. The tricell pressuremeter probe, schematically shown in Figure 3.50, is then inserted into the ground. The measuring cell inside the pressuremeter probe is connected to the water tank (volumeter in Figure 3.50) situated inside a control box on the ground surface, via high-pressure plastic tubing. The guard cells have separate pneumatic tubing to connect to the control box.

The PMT is conducted by applying air pressure to the water tank in increments. During the PMT, the pressure applied to the measuring cell is kept at a slightly higher value than the guard cells. As the pressuremeter probe expands due to increased pressure, water enters the measuring cell and water that remains in the volumeter decreases. The pressuremeter probe expansion is determined from the drop of water level in the volumeter. The purpose of guard cells is to force the measuring cell expansion close to the shape of a cylinder (i.e., uniform radius throughout the height of the measuring cell), rather than a sphere (i.e., larger radius toward the center of the measuring cell). In a typical PMT, 10 to 14 pressure increments are applied to the pressuremeter probe. A time history of pressure and volumetric expansion of the measuring cell in a PMT is conceptually described in Figure 3.51. In each increment (i), the expansion pressure (p_i) is kept constant for a period of 60 seconds, while the volumeter readings are taken at 15, 30, and 60 seconds [marked as $V_i(15)$, $V_i(30)$, and $V_i(60)$ in Figure 3.51] after the pressure increase. The main test result is a plot of the expansion pressures (p_i) versus the corresponding 60-second volumeter readings $[V_i(60)]$ from each pressure increment.

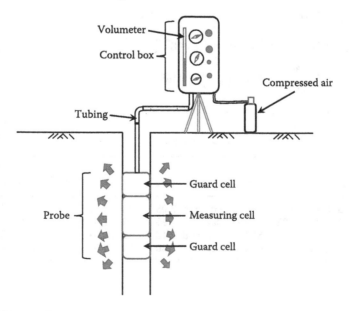

Figure 3.50 Field setup of pressuremeter test.

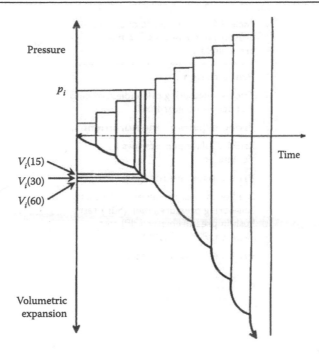

Figure 3.51 Pressure and volumetric expansion against time.

The equipment and test procedure described above are original development by Ménard and are still widely used today. This specific test procedure is referred to as the Ménard pressuremeter (MPM) test. Many modifications on the equipment and/or test procedures have been proposed in the past few decades as PMT became more popular. Table 3.6 summarizes these modifications and acronyms or abbreviations commonly used to describe them. A pressuremeter test today can involve the combination of one selection from each of the four categories shown in Table 3.6. The original Ménard development is therefore a PBP (pre-bore pressuremeter) test with a tricell probe, volumeter expansion measurement, and the test is stress controlled. Studies have indicated that if the ratio of pressuremeter probe length over its diameter is >6, then the probe expansion should be sufficiently close to a cylindrical cavity expansion (Huang et al., 1991). As long as the length to diameter ratio is large enough, the monocell design is feasible and can make the probe design and test control much more simplified by getting rid of the guard cells and thus the need to set a differential pressure between measuring cell and guard cells.

Radial displacement sensors include linear variable differential transducers (LVDTs), hall-effect gauges, and strain gauges. These sensors are installed inside the pressuremeter probe and usually have significantly higher strain resolution compared to volumeter. Unless for safety reasons, no liquid is required in the probe when using the radial displacement sensors. It is also possible to install an electric pressure transducer inside the pressuremeter probe. Figure 3.52 shows a PBP probe developed at National Chiao Tung University (the NCTU PBP) with 12 spring loaded radial displacement sensors and an internal pressure transducer. The membrane is removed to expose its radial displacement sensors (strain arms). Figure 3.53 shows the NCTU PBP in field operation.

Self-boring pressuremeter and push-in pressuremeter are not discussed herein. Interested readers are referred to Baguelin et al. (1978), Clarke (1995), and Schnaid (2009).

Table 3.6 Various types of available
pressuremeter probes and test
procedures

Probe design

Tricell—measuring cell and guard cells
Monocell—no guard cells

Probe expansion measurement

Radial displacement sensors
Volumeter

Probe insertion method

Pre-bored pressuremeter (PBP) test
Self-boring pressuremeter (SBP) test
Push-in pressuremeter (PIP) test
—partial PIP or full PIP

Probe expansion control

Stress controlled
Strain controlled

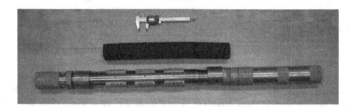

Figure 3.52 PBP with spring-loaded radial displacement sensors developed at National Chiao Tung
University. (Courtesy of An-Bin Huang, Hsinchu, Taiwan.)

The PMT probe expansion can also be strain controlled. In a strain-controlled PMT, the probe is expanded at a constant strain rate. For water-filled pressuremeter probe, a pump can be used for the purpose. A servo control system is used if the pressuremeter is expanded by air pressure. An ASTM standard (D 4719) has been established for pre-bored PMT.

3.4.8.1 Interpretation of the PMT data

Prior to field testing, the pressuremeter device should be carefully calibrated. The calibration should include the pressure and displacement-sensing elements, compliance of the measuring system, and stiffness of the membrane. Figure 3.54 shows a field test curve and the curve from system compliance and membrane stiffness calibration. Subtracting the calibration curve from the field curve yields the net pressuremeter curve that represents the pressure experienced at the borehole wall and its relationship with the cavity radial expansion.

For a stress-controlled test, the results include a creep curve in addition to the pressuremeter curve, as shown in Figure 3.55. The creep volume in Figure 3.55 is the difference between 60- and 30-second volume readings (i.e., $V_i[60] - V_i[30]$) from each pressure increment (see Figure 3.51). The volume readings are replaced with radial displacement readings when PMT probe with radial displacement measurements is used. The creep

Figure 3.53 Field operation with the NCTU PMT. (Courtesy of An-Bin Huang, Hsinchu, Taiwan.)

curve is used to identify the key readings from a PMT. Point *A* in the pressuremeter curve is identified as the point where creep volume drops to a minimum. The corresponding expansion pressure is called lift-off pressure, p_o. Point *B* corresponds to a creep pressure p_f, where the creep volume starts to increase significantly. At point *A*, the pressuremeter probe starts to make contact with the borehole wall and expands the cavity. But p_o should not be considered the same as the in-situ horizontal stress because of the disturbance associated with borehole drilling.

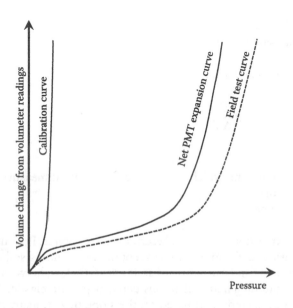

Figure 3.54 Field pressuremeter test and calibration curves.

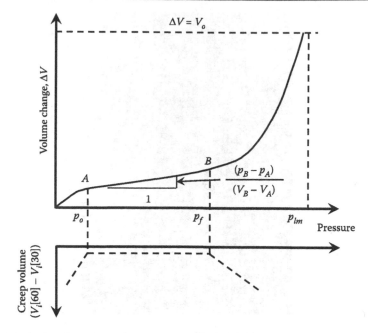

Figure 3.55 Pressuremeter and creep curves from a stress-controlled test.

For cylindrical cavity expansion in linear elastic material,

$$G = \frac{E}{2(1+v)} = \frac{0.5(p-p_o)}{\varepsilon_c} \tag{3.54}$$

$G =$ shear modulus
$E =$ Young's modulus
$v =$ Poisson's ratio
$p =$ expansion pressure
$p_o =$ initial cavity pressure
$\varepsilon_c =$ cavity expansion strain

Or, in terms of volumetric strain,

$$G = \frac{E}{2(1+v)} = (p-p_o)\frac{V_o}{\Delta V} \tag{3.55}$$

where
$V_o =$ initial cavity volume which is the same as the initial measuring cell volume of a pressuremeter probe
$\Delta V =$ volume change

Details of cavity expansion theory are described in Chapter 2. Pressuremeter expansion is similar to a cylindrical cavity expansion, except the surrounding soil is not strictly linear elastic due to disturbance and the nature of soils. Nevertheless, between points A and B in Figure 3.55, the surrounding material is considered as pseudo elastic. A pressuremeter or Ménard modulus, E_m, is computed based on the slope of AB, using a similar format as

Table 3.7 Soil types based on E_m/p_{lm}

Soil type	E_m/p_{lm}
Very loose-to-loose sand	4–7
Medium dense-to-dense sand	7–10
Sand and gravel	
Peat	8–10
Soft to firm clay	8–10
Stiff to very stiff clay	10–20
Loess	12–15
Weathered rock (depends on degree of weathering)	8–40

Source: Clarke, B.G. 1995. *Pressuremeters in Geotechnical Design*. Blackie Academic and Professional, Glasgow, 364p.

Equation 3.55 and,

$$E_m = 2(1+v)[V_o + 0.5(V_B - V_A)]\frac{(p_B - p_A)}{(V_B - V_A)} \tag{3.56}$$

where

V_o = volume of the pressuremeter probe

p_A and p_B = expansion pressure at A and B, respectively

V_A and V_B = probe volume at A and B, respectively

For PMT with radial displacement measurement, V_o is replaced with the initial probe radius, V_A and V_B with the corresponding radius values. v is typically taken as 0.33.

A limit pressure, p_{lm}, is defined as pressure corresponding to a volume increase ΔV equals to V_o, or $\Delta V/V_o = 1$. This is equivalent to 41% of radial strain. In most PBP, it is difficult to reach this strain level. Often, p_{lm} is determined by an extrapolation scheme, demonstrated in Example 3.10.

The ratio of E_m/p_{lm} is related to soil types, as shown in Table 3.7. E_m has been related to constrained modulus, M through an empirical factor, α as

$$M = E_m/\alpha \tag{3.57}$$

where α ranges between 0.25 and 1 as shown in Table 3.8.

Table 3.8 α values for estimating the constrained modulus from E_m

Soil type	Clay	Silt	Sand	Cobble
α	2/3	1/2	1/3	1/4

Source: Amar, S. et al. 1991. *The Application of Pressuremeter Test Results to Foundation Design in Europe, a State of the Art Report by the ISSFE European Technical Committee on Pressuremeters, Part 1: Predrilled Pressuremeters and Self-Boring Pressuremeters*. AA Balkema, The Netherlands, 48p.

Note
In heavily compacted fill, α can be as high as 1.

Table 3.9 Ratio between p_{lm}^* and s_u in clay

Range of $(p_{lm} - \sigma_{ho})$ (kPa)	s_u (kPa)
<300	$\dfrac{(p_{lm} - \sigma_{ho})}{5.5}$
>300	$\dfrac{(p_{lm} - \sigma_{ho})}{10} + 25$

Source: Amar, S. et al. 1991. *The Application of Pressuremeter Test Results to Foundation Design in Europe, a State of the Art Report by the ISSFE European Technical Committee on Pressuremeters, Part 1: Predrilled Pressuremeters and Self-Boring Pressuremeters.* AA Balkema, The Netherlands, 48p.

Consider PMT as a cylindrical cavity expansion. Recall in Chapter 2, for cylindrical cavity expansion in clay under undrained conditions (i.e., in Tresca soil), the ultimate expansion pressure, p_u, is related to undrained shear strength, s_u, as

$$p_u = (p_{lm} - \sigma_{ho}) = p_{lm}^* = s_u F_c' \tag{3.58}$$

and

$$F_c' = (1 + I_r) \tag{3.59}$$

where I_r is the rigidity index (see Chapter 2 on cavity expansion). For clays under undrained conditions, $\upsilon = 0.5$, $I_r = (E_u/3s_u)$ and,

$$p_u = (p_{lm} - \sigma_{ho}) = p_{lm}^* = s_u(1 + \ln E_u/3s_u) \tag{3.60}$$

where
σ_{ho} = total in-situ horizontal stress
E_u = undrained Young's modulus

Equation 3.60 is similar to Equation 2.76 except that p_o (total mean normal stress) is replaced with σ_{ho}. Table 3.9 shows the possible range of the ratio between p_{lm}^* and s_u, reported by Amar et al. (1991). An estimate of σ_{ho} is required to use Equation 3.60.

EXAMPLE 3.10

Given

Table E3.10 shows the results of a Ménard type pre-bored pressuremeter test in a weak sandstone. The initial volume of the pressuremeter measuring cell (V_o) was 997 cm³. The data have been adjusted to exclude the effects of PMT membrane stiffness and system compliance.

Required

Plot the expansion pressure versus volume change and creep volume curves and determine p_o, p_f, E_m, and p_{lm} for the PMT test result.

Table E3.10 PMT expansion readings

Expansion pressure, kPa	9.8	15	42	120	232	348	420	532
Volume change, ΔV (cm³)	20	123	180	204	223	241	253	270
Creep volume, cm³	0	43	4	4	3	4	3	4
Expansion pressure, kPa	641	744	934	1050	1165	1248	1314	
Volume change, ΔV (cm³)	291	312	373	431	516	595	690	
Creep volume, cm³	3	2	14	21	35	34	47	

Solution

Figure E3.10 shows the expansion pressure versus volume change and creep volume curves. Point A is selected as the point where the creep volume drops to a minimum. Point B is selected where the creep volume starts to increase significantly.

The expansion pressure at point A, $p_A = p_o = 45$ kPa

The expansion pressure at point B, $p_B = p_f = 747$ kPa

Volume change at point A, $\Delta V_A = 180$ cm³, and $V_A = V_o + \Delta V_A = (997) + (180) = 1177$ cm³

Volume change at point B, $\Delta V_B = 309$ cm³, and $V_B = V_o + \Delta V_B = (997) + (309) = 1306$ cm³

V_o is given as 997 cm³ and assume $v = 0.33$.

According to Equation 3.56,

$$E_m = 2(1 + v)[V_o + 0.5(V_B - V_A)]\frac{(p_B - p_A)}{(V_B - V_A)}$$

$$= 2(1 + 0.33)[(997) + (0.5)(1306 - 1177)]\frac{(747 - 45)}{(1306 - 1177)}$$ (3.56)

$$= 15,366 \text{ kPa} = 15.4 \text{ MPa}$$

The limit pressure, p_{lm}, is determined following an extrapolation scheme reported by Amar et al. (1991). A plot of expansion pressure versus $\log[1/(\Delta V + V_o)]$ using only the data points beyond the creep pressure p_f (i.e., the expansion pressure $> p_f$) is shown in Figure E3.10a. The expansion pressure $-\log[1/(\Delta V + V_o)]$ correlation is approximately a straight line. p_{lm} is selected following this straight line and extrapolated to a

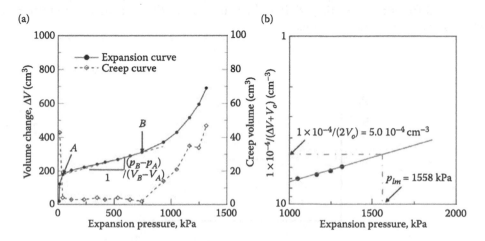

Figure E3.10 (a) PMT expansion pressure versus volume change curve and (b) PMT expansion pressure versus $\log[1/(\Delta V + V_o)]$

point where $1/(\Delta V + V_o) = 1/(2\ V_o)$ as shown in Figure E3.10a. In this case, $p_{lm} = 1558\ kPa$.

Thus for this test, $E_m/p_{lm} = (15,366)/(1558) = 9.9$, which is compatible with the lower-bound values of weathered rock shown in Table 3.7.

Hughes et al. (1977) proposed a cavity expansion-based method to determine the drained friction angle, ϕ' of sands, based on a plot of $\ln(p - u_o)/\sigma'_R$ against $\ln \varepsilon_c$ from a pressuremeter test as shown in Figure 3.56. These two terms have the following relationship:

$$\ln\frac{(p - u_o)}{\sigma'_R} = S\ \ln \varepsilon_c + A \tag{3.61}$$

where

p = expansion pressure
σ'_R = effective initial cavity yield stress
u_o = hydrostatic pressure
A = intercept of the ln-ln plot
S = slope of the ln-ln plot

$$\sin\phi' = \frac{S}{1 + (S - 1)\sin\phi'_{cs}} \tag{3.62}$$

where ϕ'_{cs} is the critical state friction angle. As described in Chapter 1, ϕ'_{cs} is typically around 30°–35° and mainly a function of mineral content, not affected by density of the sand.

Derived from theoretical modeling of pressuremeter expansion tests, Yu (1994) showed that S is also related to state parameter ξ as

$$\xi = 0.59 - 1.85S \tag{3.63}$$

PMT, especially PBP, has been applied extensively in engineering practice for shallow and deep foundation design. Results from PMT have been used in direct applications to determine the bearing capacity and for settlement analysis. Because of the similarity in stress path, PMT is especially valuable in the analysis of laterally loaded piles. Readers are referred to Baguelin et al. (1978) and Amar et al. (1991) for direct applications of PMT results in foundation designs.

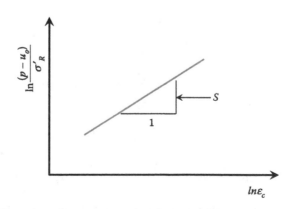

Figure 3.56 Correlation between $\ln(p - u_o)/\sigma'_R$ and $\ln \varepsilon_c$.

3.5 GEOPHYSICAL METHODS

Geophysical exploration methods can be used as part of a subsurface exploration program. Most of these methods are nonintrusive and can be applied from the ground surface (i.e., without the need of boreholes) and provide a survey of interest that covers a large area. A significant part of the geophysical exploration methods has been developed originally for geological surveys, to identify faults, bedrock, discontinuities/cavities, and groundwater. From a foundation or geotechnical engineering point of view, the geophysical exploration usually has a smaller scale but requires higher resolution in comparison with similar techniques applied in geological surveys. In foundation engineering, the objectives of geophysical exploration can include:

- Determination of in-situ elastic moduli, electrical resistivity, groundwater level and, to a lesser degree, density properties of soils and/or rocks.
- Locating hidden underground features detectable by geophysical exploration methods, such as buried underground tanks and pipes, contaminant plumes, and landfill boundaries.

The geophysical procedure measures the contrast of a physical quantity, which is not necessarily the parameter needed to provide a direct solution to the problem under consideration. The required parameters are inferred from the known geological data and the measured geophysical contrast. Usually an inverse solution is determined in geophysical exploration. Inversion implies that a cause is inferred from an effect. The physical property, the cause, is inferred from the field survey readings, the effects. The inverse solutions are not unique, and represent a most likely answer selected from multiple possibilities. The interpretation of geophysical contrasts is based on geologic assumptions and idealized ground conditions. The correlation of measured geophysical contrasts with geologic inferences most often is empirical and is dependent on the quality of both the results and the hypotheses. For these reasons, direct engineering observations such as taking samples from boreholes and performing laboratory testing are still required and not to be completely substituted by geological exploration. In fact, borings and other forms of physical experiments are often used to validate and calibrate the geophysical results and ultimately to improve the accuracy of the integrated conclusions.

A large number of geophysical exploration methods have been developed and applied successfully. The following sections introduce a few commonly used techniques that are relevant to foundation engineering applications, under the categories of seismic methods and electrical resistivity methods.

3.5.1 Seismic methods

In the seismic method, a mechanical vibration is initiated from a source and travels to the location where the vibration is sensed. The direction of travel is called the ray or raypath. A source produces motion in all directions, and the locus of first disturbances will form a spherical shell or wave front in a uniform medium. There are two major classes of seismic waves: body waves, which pass through the volume of a material, and surface waves that exist only near a boundary.

3.5.1.1 Body waves

The fastest traveling of all seismic waves is the compressional or primary wave (P-wave). The particle motion of P-waves is extension (dilation) and compression along the

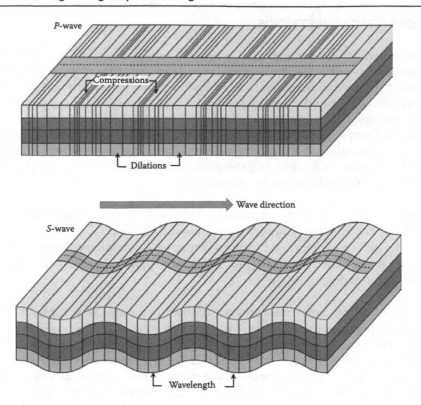

P-wave

Compressions

Dilations

Wave direction

S-wave

Wavelength

Figure 3.57 Characteristics of body waves.

propagating direction (see Figure 3.57). P-waves travel through all media that support seismic waves and include solid and various types of fluid. Compressional waves in fluids—for example, water and air—are commonly referred to as acoustic waves.

Transverse or shear wave (S-wave) is another type of body wave. S-waves travel more slowly than P-waves in solids. S-waves have particle motion perpendicular to the propagating direction, like the movement of a rope as a displacement speeds along its length (see Figure 3.57). If the dominant particle movement is vertical (i.e., polarized in the vertical plane), the S-waves are classified as SV-waves. S-waves polarized in the horizontal plane are classified as SH-waves. In soils under anisotropic stress conditions or with anisotropic soil fabric, SH-waves and SV-waves can have different velocities within the same soil mass. The S-waves can only transit material that has shear strength. S-waves do not exist in liquids and gasses, as these media have no shear strength.

Field geophysical surveys described below can be used to obtain P-wave velocity, V_p, and S-wave velocity, V_s. For an isotropic and linear elastic material, its Young's or elastic modulus (E), shear modulus (G) and either mass density (ρ) or Poisson's ratio (μ) can be determined, if V_p and V_s are known, through the following equations (Grant and West, 1965):

$$\mu = \frac{\left[\left(\frac{V_p}{V_s}\right)^2 - 2\right]}{\left\{2\left[\left(\frac{V_p}{V_s}\right)^2 - 1\right]\right\}} \tag{3.64}$$

$$E = \frac{\rho V_p^2 (1 - 2\mu)(1 + \mu)}{(1 - \mu)} \tag{3.65}$$

$$G = \frac{E}{2(1 + \mu)} \tag{3.66}$$

$$\rho = \frac{G}{V_s^2} \tag{3.67}$$

Table 3.10 shows typical values of V_p, ρ, and μ for various materials. It is clear from Table 3.10 that V_p of soil is mostly less than that of water. The V_p measurement below the water table may thus be overshadowed by the presence of water.

3.5.1.2 Surface waves

Two types of vibration-induced disturbances which exist only at "surfaces" or interfaces are Love and Rayleigh waves. These waves attenuate rapidly with distance from the surface. Surface waves travel more slowly than body waves. Love waves travel along the surfaces of layered media and are most often faster than Rayleigh waves. Love waves have particle displacement similar to shear waves in a horizontal plane (see Figure 3.58). Love waves are observed only when there is a low-velocity layer overlying a high-velocity layer/sublayers. Rayleigh waves exhibit vertical and horizontal displacement in the vertical plane of raypath. The Rayleigh wave moves back, down, forward, and up repetitively in an ellipse-like pattern, as described in Figure 3.58.

The execution of seismic geophysical exploration methods generally involves the use of a seismic source and an array or receivers to detect the arrival of seismic waves. The seismic source can be as simple as a hammer striking the ground surface, a rifle shot, a harmonic oscillator, or an explosion. The receiver is a seismometer that converts ground shaking into an electrical response. Seismic methods introduced in the following sections include reflection method, refraction method, and surface wave method. These methods can be nonintrusive where the tests are conducted from the ground surface without the

Table 3.10 Typical values of V_p, ρ, and v for various materials

Material	V_p (m/s)	ρ (Mg/m³)	μ
Air	330		
Damp loam	300–750		
Dry sand	460–900	1.8–2.0	0.3–0.35
Clay	900–1800	1.3–1.8	0.5
Water	1450–1500	1.0	
Saturated loose sand	1500		
Till	1700–2300	2.3	
Rock			0.15–0.25
Shale	800–3700		
Sandstone	2200–4000	1.9–2.7	
Dolomite and limestone	4300–6700	2.5–3.0	
Unweathered granite	4800–6700	2.6–3.1	
Unweathered basalt	2600–4300	2.2–2.3	
Metamorphic rock	2400–6600		
Steel	6000		

Source: US Army Corps of Engineers. 1995. Geophysical exploration for engineering and environmental investigations, Engineering Manual No. 1110-1-1802, Washington, DC, 208p.

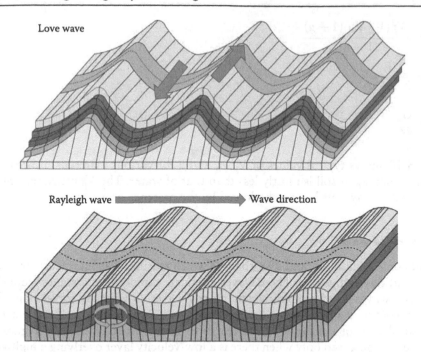

Figure 3.58 Characteristics of surface waves.

need of boreholes. Included also are the crosshole and the *P-S* logging methods, which are intrusive methods that require one or multiple boreholes to perform the test. Note that the shear wave velocity measurement in seismic cone penetration test described in Section 3.4 is also an intrusive seismic geophysical method.

3.5.2 Nonintrusive seismic methods

The raypaths in which the seismic energy follows are constructed by the method of geometrical optics. Figure 3.59 shows four rays emanating from a seismic source at the ground surface. The rays travel through a ground with two layers separated by a flat horizontal boundary. The seismic wave velocity in the upper layer or layer 1,

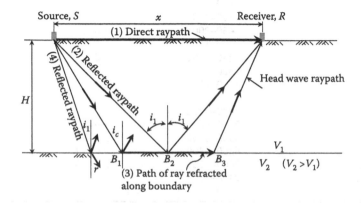

Figure 3.59 Raypaths for seismic energy generated at source and picked up at receiver.

V_1, is less than the velocity of the underlying medium or layer 2, V_2. These four rays are:

1. The direct wave that travels a horizontal path directly from source to receiver.
2. The totally reflected ray that is generated when a ray strikes the boundary between the two layers at an angle of incidence, i_1, that is greater than the critical angle of incidence, i_c, and all of the energy is reflected back toward the surface.
3. A ray striking the boundary with $i_1 = i_c$, part of the energy is reflected back toward the surface and the rest of the energy is refracted and travels along the boundary (B_1B_3) with velocity V_2.
4. A ray striking the boundary with $i_1 < i_c$, part of the energy is reflected upward and part refracted in the lower medium away from the normal to the boundary at an angle r.

The critical angle of incidence relates to the two velocities as

$$i_c = \arcsin\left(\frac{V_1}{V_2}\right) \tag{3.68}$$

For the refracted part of ray 4 ($i_1 < i_c$), according to Snell's law,

$$\frac{\sin i_1}{\sin r} = \frac{V_1}{V_2} \tag{3.69}$$

The above equations and the relationships between velocities, distance, and time constitute the basis for the interpretation of seismic test data.

3.5.2.1 Reflection method

For the simple two-layer system shown in Figure 3.59, the seismic reflection method uses the transmission time of seismic wave following the reflected raypath SB_2R (ray 2). It can be derived based on the geometry of the raypath that:

$$H = 0.5\sqrt{V_1^2 t^2 - x^2} \tag{3.70}$$

where
  $H =$ thickness of layer 1
  $t =$ travel time from source to receiver through reflected raypath SB_2R
  $x =$ horizontal distance between source and receiver

To apply Equation 3.70, t is read from the seismogram and x is predetermined when placing the receiver and seismic source in the field. To calculate H via Equation 3.70 requires that V_1 is independently measured. There are seismic reflection procedures that allow simultaneous determination of layer thickness and wave velocities but require the use of multiple seismic sources and/or receivers. However, interferences between reflected, refracted, and surface waves increase the complexity of the seismic data. Often the use of reflection method involves complex testing layouts and signal processing to enhance the signals of specific raypaths. The complexity of the wave field has limited the application of the reflection survey as part of a geotechnical exploration.

3.5.2.2 Refraction method

Refraction method uses the critically refracted wave (ray 3 of Figure 3.59) from a higher velocity layer that underlies lower-velocity material. The principles of refraction method and arrival time–distance curve for a two-layer system are shown in Figure 3.60. The seismic refraction test consists of recording the arrival time of the first impulse from a seismic excitation at a series of receivers distributed on the surface (see Figure 3.60a). The refraction method is only applicable for stiffness increasing profiles (V increases with depth). The "crossover distance" x_{c2} is defined from a plot of the first arrival time versus distance (Figure 3.60b) curve as the point where the slope of the arrival time curve changes. For the idealized two-layer system shown, the curve representing the direct wave is a straight line with a slope equal to the reciprocal of the velocity of the surface layer V_1. The head wave curve has a slope equal to the reciprocal of the velocity of the bottom layer V_2. When the distance between source and receiver x is larger than x_{c2}, the first arrival is made up of refracted energy. Depth to the first flat-lying interface, D_1, relates to the wave velocities and intercept time T_{i2} (see Figure 3.60b) as follows:

$$D_1 = \frac{(T_{i2})(V_1 V_2)}{2\sqrt{(V_2^2 - V_1^2)}} \tag{3.71}$$

The wave velocities (i.e., V_1, V_2) and time intercept, T_{i2}, can all be determined using the time–distance curve of Figure 3.60b, and that enables the calculation of D_1 following Equation 3.71.

If the ground has more flat-lying layers, such as the case of Figure 3.61, the above method can be extended and the computation begins with the first layer and progresses downward as follows:

$$D_2 = \frac{(T_{i3})(V_2 V_3)}{2\sqrt{(V_3^2 - V_2^2)}} - D_1 \left(\frac{V_2}{V_1}\right)\sqrt{\frac{(V_3^2 - V_1^2)}{(V_3^2 - V_2^2)}} \tag{3.72}$$

where D_1 is obtained from Equation 3.71, more generally, for the nth layer.

$$D_n = \frac{(T_{in+1})(V_n V_{n+1})}{2\sqrt{(V_{n+1}^2 - V_n^2)}} - \sum_{j=1}^{n-1} D_j \left(\frac{V_n}{V_j}\right)\sqrt{\frac{(V_{n+1}^2 - V_j^2)}{(V_{n+1}^2 - V_n^2)}} \tag{3.73}$$

The refraction method can be further extended for ground with multiple layers and considering inclined interfaces.

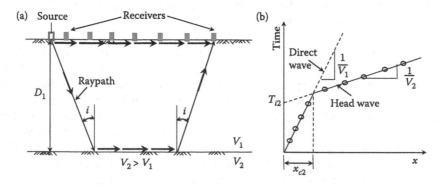

Figure 3.60 Principles of refraction survey and first arrival time–distance curve. (a) Field layout and raypath. (b) The first arrival time–distance curve.

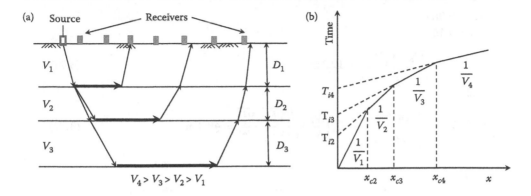

Figure 3.61 Multiple layer refraction survey. (a) Field layout and raypath. (b) The first arrival time–distance curve.

EXAMPLE 3.11

Given

An array of seismic receivers was installed in the field to perform seismic refraction test. The receivers were placed in a line and spaced at 20 m. The ground is expected to have three flat-lying layers similar to the case shown in Figure 3.61. Table E3.11 shows the record of P-wave arrival time for each of the receivers (according to their distances to the seismic source).

Table E3.11 P-wave arrival time record

Distance to Source (m)	20	40	60	80	100	120	140	160	180	200
Arrival time, s	0.04	0.08	0.12	0.16	0.188	0.212	0.238	0.262	0.288	0.312

Distance to Source (m)	220	240	260	280	300	320	340	360	380	400
Arrival time, s	0.337	0.357	0.372	0.388	0.403	0.419	0.434	0.449	0.465	0.480

Required

Determine the thickness of the top two layers (D_1 and D_2) and P-wave velocity of the three layers (V_1, V_2, and V_3).

Solution

Figure E3.11 shows the arrival time versus distance curve (time–distance curve) from the field data of Table E3.11. The slope of the time–distance curve changes at crossover distances, $x_{c2} = 80$ m (corresponding arrival time, $t_{c2} = 0.16$ s) and $x_{c3} = 220$ m ($t_{c3} = 0.337$ s). The P-wave velocity for layer 1 (V_1), layer 2 (V_2), and layer 3 (V_3) is determined respectively from the slope of the three segments of straight lines separated by x_{c2} and x_{c3} of the time–distance curve shown in Figure E3.11:

$V_1 = 500$ m/s
$V_2 = 800$ m/s
$V_3 = 1300$ m/s

The time intercepts (T_{i2} and T_{i3}) are determined by extending the respective head wave curves toward the time coordinate as shown in Figure E3.11, and

$T_{i2} = 0.06\,\text{s}$
$T_{i3} = 0.168\,\text{s}$

$$D_1 = \frac{(T_{i2})(V_1 V_2)}{2\sqrt{(V_2^2 - V_1^2)}} = 19.2\,\text{m} \tag{3.71}$$

and

$$D_2 = \frac{(T_{i3})(V_2 V_3)}{2\sqrt{(V_3^2 - V_2^2)}} - D_1\left(\frac{V_2}{V_1}\right)\sqrt{\frac{(V_3^2 - V_1^2)}{(V_3^2 - V_2^2)}} = 49.4\,\text{m} \tag{3.72}$$

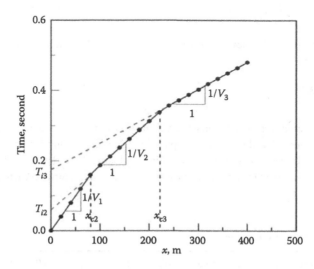

Figure E3.11 Time–distance curve.

3.5.2.3 Surface wave methods

Surface wave methods are gaining popularity as seismic methods for shallow depth (less than 30 m) profiling. The methods are fast and can cover a large sampling area. The basis of surface wave methods is the dispersive characteristics of Rayleigh waves in a layered system. The phase velocity of Rayleigh wave, V_R, depends primarily on the soil stiffness over a depth of approximately one wavelength. As a result, Rayleigh waves with different wavelengths can be used to sample different depth ranges, and the phase velocity varies accordingly. Several surface wave methods have been developed for near-surface exploration. Out of these available methods, the spectral analysis of surface waves (SASW) method (Stokoe and Nazarian, 1985) and the multi-channel analysis of surface wave (MASW) method (Park et al., 1999) are the most popular ones. In spite of differences in testing arrangements and data-processing procedures among different methods, all surface wave methods contain three stages: data acquisition, construction of field dispersion curve, and an inversion process to establish representative shear wave velocity profile. Details of surface wave methods are beyond the scope of this book. Readers interested in these test methods are referred to Stokoe et al. (2004) and Park et al. (1999).

3.5.3 Intrusive seismic methods

Intrusive seismic methods require boreholes to install the source and receivers, and the wave velocities are evaluated by measuring the travel time of specific raypaths from source to receiver(s). The seismic test included in the seismic cone penetration test described in Section 3.4 can be considered as a type of intrusive seismic geophysical test. The following sections introduce crosshole and suspension P–S logging test methods. Both are frequently used in geotechnical site characterization for foundations.

3.5.3.1 Crosshole method

The crosshole method provides a travel time measurement of seismic waves where the source and receivers are placed at the same depth in adjacent boreholes with known distances, as schematically shown in Figure 3.62. Details of field layout and test procedure of crosshole methods can be found in ASTM standard D4428. The crosshole seismic test requires an in-hole seismic source capable of generating both P-wave and S-wave propagating horizontally. The receivers must be able to record particle motions in three orthogonal directions in order to measure the P, SH, and SV wave velocities. Advantages of crosshole method include high-spatial resolution of seismic wave velocity measurements as well as measurements of P, SH, and SV wave velocities at the same depth. The receivers placed in the two boreholes record the wave arrival time difference between the two receivers, t_{12}. The distance between the two receivers, d_{12}, is known and typically at about 3 m, as shown in Figure 3.62. The seismic wave velocity, V, is then calculated as

$$V = \frac{d_{12}}{t_{12}} \tag{3.74}$$

The value obtained from Equation 3.74 reflects the velocity of the same type of waves generated by the seismic source (i.e., V_p or V_s depending on the type of waves generated by the seismic source).

3.5.3.2 P–S logging

Well logging, or borehole logging, has a long history in petroleum engineering. Logging tools can be lowered into a borehole to produce the profile of material properties. The suspension P–S logging method is a relatively new method of measuring P- and S-wave velocity profiles of soils (Nigbor and Imai, 1994) and is probably the only technique that can

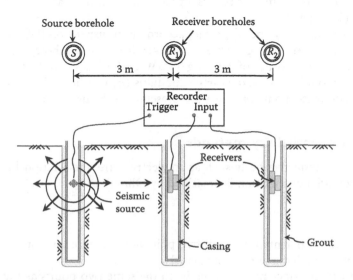

Figure 3.62 Layout of crosshole seismic testing.

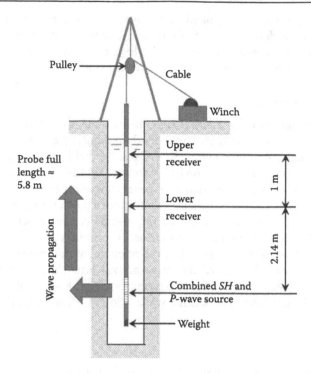

Figure 3.63 Field setup of suspension P–S logger.

provide high-resolution wave velocity profiles for deep profiles (deeper than 200 m). The setup of a suspension logger is schematically shown in Figure 3.63. The seismic source and two receivers are vertically spaced and housed in a flexible tube. The total length of the tube is approximately 5.8 m. For P–S logging test, the tube is lowered into a fluid-filled borehole. The raypath is source-fluid-surrounding material-fluid-receiver. The near and far receivers are designated, respectively, as the lower and upper receivers in Figure 3.63. The P- and S-wave velocities of the surrounding materials are inferred from the travel time between the two receivers, and the results represent the average wave velocity between the two receivers (1 m apart).

Figure 3.64 shows the V_p and V_s profiles according to suspension P–S logging tests in a sandstone formation. According to rock cores taken from the borehole, the sandstone had a mass density (ρ) on the order of 2.14 Mg/m^3. With the V_p, V_s, and ρ values, the corresponding shear modulus (G) and Young's modulus (E) can be derived using Equations 3.64 to 3.67, and these modulus values are presented in Figure 3.65.

3.5.4 Electrical resistivity methods

In an electrically conductive body such as an electric wire, the relationship between the current and potential distribution is described by Ohm's law:

$$V = IR \tag{3.75}$$

where
  V = difference of potential between two points on the wire (in volt)
  I = current through the wire (in amperes)
  R = electrical resistance measured between the same two points as the difference of potential (in Ohm, Ω)

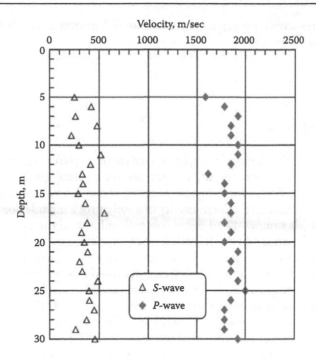

Figure 3.64 P and S wave velocity profiles from suspension P–S logging tests in a sandstone. (Courtesy of An-Bin Huang, Hsinchu, Taiwan.)

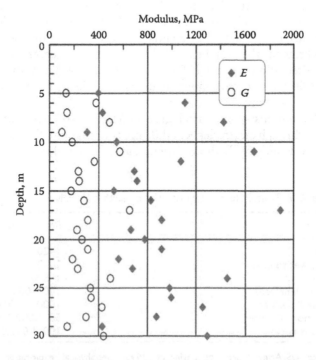

Figure 3.65 The shear and Young's modulus profiles derived from wave velocities. (Courtesy of An-Bin Huang, Hsinchu, Taiwan.)

The electrical resistivity of any conductive material having a length L and cross-sectional area of A is defined as

$$\rho = \frac{RA}{L}$$ (3.76)

where
  ρ = resistivity (in Ωm)

All materials, including soil and rock, have a resistivity as part of their intrinsic property that governs the relation between the current density and the gradient of the electrical potential. Table 3.11 shows some typical range of resistivity values of natural minerals and rocks. Properties that affect the resistivity of a soil or rock include porosity, water content, composition (clay mineral and metal content), salinity of the pore water, and grain size distribution. Variations in the resistivity of earth materials, either vertically or laterally, produce variations in the relations between the applied current and the potential distribution as measured on the surface. Using electrical methods, we can measure potentials, currents, and electromagnetic fields which occur naturally or are introduced artificially in the ground to distinguish materials through contrast of electrical properties. The enormous variation in electrical resistivity found in different soils and rocks such as those shown in Table 3.11 makes electrical methods feasible (Telford et al., 1990). A large number of electrical methods have been developed and used successfully since their early applications in late nineteenth century. The following discussion concentrates on the electrical resistivity method only.

Surface electrical resistivity survey is based on the principle that the distribution of electrical potential in the ground around a current-carrying electrode depends on the electrical resistivities and their distribution in the surrounding soils and rocks. The usual practice in the field is to apply an electrical current (I) between two electrodes implanted in the ground and to measure the difference of potential (V) between two additional electrodes that do not carry current. Usually, the potential electrodes are in line between the current electrodes, but in principle they can be located anywhere. The current used is either direct current (DC) or alternating current (AC) of low frequency, typically in 50–100 milliamperes. Figure 3.66 shows three commonly used electrode array configurations for field electrical resistivity surveying.

Consider the electrode array shown in Figure 3.66a, called the Schlumberger array. The current I is applied through electrodes A and B, and the potential difference V is measured between electrodes M and N. If the resistivity survey is conducted in a homogeneous

Table 3.11 Typical electrical resistivities of earth materials

Material	Resistivity (Ωm)
Clay	1–20
Sand, wet to moist	20–200
Shale	1–500
Porous limestone	100–1000
Dense limestone	1000–1,000,000
Metamorphic rocks	50–1,000,000
Igneous rocks	100–1,000,000

Source: US Army Corps of Engineers. 1995. Geophysical exploration for engineering and environmental investigations, Engineering Manual No. 1110-1-1802, Washington, DC, 208p.

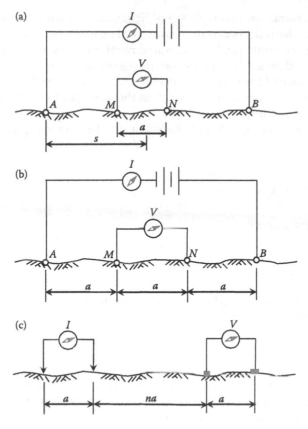

Figure 3.66 Electrode array configurations for resistivity measurements. (a) Schlumberger array. (b) Wenner array. (c) Dipole–dipole array.

half-space with a uniform resistivity ρ, then the potential difference between electrodes M and N (Keller and Frischknecht, 1966) will be:

$$V = U_M - U_N = \frac{\rho I}{2\pi}\left(\frac{1}{\overline{AM}} - \frac{1}{\overline{BM}} + \frac{1}{\overline{BN}} - \frac{1}{\overline{AN}}\right) \tag{3.77}$$

where
U_M and U_N = potential at M and N
$\overline{AM}, \overline{BM}, \overline{BN},$ and $\overline{AN}$ = distance between A and M, B and M, B and N, and A and N, respectively

The quantity inside the brackets of Equation 3.77 is denoted as $1/K$, where K is called the geometric factor and K is a function only of the geometry of the electrode arrangement. Therefore, Equation 3.77 can be rearranged to show that:

$$\rho = 2\pi K \frac{V}{I} \tag{3.78}$$

The resistivity ρ can be found from measured values of V, I, and K, the geometric factor. However, in actual field survey, the measurements are made in heterogeneous material where ρ varies. To facilitate the application of Equation 3.78 in the interpretation of field resistivity measurements, we first replace ρ of Equation 3.78 with an apparent resistivity, ρ_a. Apparent resistivity is defined as the resistivity of an electrically homogeneous and isotropic half-space that would yield the measured relationship between the applied current and the potential difference for a particular arrangement and spacing of electrodes. The real resistivity for each of the multiple layers of earth materials and their spatial boundaries present at the test site are then inferred from the ρ_a values from various locations and with various electrode configurations.

For Schlumberger array (see Figure 3.66a), it can be demonstrated that (Keller and Frischknecht, 1966):

$$\rho_a = 2\pi K \frac{V}{I} = \pi a \left[\left(\frac{s}{a}\right)^2 - \frac{1}{4} \right] \frac{V}{I} \tag{3.79}$$

and for Schlumberger array,

$$K = \frac{a}{2} \left[\left(\frac{s}{a}\right)^2 - \frac{1}{4} \right] \tag{3.80}$$

In field survey using the Schlumberger array, the ratio of s/a is set to be in a range of 3 to 30. For Wenner array (see Figure 3.66b), $K = a$ and,

$$\rho_a = 2\pi a \frac{V}{I} \tag{3.81}$$

For the dipole–dipole array shown in Figure 3.66c, where the separation between both pairs of electrodes is the same a and the separation between the centers of the dipoles is $a(n+1)$, the apparent resistivity is given by

$$\rho_a = \pi a [n(n+1)(n+2)] \frac{V}{I} \tag{3.82}$$

and for dipole–dipole array,

$$K = \frac{a[n(n+1)(n+2)]}{2} \tag{3.83}$$

where
$n = $ an integer

The basis for making an electrical sounding, irrespective of the electrode array used, is that the farther away from a current source the measurement of the potential, or the potential difference is made, the deeper the probing will be. Typically, a maximum electrode spacing of three or more times the depth of interest is necessary to assure that sufficient data are gathered.

3.5.4.1 Vertical profiling

Vertical profiling can be performed with any one of the three array configurations described above. Regardless of the electrode array used, the basic procedure is to perform a series of apparent resistivity measurements with progressively increasing electrode spacings. In electrical sounding with the Schlumberger array, $\overline{AB}/2$ ($\overline{AB} = $ distance between electrodes A and B in Figure 3.66a) is used to represent the electrode spacing. For Wenner or dipole–dipole arrays (see Figure 3.66b and c), a is used to indicate electrode spacing. For small electrode spacings, the apparent resistivity is close to

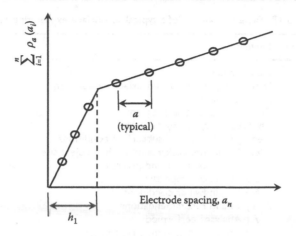

Figure 3.67 Cumulative resistivity curve for a two-layer system.

the surface layer resistivity, while at large electrode spacings, it approaches the resistivity of the bottom layer.

The interpretation for vertical profiling data is to use the relationship between apparent resistivity and electrode spacing from field measurements, to obtain the geoelectrical parameters of the tested earth material: the layer resistivities and thicknesses. This interpretation is an inverse solution, and, as in many other geophysical exploration methods, does not yield a unique answer. Curve matching is generally considered the more rigorous method of interpretation of vertical profiling data.

An empirical procedure called the Moore cumulative resistivity method has been proposed to determine the depth (but not necessarily the resistivity) to horizontal boundaries using the Wenner array (Moore, 1951). A series of apparent resistivity measurements are taken using the Wenner array; each time the electrode spacing is increased by an equal amount, a (the Wenner array electrode spacing). The cumulative resistivity curve is constructed by plotting $\sum_{i=1}^{n} \rho_a(a_i)$ versus $a_n(a_n = na)$, as conceptually described in Figure 3.67. The curve consists of straight line segments intersecting at points where the abscissa value(s) corresponds to the depth(s) of horizontal boundaries.

3.5.4.2 Horizontal profiling

The main purpose of horizontal profiling is to locate geological structures such as buried streams, veins, or dikes. In horizontal profiling, a fixed electrode spacing is chosen, and the whole electrode array is moved along a profile after each measurement is taken. Maximum apparent resistivity anomalies are obtained by orienting the profiles at right angles to the strike of the geological structure. The value of apparent resistivity is plotted, in general, at the geometric center of the electrode array. Data obtained from horizontal profiling, for engineering applications, are normally interpreted qualitatively. Apparent resistivity values are presented as either contour maps or profiles, or both.

3.6 SUBSURFACE EXPLORATION REPORT

Results of subsurface exploration should be compiled and presented in a report. Table 3.12 shows the table of contents of a typical subsurface exploration report for the purpose of foundation design. The report then becomes the basis for the design of building or structural foundations. It should not be a surprise that the sequence of the

Table 3.12 Table of contents of a typical subsurface exploration report

Item no.	Content
1.	Introduction
2.	Scope and purpose of subsurface exploration
3.	Site description
4.	The field exploration program
5.	The laboratory testing program
6.	Geological setting and subsurface conditions
	6.1 Geological background of the project site
	6.2 Soil/rock layers and profiles
	6.3 Soil/rock properties
	6.4 Groundwater conditions
7.	Analyses and recommendations
	7.1 Analyses performed
	7.2 Foundation system(s) recommended
	7.3 Expected performance of the recommended foundation system(s)
	7.4 Construction considerations
8.	References
9.	Appendices
	A. Boring location diagram
	B. Boring logs and rock core photos (if available)
	C. Laboratory test results
	D. Field (in-situ and/or geophysical) test results
	E. Test procedures (e.g., ASTM standards or Euro codes) used in the subsurface exploration
10.	Legal statements

table of contents mirrors the flow chart presented in Figure 3.1, in the beginning of this chapter, with the subsurface exploration report being the end of the flow chart. Items 1 and 2 provide the background of the subsurface exploration, its scope and purpose. Items 3–5 are results of field reconnaissance and planning of the subsurface exploration program. Item 6 describes and summarizes the results of the literature review and the subsurface exploration performed. The following chapters throughout this textbook describe, in general, the analyses involved in the selection of the foundation system(s), prediction of performance of the recommended foundation system(s), and their potential construction problems. Item 7 describes the analyses performed and recommendations made according to these analyses. All test data should be organized and presented in the appendix. All tests included in the report should either follow accepted standards or detailed procedures should be given and be included in the appendix.

3.7 REMARKS

In this chapter, we covered the following topics:

- Planning of subsurface exploration.
- Methods of borehole drilling.
- Methods to take soil and rock samples for laboratory testing.
- Five of the commonly used in-situ testing methods (SPT, CPT, FVT, DMT, and PMT).
- A brief introduction to seismic and electrical resistivity geophysical exploration method.
- Preparation of the subsurface exploration report.

All five of the in-situ testing methods described in this chapter are suitable for strength measurements (indirect interpretation), or limit state foundation analysis (direct interpretation). DMT and PMT, because of their multiple stress–displacement measurements, can also reflect the stiffness of the surrounding soil. This is an important advantage in providing the parameters for analyzing the foundation vertical/horizontal displacement under a given loading condition. Recall that stress normalization is an important part of the interpretation or application of N and q_t values. Multiple empirical equations have been developed for these normalizations and they all involve some uncertainties. In this regard, DMT is more desirable, as I_D and K_D, the two important intermediate index parameters, are normalized in the early stage of its development.

While many of the currently used subsurface exploration methods have been described in this chapter, it is by no means inclusive of all the available methods. For complicated projects, scaled model tests and field monitoring of existing structures or those under construction have also been used for the purpose of foundation engineering design. Numerical modeling can help in providing the important predictions of the proposed foundation systems under the design-loading conditions. As a natural material, variabilities in the engineering properties of soils and rocks are inevitable. The use of probability theories to quantify these variables and apply them in foundation designs is increasing. What constitutes the contents of a subsurface exploration and how the results should be analyzed are likely to change with time as various techniques evolve. In any case, the use of sound engineering judgment is always a must, regardless of the techniques applied.

HOMEWORK

3.1. Describe the objectives of literature review in the planning of subsurface exploration.

3.2. Compare the advantages and disadvantages between in-situ tests and drilling and lab testing.

3.3. Provide your recommendation of drilling/sampling, geophysical and/or in-situ testing procedures for each of the following purposes, and state your assumptions for your answer if necessary:

- Determine the soil stratigraphy within the top 5 m, at 200 test locations along a proposed 20 km-long highway.
- Determine the compressibility of an overconsolidated clay layer at 5–15 m below ground surface.
- Determine the compressibility of a sand deposit at 5–15 m below ground surface.
- Determine the groundwater table distribution within a 300-acre project site.
- Determine the type(s) and quality of rock from 50 to 100 m below ground surface.

3.4. Describe the possible procedures to determine the groundwater table in the field.

3.5. Table H3.5 shows a set of SPT data that includes the N values for every 1.5 m, and energy efficiencies (in percentage, %) from field measurements in a silty sand. The groundwater table is at 1.5 m below ground surface. The soil unit weight (γ) above groundwater table is 17 kN/m^3, saturated soil unit weight (γ_{sat}) below groundwater table is 19 kN/m^3. Calculate the effective overburden stress (σ'_{vo}) corresponding to each N value and determine the $(N_1)_{60}$ values according to energy efficiency and σ'_{vo}, plot the result of $(N_1)_{60}$ versus depth.

Table H3.5 The SPT data

Depth (m)	N blows/30 cm	Energy efficiency (%)
1.5	7	80
3.0	5	78
4.5	6	75
6.0	8	80
7.5	9	70
9.0	11	72
10.5	13	74
12.0	15	75

3.6. For the $(N_1)_{60}$ values obtained from Problem 3.5, determine the corresponding relative density, D_r (%) and peak friction angle, $\emptyset'$ (degree).

3.7. Repeat Problem 3.6 using the original N values.

3.8. Describe why we need to normalize some of the CPT and SPT test results.

3.9. Table H3.9 shows the representative q_t, f_s values for every 2.5 m from an SCPTu test in a cohesive soil. The soil total unit weight, $\gamma = 18.5 \text{ kN/m}^3$ and saturated unit weight, $\gamma_{sat} = 19.0 \text{ kN/m}^3$. The groundwater table is at 2 m below ground surface. Compute the corresponding σ_{vo}, σ'_{vo}, Q_{tn}, F_r, and I_c values and determine the soil behavior type zonal numbers according to SBTn chart.

Table H3.9 Representative q_t, f_s values

Depth (m)	2.5	5.0	7.5	10.0	12.5	15.0	17.5	20.0
q_t, MPa	3.5	1.5	1.4	1.3	0.95	0.70	0.65	0.62
f_s, kPa	180	75	32	35	20	15	7	5
u_2, MPa	2.80	0.80	0.70	0.60	0.60	0.25	0.18	0.18

3.10. Repeat Problem 3.9 using the Schneider et al. Q_t–F_r and Q_t–U_2 charts.

3.11. Discuss the effects of penetration rate and cone size on the CPTu results in saturated sand, silt, and clay, respectively.

3.12. For the data given in Problem 3.9, calculate the undrained shear strengths (Equation 3.22) and OCR (Equation 3.25) of the clay according to the cone penetration test results. Use $\emptyset' = 30°$ and $N_k = 14$.

3.13. Table H3.13 shows the q_t for every 1.5 m from CPT in a granular soil. The soil total unit weight, $\gamma = 18.5 \text{ kN/m}^3$ and saturated unit weight, $\gamma_{sat} = 19.0 \text{ kN/m}^3$. The groundwater table is at 2 m below ground surface. Compute the corresponding σ_{vo} and σ'_{vo}. Determine the relative density D_r at the corresponding depths using Equations 3.28a and b. Assume $\sigma'_{ho}/\sigma'_{vo} = 0.45$. Estimate the peak friction angle $\emptyset'$ (degree) using Equation 3.31.

Table H3.13 The q_t values

Depth (m)	q_t (MPa)
1.5	7.2
3.0	4.5
4.5	6.5
6.0	7.8
7.5	9.2
9.0	11.1
10.5	13.6
12.0	15.2

3.14. Table H3.14 shows the results of a Ménard type pre-bored pressuremeter test in a weak sandstone. The initial volume of the pressuremeter measuring cell (V_o) was 900 cm^3. The data have been adjusted to exclude the effects of PMT membrane stiffness and system compliance. Plot the expansion pressure versus volume change and creep volume curves and determine p_o, p_f, E_m, and p_{lm} for the PMT test result.

Table H3.14 The PMT test results

Expansion pressure (kPa)	19	86	139	267	376	653	1046	1404
Volume change, ΔV (cm^3)	5	10	20	48.9	60.4	78.5	90.8	98.3
Creep volume, cm^3	2	10	2	1	1	1	2	2
Expansion pressure (kPa)	1782	2578	3294	4159	4516	5252	5550	5951
Volume change, ΔV (cm^3)	104.9	116.8	128.8	142.5	148.9	161.0	168.8	178.2
Creep volume, cm^3	1	1	1	2	2	1	2	3
Expansion pressure (kPa)	6253	6555	6956	7359	7784	8100	8200	8300
Volume change, ΔV (cm^3)	187.0	195.8	208.2	224.4	280.0	350.0	490.0	650.0
Creep volume, cm^3	3	2	5	7	15	25	40	60

3.15. A vane with a height, H, of 120 mm, and a diameter, D, of 60 mm was used to perform a field vane shear test in a clay with a plasticity index, PI of 30%. The recorded peak torque, T, was 0.12 kN-m. Determine the undrained shear strength of the clay s_u using Equation 3.36, consider $a = 4$. Determine the field undrained shear strength using the Azzouz correction (Figure 3.44).

REFERENCES

Amar, S., Clarke, B.G.F., Gambin, M.P., and Orr, T.L.L. 1991. *The Application of Pressuremeter Test Results to Foundation Design in Europe, a State of the Art Report by the ISSFE European Technical Committee on Pressuremeters, Part 1: Predrilled Pressuremeters and Self-Boring Pressuremeters.* AA Balkema, The Netherlands, 48p.

Baguelin, F., Jézéquel, J.F., and Shields, D.H. 1978. *The Pressuremeter and Foundation Engineering.* Trans Tech Publications, Clausthal, Germany, 617p.

Baligh, M. 1985. Strain path method. *Journal of Geotechnical Engineering Division*, ASCE, 111(9), 1108–1136.

Barentsen, P. 1936. Short description of field testing method with cone shaped sounding apparatus. *Proceedings, First International Conference on Soil Mechanics and Foundation Engineering*, Cambridge, MA, Vol. 1(B/3), pp. 6–10.

Begemann, H.K. 1953. Improved method of determining resistance to adhesion by sounding through a loose sleeve placed behind the cone. *Proceedings, Third ICSMFE*, Zurich, Switzerland, Vol. 1, pp. 213–217.

Bjerrum, I. 1972. Embankments on soft ground. *Proceedings, ASCE Specialty Conference on Performance of Earth and Earth-Supported Structures*, Purdue University, USA, Vol. 2, pp. 1–54.

Broms, B.B. and Flodin, N. 1988. History of penetration testing. *ISOPT-1*, Orlando, pp. 157–220.

Campanella, R.G., Robertson, P.K., and Gillespie, D. 1986. Seismic cone penetration test. In: Clemence, S.P., ed., *Use of in Situ Tests in Geotechnical Engineering, Proceedings, In Situ '86*, Vol. 6, ASCE Geotechnical Specialty Publication, Blacksburg, VA, pp. 116–130.

Campanella, R.G. and Robertson, P.K. 1988. Current status of the piezocone test. *Proceedings, First International Symposium on Penetration Testing, ISOPT-1*, Orlando, pp. 93–116.

Clarke, B.G. 1995. *Pressuremeters in Geotechnical Design.* Blackie Academic and Professional, Glasgow, 364p.

Deere, D.U. 1963. Technical description of rock cores for engineering purposes. *Felsmechanik und Ingieurgeologie*, 1(1), 16–22.

Eslami, A. and Fellenius, B.H. 1997. Pile capacity by direct CPT and CPTu methods applied to 102 case histories. *Canadian Geotechnical Journal*, 34(6), 886–904.

Grant, F.S. and West, G.F. 1965. *Interpretation Theory in Applied Geophysics*. McGraw-Hill, New York.

Hatanaka, M. and Uchida, A. 1996. Empirical correlation between penetration resistance and effective friction of sandy soil. *Soils and Foundations, Japanese Geotechnical Society*, 36(4), 1–9.

Huang, A.B. and Chang, W.J. 2011. Geotechnical and geophysical site characterization oriented to seismic analysis, State of the Art Report. *Fifth International Conference on Earthquake Geotechnical Engineering, Chilean Geotechnical Society*, Santiago, Chile, pp. 63–104.

Huang, A.B., Holtz, R.D., and Chameau, J.L. 1991. A laboratory study of pressuremeter tests in clays. *Journal of Geotechnical Engineering Division, ASCE*, 117, 1549–1567.

Huang, A.B. and Hsu, H.H. 2004. Advanced calibration chambers for cone penetration testing in cohesionless soils. Keynote lecture. *Proceedings, ISC-2 on Geotechnical and Geophysical Site Characterization*, Porto, Portugal, Vol. 1, pp. 147–167.

Huang, A.B. and Hsu, H.H. 2005. Cone penetration tests under simulated field conditions. *Geotechnique*, 55(5), 345–354.

Huang, A.B. and Huang, Y.T. 2007. Undisturbed sampling and laboratory shearing tests on a sand with various fines contents. *Soils and Foundations*, 47(4), 771–781.

Huang, A.B., Tai, Y.Y., Lee, W.F., and Ishihara, K. 2008. Sampling and field characterization of the silty sand in central and southern Taiwan. *Proceedings, Third International Conference on Site Characterization*, Taipei, pp. 1457–1463.

Hughes, J.M.O., Wroth, C.P., and Windler, D. 1977. Pressuremeter tests in sands. *Geotechnique*, 27(4), 455–477.

Hvorslev, M.J. 1949. *Subsurface Exploration and Sampling of Soils for Civil Engineering Purposes*. Edited and reprinted by Waterways Experiment Station, Vicksburg, MS, 521p.

Idriss, I.M. and Bulanger, R.W. 2006. Semi-empirical procedures for evaluating liquefaction potential during earthquakes. *Soil Dynamics and Earthquake Engineering*, 26, 115–130.

Ishihara, K. 1993. Liquefaction and flow failure during earthquakes. *Geotechnique*, 43(3), 351–415.

Jamiolkowski, M., Ghionna, V., Lancellotta, R., and Paqualini, E. 1988. New correlations of penetration tests for design practice. *Proceedings, International Symposium of Penetration Testing, ISOPT-I*, Orlando, FL, AA Balkema, The Netherlands, Vol. 1, pp. 263–296.

Jamiolkowski, M., LoPresti, D.C.F., and Manassero, M. 2001. Evaluation of relative density and shear strength of sands from cone penetration test and flat dilatometer test. *Soil Behavior and Soft Ground Construction, (GSP 119)*, ASCE, Reston, VA, pp. 201–238.

Jefferies, M.G. and Been, K. 2006. *Soil Liquefaction—A Critical State Approach*. Taylor & Francis, Boca Raton, FL, 479p.

Juang, C.H., Fang, S.Y., and Khor, E.H. 2006. First-order reliability method for probabilistic liquefaction triggering analysis using CPT. *Journal of Geotechnical and Geoenvironmental Engineering*, 132(3), 337–350.

Keller, G.V. and Frischknecht, F.C. 1966. *Electrical Methods in Geophysical Prospecting*. Pergamon Press, New York.

Ladd, C.C. and DeGroot, D.J. 2003. Recommended practice for soft ground site characterization, Arthur Casagrande Lecture. *Proceedings, 12th Pan-American Conference on Soil Mechanics and Geotechnical Engineering*, Cambridge, MA, pp. 3–52.

Lee, J.T., Wang, C.C., Ho, Y.T., and Huang, A.B. 2013. Characterization of reservoir sediment with differential pressure sensored flat dilatometer and piezo-penetrometer. *Acta Geotechnica*, 8, 373–380. doi: 10.1007/s11440-012-0188-1.

Liao, S.S.C. and Whitman, R.V. 1985. Overburden correction factors for SPT in sand. *Journal of Geotechnical Engineering Division, ASCE*, 112(3), 373–377.

Lunne, T., Robertson, P.K., and Powell, J.J.M. 1997. *Cone Penetration Testing in Geotechnical Practice*. Blackie Academic/London; Routledge, New York, 312p.

Marchetti, S. and Crapps, D.K. 1981. Flat dilatometer manual, Internal Report of G.P.E. Inc.

Marchetti, S. 1980. *In situ* tests by flat dilatometer. *Journal of Geotechnical Engineering Division, ASCE*, 106(GT3), 299–321.

Marchetti, S. 1997. The Flat Dilatometer: Design applications. Keynote lecture. *Proceedings Third International Geotechnical Engineering Conference*. Cairo University, pp. 421–448.

Marchetti, S. and Totani, G. 1989. c_h evaluations from DMTA dissipation curves. *Proceedings, XII International Conference on Soil Mechanics and Foundation Engineering*, Rio de Janeiro, A.A. Balkema, Vol. 1, pp. 281–286.

Marchetti, S. 2014. The seismic dilatometer for *in situ* soil investigations. *Proceedings Indian Geotechnical Conference*, Kakinada, India.

Marchetti, S., Monaco, P., Totani, G., and Calabrese, M. 2001. The Flat Dilatometer Test (DMT) in soil investigations, A Report by the ISSMGE Committee TC16. May 2001, 41p.

Massarsch, K.R. 2014. Cone penetration testing—A historic perspective. *Proceedings, CPT14*, Paper #KN-4, Las Vegas, NV.

Mayne, P.W., Christopher, B.R., and DeJong, J. 2002. Subsurface investigations. Report No. FHWA-NHI-01-031, National Highway Institute Federal Highway Administration U.S. Department of Transportation, Washington, DC, 332p.

Mayne, P.W. 1991. Determination of OCR in clays by piezocone tests using cavity expansion and critical state concepts. *Soils and Foundations*, 31(2), 65–76.

Mayne, P.W. 2014. Interpretation of geotechnical parameters from seismic piezocone tests. *Proceedings, CPT14*, Paper #KN-2, Las Vegas, NV.

Ménard, L. 1957. An apparatus for measuring the strength of soils in place. Thesis, University of Illinois.

Moore, W. 1951. *Earth resistivity tests applied to subsurface reconnaissance surveys*. ASTM Special Technical Publication, 122, ASTM International, West Conshohocken, PA, pp. 89–103.

Nigbor, R.L. and Imai, T. 1994. The suspension P-S velocity logging method. *Geophysical Characteristics of Sites, ISSMFE, Technical Committee 10 for XIII ICSMFE*, International Science Publishers, New York, pp. 57–63.

Osterberg, J.O. 1952. New piston-type soil sampler. *Engineering News Record*, April 24.

Park, C.B., Miller, R.D., and Xia, J. 1999. Multichannel analysis of surface waves. *Geophysics*, 64(3), 800–808.

Ramsey, N. 2002. A calibrated model for the interpretation of cone penetration tests (CPTs) in North Sea Quaternary soils. Offshore site investigation and geotechnics—Diversity and sustainability. London, Society for Underwater Technology, pp. 341–356.

Robertson, P.K. 1990. Soil classification using the cone penetration test. *Canadian Geotechnical Journal*, 27(1), 151–158.

Robertson, P.K., Campanella, R.G., Gillespie, D., and Greig, J. 1986. Use of piezometer cone data. *In-Situ'86 Use of In-Situ Testing in Geotechnical Engineering, GSP 6, ASCE*, Reston, VA, Specialty Publication, SM 92, pp. 1263–1280.

Robertson, P.K. and Campanella, R.G. 1983. Interpretation of cone penetration tests: Part I - sands; Part II - clays. *Canadian Geotechnical Journal*, 20(4), 719–745.

Robertson, P.K. 2009. Interpretation of cone penetration tests—A unified approach. *Canadian Geotechnical Journal*, 49(11), 1337–1355.

Robertson, P.K. 2016. Cone penetration test (CPT)-based soil behaviour type (SBT) classification system—An update. *Canadian Geotechnical Journal*, 53, 1910–1927.

Robertson, P.K. and Wride, C.E. 1998. Evaluating cyclic liquefaction potential using the cone penetration test. *Canadian Geotechnical Journal*, 35, 442–459.

Schmertmann, J. 1977. Guidelines for cone penetration test—Performance and design, U.S. Department of Transportation, FHWA-TS-78-209, Federal Highway Administration, Washington, DC, 145p.

Schnaid, F. 2009. *In Situ Testing in Geomechanics*. Taylor & Francis, Boca Raton, FL, 329p.

Schneider, J.A., Randolph, M.F., Mayne, P.W., and Ramsey, N.R. 2008. Analysis of factors influencing soil classification using normalized piezocone tip resistance and pore pressure parameters. *Journal of Geotechnical and Geoenvironmental Engineering, ASCE*, 134(11), 1569–1586.

Schneider, J.A., Hotstream, J.N., Mayne, P.W., and Randolph, M.F. 2012. Comparing CPTU Q–F and Q–$\Delta u_2/\sigma'_{vo}$ soil classification charts. *Géotechnique Letters*, 2(4), 209–215.

Seed, H.B. and Idriss, I.M. 1982. *Ground Motions and Soil Liquefaction during Earthquakes.* Earthquake Engineering Research Center, University of California, Berkeley.

Seed, H.B., Tokimatsu, K., Harder, L.F., and Chung, R. 1985. Influence of SPT procedures in soil liquefaction resistance evaluations. *Journal of Geotechnical Engineering, ASCE*, 111(12), 1425–1445.

Stokoe, K.H. II and Nazarian, S. 1985. Use of Rayleigh waves in liquefaction studies. *Proceedings, Measurement and Use of Shear Wave Velocity for Evaluating Dynamic Soil Properties*, Geotechnical Engineering Division, ASCE, pp. 1–17.

Stokoe, K.H., Joh, S.H., and Woods, R.D. 2004. The contribution of *in situ* geophysical measurements to solving geotechnical engineering problems. *Proceedings of the Second International Conference on Site Characterization, ISC'2*, Porto, pp. 19–22.

Stroud, M.A. 1974. The Standard Penetration test in insensitive clays and soft rocks. *Proceedings, European Symposium on Penetration Testing*, Stockholm, Sweden, Vol. 2.2, pp. 367–375.

Stroud, M.A. 1989. *Standard Penetration Test: Introduction Part 2. Penetration Testing in the UK.* Thomas Telford, London, pp. 29–50.

Tanaka, H. 2008. Sampling and sample quality of soft clays. In: Huang, A.B. and Mayne, P.W., eds., *Proceedings, Geotechnical and Geophysical Site Characterization, Taipei*, Taylor & Francis Group, London, pp. 139–157.

Teh, C.I. and Houlsby, G.T. 1991. An analytical study of the cone penetration test in clay. *Geotechnique*, 41(1), 17–34.

Telford, W.M., Geldart, L.P., and Sheriff, R.E. 1990. *Applied Geophysics.* Cambridge University Press, New York.

Terzaghi, K. and Peck, R.B. 1967. *Soil Mechanics in Engineering Practice.* Wiley, New York, 729p.

Torstensson, B.A. 1975. Pore pressure sounding instrument. *Proceedings, ASCE Specialty Conference on In-Situ Measurement of Soil Properties*, Raleigh, NC, Vol. 3, pp. 48–54.

Tsai, P., Lee, D., Kung, G.T., and Juang, C.H. 2009. Simplified DMT-based methods for evaluating liquefaction resistance of soils. *Engineering Geology*, 103, 13–22.

US Army Corps of Engineers. 1995. Geophysical exploration for engineering and environmental investigations, Engineering Manual No. 1110-1-1802, Washington, DC, 208p.

Winterkorn, H.F. and Fang, H.Y. 1975. *Foundation Engineering Handbook.* Van Nostrand Reinhold Company, New York, 751p.

Wissa, A.E.Z., Martin, R.T., and Garlanger, J.E. 1975. The piezometer probe. *Proceedings, ASCE Specialty Conference on In-Situ Measurement of Soil Properties*, Raleigh, NC, Vol. 1, pp. 536–545.

Wroth, C.P. 1984. The interpretation of in-situ soil tests. *Geotechnique*, 34, 449–489.

Wroth, C.P. 1988. Penetration testing—A more rigorous approach to interpretation. *Penetration Testing 1988, Proceedings ISOPT, Orlando*, Balkema, Rotterdam, The Netherlands, Vol. 1, pp. 303–311.

Youd, T.L., Idriss, I.M., Andrus, R.D., Arango, I., Castro, G., Christian, J.T., Dobry, R. et al. 2001. Liquefaction resistance of soils: Summary report from the 1996 NCEER and 1998 NCEER/NSF workshops on evaluation of liquefaction resistance of soils. *Journal of Geotechnical and Geoenvironmental Engineering, ASCE*, 127(10), 817–833.

Yu, H.S. 1994. State parameter from self-boring pressuremeter tests in sand, *Journal of Geotechnical Engineering Division, ASCE*, 120(12), 2118–2135.

Yu, H.S. 2004. *In situ* testing: From mechanics to prediction, The James K. Mitchell Lecture. *Proceedings Second International Conference on Site Characterization*, Milpress, Porto, Vol. 1, pp. 3–38.

Chapter 4

Shallow foundations

Bearing capacity and settlement

4.1 INTRODUCTION

The design of a foundation requires the knowledge of superstructure load, ground conditions, and soil foundation interaction. Figure 4.1 shows a flow chart that describes the general procedure in the analysis/design and construction of foundations. Information regarding the structural system (i.e., a building or a bridge to be built with steel frame or reinforced concrete) and loading conditions (i.e., live and dead load in the vertical and lateral direction) at the foundation level are usually provided by the structural engineers. It should be noted that there may be occasions where we need to perform ground improvement to either densify or solidify the ground material for cost-effective construction of any kind of foundation system. Details of ground improvement are not covered in this book.

Commonly used foundations may be divided into two categories: shallow foundations and deep foundations. Typically, shallow foundations refer to those with an embedment depth equal to or less than four times the foundation width. They are suitable for the ground with relatively strong soil layers immediately below the superstructure. Otherwise, deep foundations (such as pile foundations) will be required in order to transmit the superstructure load to stronger, deep soil layers. In addition to the type(s) and dimensions of the foundation system, the design should be constructible at a reasonable cost, and the completed foundation should meet the performance requirements on a long-term basis.

To design a shallow foundation on soil, analytical or numerical methods will be required to solve soil–foundation interaction primarily to determine the bearing capacity of the foundation (i.e., the maximum pressure that the foundation can sustain before shear failure of the soil underneath the foundation occurs). In addition, the deformation or settlement of the foundation under the working load will be estimated to ensure that it is not excessive and therefore satisfies the serviceability criterion. These two analyses are commonly known as stability and deformation calculations, respectively.

This chapter is devoted to the application of the most suitable analytical methods for stability and deformation analysis of shallow foundations in soils. For all soil types, stability calculations can be carried out using limit state analysis discussed in Chapter 2. For deformation calculations, we need to treat granular soils (such as sands) and saturated clays differently. This is because for granular soil, due to its high permeability, it is common and reasonable to use theory of elasticity to estimate the foundation settlement which occurs quickly after construction. For saturated clays, however, there is additional settlement due to soil consolidation. This consolidation settlement occurs over time and can be estimated from the theory of consolidation described in Chapter 1. The contents of this chapter are as follows:

- Introduction of the theories of ultimate bearing capacity for strip foundations. By assuming the foundation as infinitely long, the bearing capacity failure can be

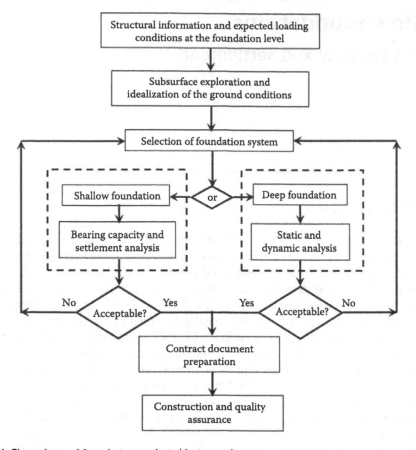

Figure 4.1 Flow chart of foundation analysis/design and construction.

simplified as a plane strain problem. Using this simplified system, we can explore the mechanisms and parameters involved in the determination of ultimate bearing capacity for shallow strip foundations.

• Derivation of the bearing capacity factors for strip foundations in cohesive and cohesive-frictional soils is introduced using the limit equilibrium and upper-bound limit analysis method that we learned in Chapter 2.

• Description of the methods to determine the ultimate bearing capacity for shallow foundations with finite dimensions such as circular or rectangular foundations. The introduction starts with cohesive soils where the problem can be approached with rigorous upper-bound limit analysis.

• A general form of ultimate bearing capacity equation is introduced. This equation can be applied to cohesive-frictional soils and considers the effects of foundation shape, depth of embedment, and inclination of the foundation load, by applying a series of empirical correction factors.

• The effects of loading eccentricity, groundwater and drainage conditions, methods to accommodate these effects, as well as the concept of safety factor are introduced. At this stage, we will have a practical and complete system to analyze shallow foundations from a stability point of view.

• The remainder of the chapter concentrates on deformation or performance analysis. We begin with elastic settlement analysis methods where the soil is assumed as linear elastic. The methods apply to both cohesive and granular soils.

- The procedure to estimate consolidation settlement for foundations underlain by compressible cohesive soils is reviewed. The settlement is a result of primary consolidation and secondary consolidation of the affected cohesive soil layer. This is a review, as details are usually covered in the soil mechanics class.
- Due to the sensitivity of soil compressibility to confining stress and drastically higher permeability, foundation settlement in granular soils is dealt with separately. A section is dedicated to the estimation of foundation settlement in granular soils.

4.2 ULTIMATE BEARING CAPACITY OF STRIP FOUNDATIONS

In practice, foundations in soils with an embedment depth, D_f, less than four times the foundation width, B, are often classified as shallow foundations. Figure 4.2a shows a cross-sectional view of a shallow strip foundation (a foundation that is infinitely long in the direction that is perpendicular to the paper) subjected to a vertical load. It is obvious that the settlement of the foundation will increase with the applied vertical load. When the vertical load is increased to certain level, the foundation will collapse due to shear failure of the soil supporting it. To ensure stability in foundation design, it is most important that for a given soil condition, we can predict or estimate the level of load, Q_u, at which the foundation collapse would occur, and the corresponding pressure (Q_u divided by the foundation area) is referred to as the ultimate bearing capacity q_u (see Figure 4.2b). Note q_u is equivalent to the collapse pressure (also called q_u) described in Chapter 2.

To simplify the problem, it is common to assume that the strip foundation is a plane strain problem and that the influence of the soil above the foundation base level is approximately represented by a surcharge q that equals the soil unit weight γ multiplied by the embedment depth D_f. This approximation is presented in Figure 4.2b.

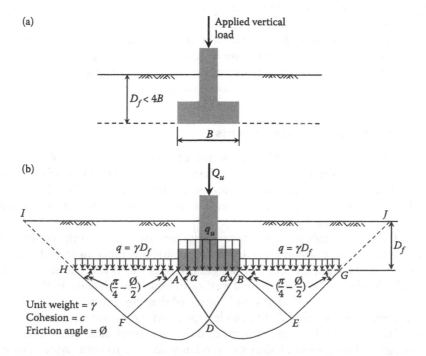

Figure 4.2 (a) A shallow foundation in soil. (b) Simplified shallow foundation failure mechanism.

For a general cohesive-frictional soil with self-weight, the ultimate bearing capacity of the shallow foundation will depend on soil strength parameters (i.e., cohesion and internal friction angle), soil unit weight, and the embedment depth (i.e., surcharge in Figure 4.2b). Strictly speaking, the solution of the problem may also be dependent on the friction of soil–foundation interface.

Terzaghi (1943) first analyzed this problem using the limit equilibrium method in which the contributions from soil cohesion, surcharge, and soil unit weight are superimposed. The α in Figure 4.2b was assumed to be equal to $\varnothing$ in Terzaghi's derivation. The contribution of shear strength from soil above the foundation base level (i.e., along the slip lines HI and GJ, represented by dash lines in Figure 4.2b) was ignored in the analysis. Mathematically, these contributions are widely expressed by the following equation, known as the ultimate bearing capacity equation:

$$q_u = cN_c + qN_q + \frac{1}{2}\gamma BN_\gamma \qquad (4.1)$$

where
   c = soil cohesion
   γ = soil unit weight
   q = surcharge = γD_f

N_c, N_q, and N_γ are known as the bearing capacity factors. Details of bearing capacity factors will be introduced later. The theoretical justification for using superposition of the three terms to obtain q_u according to Equation 4.1 has been studied by Davis and Booker (1971), who suggested that it leads to conservative estimates of the ultimate bearing capacity q_u and should therefore result in a "safe" design.

As discussed in Chapter 2, different methods may be used to carry out stability calculations, and in particular the limit equilibrium method (Terzaghi, 1943) has been widely used in geotechnical practice. However, the limit equilibrium method is only an approximate approach. In the following sections, we apply the upper-bound method of limit analysis introduced in Chapter 2 to provide a comprehensive description of the derivation of bearing capacity factors and procedure to apply the bearing capacity equation. The procedure that we describe below is an extension of the work of Chen (1975) by including a surcharge load q (see Figure 4.2b) to account for the effects of foundation embedment.

4.2.1 Kinematic failure mechanisms

As discussed in Chapter 2, in order to use the upper-bound method of limit analysis, we need to assume a kinematic failure mechanism. For a strip foundation on a cohesive-frictional soil, two failure mechanisms, separately suggested by Hill (1950) and Prandtl (1920), have been used to give approximate solutions. These two failure mechanisms are shown in Figure 4.3.

Prandtl's failure mechanism is more appropriate for cases when no slip occurs along the soil–foundation interface, while Hill's failure mechanism is more suitable for a perfectly smooth soil–foundation interface. For weightless soils, however, the ultimate bearing capacities obtained from the upper-bound method of limit analysis using both failure mechanisms are identical (Chen, 1975). It should be noted that the contributions to the ultimate bearing capacity from the self-weight of soils depend on the failure mechanism adopted.

It is noted that both of these mechanisms contain a logspiral shear zone between the triangular wedges below the foundation and surcharge load. This shear zone consists of an infinite number of infinitesimal triangular sliding blocks that are formed by a family of

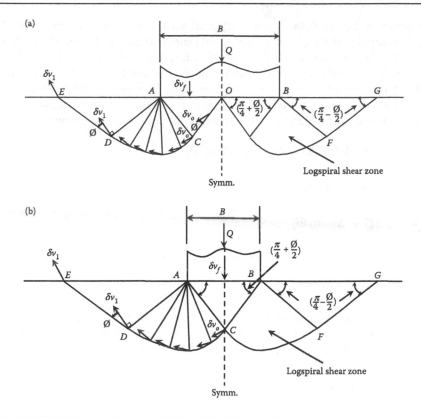

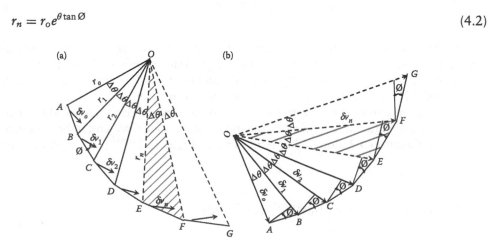

Figure 4.3 (a) Hill's failure mechanism. (b) Prandtl's failure mechanism.

concurrent straight lines (i.e., radial sliding) and logarithmic spirals (i.e., circumferential sliding).

To use upper-bound limit analysis, we need to calculate internal power dissipation within the logspiral shear zone through both radial and circumferential sliding. It is therefore necessary to discuss the procedure of this calculation.

For a logspiral shear zone, the radius–rotation relation is defined by the following equation (refer to Figure 4.4a):

$$r_n = r_o e^{\theta \tan \varnothing} \tag{4.2}$$

Figure 4.4 (a) A logspiral shear zone. (b) The velocity diagram.

For illustration, Figure 4.4a shows a logspiral shear zone that is approximated by six small rigid triangular sliding blocks with equal central angle $\Delta\theta$ (Chen, 1975). It is noted that along the shear surfaces AB, BC, CD, DE, EF, and FG, the associated plastic flow rule requires that the velocity of each rigid triangle will always be directed at a direction that makes an angle of the internal soil friction angle $\varnothing$ from the shear surface.

As illustrated by a compatible velocity diagram in Figure 4.4b (Atkinson 1981), the velocities at two adjacent triangles are related by the following equations:

$$\delta v_1 = \delta v_o + \delta v_o \Delta\theta \tan\varnothing = \delta v_o(1 + \Delta\theta \tan\varnothing) \tag{4.3}$$

$$\delta v_2 = \delta v_1(1 + \Delta\theta \tan\varnothing) \tag{4.4}$$

$$\delta v_n = \delta v_{n-1}(1 + \Delta\theta \tan\varnothing) \tag{4.5}$$

from which it follows that the velocity in the nth triangle OEF is

$$\delta v_n = \delta v_o(1 + \Delta\theta \tan\varnothing)^n \tag{4.6}$$

where δv_o denotes the velocity of the first triangle (i.e., the initial velocity).

When the number of triangles is increased to infinity, the logspiral shear zone will be recovered exactly. At the limiting case of an infinite n value, Equation 4.6 will reduce to

$$\delta v_n = \delta v_o(1 + \Delta\theta \tan\varnothing)^n = \delta v_o\left(1 + \frac{\theta}{n}\tan\varnothing\right)^n \rightarrow \delta v_o e^{\theta \tan\varnothing} \tag{4.7}$$

Now let's consider the internal power dissipation due to radial sliding along OC (see Figure 4.4a). According to Equation 2.8 described in Section 2.4, the power dissipation along OC should be as follows:

$$\delta W_{OC} = c \times r_2 \times (\delta v_1 \Delta\theta) = c r_2 \delta v_1 \Delta\theta \tag{4.8}$$

In a same way, the power dissipation along the sliding surface BC should be

$$\delta W_{BC} = c \times \frac{r_2 \Delta\theta}{\cos\varnothing} \times (\delta v_1 \cos\varnothing) = c r_2 \delta v_1 \Delta\theta \tag{4.9}$$

It is interesting to note that, for a rigid triangle, the power dissipation along the radial sliding is the same as that dissipated along the logspiral shear surface. This is a very important result that can be used to simplify upper-bound limit analysis of the ultimate bearing capacity of shallow foundations in cohesive-frictional soils.

As a result, the expression for internal power dissipation within the logspiral shear zone will be identical to the expression along the spiral shear surface, which can be obtained by integrating Equation 4.9 along a logspiral surface of $r_n = r_o e^{\theta \tan\varnothing}$ (i.e., similar to Equation 4.7):

$$\delta W_\theta = c \int_0^\theta r_n \delta v_n d\theta = c \int_0^\theta [r_0 e^{\theta \tan\varnothing}] \delta v_o e^{\theta \tan\varnothing} \, d\theta \tag{4.10}$$

$$= \frac{1}{2} c \delta v_0 r_0 \cot\varnothing (e^{2\theta \tan\varnothing} - 1)$$

4.2.2 Bearing capacity calculation neglecting self-weight of soil

For simplicity, we first consider the case when soil weight is neglected. We adopt the failure mechanism suggested by Hill (1950) in our determination of the ultimate bearing capacity. A similar procedure can be followed when Prandtl's failure mechanism is used.

If the foundation is moving downward with a velocity of δv_f, from the velocity diagram shown in Figure 4.5, we can determine the velocity of the wedge AOC of Figure 4.3a as follows:

$$\delta_{v_o} = \delta v_f \sec\left(\frac{\pi}{4} + \frac{1}{2}\varnothing\right) \tag{4.11}$$

In the logspiral shear zone ACD of Figure 4.3a, Equation 4.7 suggests that the velocity increases exponentially from the value δv_0 along the radial line CD as follows:

$$\delta v_1 = \delta v_o e^{\left(\frac{1}{2}\pi \tan \varnothing\right)} = \delta v_f \sec\left(\frac{\pi}{4} + \frac{1}{2}\varnothing\right) e^{\left(\frac{1}{2}\pi \tan \varnothing\right)} \tag{4.12}$$

In order to make use of the upper-bound method of limit analysis, we calculate the external work done by the bearing capacity of the foundation acting along OA with a vertical displacement increment of δv_f as follows:

$$\delta E_{OA} = q_u \times \frac{B}{2} \times \delta v_f = \frac{B}{2} q_u \cos\left(\frac{\pi}{4} + \frac{\varnothing}{2}\right) \delta v_o \tag{4.13}$$

In addition, the external work done by the soil surcharge acting along AE can be determined to be

$$
\begin{aligned}
\delta E_{AE} &= -q \times \left[\frac{B \cos\left(\frac{\pi}{4} - \frac{\varnothing}{2}\right)}{2 \cos\left(\frac{\pi}{4} + \frac{\varnothing}{2}\right)} e^{\left(\frac{1}{2}\pi \tan \varnothing\right)}\right] \times \left[\delta v_1 \cos\left(\frac{\pi}{4} - \frac{\varnothing}{2}\right)\right] \\
&= -\frac{qB \cos^2\left(\frac{\pi}{4} - \frac{\varnothing}{2}\right)}{2 \cos\left(\frac{\pi}{4} + \frac{\varnothing}{2}\right)} e^{(\pi \tan \varnothing)} \delta v_o
\end{aligned}
\tag{4.14}
$$

The total external work is the sum of Equations 4.13 and 4.14 and is given by

$$\delta E = \delta E_{OA} + \delta E_{AE} = \frac{Bq_u}{2} \cos\left(\frac{\pi}{4} + \frac{\varnothing}{2}\right) \delta v_o - \frac{qB \cos^2\left(\frac{\pi}{4} - \frac{\varnothing}{2}\right)}{2 \cos\left(\frac{\pi}{4} + \frac{\varnothing}{2}\right)} e^{(\pi \tan \varnothing)} \delta v_o \tag{4.15}$$

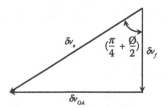

Figure 4.5 Velocity diagram based on Hill's failure mechanism.

The internal power dissipation occurs along sliding surfaces OC, CD, and DE, in which CD is the sliding surface along a logspiral shear zone.

The internal plastic power dissipation along the slip surface OC in the failure mechanism can be determined by

$$\delta W_{OC} = c \times \delta v_o \cos \varnothing \times \frac{B}{4 \cos\left(\frac{\pi}{4} + \frac{\varnothing}{2}\right)} = \frac{cB \cos \varnothing}{4 \cos\left(\frac{\pi}{4} + \frac{\varnothing}{2}\right)} \delta v_o \tag{4.16}$$

The expressions of power dissipation along the slip line CD can be obtained using Equation 4.10 as

$$\delta W_{CD} = \frac{cB \cot \varnothing}{8 \cos\left(\frac{\pi}{4} + \frac{\varnothing}{2}\right)} \left(e^{\pi \tan \varnothing} - 1\right)\delta v_o \tag{4.17}$$

As shown earlier, the power dissipation within the logspiral shear zone ACD through radial sliding is identical to that along the shear surface CD; hence

$$\delta W_{ACD} = \frac{cB \cot \varnothing}{8 \cos\left(\frac{\pi}{4} + \frac{\varnothing}{2}\right)} \left(e^{\pi \tan \varnothing} - 1\right)\delta v_o \tag{4.18}$$

Finally, the internal plastic power dissipation along the sliding surface DE can be determined as follows:

$$\delta W_{DE} = c \times \delta v_1 \cos \varnothing \times \frac{B}{4 \cos\left(\frac{\pi}{4} + \frac{\varnothing}{2}\right)} e^{\left(\frac{1}{2}\pi \tan \varnothing\right)} = c \times \delta v_o e^{\left(\frac{1}{2}\pi \tan \varnothing\right)} \cos \varnothing$$

$$\times \frac{B}{4 \cos\left(\frac{\pi}{4} + \frac{\varnothing}{2}\right)} e^{\left(\frac{1}{2}\pi \tan \varnothing\right)} = \frac{Bc \cos \varnothing}{4 \cos\left(\frac{\pi}{4} + \frac{\varnothing}{2}\right)} e^{\left(\pi \tan \varnothing\right)}\delta v_o \tag{4.19}$$

The total power dissipation is the sum of Equations 4.16 through 4.19, namely

$$\delta W = \frac{cB \cos \varnothing}{4 \cos\left(\frac{\pi}{4} + \frac{\varnothing}{2}\right)} \delta v_o + 2 \times \frac{cB \cot \varnothing}{8 \cos\left(\frac{\pi}{4} + \frac{\varnothing}{2}\right)} \left(e^{\pi \tan \varnothing} - 1\right)\delta v_o$$

$$+ \frac{Bc \cos \varnothing}{4 \cos\left(\frac{\pi}{4} + \frac{\varnothing}{2}\right)} e^{\left(\pi \tan \varnothing\right)}\delta v_o \tag{4.20}$$

According to the upper-bound theorem, we equate the external work of Equation 4.15 and the internal power dissipation of Equation 4.20, namely

$$\frac{Bq_u}{2} \cos\left(\frac{\pi}{4} + \frac{\varnothing}{2}\right)\delta v_o - \frac{qB\cos^2\left(\frac{\pi}{4} - \frac{\varnothing}{2}\right)}{2 \cos\left(\frac{\pi}{4} + \frac{\varnothing}{2}\right)} e^{\left(\pi \tan \varnothing\right)}\delta v_o = \frac{cB \cos \varnothing}{4 \cos\left(\frac{\pi}{4} + \frac{\varnothing}{2}\right)} \delta v_o$$

$$+ \frac{cB \cot \varnothing}{4 \cos\left(\frac{\pi}{4} + \frac{\varnothing}{2}\right)} \left(e^{\pi \tan \varnothing} - 1\right)\delta v_o + \frac{Bc \cos \varnothing}{4 \cos\left(\frac{\pi}{4} + \frac{\varnothing}{2}\right)} e^{\left(\pi \tan \varnothing\right)}\delta v_o \tag{4.21}$$

which gives the following expression for the ultimate bearing capacity of a shallow foundation on cohesive-frictional soils:

$$q_u = c \cot \varnothing \left[e^{(\pi \tan \varnothing)} \tan^2 \left(\frac{\pi}{4} + \frac{\varnothing}{2} \right) - 1 \right] + q e^{(\pi \tan \varnothing)} \tan^2 \left(\frac{\pi}{4} + \frac{\varnothing}{2} \right)$$
$$= c N_c + q N_q \tag{4.22}$$

where the bearing capacity factors are given by

$$N_q = e^{(\pi \tan \varnothing)} \tan^2 \left(\frac{\pi}{4} + \frac{\varnothing}{2} \right) \tag{4.23}$$

$$N_c = \cot \varnothing \left[e^{(\pi \tan \varnothing)} \tan^2 \left(\frac{\pi}{4} + \frac{\varnothing}{2} \right) - 1 \right] = \cot \varnothing (N_q - 1) \tag{4.24}$$

Note when $\varnothing = 0$, Equation 4.24 has no meaning. In this case, $N_c = (2 + \pi)$ and its derivation are described in Section 4.3. Or, if we plot N_c against $\varnothing$, you will find that $(2 + \pi)$ becomes an asymptotic value as $\varnothing$ approaches zero.

More importantly, Shield (1954) has shown that by extending satisfactorily the plastic stress field associated with the Prandtl mechanism into the remaining rigid regions below the shear surface, the upper-bound bearing capacity solution Equation 4.22 is also a lower bound for soils with an internal friction angle of less than 75°. It is therefore concluded that the bearing capacity factors of Equations 4.23 and 4.24 are exact bearing capacity solutions for shallow foundations on cohesive-frictional soils. This would be true regardless of the roughness of the soil–foundation interface.

EXAMPLE 4.1

Given

A strip foundation with a width of 3 m is located at a depth of 2 m in a cohesive-frictional soil. The cohesion of the soil c is 30 kPa and its internal friction angle $\varnothing$ is 30°. The unit weight of the soil is $\gamma = 17\,\text{kN/m}^3$. For simplicity, the overburden effect will be accounted for by using an equivalent surcharge, defined as the unit weight multiplied by embedment depth of the foundation.

Required

The ultimate bearing capacity of the strip foundation.

Solution

The surcharge is determined as follows:

$$q = \gamma D_f = 17\,\text{kN/m}^3 \times 2\,\text{m} = 34\,\text{kPa}$$

The bearing capacity factors are

$$N_q = e^{(\pi \tan \varnothing)} \tan^2 \left(\frac{\pi}{4} + \frac{\varnothing}{2} \right) = e^{(\pi \tan 30°)} \tan^2 \left(\frac{\pi}{4} + \frac{30°}{2} \right) = 18.4 \tag{4.23}$$

$$N_c = \cot \varnothing (N_q - 1) = \cot 30° (18.4 - 1) = 30.1 \tag{4.24}$$

The ultimate bearing capacity of the foundation is

$$q_u = cN_c + qN_q = (30)(30.1) + (34)(18.4) = 1530\,\text{kPa} \tag{4.22}$$

4.2.3 Bearing capacity calculation considering weight of soil

So far we have ignored the effect of soil weight in the determination of bearing capacity of shallow foundations on cohesive-frictional soils. This is partly for simplicity of illustration and partly due to the fact that the effect of soil weight is theoretically more difficult to assess and there has been no exact solution for it.

To illustrate how to include the effects of soil weight in the upper-bound calculation, we consider a case where both soil cohesion and soil surcharge are assumed to be zero, for simplicity. Once again, we use the Hill mechanism in our upper-bound analysis. Given soil cohesion is assumed to be zero, the internal power dissipation will be zero. All we need to determine will be the external work. This will include contributions from the bearing capacity pressure acting on the foundation and the weight of soil above the foundation base level.

The external work done by the bearing capacity of the foundation with a vertical displacement increment of δv_f (see Figure 4.3a) can be derived as follows:

$$\delta E_{OA} = \frac{B}{2} q_u \cos\left(\frac{\pi}{4} + \frac{\varnothing}{2}\right)\delta v_o \tag{4.25}$$

The external work done by the soil weight acting in the triangular blocks AOC and ADE is

$$\delta E_{AOC} = \frac{\gamma}{2}(r_0)^2 \cos\varnothing \cos\left(\frac{\pi}{4} + \frac{\varnothing}{2}\right)\delta v_o \tag{4.26}$$

$$\delta E_{ADE} = -\frac{\gamma}{2}(r_0)^2 e^{\left(\frac{3}{2}\pi\tan\varnothing\right)} \cos\varnothing \cos\left(\frac{\pi}{4} - \frac{\varnothing}{2}\right)\delta v_o \tag{4.27}$$

where r_0 is the length of AC and γ is the unit weight of soil. The external work done within the logspiral shear zone ACD is obtained by considering a small triangular block first and integrating throughout the whole shear zone as described above:

$$\delta E_{ACD} = \int_0^{\frac{\pi}{2}}\left\{-\frac{\gamma}{2}(r_n)^2 d\theta\left[\delta v\cos\left(\frac{3}{4}\pi - \frac{\varnothing}{2} - \theta\right)\right]\right\} \tag{4.28}$$

By using $r_n = r_0 e^{\theta\tan\varnothing}$ and $\delta v = \delta v e^{\theta\tan\varnothing}$, Equation 4.28 can be integrated to give

$$\delta E_{ACD} = \frac{-\gamma\delta v_o(r_0)^2}{2(1 + 9\tan^2\varnothing)}\left\{\left[3\tan\varnothing\sin\left(\frac{\pi}{4} + \frac{\varnothing}{2}\right) - \cos\left(\frac{\pi}{4} + \frac{\varnothing}{2}\right)\right]e^{\left(\frac{3}{2}\pi\tan\varnothing\right)}\right.$$
$$\left. + \left[3\tan\varnothing\cos\left(\frac{\pi}{4} + \frac{\varnothing}{2}\right) + \sin\left(\frac{\pi}{4} + \frac{\varnothing}{2}\right)\right]\right\} \tag{4.29}$$

Note that r_0 is linked to the foundation width B through the geometry of the failure mechanism as follows:

$$r_0 = \frac{B}{4\cos\left(\dfrac{\pi}{4} + \dfrac{\varnothing}{2}\right)} \tag{4.30}$$

The total external work is the sum of Equations 4.25 through 4.27 and 4.29. According to the upper-bound theorem, the external work is equal to internal power dissipation δW (which is zero in this case because soil cohesion is zero):

$$\delta E = \delta E_{OA} + \delta E_{AOC} + \delta E_{ADE} + \delta E_{ACD} = \delta W = 0 \tag{4.31}$$

which can be simplified to the following bearing capacity equation:

$$q_u = \frac{\gamma B}{2} N_\gamma \tag{4.32}$$

and the bearing capacity factor N_γ is

$$\begin{aligned}
N_\gamma = &\frac{1}{4}\tan\left(\frac{\pi}{4} + \frac{\varnothing}{2}\right)\left[\tan\left(\frac{\pi}{4} + \frac{\varnothing}{2}\right)e^{\left(\frac{3}{2}\pi\tan\varnothing\right)} - 1\right] \\
&+ \frac{3\sin\varnothing}{1 + 8\sin^2\varnothing}\left\{\left[\tan\left(\frac{\pi}{4} + \frac{\varnothing}{2}\right) - \frac{\cot\varnothing}{3}\right]e^{\left(\frac{3}{2}\pi\tan\varnothing\right)} + \tan\left(\frac{\pi}{4} + \frac{\varnothing}{2}\right)\frac{\cot\varnothing}{3} + 1\right\}
\end{aligned} \tag{4.33}$$

As stated by Chen (1975), the bearing capacity due to soil weight is sensitive to the roughness of soil–foundation interface. Hill's mechanism is suitable for a perfectly smooth soil–foundation interface, and Prandtl's mechanism does not allow soil–foundation slip and is therefore suitable for a perfectly rough soil–foundation interface. In fact, if we follow the above upper-bound approach but with Prandtl's mechanism, the bearing capacity obtained is exactly twice the solution of Equation 4.33 as derived from Hill's mechanism.

EXAMPLE 4.2

Given

Consider the same foundation problem as in Example 4.1. A strip foundation with a width of 3 m is located at a depth of 2 m in a cohesive-frictional soil. The cohesion of the soil c is 30 kPa and its internal friction angle $\varnothing$ is 30°. The unit weight of the soil is $\gamma = 17\ \text{kN/m}^3$.

Required

The additional bearing capacity of the foundation due to soil weight.

Solution

For the case with a perfectly smooth soil–foundation interface, we use Equation 4.33 to determine the bearing capacity factor with $\varnothing = 30°$:

$$N_\gamma = \frac{1}{4}\tan\left(\frac{\pi}{4}+\frac{\varnothing}{2}\right)\left[\tan\left(\frac{\pi}{4}+\frac{\varnothing}{2}\right)e^{(\frac{3}{2}\pi\tan\varnothing)} - 1\right]$$

$$+ \frac{3\sin\varnothing}{1+8\sin^2\varnothing}\left\{\left[\tan\left(\frac{\pi}{4}+\frac{\varnothing}{2}\right) - \frac{\cot\varnothing}{3}\right]e^{(\frac{3}{2}\pi\tan\varnothing)} + \tan\left(\frac{\pi}{4}+\frac{\varnothing}{2}\right)\frac{\cot\varnothing}{3} + 1\right\}$$

$$N_\gamma = 20.7$$

(4.33)

The ultimate bearing capacity due to soil weight can be determined using Equation 4.32:

$$q_u = \frac{\gamma B}{2}N_\gamma = \frac{(17)(3)}{2}(20.7) = 528.6 \text{ kPa}$$

(4.32)

For the case with a perfectly rough soil–footing interface, it is known that the bearing capacity factor will be twice that of a perfectly smooth soil–footing interface; therefore, we have

$$N_\gamma = 20.7 \times 2 = 41.4$$

The ultimate bearing capacity due to soil weight can then be determined using Equation 4.32 with the new N_γ:

$$q_u = \frac{\gamma B}{2}N_\gamma = \frac{(17)(3)}{2}(41.4) = 1057.2 \text{ kPa}$$

(4.32)

4.2.4 Effects of soil–foundation interface friction considering weight of soil

So far we have considered bearing capacity factors of foundations with either perfectly smooth (i.e., with no friction) or perfectly rough (i.e., with full friction) soil–foundation interfaces. In reality, most situations would be for soil–foundation interfaces with a finite friction.

As a simple solution, Chen (1975) suggests that the effect of finite soil–foundation interface friction may be taken into account by adding the internal power dissipation along the interface. Once again, we use the Hill mechanism, and the velocity along the soil–foundation interface will be

$$\delta v_{OA} = \sin\left(\frac{\pi}{4}+\frac{\varnothing}{2}\right)\delta v_o$$

(4.34)

Assume that the friction angle of the soil–foundation interface is $\varnothing_w$. The internal power dissipation due to base friction can be calculated as follows:

$$\delta W = \frac{B}{2}q_u\tan\varnothing_w\delta v_{OA} = \frac{B}{2}q_u\tan\varnothing_w\sin\left(\frac{\pi}{4}+\frac{\varnothing}{2}\right)\delta v_o$$

(4.35)

By assuming that the external work is equal to the internal power dissipation δW, we have

$$\delta E = \delta E_{OA} + \delta E_{AOC} + \delta E_{ADE} + \delta E_{ACD} = \delta W = \frac{B}{2} q_u \tan \varnothing_w \sin\left(\frac{\pi}{4} + \frac{\varnothing}{2}\right) \delta v_o \quad (4.36)$$

and this leads to the conclusion that the bearing capacity factor for a foundation with a finite base friction $\varnothing_w$ is greater than that for a perfectly smooth foundation by a factor which depends on both the base friction and soil friction, namely

$$q_u = \frac{\gamma B}{2} (N_\gamma)_f \quad (4.37)$$

where $(N_\gamma)_f$ is the bearing capacity factor for a finite base friction:

$$(N_\gamma)_f = \left\{ \frac{1}{1 - \tan \varnothing_w \tan\left(\frac{\pi}{4} + \frac{\varnothing}{2}\right)} \right\} N_\gamma \quad (4.38)$$

and N_γ is given by Equation 4.33.

It should be noted that the inclusion of a frictional interface in upper-bound analysis can be made more rigorously by using the generalized upper-bound theorem of Collins (1969). This more rigorous theorem has been applied by Yu and Sloan (1994) to solve foundation problems using a numerical formulation of upper-bound analysis. A complete lower- and upper-bound limit analysis of the bearing capacity factor N_γ for both smooth and rough foundations has been carried out by Sloan and Yu (1996) using a numerical formulation of bound theorems.

4.3 BEARING CAPACITY OF RECTANGULAR AND CIRCULAR FOUNDATIONS IN COHESIVE SOILS

The foundation problem that we have considered so far is only valid for a strip (i.e., very long rectangular) foundation. In reality, the shape of most foundations may be circular or rectangular. It is therefore important that we consider the bearing capacity of shallow foundations with limited length (L).

To illustrate the "end effects" on a foundation with limited length, we first consider an upper-bound analysis of a rectangular foundation on a purely cohesive soil $(\varnothing = 0)$. Figure 4.6 shows a shallow rectangular foundation having a width of B and length L, on a purely cohesive soil and with a surcharge q outside of the foundation (Chen, 1975).

In order to carry out upper-bound analysis, we have to assume a failure mechanism. Following Chen (1975), we assume that Hill's mechanism will be valid during the failure of the foundation. The displacement diagram at failure can be shown in Figure 4.7.

It can be seen from Figure 4.7 that the velocity of soil block abc is δv; then the velocity for the translational radial shear zone bcd will also be δv. The triangular block bde will also move along de with a velocity of δv. It is also easy to show that the downward velocity of the block acb is $\delta v/\sqrt{2}$. The upward velocity of the block bde will also be $\delta v/\sqrt{2}$.

In order to make use of the upper-bound method of limit analysis, we calculate the external work done by the bearing capacity of the foundation on line ab with a vertical

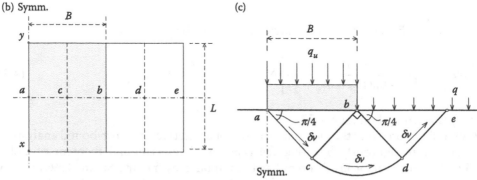

Figure 4.6 Hill's failure mechanism for a shallow foundation on cohesive soil with surcharge. (a) Failure mechanism for rectangular foundation. (b) Plan view. (c) Side view.

displacement increment of v_f as follows:

$$\delta E_{ab} = q_u \times \frac{B}{2} \times L \times \delta v_f = \frac{BL}{2} q_u \times \frac{\delta v}{\sqrt{2}} = \frac{BL}{2\sqrt{2}} q_u \delta v \qquad (4.39)$$

In addition, the external work done by the soil surcharge q acting along line be can be determined as

$$\delta E_{be} = q \times \frac{B}{2} \times L \times \left(-\frac{\delta v}{\sqrt{2}}\right) = -\frac{BL}{2\sqrt{2}} q \delta v \qquad (4.40)$$

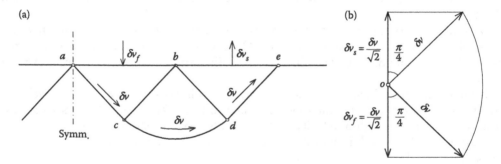

Figure 4.7 Velocity diagram based on Hill's failure mechanism. (a) Failure mechanism. (b) The velocity diagram.

The total external work is the sum of Equations 4.39 and 4.40 and is given by

$$\delta E = \delta E_{ab} + \delta E_{be} = \frac{BL}{2\sqrt{2}} q_u \delta v - \frac{BL}{2\sqrt{2}} q \delta v \tag{4.41}$$

The internal power dissipation occurs along sliding surfaces ac, cd, and de. In addition, internal plastic power will dissipate within the radial shear zone, which has been shown to be the same as that dissipated along the shear surface cd.

The internal plastic power dissipation along the slip surface ac, cd, and de and the radial shear zone bcd in the failure mechanism can be determined by

$$\delta W_{slip} = 2c\delta v \left(\frac{B}{2\sqrt{2}} \times L \right) + 2c\delta v \left(\frac{\pi}{2} \times \frac{B}{2\sqrt{2}} \times L \right) \tag{4.42}$$

The internal plastic power dissipation also takes place at the both end surfaces (at top and bottom of the foundation in plan view shown in Figure 4.6b) of the failure mechanism, which can be determined by

$$\delta W_{end} = 2c\delta v \left(\frac{B}{2\sqrt{2}} \times \frac{B}{2\sqrt{2}} \right) + 2c\delta v \left(\frac{\pi}{4} \times \frac{B^2}{8} \right) \tag{4.43}$$

If the soil–foundation interface has a friction angle of $\varnothing_w$, then the power dissipated along the soil–foundation interface will be

$$\delta W_{interface} = \frac{BL}{2} (q_u \times \tan \varnothing_w) \times \frac{\delta v}{\sqrt{2}} = \frac{BL}{2\sqrt{2}} q_u \delta v \tan \varnothing_w \tag{4.44}$$

By assuming that the external work is equal to the internal power dissipation δW, we have

$$\delta E = \delta W_{slip} + \delta W_{end} + \delta W_{interface} \tag{4.45}$$

which can be rearranged, by using Equations 4.41 through 4.44, to give the following bearing capacity expression:

$$q_u = \frac{1}{1 - \tan \varnothing_w} \left[(2 + \pi)c + q + \frac{(4 + \pi)\sqrt{2}}{8} \times \frac{B}{L} c \right] = cN_c + qN_q \tag{4.46}$$

where

$$N_c = \frac{1}{1 - \tan \varnothing_w} \left[2 + \pi + \frac{(4 + \pi)\sqrt{2}}{8} \times \frac{B}{L} \right] \tag{4.47}$$

$$N_q = \frac{1}{1 - \tan \varnothing_w} \tag{4.48}$$

For the special case of an infinitely long strip foundation (i.e., $B/L = 0$), Equation 4.47 reduces to the bearing capacity factor of strip foundations with $\varnothing = 0$ and considers the effects of soil–foundation interface friction, $\varnothing_w$, as follows:

$$N_c = \frac{(2+\pi)}{1 - \tan \varnothing_w} \tag{4.49}$$

and if $\varnothing_w = 0$, the above equation becomes

$$N_c = (2+\pi) \tag{4.50}$$

EXAMPLE 4.3

Given

Consider a rectangular foundation with a width of 3 m and a length of 6 m located at a depth of 2 m in a purely cohesive soil. The cohesion of the soil is measured to be 30 kPa and the unit weight of the soil is $\gamma = 17\,\text{kN/m}^3$.

Required

The ultimate bearing capacity of the foundation assuming a perfectly smooth soil-foundation interface.

Solution

The ultimate bearing capacity of a rectangular foundation in a cohesive soil can be determined as follows:

$$q_u = cN_c + qN_q \tag{4.46}$$

where the surcharge q can be estimated by

$$q = \gamma D_f = 17 \times 2 = 34\,\text{kPa}$$

The ultimate bearing capacity factors for a smooth soil–foundation interface ($\varnothing_w = 0$) are

$$N_c = \frac{1}{1 - \tan \varnothing_w}\left[2 + \pi + \frac{(4+\pi)\sqrt{2}}{8} \times \frac{B}{L}\right] \tag{4.47}$$

$$N_c = \frac{1}{1 - \tan(0)}\left[2 + \pi + \frac{(4+\pi)\sqrt{2}}{8}\left(\frac{3}{6}\right)\right] = 5.77$$

$$N_q = \frac{1}{1 - \tan \varnothing_w} = \frac{1}{1 - \tan(0)} = 1 \tag{4.48}$$

Therefore, the ultimate bearing capacity is

$$q_u = cN_c + qN_q = (30)(5.77) + (34)(1) = 207\,\text{kPa} \tag{4.46}$$

4.3.1 Bearing capacity of square and circular foundations on cohesive soil

For a smooth square foundation ($\emptyset_w = 0$) when $B = L$, for example, Equation 4.46 predicts a higher q_u (note qN_q is ignored for the following discussion) as follows:

$$q_u = c[3.26 + \pi] \tag{4.51}$$

It is noted that Hill's failure mechanism, shown in Figure 4.7, is less accurate for a square foundation. A better failure mechanism can be obtained by modifying Hill's mechanism, as suggested by Shield and Drucker (1953). For simplicity, we only consider the case without surcharge and soil–foundation interface friction.

The modified Hill mechanism assumed by Shield and Drucker (1953) is shown in Figure 4.8. Basically, when the foundation is moving downward with the initial velocity δv, the square is divided into four equal triangles (i.e., *col*, *con*, *cmn*, and *cml*), each of which will move downward, accommodated by lateral movement as well. The geometry of the failure mechanism is defined by angles of α and β (see Figure 4.8b).

Following the same upper-bound analysis procedure as detailed in Chen (1975), it can be shown that the bearing capacity expression is given by

$$q_u = c\left[\alpha + \beta + \sqrt{1 + (\sin\beta)^2}(\alpha + \beta + \cot\alpha + \cot\beta)\right] \tag{4.52}$$

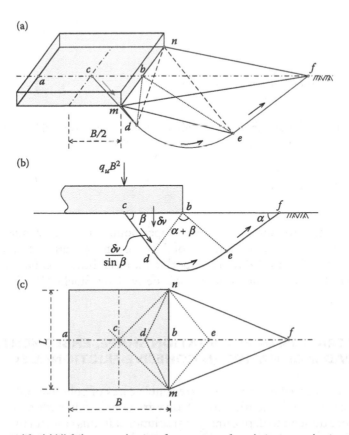

Figure 4.8 The modified Hill failure mechanism for a square foundation on cohesive soils. (a) Failure mechanism for the square foundation. (b) Vertical section. (c) Plan view.

which has a minimum q_u value of $5.8c$ when α and β are around $47°$ and $34°$, respectively. This is obviously a better (i.e., lower) upper bound than $6.4c$ as predicted from Equation 4.51 using Hill's original failure mechanism.

It may be argued that the bearing capacity of a circular foundation would be similar to that of a square foundation (i.e., $B = L =$ diameter of the foundation). This is confirmed to some extent by the fact that the exact bearing capacity of a smooth circular foundation on cohesive soil was derived as $5.69c$ by Shield (1955) using a slip line method.

By extending the square foundation failure mechanism shown in Figure 4.8, Shield and Drucker (1953) proposed an upper-bound solution for the bearing capacity of a rectangular foundation on cohesive soil. It follows the same consideration that, as the foundation is moving downward with the initial velocity δv, the rectangle is divided into four triangles, each of which will move downward, accommodated by lateral movement. The geometry of the failure mechanism along the direction of foundation width is defined by angles of α and β, as in the case of square foundations. Two additional angles, α_1 and β_1, are introduced to define the failure mechanism along the direction of foundation length, which would be different from that along the direction of foundation width.

Using the upper-bound theorem, the bearing capacity pressure for a rectangular foundation can be expressed as a function of the four unknown variables α, β, α_1, and β_1. Following this approach, the best upper bound can be defined by

$$q_u = c\left(5.24 + 0.47\frac{B}{L}\right) \quad \text{for } \frac{B}{L} \geq 0.53 \tag{4.53}$$

and

$$q_u = c\left(5.14 + 0.66\frac{B}{L}\right) \quad \text{for } \frac{B}{L} < 0.53 \tag{4.54}$$

which are lower (i.e., better) than that predicted using Equation 4.46 for a smooth rectangular foundation, namely

$$q_u = c\left(5.14 + 1.26\frac{B}{L}\right) \tag{4.55}$$

It is also interesting to note that the new upper-bound solution, defined by Equation 4.53, gives a slightly better upper solution of $5.71c$ for the special case of a square foundation on cohesive soil. This is almost identical to the exact bearing capacity of $5.69c$ for a smooth circular foundation on cohesive soil derived by Shield (1955) using the slip line method.

4.4 CONSIDERATION OF FOUNDATION SHAPE, EMBEDMENT, AND LOAD INCLINATION IN COHESIVE-FRICTIONAL SOILS

What we have considered in Section 4.3 are simple cases of cohesive soils ($\varnothing = 0$). For a more general case of a foundation in cohesive-frictional soils with surcharge and soil weight, it is more difficult to derive analytical solutions. In this case, many semi-empirical expressions of correction factors have been proposed to account for the effects of foundation shape, embedment depth, and inclination of the applied load on the foundation.

A more generalized form of ultimate bearing capacity equation evolved from the development by Meyerhof (1963) is described below:

$$q_u = cN_c F_{cs}F_{cd}F_{ci} + qN_q F_{qs}F_{qd}F_{qi} + \frac{1}{2}\gamma BN_\gamma F_{\gamma s}F_{\gamma d}F_{\gamma i} \tag{4.56}$$

The bearing capacity factors described previously for strip foundations with frictionless soil–foundation interface can be used, and are repeated as follows:

$$N_q = \tan^2\left(\frac{\pi}{4}+\frac{\varnothing}{2}\right)e^{\pi\tan\varnothing} \tag{4.23}$$

$$N_c = (N_q - 1)\cot\varnothing \tag{4.24}$$

$$N_\gamma = \frac{1}{4}\tan\left(\frac{\pi}{4}+\frac{\varnothing}{2}\right)\left[\tan\left(\frac{\pi}{4}+\frac{\varnothing}{2}\right)e^{\left(\frac{3}{2}\pi\tan\varnothing\right)} - 1\right]$$
$$+ \frac{3\sin\varnothing}{1+8\sin^2\varnothing}\left\{\left[\tan\left(\frac{\pi}{4}+\frac{\varnothing}{2}\right) - \frac{\cot\varnothing}{3}\right]e^{\left(\frac{3}{2}\pi\tan\varnothing\right)} + \tan\left(\frac{\pi}{4}+\frac{\varnothing}{2}\right)\frac{\cot\varnothing}{3}+1\right\} \tag{4.33}$$

Again, when $\varnothing = 0$, Equation 4.24 has no meaning. In this case, $N_c = (2 + \pi)$ as derived in Section 4.3. It should be noted that many equations for N_γ have been proposed. For example, an abbreviated version for N_γ has been reported by Vesic (1973) as

$$N_\gamma = 2(N_q + 1)\tan\varnothing \tag{4.57}$$

Equation 4.57 generates a slightly higher N_γ than Equation 4.33, but the difference is minimal and can be considered identical for practical purposes.

Three dimensionless correction factors are attached to each of the three bearing capacity factors in Equation 4.56 to account for the effects of foundation shape (i.e., F_{cs}, F_{qs}, and $F_{\gamma s}$), embedment depth (i.e., F_{cd}, F_{qd}, and $F_{\gamma d}$), and inclination of the load applied to the foundation (i.e., F_{ci}, F_{qi}, and $F_{\gamma i}$). Note q_u represents vertical component of the ultimate bearing capacity when the applied load is inclined. The correction factors were proposed by various researchers in the past few decades. With these correction factors, the applications of Equation 4.56 can be extended to square or circular foundations and cases with inclined loading conditions. For these reasons, Equation 4.56 is routinely used in engineering practice and is the focus of discussion in the remainder of the chapter. The following sections provide details of these correction factors.

4.4.1 Consideration of foundation shapes, inclined loading, and embedment depth

It is common to build shallow foundations with different shapes, and the load exerted on the foundation may be inclined. Many correction factors to consider these conditions have been developed based on experimental studies. The bearing capacity expression of foundations considered so far ignores the shearing resistance of the overburden soil above the foundation base level. This may be justified for cases when the overburden soil is much weaker than the bearing stratum. In some cases, however, the effect of shearing resistance of the overburden is considerable and may be taken into account in a semi-analytical manner. This is achieved by applying three further dimensionless correction factors to the bearing capacity equation to account for the effects of overburden soil.

Table 4.1 summarizes the correction factors described above. The degree of load inclination is measured by an angle α in degrees, as shown in Figure 4.9. When $\alpha = 0°$, the loading is vertical.

Where appropriate, these factors should be used when using Equation 4.56.

Note that for both square and circular foundations, the shape correction factors are obtained by simply setting $B = L$ into the above shape correction equations. It is

Table 4.1 Correction factors for foundation shapes, inclined loading, and embedment depth

Equation		Reference
Shape correction factors		
$F_{cs} = 1 + \dfrac{B}{L} \times \dfrac{N_q}{N_c}$	(4.58)	
$F_{qs} = 1 + \dfrac{B}{L} \times \tan \varnothing$	(4.59)	Vesic (1973)
$F_{\gamma s} = 1 - 0.4\dfrac{B}{L}$	(4.60)	
Load inclination correction factors (see Figure 4.9 for definition of α)		
$F_{ci} = \left(1 - \dfrac{\alpha}{90°}\right)^2$	(4.61)	
$F_{qi} = \left(1 - \dfrac{\alpha}{90°}\right)^2$	(4.62)	Meyerhof (1963)
$F_{\gamma i} = \left(1 - \dfrac{\alpha}{\varnothing}\right)^2$	(4.63)	
Embedment depth correction factors		
For cases when $(D_f/B) \leq 1$ $(D_f =$ embedment depth of foundation)		
Purely cohesive soils		
$F_{cd} = 1 + 0.4\dfrac{D_f}{B}$	(4.64)	
$F_{qd} = 1$	(4.65)	
$F_{\gamma d} = 1$	(4.66)	
Cohesive-frictional soils		
$F_{qd} = 1 + 2\tan \varnothing(1 - \sin \varnothing)^2 \dfrac{D_f}{B}$	(4.67)	
$F_{cd} = F_{qd} - \dfrac{1 - F_{qd}}{N_c \tan \varnothing}$	(4.68)	Brinch Hansen (1970)
$F_{\gamma d} = 1$	(4.69)	
For cases when $(D_f/B) > 1$		
Purely cohesive soils		
$F_{cd} = 1 + 0.4\tan^{-1}\left(\dfrac{D_f}{B}\right)$	(4.70)	
$F_{qd} = 1$	(4.71)	
$F_{\gamma d} = 1$	(4.72)	
Cohesive-frictional soils		
$F_{qd} = 1 + 2\tan \varnothing(1 - \sin \varnothing)^2 \tan^{-1}\left(\dfrac{D_f}{B}\right)$	(4.73)	
$F_{cd} = F_{qd} - \dfrac{1 - F_{qd}}{N_c \tan \varnothing}$	(4.74)	
$F_{\gamma d} = 1$	(4.75)	

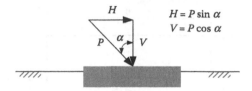

Figure 4.9 A foundation subjected to an inclined loading.

interesting to note that for the special case of cohesive soils with no soil weight and surcharge,

$$N_c F_{cs} = \left(5.14 + \frac{B}{L}\right) \qquad (4.76)$$

which is comparable to the analytical solutions defined by Equation 4.55.

4.5 BEARING CAPACITY OF ECCENTRICALLY LOADED FOUNDATIONS

The bearing capacity of foundations subject to eccentric loading is normally determined by using a reduced or effective foundation area for the calculation of the bearing capacity. Let's start with a simple case of a strip foundation that has a width B as shown in Figure 4.10a. The foundation is subjected to a design vertical load P and a moment M_B at the central location. The loading is assumed to be equivalent to a foundation subjected to a vertical load P located at a distance of e_B from the center of the foundation as shown in Figure 4.10b. The offset of the loading point or eccentricity e_B can be defined by

$$e_B = \frac{M_B}{P} \qquad (4.77)$$

The eccentricity causes a non-uniform bearing pressure distribution at the foundation base as shown in Figure 4.11. We usually assume that this bearing pressure has a linear distribution; the maximum bearing pressure q_{max} and minimum bearing pressure q_{min}

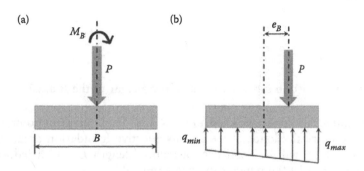

Figure 4.10 Eccentrically loaded strip foundation. (a) Eccentrically loading condition. (b) Equivalent eccentricity and non-uniform bearing pressure.

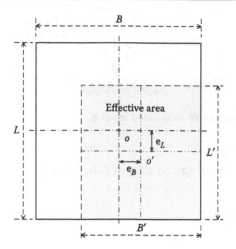

Figure 4.11 A foundation subjected to an eccentric loading.

are calculated as (consider a unit thickness of the strip foundation)

$$q_{max} = \frac{P}{(B)(1)} + \frac{e_B(P)\left(\dfrac{B}{2}\right)}{\dfrac{B^3(1)}{12}} = \frac{P}{B}\left(1 + \frac{6e_B}{B}\right)$$
(4.78)

$$q_{min} = \frac{P}{B}\left(1 - \frac{6e_B}{B}\right)$$
(4.79)

It is apparent that when $(e_B/B) > 1/6$, tensile stress will develop and $q_{min} < 0$. As soil has little tensile strength, we usually keep $(e_B/B) < 1/6$. The effective area of the strip foundation is defined by a reduced width B', which is determined by

$$B' = B - 2e_B$$
(4.80)

For a rectangular foundation (width $= B$ and length $= L$) which is also subject to an applied moment M_L, we follow a similar procedure to determine the eccentricity in the length direction,

$$e_L = \frac{M_L}{P}$$
(4.80)

where
  M_L = moment applied in a plane that is perpendicular to the B axis

Again, to avoid tensile stress, we keep $(e_L/L) < 1/6$. The equivalent point of loading for this two-way eccentrically loaded case and its effective foundation area, schematically shown in Figure 4.11, will be defined by a reduced length L' and a reduced width B' (Chen and McCarron 1991), which can be determined by

$$L' = L - 2e_L \quad \text{and} \quad B' = B - 2e_B$$
(4.81)

The ultimate bearing capacity q_u is then determined by using B' and L' as the foundation dimensions. The above case is referred to as the two-way eccentricity. For a rectangular foundation with only one-way eccentricity, for example if $e_L = 0$, then we set $L' = L$. Similarly, if $e_B = 0$, then $B' = B$.

To apply the general ultimate bearing capacity equation (Equation 4.56), we use L' and B' to determine F_{cs}, F_{qs}, and $F_{\gamma s}$ (Table 4.1). To calculate F_{cd}, F_{qd}, and $F_{\gamma d}$, we use the original B. The ultimate bearing load, Q_u, per unit length of a strip foundation can be calculated as

$$Q_u = q_u B' \tag{4.82}$$

Refer to Chen and McCarron (1991) for the determination of Q_u for rectangular foundation with two-way eccentricity. The factor of safety (FS) can be assessed as

$$FS = \frac{Q_u}{P} \tag{4.83}$$

or

$$FS = \frac{q_u}{q_{max}} \tag{4.83a}$$

where
 P = design or applied vertical load on the foundation

The foundations for reinforced concrete retaining walls to be described in Chapter 5 are typical examples of eccentrically loaded foundations.

4.6 EFFECTS OF GROUND WATER AND CONSIDERATION OF DRAINAGE CONDITIONS

So far we have assumed that the ground water level is located well below the foundation, and the bearing capacity analysis made no distinction between drained and undrained conditions. If, however, the ground water table is close to or above the foundation, we need to make certain modifications to the equations presented previously. To account for these factors, let's start by revising the bearing capacity Equation 4.56 as follows:

$$q_u = \bar{c} N_c F_{cs} F_{cd} F_{ci} + \bar{q} N_q F_{qs} F_{qd} F_{qi} + \frac{1}{2} \bar{\gamma} B N_\gamma F_{\gamma s} F_{\gamma d} F_{\gamma i} \tag{4.84}$$

Note that a "hat" has been added to the symbols of $\bar{c}$, $\bar{\varnothing}$, $\bar{q}$, and $\bar{\gamma}$ to emphasize that their values depend on the drainage and ground water conditions involved in the bearing capacity analysis. For drained or effective stress analysis, $\bar{c} = c'$, $\bar{\varnothing} = \varnothing'$, $\bar{\gamma} = \gamma'$ (buoyant unit weight), and $\bar{q} = q'$ are calculated based on γ'. For total stress or undrained analysis, $\bar{c} = c$, $\bar{\varnothing} = \varnothing$, and in the case of $\varnothing = 0$ analysis, $\bar{c} = s_u$ (undrained shear strength), $\bar{\gamma} =$ total soil unit weight γ (or γ_{sat} if saturated) and calculated based on total soil unit weight. The bearing capacity factors N_c, N_q, N_γ and the depth factors F_{cd}, F_{qd}, $F_{\gamma d}$ that involve friction angle should all be determined based on $\bar{\varnothing}$.

4.6.1 For drained analysis

Following Das (2011), we consider two different cases, as shown in Figure 4.12.

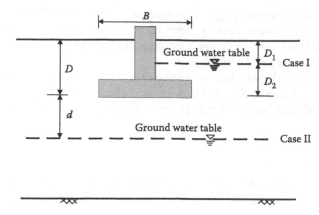

Figure 4.12 A foundation in soil with ground water level close to it.

4.6.1.1 Case I

The ground water level is above the foundation base level. In this case and in reference to Equation 4.84, the surcharge appearing in the bearing capacity equation should be determined as follows:

$$\bar{q} = q' = \gamma D_1 + (\gamma_{sat} - \gamma_w)D_2 \qquad (4.85)$$

where
  γ = soil total unit weight
  γ_{sat} = soil saturated unit weight
  γ_w = unit weight of water

In addition, the last term in the bearing capacity equation due to soil weight should be calculated using the following unit weight:

$$\bar{\gamma} = \gamma' = \gamma_{sat} - \gamma_w \qquad (4.86)$$

4.6.1.2 Case II

The distance between the foundation and water table is less than the foundation width, namely $0 \leq d \leq B$. In this case, the surcharge will be calculated as $\bar{q} = q = \gamma D$. However, the unit weight in the last term of the bearing capacity equation (i.e., Equation 4.84) should be replaced by the following expression:

$$\bar{\gamma} = \gamma' + \frac{d}{B}(\gamma - \gamma') \qquad (4.87)$$

When $d > B$, the ground water level is considered to be well below the foundation and its effect on the bearing capacity will be assumed to be negligible, then $\bar{c} = c'$, $\bar{\varnothing} = \varnothing'$, and $\bar{\gamma} = \gamma$.

4.6.2 For undrained analysis

As described above, for total stress or undrained analysis, $\bar{c} = c$, $\bar{\varnothing} = \varnothing$, and in the case of $\varnothing = 0$ analysis, $\bar{c} = s_u$ (undrained shear strength), $\bar{\gamma} =$ total soil unit weight γ (or γ_{sat} if saturated) and $\bar{q} = q$ calculated based on total soil unit weight. It should be noted that realistically, for static analysis of foundations, the undrained analysis should only be applied to cohesive soils considering $\emptyset = 0$ condition, and in this case $\bar{c} = s_u$. To have c and $\emptyset$ simultaneously could result in high (unsafe) bearing capacity values according to Equation 4.84. Judgment should be applied in this case as to whether the soil does possess the c and $\emptyset$ strength parameters in undrained loading.

4.7 FACTOR OF SAFETY AND ALLOWABLE BEARING CAPACITY

The ultimate bearing capacity that we have considered in this chapter represents the maximum load or pressure that a foundation can sustain before a shear failure in the supporting soil may occur. Given much uncertainty in ground conditions, foundations are designed to ensure an adequate margin of safety against this type of shear failure. In other words, the allowable bearing capacity will be obtained by the application of an FS as follows:

$$q_{all} = \frac{q_u}{FS} \tag{4.88}$$

Depending on the types of structure, knowledge of ground conditions, and consequence of failure, the FS used would be in the range of 2.0–4.0 (Vesic, 1975). Lower FS may be applied to temporary structures, and higher FS will be used for structures that regularly experience their maximum design load.

In practice, we often use the net allowable bearing capacity, $q_{all(net)}$, which is defined as

$$q_{all(net)} = \frac{q_u - q}{FS} \tag{4.89}$$

In this context, $q_{all(net)}$ refers to the allowable bearing pressure in excess to the surrounding overburden stress at the foundation base level.

EXAMPLE 4.4

Given

Consider a strip foundation that is embedded at a depth D_f of 2 m in a cohesive-frictional soil. The groundwater table is deep and its effects can be ignored. The cohesion of the soil is measured to be 5 kPa and its internal friction angle is 30°. The unit weight of the soil is $\gamma = 17\,\text{kN/m}^3$. The factor of safety is assumed to be 3 and the soil–foundation interface is perfectly smooth.

Required

Determine the width of the foundation B in order to support a vertical loading (q_{all}) of 400 kPa on the foundation. Assume the load is applied vertically uniformly on the foundation.

Solution

The ultimate bearing capacity of a rectangular foundation in a cohesive-frictional soil can be determined using Equation 4.84:

$$q_u = \bar{c}N_c F_{cs} F_{cd} F_{ci} + \bar{q} N_q F_{qs} F_{qd} F_{qi} + \frac{1}{2}\bar{\gamma} B N_\gamma F_{\gamma s} F_{\gamma d} F_{\gamma i} \tag{4.84}$$

where B is the width of the foundation and the surcharge can be estimated by

$$q = \gamma D_f = (17)(2) = 34\,\text{kPa}$$

Consider the groundwater table is deep, use drained analysis, $\bar{c} = c' = 5\,\text{kPa}$, $\bar{\varnothing} = \varnothing' = 30°$, and $\bar{\gamma} = \gamma = 17\,\text{kN/m}^3$.

Use trial and error to determine B, try $B = 1.5\,\text{m}$.

The ultimate bearing capacity factors can be determined according to Equations 4.23, 4.24, and 4.33 using the Excel program:

$$N_q = e^{(\pi \tan 30°)} \tan^2\left(45° + \frac{30°}{2}\right) = 18.4 \tag{4.23}$$

$$N_c = \cot 30°(N_q - 1) = 30.1 \tag{4.24}$$

$$N_\gamma = \frac{1}{4}\tan\left(\frac{\pi}{4} + \frac{\varnothing'}{2}\right)\left[\tan\left(\frac{\pi}{4} + \frac{\varnothing'}{2}\right)e^{(\frac{3}{2}\pi \tan \varnothing')} - 1\right]$$
$$+ \frac{3\sin\varnothing'}{1 + 8\sin^2\varnothing'}\left\{\left[\tan\left(\frac{\pi}{4} + \frac{\varnothing'}{2}\right) - \frac{\cot\varnothing'}{3}\right]e^{(\frac{3}{2}\pi \tan \varnothing')} + \tan\left(\frac{\pi}{4} + \frac{\varnothing'}{2}\right)\frac{\cot\varnothing'}{3} + 1\right\}$$
$$N_\gamma = 20.7 \tag{4.33}$$

For strip foundation, $F_{cs} = F_{qs} = F_{\gamma s} = 1$
For vertically applied load, $F_{ci} = F_{qi} = F_{\gamma i} = 1$
For D_f of 2 m, $(D_f/B) = (2/1.5) > 1$, in cohesive-frictional soil,

$$F_{qd} = 1 + 2\tan\varnothing'(1 - \sin\varnothing')^2 \tan^{-1}\left(\frac{D_f}{B}\right) = 1.27 \tag{4.73}$$

$$F_{cd} = F_{qd} - \frac{1 - F_{qd}}{N_c \tan\varnothing} = 1.28 \tag{4.74}$$

and

$$F_{\gamma d} = 1 \tag{4.75}$$

According to Equation 4.84,

$$q_u = \bar{c}N_c F_{cs}F_{cd}F_{ci} + \bar{q}N_q F_{qs}F_{qd}F_{qi} + \frac{1}{2}\bar{\gamma}BN_\gamma F_{\gamma s}F_{\gamma d}F_{\gamma i} \tag{4.84}$$

$$q_u = (5)(30.1)(1)(1.28)(1) + (34)(18.4)(1)(1.27)(1) + (0.5)(17)(1.5)(20.7)(1)(1)(1)$$
$$= 1250\,\text{kPa}$$

With an FS of 3, the allowable bearing capacity will be equal to

$$q_{all} = \frac{q_u}{3} = \frac{1250}{3} = 417\,\text{kPa} > 400\,\text{kPa OK}$$

Use $B = 1.5$ m.

EXAMPLE 4.5

Given

For a cohesive-frictional soil with unit weight $\gamma = 17\,\text{kN/m}^3$, soil friction angle $\varnothing = 30°$, and the strip foundation is 1.5 m wide as determined in Example 4.4. Assume the load is applied vertically and soil–foundation interface is perfectly smooth. The groundwater table is deep and its effects can be ignored.

Required

1. The strip foundation that is embedded at a depth D_f of 2 m. Determine the respective ultimate bearing capacity q_u, if cohesion of the soil $c' = 10$, 20, and 30 kPa.
2. Cohesion of the soil $c' = 5$ kPa. Determine the respective ultimate bearing capacity q_u, if the foundation embedment depth $D_f = 0$ and 4 m.

Solution

1. The soil friction angle is 30°, the same as that in Example 4.4. Thus, the bearing capacity factors are the same, as follows:

$$N_q = e^{(\pi \tan 30°)} \tan^2 \left(45° + \frac{30°}{2}\right) = 18.4 \tag{4.23}$$

$$N_c = \cot 30°(N_q - 1) = 30.1 \tag{4.24}$$

$$N_\gamma = \frac{1}{4}\tan\left(\frac{\pi}{4} + \frac{\varnothing'}{2}\right)\left[\tan\left(\frac{\pi}{4} + \frac{\varnothing'}{2}\right)e^{(\frac{3}{2}\pi \tan \varnothing')} - 1\right]$$
$$+ \frac{3\sin \varnothing'}{1 + 8\sin^2 \varnothing'}\left\{\left[\tan\left(\frac{\pi}{4} + \frac{\varnothing'}{2}\right) - \frac{\cot \varnothing'}{3}\right]e^{(\frac{3}{2}\pi \tan \varnothing')} + \tan\left(\frac{\pi}{4} + \frac{\varnothing'}{2}\right)\frac{\cot \varnothing'}{3} + 1\right\}$$
$$N_\gamma = 20.7 \tag{4.33}$$

$$q_u = \bar{c}N_c F_{cs}F_{cd}F_{ci} + \bar{q}N_q F_{qs}F_{qd}F_{qi} + \frac{1}{2}\bar{\gamma}BN_\gamma F_{\gamma s}F_{\gamma d}F_{\gamma i} \tag{4.84}$$
$$\bar{q} = \gamma D_f = (17)(2) = 34\,\text{kPa}$$

Since D_f and B are identical to those of Example 4.4, all correction factors are the same as those in Example 4.4.
When $c' = 10$ kPa,

$$q_u = (10)(30.1)(1)(1.28)(1) + (34)(18.4)(1)(1.27)(1)$$
$$+ (0.5)(17)(1.5)(20.7)(1)(1)(1) = 1444 \text{ kPa} \tag{4.84}$$

when $c' = 20$ kPa,

$$q_u = (20)(30.1)(1)(1.28)(1) + (34)(18.4)(1)(1.27)(1)$$
$$+ (0.5)(17)(1.5)(20.7)(1)(1)(1) = 1830 \text{ kPa} \tag{4.84}$$

and when $c' = 30\,\text{kPa}$,

$$q_u = (30)(30.1)(1)(1.28)(1) + (34)(18.4)(1)(1.27)(1)$$
$$+ (0.5)(17)(1.5)(20.7)(1)(1)(1) = 2218\ \text{kPa} \tag{4.84}$$

2. The bearing capacity factors remain the same. The change of D_f will affect the correction factors as follows:

When $D_f = 0\,\text{m}$, $q = 0$ and all correction factors equal to 1.

$$q_u = (5)(30.1)(1)(1)(1) + (0)(18.4)(1)(1)(1) + (0.5)(17)(1.5)(20.7)(1)(1)(1)$$
$$= 415\ \text{kPa}$$

$$\tag{4.84}$$

When $D_f = 4\,\text{m}$,

$$\bar{q} = \gamma D_f = (17)(4) = 68\,\text{kPa}$$

For D_f of 4 m, $(D_f/B) = (1/1.5) > 1$, in cohesive-frictional soil,

$$F_{qd} = 1 + 2\tan\varnothing'(1 - \sin\varnothing')^2 \tan^{-1}\left(\frac{D_f}{B}\right) = 1.35 \tag{4.73}$$

$$F_{cd} = F_{qd} - \frac{1 - F_{qd}}{N_c \tan\varnothing} = 1.37 \tag{4.74}$$

and

$$F_{\gamma d} = 1 \tag{4.75}$$

All other correction factors equal to 1.

$$q_u = (5)(30.1)(1)(1.37)(1) + (68)(18.4)(1)(1.35)(1) + (0.5)(17)(1.5)(20.7)(1)(1)(1)$$
$$= 2160\,\text{kPa}$$

$$\tag{4.84}$$

From Examples 4.4 and 4.5, we can get a feeling of how sensitive q_u is in response to the variations of c' and D_f, as the three bearing capacity factors serve as amplification factors to these parameters. With higher friction angle and hence larger bearing capacity factors, the sensitivity of q_u to these parameters increases significantly. Readers are encouraged to explore the sensitivity of q_u to various input parameters with the Excel programs associated with Examples 4.4 and 4.5, provided at the publisher's website.

EXAMPLE 4.6

Given

For a cohesive-frictional soil with unit weight $\gamma = 17\,\text{kN/m}^3$, soil cohesion $c' = 10\,\text{kPa}$, friction angle $\varnothing' = 30°$, and the strip foundation is 2 m wide. A vertical design load P of 250 kN and a moment, M_B of 50 kN-m per m of the foundation length are applied at the central location of the strip foundation. The soil–foundation interface is perfectly smooth. The strip foundation is embedded at a depth D_f of 1 m. The groundwater table is deep, and its effects can be ignored.

Required

Determine the ultimate bearing load Q_u of the strip foundation.

Solution

The soil friction angle is 30°, the same as that in Example 4.4. Thus the bearing capacity factors are the same, as follows:

$$N_q = e^{(\pi \tan 30°)} \tan^2\left(45° + \frac{30°}{2}\right) = 18.4 \tag{4.23}$$

$$N_c = \cot 30°(N_q - 1) = 30.1 \tag{4.24}$$

$$N_\gamma = 20.7 \tag{4.33}$$

The eccentricity,

$$e_B = \frac{M_B}{P} = \frac{(10)}{(30)} = 0.33\,\text{m} \tag{4.77}$$

$$\frac{e_B}{B} = \frac{0.2}{2} = 0.1 < 1/6\,\text{OK}$$

$$B' = B - 2e_B = (2) - 2(0.2) = 1.6\,\text{m} \tag{4.81}$$

when $D_f = 1$ m,

$$\bar{q} = \gamma D_f = (17)(1) = 17\,\text{kPa}$$

For D_f of 1 m and $B = 2$ m, $(D_f/B) < 1$ in cohesive-frictional soil, Equations 4.67-4.69 give

$$F_{qd} = 1 + 2 \times \tan\varphi'(1 - \sin\varphi')^2 \frac{D_f}{B}$$

$$= 1 + 2 \times \tan 30°(1 - \sin 30°)^2 \frac{1}{2} = 1.144 \tag{4.67}$$

$$F_{cd} = F_{qd} - \frac{1 - F_{qd}}{N_c \tan\varphi} = 1.153 \tag{4.68}$$

and

$$F_{\gamma d} = 1 \qquad \qquad (4.69)$$

All other correction factors equal to 1.

$$q_u = \bar{c} N_c F_{cs} F_{cd} F_{ci} + \bar{q} N_q F_{qs} F_{qd} F_{qi} + \frac{1}{2} \bar{\gamma} B N_\gamma F_{\gamma s} F_{\gamma d} F_{\gamma i} \qquad \qquad (4.84)$$

$$q_u = (10)(30.1)(1)(1.153)(1) + (17)(18.4)(1)(1.144)(1) + (0.5)(17)(1.6)(20.7)(1)(1)(1)$$
$$= 986 \, \text{kPa}$$

$$\qquad \qquad (4.84)$$

$$Q_u = q_u B' = (986)(1.6) = 1578 \, \text{kN/m}$$

4.8 FOUNDATION SETTLEMENT

It has been stressed that the foundation design has to consider two criteria relating to stability (safety) and serviceability (performance). The stability criterion ensures that shear failure does not occur under the design loading conditions. The serviceability criterion is to ensure that foundation settlement under design load is not excessive to affect its serviceability. A foundation design may be deemed not feasible because of the lack of safety (i.e., low-safety factor against bearing capacity failure) or due to poor serviceability (i.e., excessive settlement). The following sections concentrate on the issues that relate to foundation settlement.

The loading from the superstructure causes stress increase to the underlying soil stratum and that in turn induces compression of the soil and hence settlement of the foundation. The foundation settlement can generally be divided into three aspects: elastic or immediate settlement S_e, settlement due to primary consolidation of the affected soil S_c, and settlement due to secondary consolidation of soil S_s.

The elastic or immediate settlement occurs concurrently with little time delay as the superstructure is constructed and foundation loading increases. Adjustments can be made to the superstructure during construction. Therefore, the negative impact of S_e to the superstructure is usually minimal.

The primary consolidation settlement in granular soil may be considered as immediate because of the high permeability of granular soil. For clays, however, the primary consolidation settlement can be a major concern, as it can develop long after the superstructure is completed and excessive settlement can be damaging. This is especially true for foundations on normally or lightly overconsolidated clays (see Chapter 1). The high S_c values can be a reason for abandoning the use of shallow foundation.

The secondary consolidation is a long-lasting, endless phenomenon of soil structure rearrangement that continues after the primary consolidation is completed (i.e., excess pore pressure is fully dissipated). Details of secondary consolidation have been described in Chapter 1. Usually, the effects of S_s are significant for organic soil deposits due to its high-secondary compression index (C_α). Because of its powdery or fibrous characteristics, the primary consolidation in certain types of organic soils (such as peat) can occur much faster than that of inorganic cohesive soils. In any case, S_s is mostly a function of time and C_α; it is less affected by the level of stress increase.

Because of the drastically different nature among the various types of settlement described above, it may be preferable to evaluate the three components of settlement separately for foundations in cohesive soils. For foundations in granular soils, it may be advantageous to integrate the analysis for S_e, S_c, and S_s as described later.

In order to estimate foundation settlement, we often need to know the vertical stress increase in the soil due to the load applied on the foundation.

4.8.1 Vertical stress increase in soil due to external loading on foundations

It is useful to present a few fundamental elastic solutions for vertical stress increase due to external loading. These include the application of a concentrated load and a uniform pressure over a circular or rectangular area.

4.8.1.1 Stress increase due to a concentrated load

The most useful elastic solution is the vertical stress increase in soil due to a concentrated vertical load P acting on the ground surface, derived by Boussinesq (1883). The vertical stress increase solution for point A in the soil is given by

$$\Delta\sigma = \frac{3P}{2\pi z^2 \left[1 + \left(\frac{r}{z}\right)^2\right]^{5/2}} \tag{4.90}$$

where $r = \sqrt{x^2 + y^2}$ and x, y, z are the coordinates of point A.

4.8.1.2 Stress increase due to a uniformly loaded circular area

Determining the stress increase in the soil mass as a result of surface loading is often the first step in estimating the foundation settlement due to soil compressibility. Figure 4.13 describes a case of circular foundation with a radius of R. The stress increase $\Delta\sigma$ due to a uniform pressure q_o over the circular foundation can be obtained by using Equation 4.90 for a concentrated point load and integration for the entire loaded area. The resulting stress increase solution for point A, located at a depth z below the center of the circular

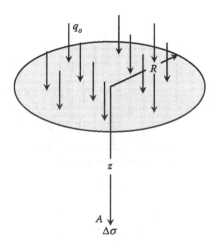

Figure 4.13 Stress increase under the center of a uniformly loaded circular area.

area, is given by

$$\Delta\sigma = q_o \left\{ 1 - \frac{1}{\left[1 + \left(\frac{R}{z}\right)^2\right]^{3/2}} \right\}$$
(4.91)

Ahlvin and Ulery (1962) showed that the stress increase for any location at a distance r from the center of the loaded area and at a depth z can also be obtained in a similar manner.

4.8.1.3 Stress increase due to a uniformly loaded rectangular area

In a similar manner, the stress increase below a uniformly loaded rectangular area as shown in Figure 4.14 can also be obtained. In particular, the stress increase solution at a point A located at a depth z below the center of the rectangular area with a length L and a width B can be shown to be

$$\Delta\sigma = q_o I_c$$
(4.92)

where I_c is called the influence factor which represents the ratio of $\Delta\sigma$ over q_o and,

$$I_c = \frac{2}{\pi} \left\{ \frac{mn}{\sqrt{1 + m^2 + n^2}} \frac{1 + m^2 + 2n^2}{(m^2 + n^2)(1 + n^2)} + \sin^{-1} \frac{m}{\sqrt{1 + n^2}\sqrt{m^2 + n^2}} \right\}$$
(4.93)

where

$$m = \frac{L}{B}$$
(4.94)

$$n = \frac{2z}{B}$$
(4.95)

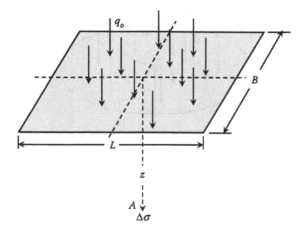

Figure 4.14 Stress increase under the center of a rectangular loaded area.

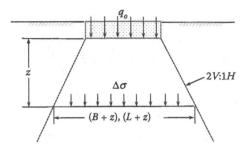

Figure 4.15 Estimate stress increase using the 2:1 method.

It should be noted that the above stress increase calculation is based on the assumption that the soil is linear elastic. The stress increase from multiple loaded areas and of various geometries can be calculated as the sum of stress increase due to individual loaded areas. For example, the stress increase below the corner of a rectangular loaded area can be calculated considering that only one quadrant of the rectangle shown in Figure 4.14 is loaded, and

$$\Delta\sigma = q_o \frac{I_c}{4} \tag{4.96}$$

where I_c is determined according to Equation 4.93.

It is useful to note that, in practice, foundation engineers often make use of an approximate solution to estimate the stress increase with depth. This approximate method is termed as the 2:1 method, as depicted in Figure 4.15. The stress increase estimated using the 2:1 method is given by

$$\Delta\sigma = q_o \frac{BL}{(B+z)(L+z)} = \left[\frac{m}{m + \dfrac{n}{2} + \dfrac{mn}{2} + \dfrac{n^2}{4}} \right] q_o \tag{4.97}$$

Through integration of the Boussinesq equation (Equation 4.90), the relationship between the vertical stress increase $\Delta\sigma$ at any location below foundation and loading on the foundation of various geometry and load configuration (e.g., an embankment) can be derived. With the derived relationship and the help of spreadsheet program, the value of $\Delta\sigma$ can be readily calculated. More details on the calculation of stress increase due to foundation loading can be found in many textbooks on soil mechanics (e.g., Holtz et al., 2011).

EXAMPLE 4.7

Given

For a flexible foundation having a shape of rectangle and a half-circle as shown in Figure E4.7. The rectangle has a width B of 3 m and length L of 5 m, and the half-circle has a radius R of 1.5 m. A uniform bearing pressure q_o of 200 kPa is applied to the foundation.

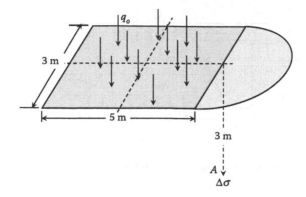

Figure E4.7 Dimensions of the flexible foundation.

Required

Determine the stress increase $\Delta\sigma$ at point A, 3 m below the center of the circle.

Solution

Divide the foundation into two parts: a rectangle and a half-circle.

For the I_c at point A under the center edge of the rectangle, I_c equals 0.5 times the I_c under the center of a foundation that has a dimensions of $L = 10$ m and $B = 3$ m.

Consider the effects of rectangular loaded area:

$$z = 3 \, \text{m}$$

$$m = \frac{L}{B} = (10)/(3) = 3.33 \tag{4.94}$$

$$n = \frac{2z}{B} = (2)(3)/(3) = 2.0 \tag{4.95}$$

$$q_o = 200 \, \text{kPa}$$

and,

$$I_c = 0.5$$
$$\times \frac{2}{\pi} \left\{ \frac{mn}{\sqrt{1 + m^2 + n^2}} \frac{1 + m^2 + 2n^2}{(m^2 + n^2)(1 + n^2)} + \sin^{-1} \frac{m}{\sqrt{1 + n^2}\sqrt{m^2 + n^2}} \right\}$$
$$= 0.266 \tag{4.93}$$

$$\Delta\sigma = q_o I_c = (200)(0.266) = 53.20 \, \text{kPa}$$

Consider the half-circular loaded area:

$$R = 1.5 \, \text{m}$$

$z = 3\,\text{m}$

$$\Delta\sigma = 0.5 \times q_o \left\{ 1 - \frac{1}{\left[1 + \left(\frac{R}{z}\right)^2 \right]^{\frac{3}{2}}} \right\} = 28.45\,\text{kPa} \tag{4.91}$$

Total stress increase $= (53.20) + (28.45) = 81.65\,\text{kPa}$.

4.9 FOUNDATION SETTLEMENT DUE TO SOIL ELASTIC DEFORMATION

This section describes some basic elastic solutions for a flexible circular and rectangular foundation under a uniform bearing pressure. Further details may be found in Davis and Selvadurai (1996).

1. *Settlement of a uniformly loaded flexible circular foundation*
   Consider a uniform pressure q_o over a circular surface foundation with a radius R. The surface settlement (i.e., vertical displacement) at a point that has a distance of ηR from the center of the foundation can be shown to be

$$S_e(r = \eta R) = \frac{2q_o(1-\mu)R}{\pi G} \int_0^{\pi/2} \sqrt{1 - \eta^2 \sin^2\Omega}\, d\Omega \tag{4.98}$$

where
$\mu = $ Poisson's ratio
$\Omega = $ is in radian, the shear modulus G is linked to Young's modulus E as follows:

$$G = \frac{E}{2(1+\mu)} \tag{4.99}$$

For the special cases of the center of the foundation ($\eta = 0$) and edge of the foundation ($\eta = 1$), we can integrate Equation 4.98 to obtain the following simple elastic surface settlement solutions:

$$S_e(\eta = 0) = \frac{q_o(1-\mu)R}{G} \tag{4.100}$$

and

$$S_e(\eta = 1) = \frac{2q_o(1-\mu)R}{\pi G} \tag{4.101}$$

2. *Settlement of a uniformly loaded surface, flexible rectangular foundation*
   Similarly, for the elastic settlement of a uniformly loaded, flexible rectangular foundation on the ground surface, with a length L (in the y direction) and width B (in the x direction). The surface settlement at the center of the rectangular foundation

can be shown to be

$$S_e = \frac{q_0(1 - \mu)B}{\pi G}\left[\sinh^{-1}(m) + m\,\sinh^{-1}\left(\frac{1}{m}\right)\right] \tag{4.102}$$

where $m = (L/B)$ and

$$\sinh^{-1}(m) = \ln\left(m + \sqrt{1 + m^2}\right) \tag{4.103}$$

The elastic settlement at the corner of the rectangular foundation can be shown to be

$$S_e = \frac{q_0(1 - \mu)B}{2\pi G}[\sinh^{-1}(m) + m\,\sinh^{-1}(\frac{1}{m})] \tag{4.104}$$

3. *Settlement of a uniformly loaded flexible rectangular foundation at depth*
    We now consider the elastic settlement of a uniformly loaded rectangular foundation buried at a depth D_f with a length L (in the y direction) and a width B (in the x direction). For this more complex problem, the average or mean settlement of the foundation has been given by Fox (1948). The final solution for the mean elastic settlement can be expressed in a closed form as follows:

$$\bar{S}_e = \frac{q_0(1 + \mu)}{4\pi E(1 - \mu)}\sum_{k=1}^{5}\beta_k\,Y_k \tag{4.105}$$

where

$$\beta_1 = 3 - 4\mu \tag{4.106}$$

$$\beta_2 = 5 - 12\mu + 8\mu^2 \tag{4.107}$$

$$\beta_3 = -4\mu(1 - 2\mu) \tag{4.108}$$

$$\beta_4 = -1 + 4\mu - 8\mu^2 \tag{4.109}$$

$$\beta_5 = -4(1 - 2\mu)^2 \tag{4.110}$$

and

$$Y_1 = B\log\frac{L + r_4}{B} + L\log\frac{B + r_4}{L} - \frac{(r_4)^3 - L^3 - B^3}{3LB} \tag{4.111}$$

$$Y_2 = B\log\frac{L + r_3}{r_1} + L\log\frac{B + r_3}{r_2} - \frac{(r_3)^3 - (r_2)^3 - (r_1)^3 + r^3}{3LB} \tag{4.112}$$

$$Y_3 = \frac{r^2}{B}\log\frac{(L + r_2)r_1}{(L + r_3)r} + \frac{r^2}{L}\log\frac{(B + r_1)r_2}{(B + r_3)r} \tag{4.113}$$

$$Y_4 = \frac{r^2(r_1 + r_2 - r_3 - r)}{LB} \tag{4.114}$$

$$Y_5 = r\tan^{-1}\left(\frac{LB}{rr_3}\right) \tag{4.115}$$

while

$$r = 2D_f \tag{4.116}$$

$$r_1 = \sqrt{B^2 + r^2} \tag{4.117}$$

$$r_2 = \sqrt{L^2 + r^2} \tag{4.118}$$

$$r_3 = \sqrt{L^2 + B^2 + r^2} \tag{4.119}$$

$$r_4 = \sqrt{L^2 + B^2} \tag{4.120}$$

For the special case of a surface foundation where $r = 2D_f = 0$, the above solution reduces to the following simple average elastic settlement:

$$\bar{S}_e = \frac{2q_0(1 - \mu^2)}{\pi E} Y_1 = m_0 \frac{q_0(1 - \mu^2)}{E} \sqrt{LB} \tag{4.121}$$

where the coefficient m_0 is only a function of $m = L/B$ and can be shown to be in the following closed form:

$$m_0 = \frac{2}{\pi} \frac{Y_1}{\sqrt{LB}} = \frac{2}{\pi} \left\{ \begin{array}{l} \dfrac{B}{\sqrt{LB}} \log \dfrac{L + \sqrt{L^2 + B^2}}{B} + \dfrac{L}{\sqrt{LB}} \log \dfrac{B + \sqrt{L^2 + B^2}}{L} \\[2mm] - \dfrac{\left(\sqrt{L^2 + B^2}\right)^3 - L^3 - B^3}{3LB\sqrt{LB}} \end{array} \right\}$$

$$= \frac{2}{\pi} \left\{ \frac{1}{\sqrt{m}} \log\left(m + \sqrt{m^2 + 1}\right) + \sqrt{m}\log\left[\frac{1}{m} + \sqrt{1 + \frac{1}{m^2}}\right] - \frac{\sqrt{m^2 + 1} - m^3 - 1}{3m\sqrt{m}} \right\} \tag{4.122}$$

The procedures for the estimation of elastic settlement are developed based on elasticity theories, and the soil is assumed as linear elastic with a constant Young's modulus E. Soil is known to have a rather non-linear stress–strain relationship. A reasonable estimate of the E value should consider the expected strain level to be experienced by the soil under the loading conditions. Schnaid (2009) recommended empirical equations to estimate E based on SPT N_{60} (SPT N value corrected to an energy ratio of 60%) that consider the foundation allowable bearing pressure q_{all} determined with an FS of 3 (i.e., $q_{all} = q_u/3$) as follows.

For granular soils:

$$\frac{E}{N_{60}} = 1 \text{ (MPa)} \tag{4.123}$$

and for undrained cohesive soils:

$$\frac{E}{N_{60}} = 1 \text{ to } 1.2 \text{ (MPa)} \tag{4.124}$$

EXAMPLE 4.8

Given

For a flexible rectangle foundation having a width B of 3 m and length L of 5 m. A uniform bearing pressure q_o of 200 kPa is applied to the foundation. The underlying soil has a shear modulus G of 5000 kPa and Poisson's ratio μ of 0.3.

Required

a. Consider the foundation is at the ground surface. Determine the elastic settlement at the center and corner of the rectangular foundation using Equations 4.102 and 4.104, respectively (that is, the Davis and Selvadurai (1996) solution).
b. Consider the same foundation embedded (D_f) at 2 and 0 m below ground surface. Determine the respective average elastic settlement of the rectangular foundation using the Fox (1948) solution (i.e., Equations 4.105 and 4.121, respectively).

Solution

$B = 3$ m. $L = 5$ m, $q_o = 200$ kPa, $G = 5000$ kPa, and $\mu = 0.3$
$m = L/B = 5/3 = 1.67$

a. Elastic settlement using the Davis and Selvadurai (1996) solution for the surface foundation:

To determine the elastic settlement at the center of the foundation,

$$\sinh^{-1}(m) = \ln\left(m + \sqrt{1+m^2}\right) = 1.695 \tag{4.103}$$

$$\sinh^{-1}\left(\frac{1}{m}\right) = \ln\left(\frac{1}{m} + \sqrt{1 + \frac{1}{m^2}}\right) = 0.673 \tag{4.103}$$

$$S_{e(center)} = \frac{q_o(1-\mu)B}{\pi G}\left[\sinh^{-1}(m) + m\sinh^{-1}\left(\frac{1}{m}\right)\right] = 75\,\text{mm} \tag{4.104}$$

At corner of the foundation,

$S_{e(corner)} = S_{e(center)}/2 = 37.5$ mm according to Equation 4.104

b. Elastic settlement using the Fox (1948) solution:
Consider $D_f = 2$ m

$$\beta_1 = 3 - 4\mu = 1.80 \tag{4.106}$$

$$\beta_2 = 5 - 12\mu + 8\mu^2 = 2.12 \tag{4.107}$$

$$\beta_3 = -4\mu(1 - 2\mu) = -0.48 \tag{4.108}$$

$$\beta_4 = -1 + 4\mu - 8\mu^2 = -0.52 \tag{4.109}$$

$$\beta_5 = -4(1 - 2\mu)^2 = -0.64 \tag{4.110}$$

$$r = 2D_f = 4.0 \tag{4.116}$$

$$r_1 = \sqrt{B^2 + r^2} = 5.0 \tag{4.117}$$

$$r_2 = \sqrt{L^2 + r^2} = 6.40 \tag{4.118}$$

$$r_3 = \sqrt{L^2 + B^2 + r^2} = 7.07 \tag{4.119}$$

$$r_4 = \sqrt{L^2 + B^2} = 5.83 \tag{4.120}$$

$$Y_1 = B \log \frac{L + r_4}{B} + L \log \frac{D + r_4}{L} - \frac{(r_4)^3 - r_?^3 - B^3}{3LB} = 1.00 \tag{4.111}$$

$$Y_2 = B \log \frac{L + r_3}{r_1} + L \log \frac{B + r_3}{r_2} - \frac{(r_3)^3 - (r_2)^3 - (r_1)^3 + r^3}{3LB} = 1.47 \tag{4.112}$$

$$Y_3 = \frac{r^2}{B} \log \frac{(L + r_2)r_1}{(L + r_3)r} + \frac{r^2}{L} \log \frac{(B + r_1)r_2}{(B + r_3)r} = 0.39 \tag{4.113}$$

$$Y_4 = \frac{r^2(r_1 + r_2 - r_3 - r)}{LB} = 0.35 \tag{4.114}$$

$$Y_5 = r \tan^{-1}\left(\frac{LB}{rr_3}\right) = 1.95 \tag{4.115}$$

$$\bar{S}_e = \frac{q_0(1 + \mu)}{4\pi E(1 - \mu)} \sum_{k=1}^{5} \beta_k \, Y_k = 11 \, \text{mm} \tag{4.105}$$

Consider $D_f = 0$ m

$$\bar{S}_e = \frac{2q_0(1 - \mu^2)}{\pi E} Y_1 = 15.1 \, \text{mm} \tag{4.121}$$

4.10 FOUNDATION SETTLEMENT DUE TO PRIMARY CONSOLIDATION IN COHESIVE SOILS

In Chapter 1, the concept of consolidation and time-dependent consolidation deformation for saturated clay soils under effective stress increments was presented. This is particularly applicable for estimation of consolidation settlement of a foundation over saturated clay soil.

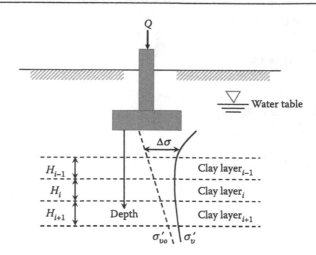

Figure 4.16 Consolidation settlement calculations.

The foundation settlement over a saturated clay stratum, as shown in Figure 4.16, due to one-dimensional consolidation can be calculated by dividing the clay stratum into a number of layers each having a thickness of H_i. For each clay layer i, the settlement due to primary consolidation ΔSc_i is related to the change of void ratio Δe_i and initial thickness of the layer H_i as follows:

$$\Delta S_{ci} = \frac{\Delta e_i}{1 + e_{oi}} H_i \tag{4.125}$$

where
  e_{oi} = initial void ratio of layer i

The change (reduction) of void ratio caused by an increase of the effective vertical stress from σ'_{vo} to $\sigma'_v = \sigma'_{vo} + \Delta\sigma$ for each layer depends on the relative values of the initial effective vertical stress, final effective vertical stress, and the preconsolidation pressure σ'_p. It is noted that $\Delta\sigma$ denotes the average vertical stress increase over each clay layer due to the construction of the foundation. e_{oi} is the initial void ratio of the clay layer i.

The total consolidation settlement of the clay stratum will be the sum of the consolidation settlement for each clay layer, namely

$$S_c = \sum_{i=1}^{n} \Delta S_{ci} = \sum_{i=1}^{n} \frac{\Delta e_i}{1 + e_{oi}} H_i \tag{4.126}$$

where n is the number of clay layers over the clay stratum. Apparently, the number of layers that we use to divide the clay stratum can affect the Δe_i for each clay layer and thus the S_c. Judgment is required to reach a balance between the accuracy of computation and accuracy of the input parameters, especially the clay compressibility.

As shown in Figure 1.3, we may encounter three possible cases for which different equations would need to be used to determine the void ratio reduction and consolidation settlement for each clay layer (note that the subscript i is omitted below for brevity):

a. The case of normal consolidation when $\sigma'_{vo} = \sigma'_p$:

$$\Delta e = C_c \log \frac{\sigma'_v}{\sigma'_{vo}} \tag{4.127}$$

$$\Delta S_c = \frac{HC_c}{1 + e_o} \log \frac{\sigma'_{vo} + \Delta\sigma}{\sigma'_{vo}} \tag{4.128}$$

b. The case of overconsolidation when $\sigma'_{vo} < \sigma'_v \le \sigma'_p$:

$$\Delta e = C_s \log \frac{\sigma'_v}{\sigma'_{vo}} \tag{4.129}$$

$$\Delta S_c = \frac{HC_s}{1 + e_o} \log \frac{\sigma'_{vo} + \Delta\sigma}{\sigma'_{vo}} \tag{4.130}$$

c. The case of overconsolidation followed by normal consolidation when $\sigma'_{vo} < \sigma'_p \le \sigma'_v$:

$$\Delta e = C_s \log \frac{\sigma'_p}{\sigma'_{vo}} + C_c \log \frac{\sigma'_v}{\sigma'_p} \tag{4.131}$$

$$\Delta S_c = \frac{HC_s}{1 + e_o} \log \frac{\sigma'_p}{\sigma'_{vo}} + \frac{HC_c}{1 + e_o} \log \frac{\sigma'_{vo} + \Delta\sigma}{\sigma'_p} \tag{4.132}$$

where C_c is the compression index and C_s is the swelling index of the clay layer.

It must be stressed that the above consolidation settlement is based on the assumption of one-dimensional (i.e., the vertical displacement) consolidation theory. In other words, it is assumed that the soil will only move in the vertical direction under the external loading applied from the construction of a foundation. In reality, of course, soils will also be able to move sideways, although their magnitude may be much smaller than that of the vertical direction. A more rigorous treatment and quantification would require a numerical-based, three-dimensional consolidation analysis. However, it may be reasonable to state that the simplified, one-dimensional consolidation analysis described above would likely overestimate the consolidation settlement of a foundation when compared with a three-dimensional consolidation analysis, and therefore is conservative for foundation design.

EXAMPLE 4.9

Given

A rectangular foundation with a width of 3 m and a length of 6 m is located at a depth of 2 m in a soil profile with a sand layer of 4 m from the ground surface underlain by a normally consolidated clay layer of 4 m, as shown in Figure E4.9. The increased pressure at the foundation base caused by the construction of the foundation is 200 kN/m². The initial void ratio e_o of the normally consolidated clay is 1.35. The compression index C_c of the clay is 0.35. The water table is at the top of the clay layer. The total unit weight of sand $\gamma = 16.5 \text{ kN/m}^3$. The saturated unit weight of the clay $\gamma_{sat} = 18.5 \text{ kN/m}^3$. The unit weight of water $\gamma_w = 10 \text{ kN/m}^3$.

Required

Determine the expected consolidation settlement due to the construction of the foundation.

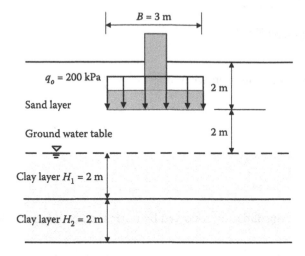

Figure E4.9 The foundation and multi-layer soil profile.

Solution

For simplicity, we divide the clay layer into two thinner layers of 2 m each.

First consider the center of the upper-clay layer, which is 5 m deep from the ground surface, including 4 m of sand and 1 m of clay. Before the construction of foundation, its initial effective vertical stress is

$$\sigma'_{vo1} = \gamma \times 4\,\text{m} + (\gamma_{sat} - \gamma_w) \times 1\,\text{m} = (16.5)(4) + (18.5 - 10)(1) = 74.5\ \text{kN/m}^2$$

Then consider the center of the lower-clay layer, which is 7 m deep from the surface, including 4 m of sand and 3 m of clay. Before the construction of foundation, its initial effective vertical stress is

$$\sigma'_{vo2} = \gamma \times 4\,\text{m} + (\gamma_{sat} - \gamma_w) \times 3\,\text{m} = (16.5)(4) + (18.5 - 10)(3) = 91.5\ \text{kN/m}^2$$

Due to the construction of the foundation, there is an additional stress increment at each of the above two depths, which can be determined by Equations 4.93 through 4.95. Alternatively, it can be estimated using the simpler Equation 4.97.

If Equation 4.97 is used, then the vertical stress increment at a depth z from the bottom of the foundation with a length of L and a width of B is

$$\Delta\sigma = q_o \frac{BL}{(B+z)(L+z)} = \left[\frac{m}{m + \frac{n}{2} + \frac{mn}{2} + \frac{n^2}{4}} \right] q_o \tag{4.97}$$

where $m = L/B$, $n = 2z/B$, and q_o is the additional pressure exerted at the bottom of the foundation.

At the center of the upper-clay layer, we have $L = 6$ m, $B = 3$ m, and $z = 3$ m $q_o = 200$ kPa, and therefore

$$\Delta\sigma_1 = q_o \frac{BL}{(B+z)(L+z)} = 200 \times \frac{3 \times 6}{(3+3) \times (6+3)} = 66.7\ \text{kN/m}^2$$

At the center of the lower-clay layer, we have $L = 6$ m, $B = 3$ m, and $z = 5$ m $q_o = 200$ kPa, and therefore

$$\Delta\sigma_2 = q_o \frac{BL}{(B+z)(L+z)} = 200 \times \frac{3 \times 6}{(3+5) \times (6+5)} = 40.9 \, \text{kN/m}^2$$

The consolidation settlement of the upper- and lower-clay layers can be estimated using Equation 4.128. For the upper-clay layer, we have

$$\Delta S_{c1} = \frac{H_1 C_c}{1+e_o} \log \frac{\sigma'_{vo1} + \Delta\sigma_1}{\sigma'_{vo1}} = \frac{2 \times 0.35}{1+1.35} \log \left(\frac{74.5 + 66.7}{74.5} \right) = 0.0827 \, \text{m} \quad (4.128)$$

For the lower-clay layer, we have

$$\Delta S_{c2} = \frac{H_2 C_c}{1+e_o} \log \frac{\sigma'_{vo2} + \Delta\sigma_2}{\sigma'_{vo2}} = \frac{2 \times 0.35}{1+1.35} \log \left(\frac{91.5 + 40.9}{91.5} \right) = 0.0478 \, \text{m} \quad (4.128)$$

The total consolidation settlement is therefore

$$\Delta S_c = \Delta S_{c1} + \Delta S_{c2} = 0.0827 + 0.0478 = 0.1305 \, \text{m} \approx 131 \, \text{mm}$$

4.11 FOUNDATIONS SETTLEMENT DUE TO SECONDARY CONSOLIDATION IN COHESIVE SOILS

The secondary consolidation is mostly a function of time. The settlement due to secondary consolidation S_s can be estimated as follows:

$$S_s = \frac{C_\alpha H_c}{1+e_p} \log \frac{t_p + \Delta t}{t_p} \tag{4.133}$$

where
C_α = secondary compression index defined by Equation (1.21)
H_c = thickness of the clay layer
e_p = void ratio of the clay layer at the end of primary consolidation
t_p = time required to reach the end of primary consolidation after the foundation loading is applied
Δt = time duration (in the same unit as t_p) after primary consolidation for the secondary consolidation analysis

If the e_p varies significantly due to the thickness of the clay layer, the clay layer can be divided into multiple layers for the analysis similar to that for primary consolidation analysis.

EXAMPLE 4.10

Given

For the same foundation and soil profile described in Example 4.9. The time duration to complete the primary consolidation t_p in the clay layer is expected to be 5 years.

Required

1. Compute the end of primary consolidation void ratio e_p for the clay layer.
2. Estimate secondary compression index $C_\alpha = 0.04 \times C_c$.
3. Estimate the settlement in the clay layer due to its secondary consolidation 20 years after the end of primary consolidation ($\Delta t = 20$ years).

Solution

As described in Example 4.9, $e_o = 1.35$

In clay layer 1:

Initial effective overburden stress, $\sigma'_{vo1} = 74.5$ kPa

Effective overburden stress increase, $\Delta\sigma_1 = 66.7$ kPa

End of primary consolidation effective overburden stress, $\sigma'_{v1} = (74.5) + (66.7) = 141.2$ kPa

$$\Delta e = C_c \log \frac{\sigma'_v}{\sigma'_{vo}} = 0.097 \tag{4.127}$$

$$e_p = e_o - \Delta e = (1.35) - (0.097) = 1.25$$

$$S_s = \frac{C_\alpha H_c}{1 + e_p} \log \frac{t_p + \Delta t}{t_p} = 8.7 \, \text{mm} \tag{4.133}$$

In clay layer 2:

Initial effective overburden stress, $\sigma'_{vo2} = 91.5$ kPa

Effective overburden stress increase, $\Delta\sigma_2 = 40.9$ kPa

End of primary consolidation effective overburden stress, $\sigma'_{v2} = (91.5) + (40.9) = 132.4$ kPa

$$\Delta e = C_c \log \frac{\sigma'_v}{\sigma'_{vo}} = 0.056 \tag{4.127}$$

$$e_p = e_o - \Delta e = (1.35) - (0.056) = 1.29$$

$$S_s = \frac{C_\alpha H_c}{1 + e_p} \log \frac{t_p + \Delta t}{t_p} = 8.5 \, \text{mm} \tag{4.133}$$

Total settlement due to secondary consolidation $= (8.7) + (8.5) = 17.2$ mm

4.12 FOUNDATION SETTLEMENT IN GRANULAR SOILS USING THE STRAIN INFLUENCE FACTOR METHOD

The strength of granular soil, as a frictional material, increases significantly with lateral or confining stress. Its stiffness also increases significantly with lateral stress. The foundation loading causes the vertical as well as lateral stress to increase in the soil below the foundation. Had the soil stiffness remained constant, the vertical strains in the soil below the foundation would have decreased monotonically with depth, as the vertical stress increase diminishes. For granular soil immediately below the foundation, the stiffness increase due to lateral stress increases more than offsets the vertical stress increase. As a result, the vertical strain starts low at the foundation base, and increases with depth to a maximum value before starting its monotonic decreasing trend with depth. Recognizing the characteristics of foundation settlement in granular soils, Schmertmann et al. (1978) proposed a method to estimate the foundation settlement S_e using a strain influence factor, as follows:

$$S_e = C_1 C_2 (q_{all} - q) \sum_0^{z_2} \frac{I_z}{E} \Delta z \tag{4.134}$$

where
   q_{all} = the allowable or design bearing pressure applied to the foundation
   q = effective overburden stress at the foundation base level
   $C_1 = 1 - 0.5[q/(q_{all} - q)]$
   $C_2 = 1 + 0.2 \log(\text{time in years}/0.1)$
   E = Young's modulus of the soil
   I_z = strain influence factor

Figure 4.17 shows the variation of the strain influence factor with depth according to a modified version reported by Salgado (2008). B is the width and L is the length of the foundation. The strain influence factor I_z is used to adjust the strain values (i.e., $(q_{all} - q)/E$) to account for the combined effects of vertical and lateral stress at different depths below the foundation and impose the characteristics of strain distribution in granular soil below the foundation as described above.

The variation of I_z can be described by the following linear equations according to Salgado (2008). I_z at $z = 0$, $I_{z(0)}$:

$$I_{z(0)} = 0.1 + 0.011 \left(\frac{L}{B} - 1 \right) \le 0.2 \tag{4.135}$$

Variation of (z_1/B) (z_1 = the depth of maximum I_z from below the foundation base):

$$\frac{z_1}{B} = 0.5 + 0.0555 \left(\frac{L}{B} - 1 \right) \le 1 \tag{4.136}$$

Variation of (z_2/B) (z_2 = depth where I_z reduces to 0, from below the foundation base):

$$\frac{z_2}{B} = 2 + 0.222 \left(\frac{L}{B} - 1 \right) \le 4 \tag{4.137}$$

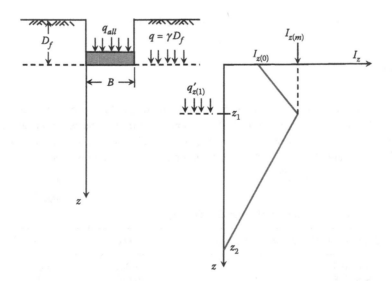

Figure 4.17 Variation of influence factor with depth.

and the maximum influence factor, $I_{z(m)}$ (I_z at depth z_1):

$$I_{z(m)} = 0.5 + 0.1 \sqrt{\frac{(q_{all} - q)}{q'_{z(1)}}} \qquad (4.138)$$

where

$q'_{z(1)}$ = effective vertical stress at z_1 where influence factor reaches its maximum value $I_{z(m)}$ before foundation construction

Equation 4.134 provides a rather comprehensive way to estimate the foundation settlement in granular soils. The effects of foundation embedment on foundation settlement are included in C_1. Additional settlement due to secondary consolidation is considered in C_2. The term S_e is used to represent foundation settlement in granular soils because the settlement (or the majority of it) is immediate and similar to the case of elastic settlement. The summation in Equation 4.134 is used to accommodate the variations of I_z and possible variations of E with depth.

To apply the strain influence factor method, the following procedure is recommended:

1. Establish the soil profile that includes the variation of soil unit weight γ and Young's modulus E with depth z and ground water level if present. The E values can be estimated using Equation 4.123 according to Standard Penetration Test (SPT).
2. Collect the foundation design parameters that include foundation width B, foundation length L, allowable bearing pressure q_{all}, and depth of embedment D_f. Determine q at the foundation base level.
3. Determine z_1, z_2, $I_{z(0)}$, $q'_{z(1)}$, and $I_{z(m)}$ using Equations 4.135 through 4.138.
4. Determine C_1 and C_2 according to Equation 4.134 and the time period to be considered in the settlement analysis.
5. Divide the soil stratum below the foundation into layers to accommodate the variations of I_z and E. The range of I_z and E values in each of the divided layers should be reasonably represented by the corresponding values at the center of that layer.
6. Establish a spreadsheet that includes the depth range of each layer, representative I_z and E values, and execute the computation according to Equation 4.134.

Example 4.11 demonstrates how the procedure is applied.

EXAMPLE 4.11

Given

A shallow foundation that is 2 m wide (B) and 3 m long (L) is to be embedded (D_f) at 1.5 m below ground surface. Consider the groundwater table to be very deep, and it has no effect on the foundation settlement calculations. The variations of γ and E with depth are presented in Table E4.11. The foundation is designed for an allowable bearing pressure q_{all} of 250 kPa.

Table E4.11 Variations of γ and E with depth

Depth range (From foundation base), m	E, kPa	γ, kN/m³
1.5 to 0.5	12,000	18.0
0.5 to 3.5	10,000	17.0
3.5 and below	15,000	19.0

Required

Predict the foundation settlement 10 years after the construction.

Solution

Following the procedure described above for the execution of Equation 4.134:

1. Variations of γ and E with depth z are given in Table E4.11.
2. $B = 2$ m, $L = 3$ m, $D_f = 1.5$ m, $q = \gamma D_f = (18.0)(1.5) = 27$ kPa and $q_{all} = 250$ kPa as given.
3. Determine z_1, z_2, $I_{z(0)}$, $q'_{z(1)}$, and $I_{z(m)}$

$$I_{z(0)} = 0.1 + 0.011\left(\frac{L}{B} - 1\right) = 0.1 + 0.011((3)/(2) - 1) = 0.106 < 0.2 \quad (4.135)$$

$$z_1 = B\left[0.5 + 0.0555\left(\frac{L}{B} - 1\right)\right] = (2)[0.5 + 0.0555((3)/(2) - 1)]$$
$$= 1.06\,\text{m} < B \quad (4.136)$$

$$z_2 = B\left[2 + 0.222\left(\frac{L}{B} - 1\right)\right] = (2)[2 + 0.222((3)/(2) - 1)] = 4.22\,\text{m} < 4B \quad (4.137)$$

$$q'_{z(1)} = (2)(18) + (1.5 + 1.06 - 2)*(17) = 45.44\,\text{kPa}$$

$$I_{z(m)} = 0.5 + 0.1\sqrt{\frac{(q_{all} - q)}{q'_{z(1)}}} = 0.722 \quad (4.138)$$

4. Determine C_1 and C_2.

   $C_1 = 0.94$ according to Equation 4.134
   $C_2 = 1.40$ according to Equation 4.134

5. Divide the soil stratum into four layers, as shown in Figure E4.11, and representative E and I_z are shown in Table E4.11a.

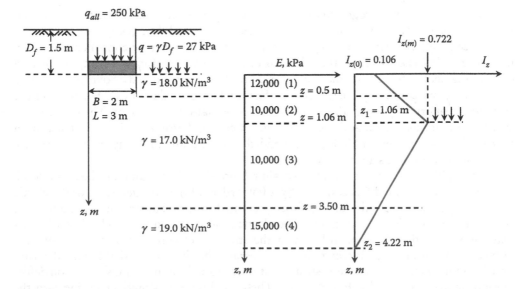

Figure E4.11 Multi-layer soil profile.

Table E4.11a Representative I_z and E values

Layer no.	Δz, m	E, kPa	I_z	$\frac{I_z}{E}\Delta z$, m^2/kN
1	0.50	12,000	0.25	1.05E−05
2	0.56	10,000	0.56	3.13E−05
3	2.44	10,000	0.44	10.81E−05
4	0.72	15,000	0.08	0.39E−06
			$\sum_0^{z_2}\frac{I_z}{E}\Delta z$	15.38E−05

6. Based on Table E4.11 a, Equation 4.134 gives the total settlement as

$$S_e = C_1 C_2(q_{all} - q) \sum_0^{z_2}\frac{I_z}{E}\Delta z = (0.94)(1.40)(250 - 27)(0.0001538)$$

$$= 45.1 \, mm \tag{4.134}$$

4.13 REMARKS

In this chapter, we reviewed the theories and methods for the stability (safety) and deformation (performance) analysis of shallow foundations. The bearing capacity factors were derived using the upper-bound limit analysis. The bearing capacity factors are sensitive to the variations of friction angle $\emptyset'$ and can be substantial when $\emptyset'$ approaches or exceeds 30°. The ultimate bearing capacity, q_u, can be unsafely high if both c' and $\emptyset'$ are involved and c' is significant, because of the high value of N_c. It is thus imperative to verify if there is a justification for using c'. This is also the reason we emphasize that for undrained shearing (including bearing capacity failure) in cohesive soils, the undrained shear strength s_u should be used; in this case $\emptyset = 0$. Readers are encouraged to use the Excel program provided on the publisher's website to explore the sensitivity of q_u to the various input parameters such as foundation dimensions, embedment, and strength parameters c' and $\emptyset'$.

Readers are also encouraged to explore the sensitivity of foundation settlement calculation to the various input parameters such as elastic modulus, compression indexes, and preconsolidation stress σ'_p. Experience shows that the value of consolidation settlement prediction is sensitive to the selection of σ'_p. The value of σ'_p is an important part of our judgment as to whether the clay layer is normally consolidated or overconsolidated (i.e., whether we can justify the C_s instead of a march larger value of C_c). The time for the completion of primary consolidation is another potential source of error. The result is sensitive to the drainage path and the coefficient of consolidation used in the prediction of time period. A thin permeable layer embedded in the middle of a clay layer can reduce the consolidation time significantly.

Soil is highly non-linear in its stress–strain relationship. Depending on the strain level, the values of elastic modulus can vary by a few orders of magnitude. Most of the elastic settlement analysis methods such as those presented in this chapter unfortunately use a constant shear or Young's modulus. The elastic modulus used in the elastic settlement analysis should thus be compatible with the strain level expected in the soil stratum induced by the foundation loading. Intensive research has been carried out in the determination of the soil non-linear stress–strain relationship and its applications in the soil deformation analysis in the past few decades. These methods are slowly emerging into the mainstream of engineering practice.

HOMEWORK

4.1. A strip foundation with a width of 4 m is located at a depth of 2 m in a dry cohesive-frictional soil. The cohesion of the soil is measured to be 10 kPa and its internal friction angle is 28°. The unit weight of the soil is $\gamma = 17.5$ kN/m^3. The friction angle of the soil–foundation interface is 30°. Calculate the ultimate bearing capacity of the footing.

4.2. A rectangular foundation with a width of 3 m and a length of 6 m that is located at a depth of 2 m in a cohesive-frictional soil above water table. The cohesion of the soil is measured to be 10 kPa and its internal friction angle is 28°. The unit weight of the soil is $\gamma = 17$ kN/m^3. Estimate the ultimate bearing capacity of the foundation assuming that soil–foundation interface is perfectly smooth.

4.3. Repeat Problem H4.2 for a circular foundation with a diameter of 4 m located at a depth of 2 m in the same soil mass.

4.4. A strip foundation is designed to be located at a depth of 2 m in a cohesive-frictional soil with the water table at a depth of 0.5 m. The cohesion of the soil is measured to be 10 kPa and its internal friction angle is 28°. The total unit weight of the soil is = 17 kN/m^3. The saturated unit weight of soil is $\gamma_{sat} = 19$ kN/m^3 and the unit weight of water is $\gamma = 10$ kN/m^3. The FS is assumed to be 3 and the soil–foundation interface is perfectly smooth. Design the width of the foundation subjected to a vertical loading of 250 kN/m of footing length.

4.5. Consider a rectangular foundation, with a width of 3 m and a length of 6 m, located at a depth of 2 m in a soil profile with a sand layer of 4 m from the ground surface overlain by an overconsolidated clay layer of 4.5 m. The increased pressure at the foundation base caused by the construction of the foundation is 220 kPa. The initial void ratio of the overconsolidated clay is 1.1. The compression index of the clay is 0.35 and the swelling index of the clay is 0.05. The preconsolidation pressure is measured to be 220 kPa. The water table is at the top of the clay. The total unit weight of sand is $\gamma = 17$ kN/m^3. The saturated unit weight of the clay is $\gamma_{sat} = 19$ kN/m^3. The unit weight of water is $\gamma_w = 10$ kN/m^3. Determinate the final consolidation settlement due to the construction of the foundation.

4.6. Repeat Example 4.9 but use the Boussinesq (1883) method (Equation 4.93) to compute the stress increase. The foundation and soil profile are repeated in Figure H4.6.

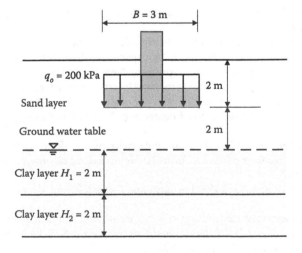

$B = 3$ m

$q_o = 200$ kPa

2 m

Sand layer

Ground water table

2 m

Clay layer $H_1 = 2$ m

Clay layer $H_2 = 2$ m

Figure H4.6 The foundation and multi-layer soil profile.

4.7.   If the clay in Problem H4.6 has a coefficient of consolidation c_v of 10 mm^2/m, and the lower boundary of the clay layer is not permeable (i.e., bounded by solid rock), determine the time required to reach 95% consolidation for the clay layer, t_{95}.

4.8.   Consider the t_{95} obtained from Problem H4.7 the same as the time required to complete the primary consolidation, t_p. Repeat Example 4.10 and predict the settlement due to secondary consolidation 20 years after the end of primary consolidation, using the t_p obtained herein.

4.9.   Repeat Problems H4.7 and 4.8 if there is a thin but permeable sand layer that can provide drainage for pore pressure dissipation in the middle of the clay layer (i.e., at 4 m below the foundation base).

4.10. For a flexible circular foundation having a radius R of 3 m, a uniform bearing pressure q_o of 200 kPa is applied to the foundation. The underlying soil has a shear modulus G of 5000 kPa and Poisson's ratio μ of 0.3. Estimate the elastic settlement of the foundation at the center and edge of the foundation using Equation 4.98.

4.11. Repeat Example 4.8 for a flexible rectangle foundation having a width B of 2 m and length L of 4 m. A uniform bearing pressure q_o of 200 kPa is applied to the foundation. The underlying soil has a shear modulus G of 4000 kPa and Poisson's ratio μ of 0.5.

4.12. Repeat Example 4.11 soil profile remains the same, but consider the following conditions:
   a.   The foundation is 2 m wide (B) and 3 m long (L) embedded (D_f) at 0.5 m, 1.5 m, and 3.0 m below ground surface.
   b.   The foundation is 2 m wide (B) and 10 m long (L) embedded (D_f) at 0.5, 1.5, and 3.0 m below ground surface.

REFERENCES

Ahlvin, R.G. and Ulery, H.H. 1962. Tabulated values of determining the composite pattern of stresses, strains and deflections beneath a uniform load on a homogeneous half space. *Highway Research Board Bulletin*, 342, 1–13.

Atkinson, J.H. 1981. *Foundations and Slopes: An Introduction to Applications of Critical State Soil Mechanics*. Wiley, New York, 382pp.

Boussinesq, J. 1883. *Application des potentials a L'Etude de L'Equilibre et du Mouvement des Solides Elastiques*, Gauthier-Villars, Paris.

Brinch Hansen, J. 1970. A revised and extended formula for bearing capacity. *Bulletin of Danish Geotechnical Institute*, 28, 5–11.

Chen, W.F. 1975. *Limit Analysis and Soil Plasticity*. Elsevier, New York, NY.

Chen, W.F. and McCarron, W.O. 1991. Bearing capacity of shallow foundations. In Fang, H.Y., ed., *Foundation Engineering Handbook*. Springer, New York, USA, pp. 144–165.

Collins, I.F. 1969. The upper bound theorem for rigid/plastic solids generalised to include Coulomb friction. *Journal of the Mechanics and Physics of Solids*, 17, 323–338.

Das, B.M. 2011. *Principles of Foundation Engineering*, 7th Edition. Cengage Learning, Stamford, CT, USA.

Davis, E.H. and Booker, J.R. 1971. The bearing capacity of strip footings from the standpoint of plasticity theory. *Proceedings of the 1st Australia-New Zealand Geomechanics Conference*, Melbourne, pp. 275–282.

Davis, R.O. and Selvadurai, A.P.S. 1996. *Elasticity and Geomechanics*. Cambridge University Press, Cambridge, UK.

Fox, E.N. 1948. The mean elastic settlement of a uniformly loaded area at a depth below the ground surface. *Proceedings of the 2nd International Conference on Soil Mechanics and Foundation Engineering*, Rotterdam, Vol. 1, pp. 129–132.

Hill, R. 1950. *The Mathematical Theory of Plasticity*. Clarendon Press, Oxford, UK.

Holtz, R.D., Kovacs, W.D., and Sheahan, T.C. 2011. *An Introduction to Geotechnical Engineering*, 2nd Edition. Prentice Hall, Englewood Cliffs, NJ, USA.

Meyerhof, G.G. 1963. Some recent research on the bearing capacity of foundations. *Canadian Geotechnical Journal*, 1, 16–26.

Prandtl, L. 1920. Ueber die haerte plastischer koerper. *Geottinger Nachr Math Phys*, K1, 74–85.

Salgado, R. 2008. *The Engineering of Foundations*. McGraw-Hill, New York.

Schmertmann, J.H., Hartman, J.P., and Brown, P.R. 1978. Improved strain influence factor diagrams. *Journal of The Geotechnical Engineering Division, ASCE*, 104(GT8), 1131–1135.

Schnaid, F. 2009. *In Situ Testing in Geomechanics*. Taylor and Francis, London, 329pp.

Shield, R.T. 1954. On the plastic flow of metals under conditions of axial symmetry. *Proceedings of the Royal Society, London, Series A*, 233, 267–287.

Shield, R.T. 1955. Stress and velocity fields in soil mechanics. *Quarterly Applied Mathematics*, 12, 144–156.

Shield, R.T. and Drucker, D.C. 1953. The application of limit analysis to punch-indentation problems. *Journal of Applied Mechanics, ASME*, 75, 453–460.

Sloan, S.W. and Yu, H.S. 1996. Rigorous plasticity solutions for the bearing capacity factor $N\gamma$. *Proceedings of the 7th Australia-New Zealand Geomechanics Conference*, Adelaide, 544–550.

Terzaghi, K. 1943. *Theoretical Soil Mechanics*. Wiley, New York.

Vesic, A.S. 1973. Analysis of ultimate loads of shallow foundations. *Journal of the Soil Mechanics and Foundations Division, ASCE*, 99, 45–73.

Vesic, A.S. 1975. Bearing capacity of shallow foundations. In Winterkorn, H.F. and Fang, H.Y., eds., *Foundation Engineering Handbook*, pp. 121–145, Van Nostrand Reinhold, New York.

Yu, H.S. and Sloan, S.W. 1994. Upper-bound limit analysis of a rigid-plastic body with frictional interfaces. *International Journal of Mechanical Sciences*, 36, 219–229.

Chapter 5

Lateral earth pressure and retaining structures

5.1 INTRODUCTION

The amount of lateral earth pressure exerted on a retaining structure (e.g., a retaining wall) is a classic soil mechanics problem that dates back hundreds of years. When we design a retaining wall to hold the soil behind it, the first question we face is how much lateral earth pressure, or lateral force, the wall is subjected to. We need the lateral earth pressure/force (i.e., the demand) to determine the dimensions and other structural details of the retaining wall. Sometimes we build a retaining structure as a supporting element to resist other structural loads and use soil behind the retaining wall as part of the resistance. In this case, we need to know the lateral resistance (i.e., the supply) that we can rely on using the weight and strength of the soil behind the wall. Lateral earth pressure/force is part of a soil–structure interaction problem. The magnitude of lateral earth pressure depends on the direction and amount of lateral movement of the retaining wall against the soil. We generally divide the conditions into three categories, depending on how the wall moves in reference to the soil, as follows (see Figure 5.1):

- At-rest condition: when the retaining wall remains fixed against the soil
- Active condition: when the retaining wall moves away from the soil
- Passive condition: when the retaining wall pushes toward the soil

Figure 5.1a shows the schematic view of a laboratory test, reported by Chen (1975), that involves a large container with a movable wall. By filling the container with sand behind the wall, a lateral earth pressure is developed against the wall. A horizontal force P_n normal to the wall must be applied to maintain stability. Figure 5.1b shows the load displacement relationship depicting the soil behavior under active, at-rest, and passive earth pressure. If the wall is at rest, then $P_n = P_o$ (at-rest condition). By moving the wall gradually outward, P_n will reduce accordingly, the soil goes through elastic movement, into plastic, and eventually reaches an unrestricted plastic flow (failure) and thus defines the active collapse load, P_a. Conversely, by moving the wall gradually inward, P_n will increase accordingly and eventually reach an unrestricted plastic flow (failure), and that defines the passive collapse load, P_p. Table 5.1 shows the magnitudes of wall movement to reach active and passive failure according to previous studies (DM7.2, 1982).

Details of the theories describing the lateral earth pressure/force under these conditions are presented in the first part of this chapter. It will soon become obvious that it is imperative to distinguish the differences between pressure and force. Although a seemingly simple issue, failure to recognize the difference can lead to misunderstandings in the development of the earth pressure theories and errors in the design of the retaining structures.

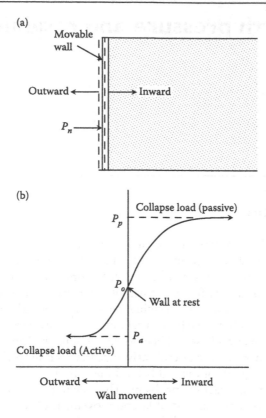

Figure 5.1 Relationship between horizontal force experienced by the wall and wall movement. (a) Section view of the test wall. (b) Load displacement relationship.

No significant changes have been made in the theories of lateral earth pressure analysis in recent history. The materials and methods used in the construction of retaining structures, on the other hand, have been improved tremendously. Concrete or masonry retaining walls, with or without reinforcement, have been used for centuries and are still used today. The reinforced earth is likely the newest addition to the family of retaining wall structures. By embedding manufactured materials such as steel strips or plastic sheets into an earth structure, the overall performance of the composite material is significantly enhanced. The reinforced earth walls are becoming more popular as retaining structures

Table 5.1 Magnitudes of wall movement to reach failure

Soil type and condition	Wall movement, y/H[a]	
	Active	Passive
Dense cohesionless	0.0005	0.002
Loose cohesionless	0.002	0.006
Stiff cohesive	0.01	0.02
Soft cohesive	0.02	0.04

Source: DM7.2. 1982. *Foundations and Earth Structures Design Manual 7.2.* Naval Facilities Engineering Command, Alexandria, VA, Department of Navy.

Note:

[a] y = horizontal movement, H = height of the wall.

due to their competiveness in cost, performance, and versatility. Prefabricated, interlocking sheets (i.e., thin plates) that can be made of metal, concrete, or timber driven into the ground have been used to retain soil. These types of retaining structures are referred to as sheet pile walls. The second part of this chapter is dedicated to the design of the various types of retaining walls.

The calculations described in this chapter may involve elaborate equations which are cumbersome to use without the help of a computer program, and charts and tables as typically offered in many textbooks to help interpolate the necessary values for the equations. A series of Excel spreadsheet programs are provided on the publisher's website for registered users. With the help of these spreadsheet computer programs, the calculations can be significantly simplified and expedited. The charts and tables are kept to a minimum.

5.2 AT-REST LATERAL EARTH PRESSURE

Under at-rest condition, the soil behind the retaining wall is at a static equilibrium with zero lateral strain. For a soil element (see Figure 5.2a) behind the retaining wall at a given depth z (or given vertical effective stress, σ'_v), the Mohr circle (circle ab in Figure 5.2b) that represents the state of stress for that soil element is below the failure envelope, as shown in Figure 5.2b. It is assumed here that the soil behind the retaining wall is normally

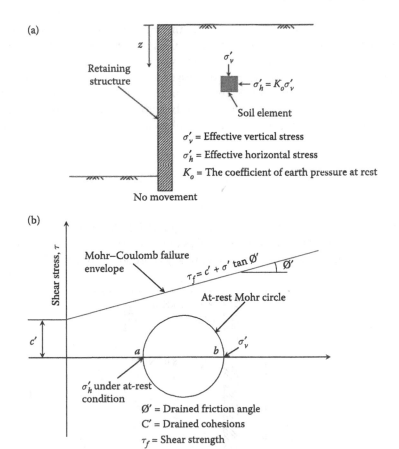

Figure 5.2 Vertical and horizontal stresses under at-rest condition. (a) Soil element behind the retaining wall. (b) The Mohr circle.

consolidated where the effective horizontal stress, σ_h', is the minor principal stress (i.e., $\sigma_h' = \sigma_3'$) and the effective vertical stress, σ_v', is the major principal stress (i.e., $\sigma_v' = \sigma_1'$). If the soil behind the retaining wall is overconsolidated, then it is possible that $\sigma_h' = \sigma_1'$ and $\sigma_v' = \sigma_3'$ (note σ_1' is always larger than σ_3'). In any case, the Mohr circle under at-rest condition is below the failure envelope.

The effective horizontal stress σ_h' can be related to effective vertical stress, σ_v', as

$$\sigma_h' = K_o \sigma_v' \tag{5.1}$$

where
K_o = the coefficient of earth pressure at rest

The empirical equation proposed by Jaky (1944) is widely used for the estimation of K_o for normally consolidated soil where K_o relates to drained soil friction angle, $\emptyset'$, as

$$K_o = 1 - \sin \emptyset' \tag{5.2}$$

For overconsolidated soil, K_o may be expressed as (Mayne and Kulhawy, 1982)

$$K_o = \left(1 - \sin \emptyset'\right) \mathrm{OCR}^{\sin \emptyset'} \tag{5.3}$$

where
OCR = overconsolidation ratio

Note that water has no strength (i.e., $\emptyset' = 0$) and thus pore water pressure, u, is isotropic, or its $K_o = 1$. For this reason, where water table is present behind a retaining wall, the lateral soil pressure (based on effective stress) and water pressure are calculated separately and then combined. The combined lateral pressure is higher than that without water. Also, water freezes and expands in cold temperatures. To minimize these negative effects, we usually provide a filter and drainage system behind the retaining wall to assure that there is no accumulation of water. Details of filter designs are discussed in the latter part of this chapter.

EXAMPLE 5.1

Given

For the rigid and fixed retaining wall shown in Figure E5.1, the groundwater table is at 3 m below the ground surface. Consider soil as dry above the groundwater table. The granular soil behind the retaining wall is normally consolidated (OCR = 1).

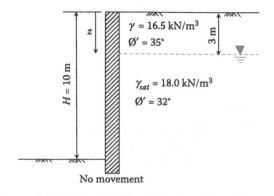

z

$\gamma = 16.5 \ \mathrm{kN/m^3}$
$\emptyset' = 35°$
3 m

$\gamma_{sat} = 18.0 \ \mathrm{kN/m^3}$
$\emptyset' = 32°$

$H = 10 \ \mathrm{m}$

No movement

Figure E5.1 The rigid and fixed retaining wall.

Required

Calculate and plot the lateral pressure distribution on the back of the retaining wall.

Solution

For soil layer 1 (0 to 3 m):

$$\gamma' = \gamma = 16.5 \, \text{kN/m}^3$$

For normally consolidated soil,

$$K_o = (1 - \sin \varnothing') = 1 - \sin (35°) = 0.43 \tag{5.2}$$

For soil layer 2 (3 to 10 m),

$$\gamma' = \gamma_{sat} - \gamma_w = (18.0) - (9.81) = 8.19 \, \text{kN/m}^3$$

$$K_o = (1 - \sin \varnothing') = 1 - \sin (32°) = 0.47 \tag{5.2}$$

Calculations of the effective lateral stress, σ'_h, and pore water pressure, u, at $z = 0, 3,$ and 10 m are shown in Table E5.1. The pressure profiles depicted in Figure E5.1a are based on the results in Table E5.1.

Table E5.1 Computations of σ'_h and u

z, m	σ'_v, kPa	$\sigma'_h = K_o \sigma'_v$, kPa	u, kPa	$\sigma'_h + u$, kPa
0	0	0	0	0
3	49.50	21.11	0	21.11
3	49.50	23.27	0	23.27
10	106.83	50.22	68.67	118.89

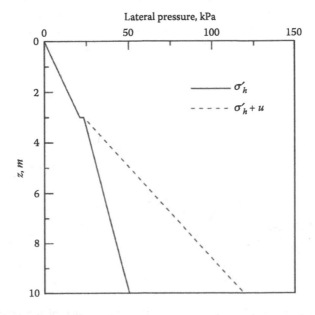

Figure E5.1a Lateral pressure profiles.

5.3 LATERAL EARTH PRESSURE—COULOMB'S METHOD

Coulomb (1776) approached the problem of calculating the active and passive lateral pressure on a retaining wall using the concept of limit (or force) equilibrium. The method involves the following assumptions:

- The soil behind the retaining wall is uniform and can have both friction and cohesion.
- The soil is assumed to fail by sliding along a plane.
- The frictional resistance is distributed uniformly along the failure plane.
- The wedge bounded by the wall, the backfill surface, and failure plane moves as a rigid body.
- There can be friction between the wall and soil, represented by a soil–wall interface friction angle, δ'.
- The failure is a plane strain problem. The analysis considers a unit thickness of the soil and wall system.

It is important to note that Coulomb's method is based on forces. The direct outcome of the analysis is lateral force. The lateral earth pressure is derived based on an assumption that it increases linearly with depth and that the integration of the earth pressure along the wall gives the same total force as that from the force-based analysis.

5.3.1 Coulomb active earth pressure

To describe the method of calculating the active earth pressure, let's start with a simple case of a smooth, vertical wall and a flat granular soil ($c' = 0$) backfill (see Figure 5.3a). In this case, the soil–wall interface friction angle, $\delta' = 0$. The drained friction angle of the granular soil is $\emptyset'$. The height of the wall, H, and unit weight of the granular backfill (γ) are given. A failure plane inclined at an angle Ω from the vertical as shown in Figure 5.3a is assumed. The angle Ω is arbitrarily chosen initially. Since the wedge ABC in Figure 5.3a is in a state of equilibrium, the force polygon indicated in Figure 5.3b must close. With H, γ, and Ω either given or assumed, the weight of the wedge ABC, W, can be calculated as

$$W = \frac{1}{2}\gamma' H^2 \tan \Omega \tag{5.4}$$

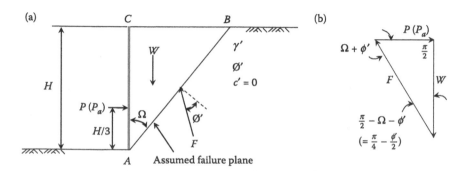

Figure 5.3 Coulomb's limit equilibrium approach. (a) Sectional view of the retaining wall and the assumed failure plane. (b) The force polygon.

F is the reaction force along the face AB, at an angle $\varnothing'$ to the normal (i.e., known direction). The lateral reaction force P in horizontal direction (see Figure 5.3b) can be determined using the force polygon.

$$P = W \tan\left(\frac{\pi}{2} - \Omega - \phi'\right) = \frac{1}{2}\gamma' H^2 \tan\Omega \cdot \tan\left(\frac{\pi}{2} - \Omega - \phi'\right) \tag{5.5}$$

The lateral active force, P_a, corresponds to the case where Ω is chosen so that P reaches its maximum value. It can be demonstrated mathematically (i.e., set derivative $dP/d\Omega = 0$ and determine Ω) that when $\Omega = \frac{\pi}{4} - \frac{\phi'}{2}, P = P_a$ (maximum P).
Or

$$P_a = \frac{1}{2}\gamma' H^2 \tan^2\left(\frac{\pi}{4} - \frac{\phi'}{2}\right) \tag{5.6}$$

We lump the latter part of Equation (5.6) into one variable, K_a, so that

$$P_a = \frac{1}{2}\gamma' H^2 K_a \tag{5.7}$$

where
$K_a = $ Coulomb's active earth pressure coefficient

and

$$K_a = \tan^2\left(\frac{\pi}{4} - \frac{\phi'}{2}\right) = \frac{1 - \sin\phi'}{1 + \sin\phi'} \tag{5.8}$$

Assuming lateral earth pressure increases linearly with depth, P_a acts at $H/3$ from the base of the wall (see Figure 5.3a).

Using a similar approach, we can calculate the active lateral force P_a for more general conditions where the retaining wall has a back face inclined at an angle of β with the vertical, and the granular backfill slopes at an angle of α with the horizontal, as shown in Figure 5.4. The angle of friction between the soil and the wall is δ'. In this case, P or P_a is inclined at an angle δ' to the normal of the back face, and

$$K_a = \frac{\cos^2(\beta - \varnothing')}{\cos^2(\beta)\cos(\beta + \delta')\left[1 + \sqrt{\dfrac{\sin(\varnothing' + \delta')\sin(\varnothing' - \alpha)}{\cos(\beta + \delta')\cos(\beta - \alpha)}}\right]^2} \tag{5.9}$$

In practice, it is generally assumed that δ' ranges from $\frac{1}{2}\phi'$ to $\frac{2}{3}\phi'$. In active condition, the failure wedge moves away from the backfill and with a downward vertical component.

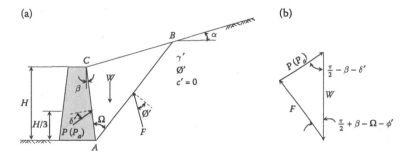

(a)   (b)

Figure 5.4 Coulomb's active force analysis for general case. (a) Sectional view of the retaining wall and the assumed failure plane. (b) The force polygon.

When there is friction between the wall and soil (i.e., $\delta' \neq 0$), P_a reacts to the failure wedge movement and therefore points upward toward the failure wedge, as shown in Figure 5.4.

The Excel spreadsheet program available on the publisher's website is capable of calculating the Coulomb and Rankine (described later) earth pressure coefficients. With the help of this spreadsheet computer program, the coefficients can be readily calculated with any reasonable combinations of ϕ', δ', α, and β.

5.3.2 Coulomb passive earth pressure

The derivation for the Coulomb passive earth pressure coefficient is similar to the active case presented above. The main difference is that in passive condition, the failure wedge is pushed laterally toward the backfill with an upward component, by the lateral force stemming from the retaining wall. When there is friction between the wall and soil (i.e., $\delta' \neq 0$), the passive lateral force P_p reacts to the failure wedge movement and points downward, as shown in Figure 5.5a. For an arbitrarily chosen Ω, the lateral force P is calculated using the force polygon shown in Figure 5.5b. The lateral passive force P_p corresponds to the case where Ω is chosen so that P reaches its minimum value. By setting the derivative $dP/d\Omega = 0$, $\Omega = \frac{\pi}{4} + \frac{\phi'}{2}$, and $P = P_p$.

In this case,

$$P_p = \frac{1}{2}\gamma' H^2 K_p \tag{5.10}$$

Where K_p = Coulomb's passive earth pressure coefficient, and

$$K_p = \frac{\cos^2(\beta + \varnothing')}{\cos^2(\beta)\cos(\beta - \delta')\left[1 - \sqrt{\dfrac{\sin(\varnothing' + \delta')\sin(\varnothing' + \alpha)}{\cos(\beta - \delta')\cos(\beta - \alpha)}}\right]^2} \tag{5.11}$$

Again, assume passive earth pressure increases linearly with depth; P_p therefore acts at $H/3$ from the base of the wall (point A of Figure 5.5a). For a simple case of smooth ($\delta' = 0$), vertical wall ($\beta = 0$), and a flat ($\alpha = 0$) backfill, Equation (5.11) is simplified as

$$K_p = \frac{\cos^2(\varnothing')}{[1 - \sin(\varnothing')]^2} = \tan^2\left(\frac{\pi}{4} + \frac{\phi'}{2}\right) = \frac{1 + \sin\phi'}{1 - \sin\phi'} \tag{5.12}$$

Thus for the same simple case ($\delta' = 0$, $\beta = 0$, and $\alpha = 0$), $K_p = 1/K_a$.

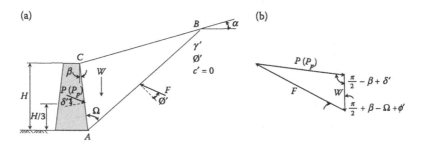

Figure 5.5 Coulomb's passive force analysis. (a) Sectional view of the retaining wall and the assumed failure plane. (b) The force polygon.

EXAMPLE 5.2

Given

For the wall described in Figure E5.2, the backfill is granular material. Consider the following soil parameters, wall, and backfill configurations:

$\emptyset' = 30°, 35°,$ and $40°$
$c' = 0$
$\delta = (2/3)\,\emptyset'$
$\alpha = 0°, 5°,$ and $10°$
$\beta = 0°, 10°,$ and $20°$

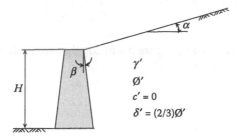

Figure E5.2 Wall and backfill conditions for Example 5.2.

Required

Compute the Coulomb K_a and K_p coefficients under these conditions.

Solution

K_a of Coulomb's method is

$$K_a = \frac{\cos^2(\beta - \emptyset')}{\cos^2(\beta)\,\cos(\beta + \delta')\left[1 + \sqrt{\dfrac{\sin(\emptyset' + \delta')\,\sin(\emptyset' - \alpha)}{\cos(\beta + \delta')\,\cos(\beta - \alpha)}}\right]^2} \tag{5.9}$$

and K_p of Coulomb's method is

$$K_p = \frac{\cos^2(\beta + \emptyset')}{\cos^2(\beta)\,\cos(\beta - \delta')\left[1 - \sqrt{\dfrac{\sin(\emptyset' + \delta')\,\sin(\emptyset' + \alpha)}{\cos(\beta - \delta')\,\cos(\beta - \alpha)}}\right]^2} \tag{5.11}$$

Using the spreadsheet program provided at the publisher's website and inserting the given parameters, the following table can be prepared.

$\emptyset'$		$\beta = 0°$ α 0°	5°	10°	$\beta = 10°$ α 0°	5°	10°	$\beta = 20°$ α 0°	5°	10°
30°	K_a	0.30	0.32	0.34	0.38	0.40	0.44	0.48	0.52	0.57
	K_p	6.11	8.05	10.90	4.45	5.62	7.16	3.58	4.42	5.45
35°	K_a	0.24	0.26	0.27	0.32	0.34	0.37	0.43	0.46	0.50
	K_p	9.96	14.23	21.54	6.47	8.56	11.62	4.81	6.13	7.88
40°	K_a	0.20	0.21	0.22	0.28	0.29	0.31	0.38	0.41	0.44
	K_p	18.72	30.81	58.46	10.18	14.49	21.71	6.80	9.06	12.34

5.4 LATERAL EARTH PRESSURE—RANKINE'S METHOD

Rankine (1857) developed the lateral earth pressure theory using the concept of stress. It is assumed that the backfill behind the retaining wall is granular (i.e., no cohesion, or $c' = 0$) and the soil within the failure zone is in a state of plastic equilibrium.

5.4.1 Rankine active earth pressure

We again start with a simple case of vertical wall ($\beta = 0$) and flat backfill ($\alpha = 0$), as shown in Figure 5.6a. Within the active failure zone, there are numerous failure planes inclined at an angle of $\left(\frac{\pi}{4}\right) + \left(\frac{\phi'}{2}\right)$ with the horizontal as shown in Figure 5.6a, as predicted by the Mohr–Coulomb failure envelope described in Figure 5.6b. When the soil element shown in Figure 5.6a is under active condition, the wall movement causes σ'_h to be low enough that the Mohr circle (circle AB in Figure 5.6b) touches the failure envelope and $\sigma'_h = \sigma'_3$ (i.e., the minor principal stress), as shown in Figure 5.6b. The Mohr–Coulomb failure envelope (note it is assumed that $c' = 0$) is defined as

$$\tau_f = \sigma' \tan \phi' \tag{5.13}$$

In this case, the major and minor principal stresses are related to each other as follows:

$$\sigma'_1 = \sigma'_3 \tan^2\left(\frac{\pi}{4} + \frac{\phi'}{2}\right) \tag{5.14}$$

where
$\sigma'_1 = \sigma'_v =$ major principal stress and $\sigma'_3 = \sigma'_h = \sigma'_a =$ minor principal stress.

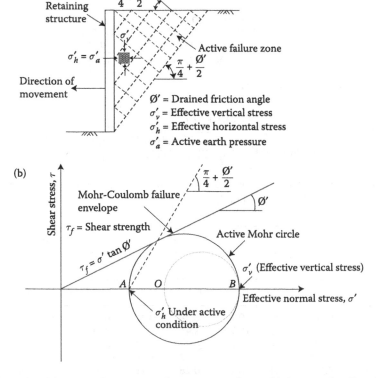

Figure 5.6 Vertical and horizontal stresses under active condition. (a) A vertical wall with flat backfill. (b) The Mohr circle.

Replace σ_v' and σ_a' into Equation (5.14),

$$\sigma_v' = \sigma_a' \tan^2\left(\frac{\pi}{4} + \frac{\varnothing'}{2}\right) \tag{5.15}$$

Reorganizing Equation (5.15) yields

$$\sigma_a' = \sigma_v' \tan^2\left(\frac{\pi}{4} - \frac{\varnothing'}{2}\right) \tag{5.16}$$

$$= \sigma_v' K_a$$

where
$$K_a = \tan^2\left(\frac{\pi}{4} - \frac{\varnothing'}{2}\right) = \frac{1 - \sin\varnothing'}{1 + \sin\varnothing'} = \text{Rankine active earth pressure coefficient.}$$

For the case shown in Figure 5.6, the wall face is vertical ($\beta = 0°$), frictionless ($\delta' = 0$), the backfill is flat ($\alpha = 0$), and K_a is the same as that of Coulomb.

Using a similar approach, the Rankine active earth pressure problem can be extended to a more general case with and inclined backfill, α (from horizontal). The active earth pressure coefficient, K_a, can be expressed as

$$K_a = \cos\alpha \frac{\cos\alpha - \sqrt{\cos^2\alpha - \cos^2\varnothing'}}{\cos\alpha + \sqrt{\cos^2\alpha - \cos^2\varnothing'}} \tag{5.17}$$

Note that for the case of Equation (5.17), the wall face is vertical ($\beta = 0°$) and frictionless ($\delta' = 0$). At any depth, z, the Rankine active earth pressure σ_a' is

$$\sigma_a' = \sigma_v' K_a = \gamma' z K_a \tag{5.18}$$

and the active lateral force on the wall is

$$P_a = \frac{1}{2}\gamma' H^2 K_a \tag{5.19}$$

The P_a and σ_a' are parallel with the surface of the backfill and inclined at an angle of α from the horizontal, as shown in Figure 5.7. P_a acts at $H/3$ from the bottom of the wall.

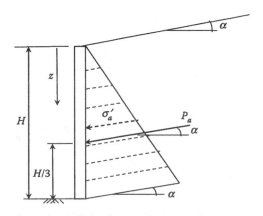

Figure 5.7 Direction of Rankine active pressure/force.

5.4.2 Rankine passive earth pressure

If we follow the simple case of vertical wall ($\beta = 0$) and flat backfill ($\alpha = 0$) but consider the passive condition as shown in Figure 5.8, the failure planes will incline at an angle of $\left(\frac{\pi}{4} - \frac{\phi'}{2}\right)$ with the horizontal as shown in Figure 5.8a, as predicted by the Mohr–Coulomb failure envelope described in Figure 5.8b. When the soil element shown in Figure 5.8a is under passive condition, the wall movement causes σ'_h to be high enough that it becomes the major principal stress before the Mohr circle (circle BC in Figure 5.8b) touches the failure envelope (i.e., $\sigma'_h = \sigma'_1 = \sigma'_p$ and $\sigma'_v = \sigma'_3$). The Mohr–Coulomb failure envelope (note it is assumed that $c' = 0$) remains the same as described by Equation (5.14). By replacing σ'_h and σ'_v into Equation (5.14), we have

$$\sigma'_p = \sigma'_v \tan^2\left(\frac{\pi}{4} + \frac{\phi'}{2}\right) \tag{5.20}$$

$$= \sigma'_v K_p$$

where
$$K_p = \tan^2\left(\frac{\pi}{4} + \frac{\phi'}{2}\right) = \frac{1 + \sin\phi'}{1 - \sin\phi'} = \text{Rankine passive earth pressure coefficient.}$$

Similarly, the Rankine passive earth pressure problem for a vertical and frictionless wall ($\beta = 0$ and $\delta' = 0$) can be extended to a more general case with and inclined backfill, α (to the horizontal). The passive earth pressure coefficient, K_p, can be expressed as

$$K_p = \cos\alpha \frac{\cos\alpha + \sqrt{\cos^2\alpha - \cos^2\phi'}}{\cos\alpha - \sqrt{\cos^2\alpha - \cos^2\phi'}} \tag{5.21}$$

(a)

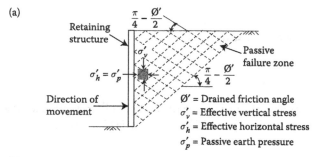

ϕ' = Drained friction angle
σ'_v = Effective vertical stress
σ'_h = Effective horizontal stress
σ'_p = Passive earth pressure

(b)

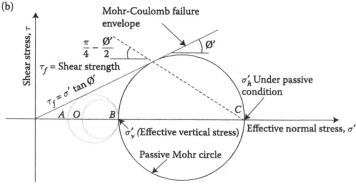

Figure 5.8 Vertical and horizontal stresses under passive condition. (a) Passive conditions of a vertical wall with flat backfill. (b) The Mohr–Coulomb failure envelope.

EXAMPLE 5.3

Given

For the same backfill conditions described in Example 5.2 with the following soil parameters:

$\varnothing' = 30°, 35°,$ and $40°$
$c' = 0$
$\alpha = 0°, 5°,$ and $10°$

But,

$\delta' = 0°,$ and
$\beta = 0°$

Required

Compute the Rankine K_a and K_p under these conditions.

Solution

According to Equation (5.17), K_a of Rankine's method is

$$K_a = \cos \alpha \frac{\cos \alpha - \sqrt{\cos^2 \alpha - \cos^2 \varnothing'}}{\cos \alpha + \sqrt{\cos^2 \alpha - \cos^2 \varnothing'}} \qquad (5.17)$$

and K_p of Rankine's method is

$$K_p = \cos \alpha \frac{\cos \alpha + \sqrt{\cos^2 \alpha - \cos^2 \varnothing'}}{\cos \alpha - \sqrt{\cos^2 \alpha - \cos^2 \varnothing'}} \qquad (5.21)$$

Using the spreadsheet program provided at the publisher's website and inserting the given parameters, the following table can be prepared.

		$\beta = 0°$		
		α		
$\varnothing'$		0°	5°	10°
30°	K_a	0.33	0.34	0.35
	K_p	3.00	2.94	2.77
35°	K_a	0.27	0.27	0.28
	K_p	3.69	3.63	3.44
40°	K_a	0.22	0.22	0.22
	K_p	4.60	4.53	4.32

The Rankine K_p actually decreases with α. This trend is not reasonable.

5.5 LATERAL EARTH PRESSURE—LIMIT ANALYSIS

As stated in Chapter 2, the upper- and lower-bound methods are based in the rigorous theory of plasticity and are theoretically correct for perfectly plastic materials. The limit equilibrium method (Terzaghi, 1943) makes use of a failure mechanism and ensures

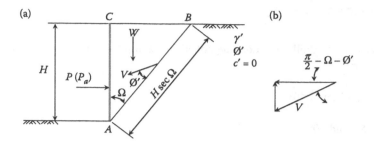

Figure 5.9 Limit analysis for active earth pressure. (a) Failure mechanism. (b) Velocity field.

overall force equilibrium of soil structures. Limit equilibrium analysis is similar to upper-bound limit analysis in that both approaches make use of a failure mechanism. The upper-bound method is based on an energy balance equation to derive an upper-collapse load, while the limit equilibrium method is based on force equilibrium to derive an estimated collapse load. For a class of translational failure mechanism, the solution from a limit equilibrium calculation may be regarded as an upper bound. For more general analysis, however, the limit equilibrium method would not satisfy all the requirements of either an upper-bound or a lower-bound theorem and therefore would not provide a bound solution on the exact collapse load.

To demonstrate the limit analysis for active and passive lateral earth pressure, we again start with a simple case of frictionless ($\delta' = 0$) and vertical ($\beta = 0$) wall with a flat backfill ($\alpha = 0$) of cohesionless soil ($c' = 0$). Figure 5.9 shows the case of upper-bound (based on kinematically admissible velocity field) analysis for the active earth pressure. The failure mechanism shown in Figure 5.9a consists of two rigid bodies (i.e., soil wedge ABC and remainder of the backfill) and a plane sliding surface (AB). The angle Ω is again arbitrarily chosen initially. The compatible velocity field is depicted in Figure 5.9b where velocity V of the soil wedge ABC is at an angle $\emptyset'$ to the sliding plane AB.

The rate of internal energy dissipation occurs along AC and AB. As the wall is frictionless, no energy is dissipated along AC. The rate of internal energy, W_i, dissipation along AB is $W_i = c'V\cos\emptyset'(H\sec\Omega) = 0$ since $c' = 0$. The rate of external work, W_e, consists of the work done by soil wedge moving downward and that by the force P moving horizontally:

$$W_e = \left(\frac{1}{2}\gamma'H^2\tan\Omega\right)\left[V\cos(\Omega+\emptyset')\right] - PV\sin(\Omega+\emptyset') \tag{5.22}$$

Equating the internal and external work $W_i = W_e$, thus

$$\left(\frac{1}{2}\gamma'H^2\tan\Omega\right)\left[V\cos(\Omega+\emptyset')\right] - PV\sin(\Omega+\emptyset') = 0$$

or

$$P = \left(\frac{1}{2}\gamma'H^2\tan\Omega\right)\cot(\Omega+\emptyset') \tag{5.23}$$

When $\Omega = \frac{\pi}{4} - \frac{\emptyset'}{2}$, P reaches its maximum value or $P = P_a = \frac{1}{2}\gamma'H^2\tan^2\left(\frac{\pi}{4} - \frac{\emptyset'}{2}\right)$, which is the same as Equation (5.6).

For the passive earth pressure case, the mechanism is very similar. The main difference is that V is now pointing upward but still inclined at an angle $\emptyset'$ to the sliding plane. The rate of internal and external work is found in the same manner as for the active earth pressure; the upper bound is minimized to obtain the passive earth pressure. When $\Omega = \frac{\pi}{4} + \frac{\varnothing'}{2}$, P reaches its minimum value, or $P = P_p = \frac{1}{2}\gamma'H^2 \tan^2(\frac{\pi}{4} + \frac{\varnothing'}{2})$, which is the same as Equation (5.12).

Figure 5.10 shows the discontinuous equilibrium solution to obtain the lower bounds (based on statically admissible stress field) in which K is a chosen parameter such that the Coulomb yield criterion will be satisfied in region I. The Mohr circles in the figure show that the two K values that correspond to K_a and K_p, respectively, can furnish the needed lower-bound solutions of the lateral earth pressure problem, and for active earth pressure:

$$K = K_a = \tan^2\left(\frac{\pi}{4} - \frac{\varnothing'}{2}\right) \tag{5.24}$$

for passive earth pressure:

$$K = K_p = \tan^2\left(\frac{\pi}{4} + \frac{\varnothing'}{2}\right) \tag{5.25}$$

In this case the upper and the lower bounds for the active and passive earth pressure, respectively, are identical, indicating that these are exact solutions.

The Coulomb and Rankine methods consistently overestimate the passive earth pressures when compared with those measured in the field and laboratory model tests (Chen, 1975) when $\varnothing'$ becomes larger than 35°. Chen (1975) tried six different possible failure mechanisms in his investigation in the use of upper-bound solution for active and

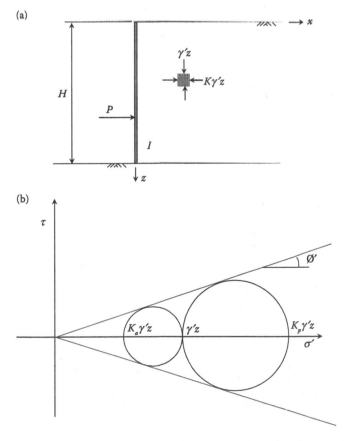

Figure 5.10 Lower-bound stress field for lateral earth pressure. (a) An equilibrium solution. (b) Mohr's circles of region I.

passive earth pressures. It was concluded that the so-called log-sandwich mechanism as shown in Figure 5.11a provided the best results in many aspects. In the log-sandwich mechanism, the nonlinear failure surface is defined by two parameters, ρ and Ψ. (Note: In Coulomb and Rankine methods, a single parameter such as Ω in Figure 5.5a was used to define the sliding plane.) The angles ρ and Ψ in Figure 5.11a are arbitrarily chosen initially; the corresponding lateral earth pressure coefficients termed K_{at} and K_{pt} are temporary values for active and passive conditions, respectively, and based on the trial ρ and Ψ. For cohesionless backfill and a smooth wall ($\delta' < \varnothing'$):

$$\left\{ \begin{matrix} K_{at} \\ K_{pt} \end{matrix} \right\} = \frac{\mp \sec \delta'}{A} \times \{B + C[D+E] + F\} \tag{5.26}$$

and

$$A = \mp \cos + \tan \delta' \sin \beta - \frac{\tan \delta' \sin(\rho+\beta)}{\cos \rho} \tag{5.26a}$$

$$B = \frac{\tan \rho \cos(\rho \pm \varnothing') \sin(\rho+\beta)}{\cos \beta \cos \varnothing'} \tag{5.26b}$$

$$C = \frac{\cos^2(\rho \pm \varnothing')}{\cos \rho \cos \beta \cos^2 \varnothing' \left(1 + 9 \tan^2 \varnothing'\right)} \tag{5.26c}$$

$$D = \sin(\beta+\rho)\big[\pm 3 \tan \varnothing' + (\mp 3 \tan \varnothing' \cos \Psi + \sin \Psi) \exp(\mp 3 \Psi \tan \varnothing')\big] \tag{5.26d}$$

$$E = \cos(\beta+\rho)\big[1 + (\mp 3 \tan \varnothing' \sin \Psi - \cos \Psi) \exp(\mp 3 \Psi \tan \varnothing')\big] \tag{5.26e}$$

$$F = \frac{\cos^2(\rho \pm \varnothing') \cos(\beta+\rho+\Psi-\alpha) \sin(\beta+\rho+\Psi) \exp(\mp 3 \Psi \tan \varnothing')}{\cos \varnothing' \cos \beta \sin(\beta+\rho+\Psi \pm \varnothing'-\alpha) \cos \rho} \tag{5.26f}$$

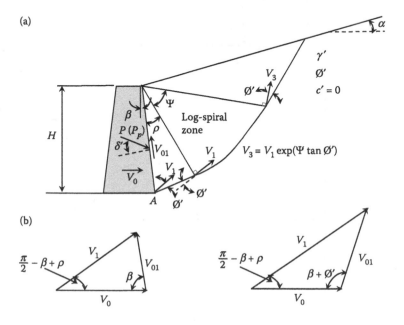

Figure 5.11 The log-sandwich mechanism. (a) Failure mechanism. Smooth wall $\delta' < \varnothing'$. Rough Wall $\delta' \geq \varnothing'$. (b) Velocity field.

For cohesionless backfill and a rough wall ($\delta' \geq \varnothing'$):

$$\begin{Bmatrix} K_{at} \\ K_{pt} \end{Bmatrix} = \frac{\mp \sec \delta'}{A} \times \{B \mp C[D + E] + F\} \tag{5.27}$$

and

$$A = \mp \cos \beta + \tan \delta' \sin \beta \tag{5.27a}$$

$$B = \frac{\sin^2 \rho \cos(\rho \pm \varnothing') \sin(\beta + \rho) \cos(\beta \pm \varnothing')}{\cos^2 \beta \cos \varnothing' \cos(\rho \mp \varnothing')} \tag{5.27b}$$

$$C = \frac{\cos^2(\rho \pm \varnothing') \cos(\beta \pm \varnothing')}{\cos^2 \beta \cos^2 \varnothing'(1 + 9\tan^2 \varnothing') \cos(\rho \mp \varnothing')} \tag{5.27c}$$

$$D = \sin(\beta + \rho) \\ \times \left[\pm 3 \tan \varnothing' + (\mp 3 \tan \varnothing' \cos \Psi + \sin \Psi) \exp(\mp 3\Psi \tan \varnothing')\right] \tag{5.27d}$$

$$E = \cos n(\beta + \rho)\left[1 + (\mp 3 \tan \varnothing' \sin \Psi - \cos \Psi) \exp(\mp 3 \Psi \tan \varnothing')\right] \tag{5.27e}$$

$$F = \frac{\cos^2(\rho \pm \varnothing') \cos(\beta + \rho + \Psi - \alpha) \sin(\beta + \rho + \Psi) \cos n(\beta \pm \varnothing') \exp(\mp 3 \Psi \tan \varnothing')}{\cos^2 \beta \cos \varnothing' \sin(\beta + \rho + \Psi - \alpha \pm \varnothing') \cos(\rho \mp \varnothing')} \tag{5.27f}$$

To apply these solutions to obtain K_a and K_p, an iterative procedure is required to select ρ and Ψ such that the maximum $K_{at} = K_a$ and minimum $K_{pt} = K_p$. To expedite the computations, the following initial values for ρ and Ψ (in degrees) may be considered (Bowles, 1996) in the search routine:

$$\rho \approx 0.5(90 + \alpha - \beta)$$

$$\Psi \approx 0.2(90 + \alpha - \beta)$$

The computation may not converge if extreme initial values of ρ and Ψ (i.e., close to 0 or $\alpha + \beta$) are used. It should be noted that in comparison with those of Coulomb and Rankine methods, significant differences exist mostly in passive earth pressure when using the log-sandwich method. The passive earth pressures from log-sandwich method compare favorably with results from rigorous numerical analysis (Chen, 1975). Table 5.2 shows selected K_p values from log-sandwich solution for flat, cohesionless backfill ($\alpha = 0$).

Table 5.2 Selected K_p from log-sandwich solution for flat, cohesionless backfill ($\alpha = 0$)

$\varnothing'$ (degree)	δ' (degree)	$\beta = 0°$	$\beta = 20°$
20	0	2.11	1.82
	10	2.61	2.11
	20	3.54	2.60
30	0	3.23	2.53
	15	4.71	3.22
	30	9.11	4.90
40	0	6.98	3.55
	20	10.70	5.44
	40	30.81	15.66

A spreadsheet program is provided at the publisher's website for the computation of K_p using the upper-bound log-sandwich method.

EXAMPLE 5.4

Given

For the same backfill conditions described in Examples 5.2 and 5.3 with the following soil parameters:

$\emptyset' = 30°, 35°,$ and $40°$
$c' = 0$
$\alpha = 0°,$ and $5°$

But,

$\delta = 0°,$ and $(2/3)\emptyset'$
$\beta = 0°,$ and $20°$ (wall inclination)

Required

Compute the K_p values under these conditions using the upper-bound log-sandwich method.

Solution

Use the spreadsheet program provided at the publisher's website and insert the given parameters, the following table for K_p can be prepared:

| | | α | | | |
| | | $\beta = 0°$ | | $\beta = 20°$ | |
$\emptyset'$		$0°$	$5°$	$0°$	$5°$
$30°$	$\delta = 0°$	3.34	3.88	2.53	2.93
	$\delta = (2/3)\emptyset'$	5.58	6.84	3.61	4.33
$35°$	$\delta = 0°$	4.38	5.20	3.06	3.61
	$\delta = (2/3)\emptyset'$	8.39	10.70	4.77	5.88
$40°$	$\delta = 0°$	6.04	7.36	3.83	4.62
	$\delta = (2/3)\emptyset'$	13.57	18.21	6.61	8.40

The K_p values are smaller than those using the Coulomb method under similar conditions (Example 5.2).

5.6 EFFECTS OF COHESION ON LATERAL EARTH PRESSURE

The original development of K_a and K_p by Coulomb and Rankine did not directly consider cohesion. So far, the description on active and passive earth pressure involves $\emptyset'$ only. Cohesion, c', has been left out deliberately because it has to be treated with caution. To start with, let's consider the active case shown in Figure 5.12 where the Mohr–Coulomb failure envelope involves both c' and $\emptyset'$ as

$$\tau_f = c' + \sigma' \tan \emptyset' \tag{5.28}$$

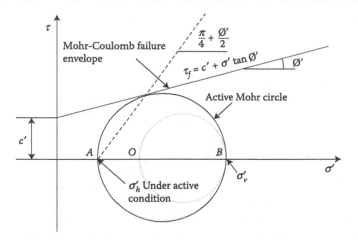

Figure 5.12 Active failure envelope with cohesion.

This is the same as Rankine's stress-based approach. The direction of the failure planes (dashed line in Figure 5.12) remains the same. The major and minor principal stresses on the Mohr circle are now related to each other as follows:

$$\sigma_1' = \sigma' \tan^2\left(\frac{\pi}{4} + \frac{\varnothing'}{2}\right) + 2c' \tan\left(\frac{\pi}{4} + \frac{\varnothing'}{2}\right) \tag{5.29}$$

where $\sigma_1' = \sigma_v' =$ major principal stress and $\sigma_3' = \sigma_h' = \sigma_a' =$ minor principal stress.
Replace σ_v' and σ_a' into Equation (5.29),

$$\sigma_v' = \sigma_a' \tan^2\left(\frac{\pi}{4} + \frac{\varnothing'}{2}\right) + 2c' \tan\left(\frac{\pi}{4} + \frac{\varnothing'}{2}\right) \tag{5.30}$$

Reorganizing Equation (5.30) yields

$$\sigma_a' = \sigma_v' \tan^2\left(\frac{\pi}{4} - \frac{\varnothing'}{2}\right) - 2c' \tan\left(\frac{\pi}{4} - \frac{\varnothing'}{2}\right) \tag{5.31}$$

The general practice is to take $K_a = \tan^2\left(\frac{\pi}{4} - \frac{\varnothing'}{2}\right)$ as the Rankine active earth pressure coefficient. Through analogy, Equation (5.31) can now be expressed as

$$\sigma_a' = \sigma_v' K_a - 2c'\sqrt{K_a} \tag{5.32}$$

Therefore K_a is not affected by the consideration of c', but σ_a' is reduced because of the second term in Equation (5.32).

Similarly, c' can be added to the calculation of shear resistance along the failure plane in Coulomb's force-based analysis, and P_a will also be reduced. In any case, it is clear that c' causes P_a to be lowered, or the demand for the retaining structural capacity is reduced. The question is, can we rely on the reduced P_a in our design?

At ground surface ($z = 0$), $\sigma_v' = 0$ and $\sigma_a' = -2c'\sqrt{K_a}$, a negative value or σ_a' is a tensile stress as indicated in Figure 5.13. The tensile stress can cause cracks within soil and soil-wall interface because soil has practically zero tensile strength. This tensile stress decreases with depth until it reaches a depth where σ_a' becomes zero as

$$\sigma_a' = \gamma' z_c K_a - 2c'\sqrt{K_a} = 0$$

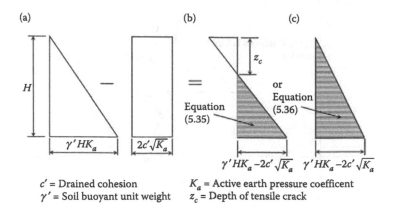

c' = Drained cohesion K_a = Active earth pressure coefficent
γ' = Soil buoyant unit weight z_c = Depth of tensile crack

Figure 5.13 Rankine active pressure with tensile stress. (a) Active pressure with tensile stress. (b) Active pressure below the tensile crack only. (c) Active pressure from ground surface.

thus

$$z_c = \frac{2c'}{\gamma'\sqrt{K_a}} \tag{5.33}$$

z_c is called the depth of tensile crack. If the effects of the tensile stress are considered (assuming no cracks), then the total Rankine active force per unit length of the wall is

$$P_a = \frac{1}{2}\gamma'H^2K_a - 2c'H\sqrt{K_a} \tag{5.34}$$

A higher c' causes lower P_a (i.e., less demand on retaining structure). Thus an overestimation of c' and/or its effects can lead to unsafe design. In practice, P_a is usually calculated considering that the tensile crack has appeared and only the earth pressures between $z = z_c$ and $z = H$ are included, as shown in the shaded area of Figure 5.13b, where

$$P_a = \frac{1}{2}(H - z_c)\left(\gamma'HK_a - 2c'\sqrt{K_a}\right) \tag{5.35}$$

Or, assume that the crack is also filled with soil, and

$$P_a = \frac{1}{2}H\left(\gamma'HK_a - 2c'\sqrt{K_a}\right) \tag{5.36}$$

As shown in the shaded area of Figure 5.13c.

Theoretically, the Rankine and Coulomb earth pressure theories can also be formulated in total stress (i.e., in terms of γ, $\varnothing$, and c) or $\varnothing = 0$ conditions in saturated clays (i.e., $\gamma = \gamma_{sat}$ and $c = s_u$). After all, we build earth-retaining structures mostly for long-term applications, and the backfill will eventually become drained (excess pore water pressure fully dissipated). Unless it can be demonstrated that the earth pressure is more critical in undrained conditions, the earth pressures are usually calculated with drained soil parameters in the design of retaining structures.

EXAMPLE 5.5

Given

Figure E5.5 shows a case of vertical wall ($\beta = 0$) with flat ($\alpha = 0$) cohesive frictional backfill. The soil parameters are included in Figure E5.5.

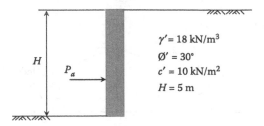

Figure E5.5 Vertical wall with a flat, cohesive frictional backfill.

Required

Determine the depth of tensile crack, z_c, and Rankine lateral earth force (P_a) considering the effects of tensile crack according to Equation (5.35) and Equation (5.36).

Solution

For the soil conditions given, $K_a = \tan^2\left(\dfrac{\pi}{4} - \dfrac{\varnothing'}{2}\right) = 0.33$

$$z_c = \frac{2c'}{\gamma'\sqrt{K_a}} = (2)(10)/\left[(18)(0.33)^{0.5}\right] = 1.93\,\mathrm{m} \tag{5.33}$$

Before tensile crack develops, compute P_a using Equation (5.34)

$$P_a = \frac{1}{2}\gamma' H^2 K_a - 2c'H\sqrt{K_a} = (0.5)(18)(5^2)(0.33) - (2)(10)(5)(0.33^{0.5})$$
$$= 17.27\,\mathrm{kN/m}$$

Consider the tensile crack and use Equation (5.35):

$$P_a = \frac{1}{2}(H - z_c)\left(\gamma' H K_a - 2c'\sqrt{K_a}\right)$$
$$= (0.5)(5 - 1.93)\left[(18)(5)(0.33) - (2)(10)(0.33^{0.5})\right] = 28.38\,\mathrm{kN/m}$$

Consider the tensile crack and use Equation (5.36):

$$P_a = \frac{1}{2}H\left(\gamma' H K_a - 2c'\sqrt{K_a}\right) = (0.5)(5)\left[(18)(5)(0.33) - (2)(10)(0.33^{0.5})\right]$$
$$= 46.13\,\mathrm{kN/m}$$

For the passive earth pressure, we can follow Equation (5.29) again, except for the passive case, $\sigma'_1 = \sigma'_h = \sigma'_p$ = major principal stress and $\sigma'_3 = \sigma'_v$ = minor principal stress as shown in Figure 5.14.

Replace σ'_v and σ'_p into Equation (5.29),

$$\sigma'_p = \sigma'_v \tan^2\left(\frac{\pi}{4} + \frac{\varnothing'}{2}\right) + 2c' \tan\left(\frac{\pi}{4} + \frac{\varnothing'}{2}\right) \tag{5.37}$$

Take $K_p = \tan^2\left(\frac{\pi}{4} + \frac{\varnothing'}{2}\right)$ as the Rankine passive earth pressure coefficient. Through analogy, Equation (5.37) can now be expressed as

$$\sigma'_p = \sigma'_v K_p + 2c'\sqrt{K_p} \tag{5.38}$$

Rankine passive force per unit length of the wall is

$$P_p = \frac{1}{2}\gamma' H^2 K_p + 2c'H\sqrt{K_p} \tag{5.39}$$

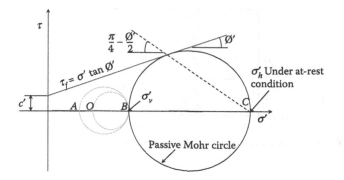

Figure 5.14 Passive Mohr circle and failure envelope with cohesion.

Therefore K_p is not affected by the consideration of c', but σ_p' and P_p are increased because of the second term in Equations (5.38) and (5.39). Similarly, c' can be added to the calculation of shear resistance along the failure plane in Coulomb's force-based analysis, and P_p will also be increased.

5.7 CONSIDERATION OF SURCHARGE

Using Coulomb's limit equilibrium approach, Okabe (1924) proposed a general method that considers the effects of surcharge on the surface of the backfill. Figure 5.15 shows the failure wedge and force polygon for the active earth pressure analysis, considering a uniform surcharge distribution and the backfill is cohesive-frictional ($c' \neq 0$). For wedge ABD with a trial angle Ω, the lateral force, P, can be calculated as

$$P = \frac{AB - C}{D} \tag{5.40}$$

and

$$A = \left[\frac{\gamma' h^2}{2} \frac{\cos(\beta - \alpha)}{\cos(\beta)} + qh\right] \tag{5.40a}$$

$$B = \cos(\beta - \Omega - \varnothing') \sin(\Omega) \tag{5.40b}$$

$$C = c'h \, \cos(\beta - \alpha) \cos\varnothing' \tag{5.40c}$$

$$D = \cos(\beta) \sin(\Omega + \varnothing' + \delta') \, \cos(\beta - \Omega - \alpha) \tag{5.40d}$$

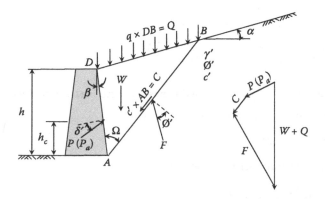

Figure 5.15 Coulomb's active earth pressure considering cohesion and surcharge.

The active earth force, P_a, corresponds to the maximum value of P when

$$\Omega = \frac{1}{2}\left(\frac{\pi}{2} + \beta - \varnothing' + \tan^{-1}\frac{bc + a\sqrt{b^2 - a^2 + c^2}}{b^2 - a^2}\right) \tag{5.41}$$

and

$$a = \sin(\delta' + \alpha) \tag{5.41a}$$

$$b = \sin(\alpha - \beta - \varnothing' - \delta') - \sin(\varnothing' - \beta)\cos(\delta' + \alpha)$$
$$+ \frac{2c'\cos(\alpha - \beta)\cos\varnothing'}{\left\{\dfrac{\gamma' h \cos(\alpha - \beta)}{\cos(-\beta)} + q\right\}}\cos(\delta' + \alpha) \tag{5.41b}$$

$$c = \sin(\varnothing' - \beta)\sin(\delta' + \alpha) - \frac{2c'\cos(\alpha - \beta)\cos\varnothing'}{\left\{\dfrac{\gamma' h \cos(\alpha - \beta)}{2\cos(-\beta)} + q\right\}}\sin(\delta' + \alpha) \tag{5.41c}$$

The lateral force P_a is acting at h_c from the bottom of the wall (point A), where

$$h_c = h\frac{\left(\dfrac{W}{3} + \dfrac{Q}{2}\right)\cos(\Omega + \varnothing' - \beta) - \dfrac{C}{2}\cos\varnothing'}{(W + Q)\cos(\Omega + \varnothing' - \beta) - C\cos\varnothing'} \tag{5.42}$$

and

$$W = \frac{\gamma' h^2 \cos(\alpha - \beta)\sin\Omega}{2\cos^2(-\beta)\cos(\Omega + \alpha - \beta)} \tag{5.42a}$$

$$Q = qh\frac{\sin\Omega}{\cos(-\beta)\cos(\Omega + \alpha - \beta)} \tag{5.42b}$$

$$C = \frac{c'h\cos(\alpha - \beta)}{\cos(-\beta)\cos(\Omega + \alpha - \beta)} \tag{5.42c}$$

where (see Figure 5.15)

W = weight of the soil within the failure wedge (ABD)
Q = total force from the surcharge acting on DB
C = resistance force along the shear plane AB due to soil cohesion

EXAMPLE 5.6

Given

For the wall and backfill conditions described in Figure E5.2. The backfill is granular material. Consider the following soil parameters, wall, and backfill configurations:

$\varnothing' = 30°$, $35°$, and $40°$
$c' = 0$
$\gamma = 18$ kN/m^3
$\delta' = (2/3)\varnothing'$
$\alpha = 0°$, $5°$, and $10°$
$\beta = 0°$, $10°$, and $20°$
H = height of the retaining wall = 5 m
q = surcharge on the backfill = 5 kN/m^2

Required

Calculate the P_a per unit thickness of the wall (in kN/m) and h_c (in m) values for the given conditions.

Solution

Use the spreadsheet program provided at the publisher's website and insert the given parameters, the following table for P_a [Equation (5.40)] and h_c [Equation (5.42)] can be prepared:

		$\beta = 0°$			$\beta = 10°$			$\beta = 20°$		
		α			α			α		
$\varnothing'$		0°	5°	10°	0°	5°	10°	0°	5°	10°
30°	P_a, kN	74.33	79.15	85.14	94.22	100.96	109.23	119.84	129.50	141.25
	h_c, m	1.75	1.75	1.75	1.75	1.75	1.75	1.75	1.75	1.75
35°	P_a, kN	61.10	64.59	68.81	80.87	86.06	92.26	106.67	114.53	123.83
	h_c, m	1.75	1.75	1.75	1.75	1.75	1.75	1.75	1.75	1.75
40°	P_a, kN	49.96	52.47	54.43	69.34	73.34	78.00	95.15	101.56	108.98
	h_c, m	1.75	1.75	1.75	1.75	1.75	1.75	1.75	1.75	1.75

h_c is largely controlled by the magnitude of q in this example. As the same q is used for the example, h_c is practically unchanged for all the applied parameters.

For surcharges occupying a limited area, the Boussinesq equation of theories of elasticity that relates lateral stress, σ_r, in the ground to a point load P on the ground surface (see Figure 5.16) can be used as a basis to estimate the lateral stress profile against the retaining wall. The Boussinesq equation states that

$$\sigma_r = \frac{P}{2\pi}\left[\frac{3r^2 z}{R^5} - \frac{1 - 2\mu}{R(R + z)}\right] \tag{5.43}$$

where

$\mu = $ Poisson's ratio
$r = \sqrt{x^2 + y^2}$
$R = \sqrt{r^2 + z^2}$

By integrating σ_r considering P located along a line (line load), a band (strip load) or with various intensity (embankment load), a wide variety of surcharge loading conditions

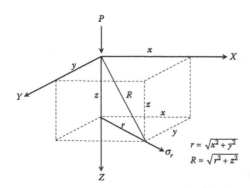

Figure 5.16 Notations of the terms in Equation (5.43).

and their locations can be considered. A lateral stress profile can be obtained by calculating σ_r along the back of the retaining wall and assuming that $\sigma_h = \sigma_r$.

5.7.1 Point load

According to his studies, Terzaghi (1954) showed that the point load-induced lateral earth pressure along the line ab, $\sigma_{h,ab}$ (see Figure 5.17) on the retaining wall can be estimated using the following empirical equations:

If $m > 0.4$
$$\sigma_{h,ab} = \left(\frac{1.77P}{H^2}\right)\left(\frac{m^2 n^2}{(m^2 + n^2)^3}\right) = I_p \frac{P}{H^2} \tag{5.44}$$

If $m \le 0.4$
$$\sigma_{h,ab} = \left(\frac{0.28P}{H^2}\right)\left(\frac{n^2}{(0.16 + n^2)^3}\right) = I_p \frac{P}{H^2} \tag{5.45}$$

The surcharge induced lateral earth pressure off the line of ab, at depth comparable to those obtained in Equations (5.44) or (5.45), can be estimated as (Terzaghi, 1954)

$$\sigma_h = \sigma_{h,ab} \cos^2(1.1\theta) \tag{5.46}$$

where

$\sigma_{h,ab}$ = point load-induced lateral stress on the retaining wall along line ab
I_p = coefficient of pressure
σ_h = point load-induced lateral stress at any point on the retaining wall
θ = angle of deviation from line ab

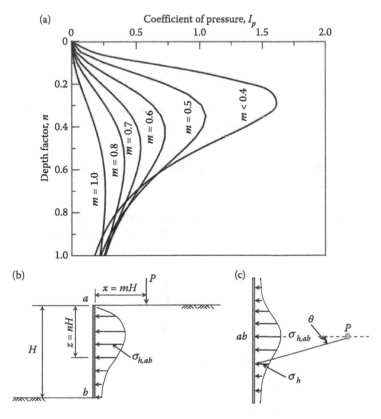

Figure 5.17 Lateral earth pressure due to a point load. (a) Lateral stress along line ab. (b) Elevation. (c) Plane.

5.7.2 Line load

Integrating Equation (5.43) along a line on the ground surface located at mH behind the wall (see Figure 5.17b) and considering $\mu = 0.5$, we can derive an equation that relates σ_h to the line load of P/unit length as

$$\sigma_h = \left(\frac{2P}{\pi H}\right)\left(\frac{m^2 n}{(m^2 + n^2)^3}\right) \tag{5.47}$$

The definitions of m and n are identical as those shown in Figure 5.17b. Studies showed that the measured σ_h values were approximately twice as much as that predicted by Equation (5.47). Terzaghi (1954) proposed the following modified equations for the line load-induced lateral stress as follows:

$$\text{If } m > 0.4 \qquad \sigma_h = \left(\frac{4P}{\pi H}\right)\left(\frac{m^2 n}{(m^2 + n^2)^3}\right) \tag{5.48}$$

$$\text{If } m \le 0.4 \qquad \sigma_{h,ab} = \left(\frac{0.203P}{H}\right)\left(\frac{n}{(0.16 + n^2)^2}\right) \tag{5.49}$$

5.7.3 Strip load

This is the case where the load is applied over a finite width, such as a highway, railroad, or embankment placed in parallel to the retaining wall. The induced lateral stress can be obtained by integrating the equation of a line load over the given width of the strip load. The derived equation is:

$$\sigma_h = \left(\frac{2P}{\pi}\right)(\beta - \sin\beta \cos 2\alpha) \tag{5.50}$$

The definition of α and β are shown in Figure 5.18. α and β are in radians and P represents applied load per unit area.

The above equations for the calculation of surcharge load-induced lateral stress can be readily executed with the help of a spreadsheet program.

5.8 ACTIVE LATERAL EARTH PRESSURE FOR EARTHQUAKE CONDITIONS

Okabe (1924) and Mononobe (1929) reported a method to consider the effects of earthquake loading for retaining walls with granular backfill. Let g be the gravity

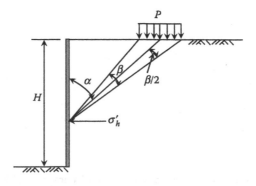

Figure 5.18 Strip load behind a retaining wall.

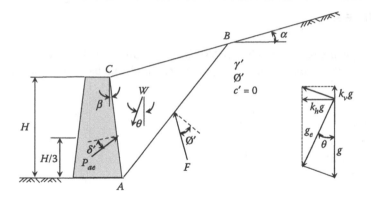

Figure 5.19 Coulomb's active force for earthquake conditions.

acceleration, $k_h g$ = maximum horizontal acceleration, and $k_v g$ = maximum upward vertical acceleration during an earthquake event. The analysis was developed considering that the direction of gravity (thus direction of soil weight, W) was rotated by an angle, θ, as shown in Figure 5.19. The magnitude of θ was calculated according to the values of k_h and k_v as

$$\tan^{-1}\theta = \frac{k_h}{1 - k_v} \tag{5.51}$$

The lateral force exerted on the retaining wall for the earthquake conditions, P_{ae}, was then derived following Coulomb's limit equilibrium procedure as

$$P_{ae} = \frac{1}{2}\gamma'H^2(1 - k_v)\frac{\cos^2(\beta - \varnothing' + \theta)}{\cos^2(\beta)\,\cos(\beta + \delta' + \theta)\left[1 + \sqrt{\frac{\sin(\varnothing' + \delta')\sin(\varnothing' - \theta - \alpha)}{\cos(\beta + \delta' + \theta)\cos(\beta - \alpha)}}\right]^2} \tag{5.52}$$

And thus an active lateral earth pressure coefficient, K_{ae}, for the earthquake conditions can be identified as

$$K_{ae} = \frac{\cos^2(\beta - \varnothing' + \theta)}{\cos\theta\cos^2(\beta)\,\cos(\beta + \delta' + \theta)\left[1 + \sqrt{\frac{\sin(\varnothing' + \delta')\sin(\varnothing' - \theta - \alpha)}{\cos(\beta + \delta' + \theta)\cos(\beta - \alpha)}}\right]^2} \tag{5.53}$$

Equation (5.53) is often referred to as the Mononobe–Okabe Equation. Mononobe (1929) suggested considering the point of application for P_{ae} at $H/3$ from A (the bottom of the wall) in Figure 5.19.

Whitman (1990) suggested a simple estimate of the K_{ae} as

$$K_{ae} = K_a + 0.75k_h \tag{5.54}$$

where K_a is the active lateral earth pressure coefficient from Coulomb's method [Equation (5.9)].

EXAMPLE 5.7

Given

For the wall and backfill conditions described in Figure E5.2. The backfill is granular material. Consider the following soil parameters, wall, and backfill configurations:

$\emptyset' = 30°$
$c' = 0$
$\delta' = (2/3)\emptyset'$
$\alpha = 0°$, and $10°$
$\beta = 0°$, and $20°$
$k_h = 0.1$, and 0.2
$k_v = 0$

Required

Calculate the active lateral earth pressure coefficient, K_{ae}, for the given earthquake conditions and compare values with Whitman's equation.

Solution

According to Mononobe–Okabe equation,

$$K_{ae} = \frac{\cos^2(\beta - \emptyset' + \theta)}{\cos\theta \cos^2(\beta)\, \cos(\beta + \delta' + \theta)\left[1 + \sqrt{\dfrac{\sin(\emptyset' + \delta')\sin(\emptyset' - \theta - \alpha)}{\cos(\beta + \delta' + \theta)\cos(\beta - \alpha)}}\right]^2}$$

(5.53)

With Whitman's equation,

$$K_{ae} = K_a + 0.75k_h$$

(5.54)

To apply Whitman's equation, K_a is computed using Coulomb's method [Equation (5.9)]. Use the spreadsheet program provided at the publisher's website and insert the given parameters, the following table for K_{ae} can be prepared:

	$\beta = 0°$		$\beta = 20°$	
K_a and K_{ae}	$\alpha = 0°$	$\alpha = 10°$	$\alpha = 0°$	$\alpha = 10°$
K_a (Coulomb's method)	0.30	0.34	0.48	0.57
K_{ae} ($k_h = 0.1$) (MOE)[a]	0.37	0.43	0.57	0.70
K_{ae} ($k_h = 0.1$) (WE)[b]	0.37	0.41	0.55	0.64
K_{ae} ($k_h = 0.2$) (MOE)[a]	0.45	0.57	0.68	0.90
K_{ae} ($k_h = 0.2$) (WE)[b]	0.44	0.48	0.62	0.71

[a] MOE: Mononobe–Okabe equation.

[b] WE: Whitman's equation.

5.9 COMMENTS ON THE MERITS OF THE LATERAL EARTH PRESSURE THEORIES

Variations of active earth pressure coefficients among different methods are limited. It is apparent that Rankine's method has more restrictions, such as that the wall has to be frictionless ($\delta' = 0$). However, the nature of tension crack can be effectively described only under Rankine's stress-based analysis. It should be noted that the Rankine K_p values tend to decrease with backfill inclination angle, α. This is obviously not reasonable.

Coulomb's method is more flexible in its acceptable conditions that can be analyzed. Also, the limit (force) equilibrium approach is easier to formulate and execute. It is no surprise that the more recent developments in the consideration of earthquake conditions and nonlinear shear surfaces are mostly based on limit equilibrium. However, when friction angle, $\emptyset'$, exceeds 35° (or with high soil wall interface friction angle, δ'), both Coulomb and Rankine passive earth pressure becomes consistently high (i.e., can be not conservative). This is generally believed to be caused by the use of a planar shear surface in the approach.

Several nonplanar failure surface methods using the limit equilibrium approach have been proposed (Terzaghi, 1943; Caquot and Kerisel, 1948; Janbu, 1957; Shields and Tolunay, 1973). Some of these methods are similar to the method of slices used in slope stability analysis (Chapter 9). More reasonable passive earth pressures can be obtained with the analysis that considers a nonplane failure surface. The upper-bound limit analysis (energy based, and see Chapter 2) has a sound theoretical basis and it is versatile. Calculations that consider nonplane failure surfaces can readily be executed.

With the help of spreadsheet computer program, the above-mentioned close form methods can all be applied efficiently.

5.10 EFFECTS OF WATER IN BACKFILL

Water has no strength ($\emptyset' = 0$), its $K_o = K_a = K_p = 1$. Because of this, when calculating the lateral earth pressure below a water table it is necessary to separate water pressure from that of the soil. As demonstrated in Example 5.1, when water is present, the total lateral pressure can be significantly higher than soil pressure that is calculated using the effective stress. In addition, water can freeze and expand in cold regions, and that further increases the loading on the retaining wall. Thus in practice we normally would install a drainage system at the base of the retaining wall to make sure that no water can accumulate behind the retaining wall. Details for the design of the drainage system are described in the following sections.

5.11 DESIGN OF RETAINING STRUCTURES

Concrete or masonry retaining walls, with or without reinforcement, have been used for centuries. Figure 5.20 provides schematic views of different types of concrete retaining walls. The gravity wall (Figure 5.20a) relies on its heavy weight (thus, gravity) to resist the lateral earth pressure. Little or no reinforcement is required in the gravity wall design.

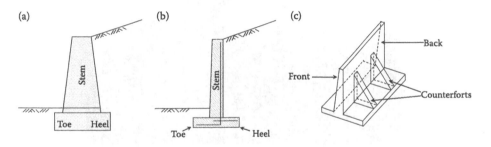

Figure 5.20 Various types of concrete retaining walls. (a) Gravity retaining wall. (b) Cantilever retaining wall. (c) Cantilever retaining wall with counterfort.

Figure 5.21 A geosynthetic reinforced soil-retaining wall in Jodhpur, India. (Courtesy of An-Bin Huang, Hsinchu, Taiwan.)

The wall can be built with concrete or masonry. Because of the large volume of concrete involved, gravity walls are rarely used today. The cantilever retaining wall (Figure 5.20b) takes advantage of the steel reinforcement; the wall thickness is much reduced and thus more material efficient. For excessively high (wall height exceeds 6 or 7 m) cantilever walls, a series of counterforts (Figure 5.20c) can be placed on the back of the wall to enhance its structural capacity. This class of retaining wall is usually fabricated in the field and is rigid.

The reinforced soil structure is a relatively new addition to the family of retaining walls. Although the history of reinforced soil can be traced back for thousands of years, the development of rigorous design, material fabrication, and construction procedures for reinforced soil structures started in the mid-twentieth century. Figure 5.21 shows the picture of a geosynthetic reinforced retaining wall. The geosynthetic is a manmade material (mostly polymers) with suitable tensile strength. The overall stability of the reinforced soil structure is significantly enhanced with the inclusion of the reinforcement elements such as the geosynthetic. The reinforced soil-retaining wall is gaining popularity because of its superior performance and relatively low cost.

Sheet pile walls are another class of retaining structures. Prefabricated, interlocking sheets (i.e., thin plates) that can be made of metal (mostly steel), timber, or concrete are driven side by side into the ground to form a continuous vertical wall. Figure 5.22 shows

Figure 5.22 Field installation of a sheet pile wall. (Courtesy of Trinity Foundation Engineering Consultants, Co. Ltd., Taipei, Taiwan.)

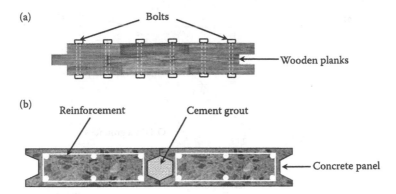

Figure 5.23 Cross-sectional view of sheet piles of other types of material. (a) Timber sheet pile. (b) Concrete sheet pile.

the field installation of a sheet pile wall. The sheet piles are usually driven with a vibratory hammer (see Section 8.2). The alignment and resistance or thrusts are normally provided by horizontal wales, braces, or anchors.

Wooden planks pinned together side by side have been used to make sheet piles. If water tightness is desired, Wakefield or tongue-and-groove sheeting is generally used. Wakefield piles are made with three layers of wooden planks, 5 cm to 10 cm in thickness. The planks are bolted together with the middle plank offset, forming a tongue on one edge and a groove on the other, as shown in Figure 5.23a. Timber sheet piles have light weight, and as such the equipment required for pile driving is also light.

Precast concrete piles (or panels) made in rectangular cross-section driven side by side can also be used as sheet piles. The interlock between two piles can be provided with tongue and groove, as in the case of wooden sheet piles, or after the piles are driven to the required depth, the joint is grouted, as shown in Figure 5.23b. The concrete piles are usually reinforced and often pre-stressed. The piles are normally beveled at their feet to facilitate tight close driving of a pile against the already driven one.

The prefabricated sheet piles can be driven under water and/or directly into the soil to be retained (i.e., no backfill). Because of these characteristics, sheet pile retaining walls are commonly used for waterfront structures such as quay walls or ship docks, where a significant part of the construction takes place under water. Or, the sheet pile forms an important part of a braced cut system, described in Chapter 6. An important difference in the braced cut system is that the soil to be retained is usually not backfilled. This change in construction sequence causes significant differences in the lateral earth pressures involved. Details of the construction sequence effects are described in Chapter 6.

The following sections provide details in the analysis and design of concrete retaining walls, reinforced soil-retaining walls, and sheet pile walls.

5.12 CONCRETE RETAINING WALLS

Regardless of the details of their structural designs, the construction of concrete retaining walls often requires the fabrication of forms, followed by pouring of concrete. Figure 5.24 provides a cross-sectional view of the sequence of constructing a typical concrete retaining wall. The ground is excavated to the desired depth where the retaining wall foundation is to be placed (see Figures 5.24a). Additional excavation to shape the ground surface

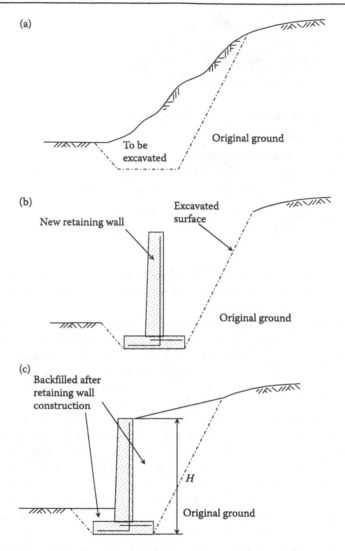

Figure 5.24 Typical sequence of a concrete retaining wall construction. (a) Original ground profile. (b) Ground excavation and installation of retaining wall. (c) Backfill.

behind the proposed retaining wall may be necessary to maintain stability of the excavated slope surface and create room for retaining wall construction work (see Figure 5.24b). The excavated space is backfilled with compacted soil to the final design grade after completion of the retaining wall (Figure 5.24c).

For the design of a concrete retaining wall, we need to collect the following information first:

*   Conditions of the original ground material that should include ground surface profile, soil type(s), shear strength and unit weight of the ground material(s), and groundwater table (if present).
*   Total height of the proposed retaining wall, from bottom of the foundation to top of the wall (H in Figure 5.24c).
*   Lateral earth pressure to be retained by the retaining wall. This can be determined based on the information provided in the above two items.

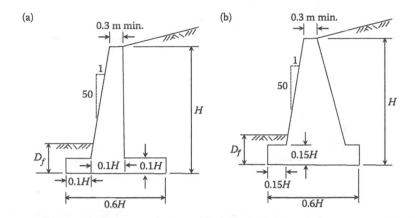

Figure 5.25 Proportioning of the concrete retaining walls. (Note: the drawings are not to scale and the slope of the front face of the retaining wall is not necessary.) (a) Tentative proportions for cantilever retaining wall. (b) Tentative proportions for gravity retaining wall.

There are two major aspects to be considered in the design of a concrete retaining wall: (1) overall stability of the wall/soil system, which is mostly a geotechnical problem; and (2) structural design of the retaining wall that involves detailed dimensioning of the various structural elements, material (i.e., concrete) properties, and reinforcement (if used). This textbook concentrates on the overall stability analysis only. We do not discuss the structural designs of the retaining wall, as this information can be found in manuals or textbooks on structural designs.

As in many civil engineering problems, the system is indeterminate, meaning that there is more than one acceptable answer or retaining wall design for a given set of conditions. In these situations, we assume a set of tentative dimensions and then verify if the dimensions are acceptable. Adjustments are made if necessary to assure that all safety requirements are met. Figures 5.25a and 5.25b show the initial tentative proportions of the cantilever and gravity retaining walls, respectively. These proportions are likely to meet the stability requirements based on previous experience. The geometry is reasonable in its aesthetic appearance and likely to meet the requirements for structural design and construction. The foundation embedment, D_f, should be deep enough that the foundation is not affected by temperature and moisture fluctuations.

As indicated in Figures 5.25a and 5.25b, all dimensions are based on the total height of the retaining wall, H, which is known or is part of the information to be collected as described above. Therefore all key dimensions for the proposed retaining wall can be established, as a start, with the help of Figures 5.25a, 5.25b, and the known magnitude of H. The following sections provide the procedures for the stability analysis.

5.12.1 Stability analysis of cantilever retaining walls

We start with the relatively popular cantilever concrete retaining walls. The stability analysis considers four possible failure modes of the soil/structural system, as shown in Figure 5.26. In analyzing the potential failures, the retaining wall itself is assumed to be intact and remains rigid. For the analysis of deep-seated shear failure (Figure 5.26d), we use the same techniques of slope stability analysis described in Chapter 9. We concentrate on the first three modes of failure in this section: rotation, sliding, and bearing capacity failure. The following is a step-by-step description of the procedure.

Step 1: Establish the earth pressures (forces) to be experienced by the retaining wall.

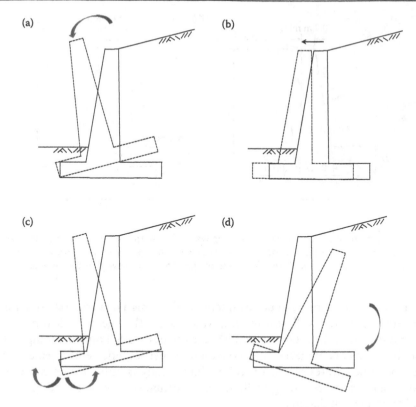

Figure 5.26 Potential failure modes to be considered in the stability analysis. (a) Rotation failure. (b) Sliding failure. (c) Bearing capacity failure. (d) Deep-seated shear failure.

Figure 5.27 shows a general case of a cantilever retaining wall with a sloped backfill. For simplicity, it is assumed that the soil above the foundation is part of the retaining wall. With this assumption, the lateral earth pressures act within the soil mass and therefore avoid the issue of dealing with the soil/structure interface friction angle, δ', and this allows Rankine's earth pressure theory be applied. The active earth pressure (and thus the active force, P_a) acts on the surface of a fictitious wall represented by the dash line in

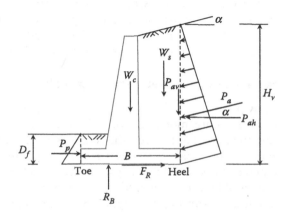

Figure 5.27 Loading and reaction forces experienced by a retaining wall.

Figure 5.27 with a total height of H_v. P_a and P_p are parallel to the surface of the backfill, and according to Rankine's theory,

$$P_a = \frac{1}{2}\gamma' H_v^2 K_a - 2c' H_v \sqrt{K_a} \qquad (5.34)$$

and

$$P_p = \frac{1}{2}\gamma' D_f^2 K_p + 2c' D_f \sqrt{K_p} \qquad (5.39)$$

The horizontal component of P_a, P_{ah} is the major driving force that is responsible for causing rotation and sliding failure. P_{ah} also causes the bearing pressure at the retaining wall foundation to be eccentric. P_{av}, the vertical component of P_a and P_p, are important resisting forces against rotation and sliding failure.

Note that the earth pressures in Figure 5.27 consider that the related soils are cohesionless (i.e., $c' = 0$). As described in Section 5.4, the inclusion of c' makes P_p higher and P_a lower, and both of these conditions can lead to unsafe design. If c' is to be included as part of the soil strength parameters, it should be used with caution. In any case, the active force, P_a, should be computed considering the post-tension crack conditions (i.e., use Equation 5.35 or 5.36). For P_p, it is advisable to assume that $c' = 0$. P_p acts at $D_f/3$ and P_a at $H_v/3$ from the bottom of the retaining wall with $c' = 0$. P_a and P_p are calculated considering a unit thickness of the retaining wall.

Step 2: Determine the weight of the concrete structure, W_c and soils, W_s above the heel.

The weight of concrete (W_c) and soil (W_s) are part of the driving forces that induce bearing capacity failure. At the same time, they are responsible for the development of frictional resistance (F_R in Figure 5.27) against sliding and moment against rotation failure. W_c and W_s are calculated considering a unit thickness of the retaining wall by multiplying the respective unit weight and its cross-sectional area (and unit thickness).

Step 3: Reaction forces in response to the applied loads.

The reaction forces induced by the loading conditions described in Step 1 include frictional resistance, F_R, and bearing pressure (represented by its resultant R_B in Figure 5.27) at the base of the retaining wall foundation. In a way, P_p, which is developed passively due to movement of the retaining wall against the soil, should also be considered as a reaction force. F_R is defined as

$$\begin{aligned} F_R &= (W_c + W_s + P_{av}) \tan \delta' + Ba' \\ &= (\Sigma F_v) \tan \delta' + Ba' \end{aligned} \qquad (5.55)$$

where

δ' = soil–wall interface friction, $\phi'/2$ to $\phi'2/3$
ϕ' = drained friction angle of the foundation soil
a' = soil–wall interface adhesion, $c'/2$ to $c'2/3$
c' = drained cohesion of the foundation soil
P_{av} = vertical component of P_a

and

$$R_B = \Sigma F_v = (W_c + W_s + P_{av}) \qquad (5.56)$$

Step 4: Check FS against overturning failure.

Refer to Figure 5.28, by taking moment about point O, the FS against overturning is

$$FS = \frac{\Sigma M_R}{\Sigma M_O} \qquad (5.57)$$

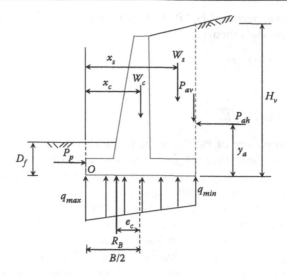

Figure 5.28 The eccentrically loaded retaining wall foundation.

where

ΣM_R = sum of the moment-resisting overturning

ΣM_O = sun of moment causing overturning of the wall

The effects of passive force (P_p) and weight of soil above the toe (see Figure 5.27) are ignored in the calculation of ΣM_R.

For the case shown in Figure 5.28,

$$\Sigma M_R = x_c W_c + x_s W_s + B P_{av} \quad (5.58)$$

and

$$\Sigma M_O = y_a P_{ah} \tag{5.59}$$

where

x_c, x_s, and y_a = moment arm corresponds to W_c, W_s, and P_{ah}, respectively

P_{ah} = horizontal component of P_a

P_{av} = vertical component of P_a

The moment arm for P_{av} is B, width of the retaining wall foundation. A minimum FS of 2 against overturning is generally required.

Step 5: Check FS against sliding failure.

As P_{ah} is the only driving force causing the sliding failure, the FS against sliding is

$$FS = \frac{F_R + P_p}{P_{ah}} \tag{5.60}$$

where F_R is the frictional resistance shown in Figure 5.27 and defined by Equation (5.55). A minimum FS of 1.5 against sliding is generally required.

Step 6: Check FS against bearing capacity failure.

The retaining wall foundation is likely to be eccentrically loaded because of the off-centered (i.e., W_c and W_s) and lateral (such as P_a) forces involved, thus the bearing pressure is likely to be non-uniform. The eccentricity of the resultant R_B can be calculated by taking moment about O, and

$$R_B\left(\frac{B}{2} - e_c\right) = \Sigma M_R - \Sigma M_O \tag{5.61}$$

or

$$e_c = \frac{B}{2} - \frac{(\Sigma M_R - \Sigma M_O)}{R_B} = \frac{B}{2} - \frac{(\Sigma M_R - \Sigma M_O)}{\Sigma F_v} \tag{5.62}$$

Assume that the bearing pressure is linearly distributed as shown in Figure 5.28, the maximum (q_{max}) and minimum bearing pressure (q_{min}) can be calculated using the concept in the mechanics of materials as

$$q_{max} = \frac{\Sigma F_v}{(B)(1)} + \frac{(\Sigma M_R - \Sigma M_O)\left(\frac{B}{2}\right)}{I} \tag{5.63}$$

and

$$q_{min} = \frac{\sum F_v}{(B)(1)} - \frac{\left(\sum M_R - \sum M_O\right)\left(\frac{B}{2}\right)}{I} \tag{5.64}$$

where
I = moment of inertia per unit length of the foundation section = $\dfrac{B^3(1)}{12}$

$$\left(\sum M_R - \sum M_O\right) = e_c R_B = e_c \sum F_v \tag{5.65}$$

Substitute I and $\left(\sum M_R - \sum M_O\right)$ into Equation (5.63) and reorganize

$$q_{max} = \frac{\sum F_v}{(B)(1)} + \frac{e_c(\sum F_v)\left(\frac{B}{2}\right)}{\dfrac{B^3(1)}{12}} = \frac{\sum F_v}{B}\left(1 + \frac{6e_c}{B}\right) \tag{5.66}$$

In a similar manner,

$$q_{min} = \frac{\sum F_v}{B}\left(1 - \frac{6e_c}{B}\right) \tag{5.67}$$

It is evident that when $e_c/B > 1/6$, q_{min} becomes negative or tension exists at the base of the foundation. As soil has little tensile strength, we usually keep $e_c/B \le 1/6$ when proportioning a retaining wall. The retaining wall foundation is an eccentrically loaded, strip (or continuous) shallow foundation where the applied load is also inclined. Consider that the groundwater table is significantly below the foundation level and for drained analysis, as described in Chapter 4, the ultimate bearing capacity is

$$q_u = \bar{c}N_c F_{cs}F_{cd}F_{ci} + \bar{q}N_q F_{qs}F_{qd}F_{qi} + \frac{1}{2}\bar{\gamma}BN_\gamma F_{\gamma s}F_{\gamma d}F_{\gamma i} \tag{4.84}$$

where

$\bar{c} = c'$ drained cohesion of the soil

$\bar{\gamma} = \gamma$ unit weight of the soil

$q = \gamma D_f$

$B' = B - 2e_c$

$F_{cs}, F_{qs}, F_{\gamma s}$ = shape factors = 1 for continuous foundation ($B/L \approx 0$)

F_{qd} = depth factor = $1 + 2\tan\emptyset'(1 - \sin\emptyset')^2 \left(\frac{D_f}{B}\right)$

F_{cd} = depth factor = $d_q - \frac{1-d_q}{N_c \tan\emptyset'}$

$F_{\gamma d}$ = depth factor = 1

F_{ci} = inclination factor = $\left(1 - \frac{\alpha}{90^0}\right)^2$

F_{qi} = inclination factor = i_c

$F_{\gamma i}$ = inclination factor = $\left(1 - \frac{\alpha}{\emptyset'}\right)^2$ ($\emptyset'$ and α in degrees)

α = inclination of the applied load with respect to vertical = $\tan^{-1}\left(\frac{P_{ah}}{\Sigma F_v}\right)$, in degrees

N_c, N_q, N_γ = bearing capacity factors

Because a drainage system (described later) is usually installed behind the retaining wall to keep the water table low, we normally will use drained or effective stress analysis for concrete retaining walls. The factor of safety (FS) against bearing capacity failure can be calculated as

$$FS = \frac{q_u}{q_{max}} \tag{5.68}$$

A minimum FS of 3 against bearing capacity failure is generally required.

EXAMPLE 5.8

Given

Consider the dimensions of a concrete cantilever retaining wall and the soil profile shown in Figure E5.8. For the calculation of the shear resistance at the base of the wall foundation, assume $\delta' = (2/3)\emptyset' = 20°$.

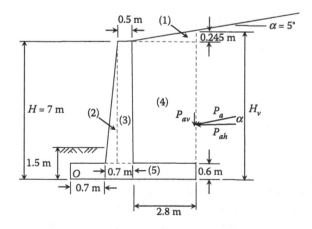

Figure E5.8 Dimensions and soil profile of the retaining wall for Rankine's method.

Required

Determine the FS of the retaining wall design against overturning, sliding, and bearing capacity failure. Use Rankine's method for the calculation of lateral earth pressure.

Solution

Computations involved in this example take advantage of the spreadsheet program provided on the publisher's website.

Step 1: K_a of Rankine's method is

$$K_a = \cos\alpha \frac{\cos\alpha - \sqrt{\cos^2\alpha - \cos^2\varnothing'}}{\cos\alpha + \sqrt{\cos^2\alpha - \cos^2\varnothing'}} = 0.34 \quad \text{for } \varnothing' = 30° \text{ and } \alpha = 5° \quad (5.17)$$

and K_p of Rankine's method is

$$K_p = \cos\alpha \frac{\cos\alpha + \sqrt{\cos^2\alpha - \cos^2\varnothing'}}{\cos\alpha - \sqrt{\cos^2\alpha - \cos^2\varnothing'}} = 3 \quad \text{for } \varnothing' = 30° \text{ and } \alpha = 0° \quad (5.21)$$

and per-unit thickness of the wall,

$$P_a = \frac{1}{2}\gamma H_v^2 K_a = (0.5)(19)(7 + 0.245)^2(0.34) = 168.14\,\text{kN}$$

$$P_{ah} = P_a \cos\alpha = (168.14)(\cos 5°) = 167.50\,\text{kN}$$

$$P_{av} = P_a \sin\alpha = (168.14)(\sin 5°) = 14.65\,\text{kN}$$

$$P_p = \frac{1}{2}\gamma D_f^2 K_p = (0.5)(19)(1.5)(3.0) = 64.13\,\text{kN}$$

Step 2: Divide the soil above the heel and concrete into 5 rectangular and triangular sections (as marked in Figure E5.8) for the calculation of W_s, W_c per unit thickness of the wall, and their respective moment arms.

Unit weight of concrete, $\gamma_{concrete} = 24\,\text{kN/m}^3$

For the preparation of the following table,

$W_i = (A_i)(\text{unit weight of soil or concrete})$

$x_i =$ horizontal distance between centroid of the respective section to point O of Figure E5.8

$M_i = (A_i)(x_i)$

Section no.	Area (A_i), m^2	Weight (W_i), kN	Moment arm (x_i), m	Moment (M_i), kN-m
1 (soil)	0.34	6.52	3.27	21.29
2 (concrete)	0.64	15.36	0.83	12.80
3 (concrete)	3.2	76.80	1.15	88.32
4 (soil)	17.92	340.48	2.80	953.34
5 (concrete)	2.52	60.48	2.10	127.01
		$\sum W_i = 499.64$		$\sum M_i = 1202.76$

Step 3: $\sum W_i$ in the above table $= W_c + W_s$

For the calculation of shear resistance at the base, F_R, c' and thus soil–wall interface adhesion $a' = 0$,

$$F_R = (W_c + W_s + P_{av})\tan\delta' + Ba' = (\sum W_i + P_{av})\tan\delta'$$
$$= (499.64 + 14.65)(\tan 20°) = 187.19\,\text{kN} \quad (5.55)$$

and

$$R_B = \Sigma F_v = (W_c + W_s + P_{av}) = (499.64 + 14.80) = 514.44\,\text{kN} \quad (5.56)$$

Step 4: Take moment about point O, the FS against overturning is

$$FS = \frac{\Sigma M_R}{\Sigma M_O} \qquad (5.57)$$

$$\Sigma M_R = \Sigma M_i + BP_{av} = 1202.76 + (2.8 + 0.7 + 0.7)(14.65)$$
$$= 1263.80 \, \text{kN-m}$$

where $B = (2.8 + 0.7 + 0.7) = 4.2 \, \text{m}$

$$\Sigma M_O = \frac{1}{3} H_v P_{ab} = \left(\frac{1}{3}\right)(7 + 0.245)(167.50) = 404.52 \, \text{kN-m}$$

Thus, FS against overturning $= (1263.80)/(404.52) = 3.1 > 3.0$

Step 5: According to Equation (5.60),

$$\text{FS against sliding} = \frac{F_R + P_p}{P_{ab}} = (187.19 + 64.13)/(167.50) \approx 1.50, \text{ OK} \qquad (5.60)$$

Step 6: According to Equation (5.62)

$$e_c = \frac{B}{2} - \frac{\left(\Sigma M_R - \Sigma M_O\right)}{R_B} = (4.2)(0.5) - (1264.31 - 404.52)/514.29 = 0.43 \text{m} < B/6$$

Following Equations (5.63) and (5.64),

$$q_{max} = \frac{\Sigma F_v}{B}\left(1 + \frac{6e_c}{B}\right) = \frac{(514.29)}{(4.2)}\left[1 + \frac{(6)(0.43)}{(4.2)}\right] = 197.36 \, \text{kN/m}^2 \qquad (5.63)$$

$$q_{min} = \frac{\Sigma F_v}{B}\left(1 - \frac{6e_c}{B}\right) = \frac{(514.29)}{(4.2)}\left[1 - \frac{(6)(0.43)}{(4.2)}\right] = 47.54 \, \text{kN/m}^2 \qquad (5.64)$$

Consider the wall foundation as a continuous foundation and $c' = 0$,

$$q_u = \bar{q} N_q F_{qs} F_{qd} F_{qi} + \frac{1}{2}\bar{\gamma} B' N_\gamma F_{\gamma s} F_{\gamma d} F_{\gamma i} \qquad (4.84)$$

$$F_{qs} = F_{\gamma s} = 1$$

$$\bar{q} = \gamma D_f = (19)(1.5) = 28.5 \, \text{kN/m}^2$$

$$B' = B - 2e_c = (4.2) - (2)(0.44) = 3.34\text{m}$$

$$F_{qd} = 1 + 2\tan\varnothing'(1 - \sin\varnothing')^2\left(\frac{D_f}{B}\right) = (1) + (2)(\tan 30°)(1 - \sin 30°)^2(1.5/4.2)$$
$$= 1.10$$

$$F_{\gamma d} = 1$$

$$\alpha = \tan^{-1}\left(\frac{P_{ab}}{\Sigma F_v}\right) = \tan^{-1}\left(\frac{167.50}{514.29}\right) = 18.04°$$

$$F_{qi} = \left(1 - \frac{\alpha}{90°}\right)^2 = [(1) - (18.04)/(90)]^2 = 0.64$$

$$F_{\gamma i} = \left(1 - \frac{\alpha}{\varnothing'}\right)^2 = [(1) - (18.04)/(30)]^2 = 0.16$$

For $\emptyset' = 30°$,

$$N_q = \tan^2\left(45 + \frac{\emptyset'}{2}\right)e^{\pi \tan \emptyset'} = \tan^2\left(45 + \frac{(30)}{2}\right)e^{(3.1416)(\tan (30))} = 18.40 \quad (4.23)$$

$$N_\gamma = 2(N_q + 1)\tan \emptyset' = (2)(18.40 + 1)(\tan (30)) = 22.40 \quad (4.57)$$

$$\begin{aligned} q_u &= \bar{q}N_q F_{qs}F_{qd}F_{qi} + \frac{1}{2}\gamma B' N_\gamma F_{\gamma s}F_{\gamma d}F_{\gamma i} \\ &= (28.5)(18.40)(1)(0.64)(1.10) + (0.5)(19)(3.34)(22.40)(1)(0.16)(1) \\ &= 482.93\,\text{kN/m}^2 \end{aligned} \quad (4.84)$$

Thus, FS against bearing capacity failure is

$$FS = \frac{q_u}{q_{max}} = (482.93)/(197.36) = 2.45 \text{ which is less than } 3.0. \quad (5.68)$$

Coulomb's lateral earth pressure theory can also be used to analyze the stability of a retaining wall. Because, the method allows wall friction δ and inclination of the back face β to be included. The active lateral force can be applied directly to the surface of the wall. We use the following example to demonstrate its application.

EXAMPLE 5.9

Given

Consider the same cantilever retaining wall and the soil profile in Example 5.8. At the back and base of the retaining wall, assume the soil–wall interface friction, $\delta' = (2/3)\emptyset' = 20°$.

Required

Determine the FS of the retaining wall design against overturning. Use Coulomb's method for the calculation of lateral earth pressure.

Solution

Replot the retaining wall and the soil profile as shown in Figure E5.9, but now with the lateral active force applied directly to the back of the wall. The following computations take advantage of the spreadsheet program provided on the publisher's website.

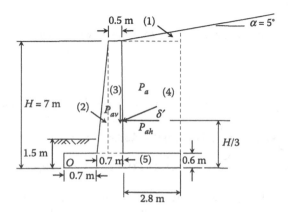

Figure E5.9 Dimensions and soil profile of the retaining wall for Coulomb's method.

Step 1: For $\emptyset' = 30°$, $\alpha = 5°$, and $\beta = 0°$, K_a of Coulomb's method is

$$K_a = \frac{\cos^2(\beta - \emptyset')}{\cos^2(\beta)\,\cos(\beta + \delta')\left[1 + \sqrt{\dfrac{\sin(\emptyset' + \delta')\sin(\emptyset' - \alpha)}{\cos(\beta + \delta')\cos(\beta - \alpha)}}\right]^2} = 0.32 \qquad (5.9)$$

$P_a = \frac{1}{2}\gamma' H^2 K_a = (0.5)(19)(7)^2(0.32) = 147.32\ \text{kN}$

$P_{ah} = P_a \cos \delta' = (147.32)(\cos 20°) = 138.43\ \text{kN}$

$P_{av} = P_a \sin \delta' = (147.32)(\sin 20°) = 50.39\ \text{kN}$

Step 2: Divide the soil above the heel and concrete into 5 rectangular and triangular sections (as marked in Figure E5.9) for the calculation of W_s, W_c per unit thickness of the wall and their respective moment arms.

Unit weight of concrete, $\gamma_{concrete} = 24\ \text{kN/m}^3$

For the preparation of the following table,

$W_i = (A_i)(\text{unit weight of soil or concrete})$

$x_i =$ horizontal distance between centroid of the respective section to point O of Figure E5.9

$M_i = (A_i)(x_i)$

The result is the same as that shown in Example 5.8.

Section no.	Area (A_i), m^2	Weight (W_i), kN	Moment arm (x_i), m	Moment (M_i), kN-m
I (soil)	0.34	6.52	3.27	21.29
2 (concrete)	0.64	15.36	0.83	12.80
3 (concrete)	3.2	76.80	1.15	88.32
4 (soil)	17.92	340.48	2.80	953.34
5 (concrete)	2.52	60.48	2.10	127.01
		$\sum W_i = 499.64$		$\sum M_i = 1202.76$

Step 3: $\sum W_i$ in the above table $= W_s + W_c = 499.64\ \text{kN}$

According to Equation (5.56),

$$R_B = \sum F_v = (W_c + W_s + P_{av}) = (499.64 + 50.39) = 550.03\ \text{kN} \qquad (5.56)$$

Step 4: Take moment about point O, the FS against overturning is

$$FS = \frac{\sum M_R}{\sum M_O} \qquad (5.55)$$

$$\sum M_R = \sum M_i + (0.7 + 0.7)P_{av} = 1202.76 + (1.4)(50.39) = 1273.30\ \text{kN-m}$$

$$\sum M_O = \frac{1}{3}HP_{ah} = \left(\frac{1}{3}\right)(7)(138.43) = 138.42\ \text{kN-m}$$

Thus, FS against overturning $= (1273.30)/(138.42) = 3.9 > 3.0$, OK

The gravity retaining walls are rarely used, as they are much more costly to build in comparison with other types of retaining walls currently available. No further discussion will be made regarding the gravity retaining wall designs.

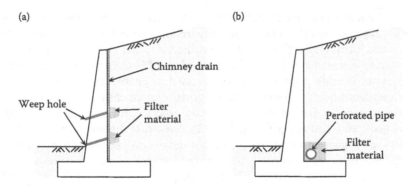

Figure 5.29 Possible elements in a drainage system behind a retaining wall. (a) Weep holes and chimney drain. (b) Perforated drainage pipe.

5.12.2 Drainage for the retaining walls

As described previously, water has no shear strength. Significantly higher lateral pressure against a retaining structure can develop if water is allowed to accumulate as demonstrated in Example 5.1. It is a general practice to drain the water away from the backfill rather than to design the wall to resist the water pressure. Figure 5.29 shows the elements that may be used as part of a drainage system behind a retaining wall. The drainage pipes or weep holes as well as the chimney drain, if used, should be protected by a filter system.

The filter must retain the backfill soil particles from entering the filter (i.e., the retention criterion to prevent clogging of the drainage system) while allowing water to pass (i.e., the permeability criterion to prevent buildup of pore water pressure). The filter can be made of granular material with gradations that meet the following criteria (Terzaghi and Peck, 1967):

$$\frac{D_{15(F)}}{D_{85(B)}} < 5 \qquad \text{(for retention criterion)} \tag{5.69}$$

and

$$\frac{D_{15(F)}}{D_{15(B)}} > 4 \qquad \text{(for permeability criterion)} \tag{5.70}$$

where
$D_{15(F)}$ = grain size of the filter material for which 15% are smaller
$D_{15(B)}$ = grain size of the backfill for which 15% are smaller
$D_{85(B)}$ = grain size of the backfill for which 85% are smaller

Woven or nonwoven geotextile has been used successfully as a filter. Readers are referred to Holtz et al. (1997) for details of the geotextile filter design.

5.13 MECHANICALLY STABILIZED EARTH-RETAINING WALLS

The modern invention of reinforcing the soil backfill behind the retaining walls was developed by a French Engineer, Henri Vidal (Vidal, 1966). The Vidal system, trademarked as Reinforced Earth™, uses metal strips for reinforcement, as schematically shown in

Figure 5.30. The Reinforced Earth walls have been used extensively throughout the world. Various other types of metallic and nonmetallic reinforcement have also been introduced since Reinforced Earth walls. The use of geosynthetics rather than metallic strips for reinforcement started in the early 1970s in France, and the U.S. Geosynthetics is a generic term that encompasses flexible polymeric materials used in geotechnical engineering such as geotextiles, geomembranes, geonets, and geogrids. Out of these geosynthetic materials, geogrids and geotextiles are the most commonly used reinforcing elements. Figure 5.31 shows a few geotextile and geogrid samples to demonstrate the appearance of these materials. Readers are referred to Koerner (2012) for details on geosynthetics and their applications in geotechnical engineering. The geosynthetics are much more stretchable or extensible than metal strips. The tensile strength per unit width of the geosynthetics is much weaker than typical metal strips. Metal strips are installed as individual reinforcement elements with horizontal and vertical spacings as shown in Figure 5.30b. The geosynthetics are deployed as sheets of reinforcement with vertical spacing only. Details of the

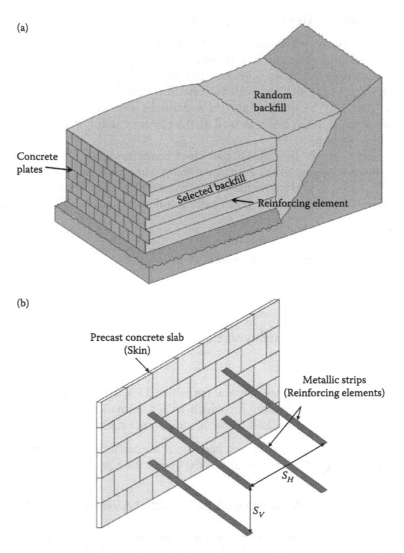

Figure 5.30 The reinforced earth wall system. (a) Reinforced earth-retaining wall. (b) Use of metal strips as reinforcement.

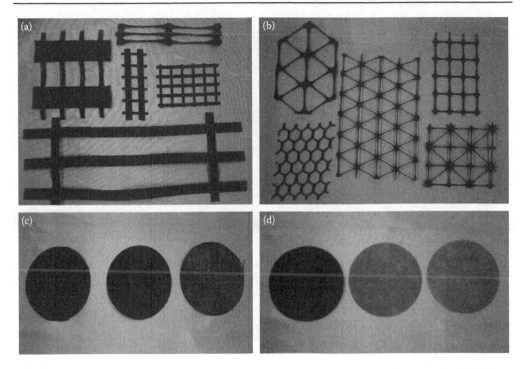

Figure 5.31 Photographs of geosynthetics. (a) Uniaxial geogrid. (b) Biaxial geogrid. (c) Woven geotextile. (d) Nonwoven geotextile.

geosynthetic reinforced retaining wall construction are described later. Since expiration of the early patents, the engineering community has adopted a generic term, mechanically stabilized earth (MSE), to describe this type of reinforced soil-retaining wall construction. Reinforced soil structures with their face inclined at 70–90° from horizontal are considered as the MSE retaining walls.

5.13.1 Mechanics of soil reinforcement

McKittrick (1978) provided a simple explanation for the basic mechanics of MSE. Consider a granular soil specimen encased in a rubber membrane and placed in a triaxial cell under a given confining stress (σ'_c) as shown in Figure 5.32a. If we increase the vertical stress, $\Delta\sigma'_v$, while maintaining the same σ'_c (i.e., $\Delta\sigma'_h = 0$), the Mohr circle shown in Figure 5.32b will enlarge and eventually touch the failure envelope (meaning the specimen is failed). In the process, the specimen expands laterally while shortening in the vertical direction, as we learned in typical triaxial tests. On the other hand, if horizontal reinforcing elements are placed within the soil mass, as shown in Figure 5.32c, these reinforcements will prevent or reduce lateral as well as vertical strain because of friction between the reinforcing elements and the soil. The behavior will be as if a lateral restraining force or load is imposed on the soil specimen. In this case, when σ'_v is applied, the horizontal stress σ'_h maintains a consistent relationship with the vertical stress so that $\sigma'_h = K\sigma'_v$. For inextensible horizontal reinforcing elements (such as metal strips), $K \approx K_o$ (zero lateral strain). For relatively extensible reinforcing elements such as geotextiles and geogrids, $K < K_o$. Therefore as the vertical stress increases, the horizontal restraining stress also increases in direct proportion, keeping the Mohr circle below the failure envelope (no failure) as shown in Figure 5.32d.

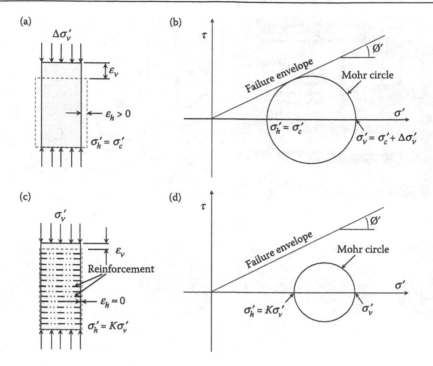

Figure 5.32 Mechanics of soil reinforcement. (a) Soil specimen with no reinforcement. (b) Mohr's circle for the case with no reinforcement. (c) Soil specimen with reinforcement. (d) Mohr's circle for the case with reinforcement.

MSE is therefore a composite material, combining the compressive and shear strengths of compacted granular fill with the tensile strength of horizontal reinforcements. With the addition of a facing system, MSE structures are well suited for use as retaining walls, bridge abutments, and other heavily loaded structures. The vertical face of the soil/reinforcement system is essentially self-supporting, allowing vertical walls to be constructed safely. MSE walls are cost-effective soil-retaining structures that can tolerate much larger settlements than reinforced concrete walls.

The same reinforced soil specimen described in Figure 5.32 could fail if the reinforcements are pulled out, or the reinforcement yielded, and cease to provide additional horizontal stress to maintain stability. These two factors are also the major concerns in the design of MSE walls described in the following sections.

5.13.2 Stress transfer at the reinforcement–soil interface

Stresses are transferred between soil and reinforcement by friction (Figure 5.33a) and/or passive resistance (Figure 5.33b), depending on reinforcement geometry. Frictional resistance develops at locations where there is a relative shear displacement and corresponding shear stress between soil and reinforcement surface. Reinforcing elements with primarily planar contact with the soil such as steel strips, longitudinal bars in geogrids, and geotextile are likely to generate frictional resistance. Passive resistance occurs through the development of bearing stresses on the transverse reinforcement elements normal to the direction of soil reinforcement relative movement. Passive resistance is a primary interaction for reinforcements such as rigid geogrids, wire mesh, and strips with transverse ribs.

The contribution of each transfer mechanism for a particular reinforcement will depend on the roughness of the contact surface, normal effective stress, grid opening dimensions,

(a)

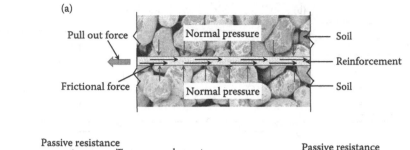

(b)

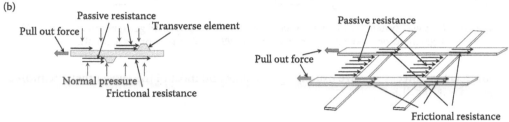

Figure 5.33 Stress transfer mechanism for soil reinforcement. (a) Frictional stress transfer between soil and reinforcement surface. (b) Soil passive (bearing) resistance on transverse reinforcement elements.

thickness of the transverse members, and elongation characteristics of the reinforcement. Equally important for interaction development are the soil characteristics, including grain size distribution, grain characteristics, density, water content, cohesion, and stiffness. Understanding the basic mechanics of MSE is crucial to the correct use of this composite construction material.

5.13.3 Types of MSE wall systems

An MSE wall system can be described by its reinforcement geometry, reinforcement material, extensibility of the reinforcement material, and the type of facing, among others. The following is a list of possible ways to classify MSE wall systems.

5.13.3.1 By reinforcement geometry

The reinforcement can be unidirectional or planar bidirectional. The unidirectional reinforcement can include steel or geogrids (see Figure 5.31a) with grid spacing greater than 150 mm in one direction. Typical planar bidirectional reinforcement includes continuous sheets of geosynthetics (see Figure 5.31b) with element spacing less than 150 mm in both directions.

5.13.3.2 By reinforcement material

The material can generally be divided in two major categories: metallic and nonmetallic. Metallic reinforcements are typically made of mild steel, usually galvanized or epoxy coated. Nonmetallic reinforcements include polymeric materials consisting of polypropylene, polyethylene, or polyester.

5.13.3.3 By extensibility of the reinforcement

For the inextensible reinforcement, the deformation of the reinforcement at failure is much less than the deformability of the soil. The deformation of the reinforcement at failure is comparable to or even greater than the deformability of the soil for the extensible

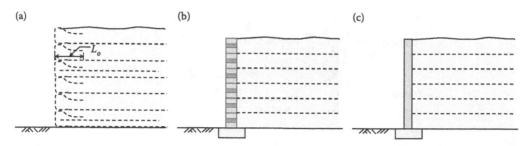

Figure 5.34 MSE wall systems with different facing. (a) With wrap-around geosynthetic facing. (b) With segmented or modular concrete block. (c) With full-height precast panels.

reinforcement. Steel strips are usually inextensible and most of the geosynthetic reinforcements are extensible.

5.13.3.4 By facing system

The types of facing elements used in the different MSE walls are the only visible parts of the completed structure. A wide range of finishes and colors can be provided in the facing. In addition, the facing provides protection against backfill sloughing and erosion, and provides in certain cases drainage paths. Figure 5.34 depicts in concept three types of MSE wall facings. Figure 5.34a shows a wraparound facing which is made simply by looping around the geosynthetic at the facing. Facing can also be made with segmented or modular concrete blocks as shown in Figure 5.34b or full-height precast panels as depicted in Figure 5.34c.

Regardless of the type of reinforcement used, the MSE walls are built in layers of compacted soil. The reinforcement is installed at the designated locations or depth intervals after each layer of backfill is placed and properly compacted. Figure 5.35 shows the field construction of a geogrid reinforced MSE wall with a wraparound facing. MSE walls with heights in excess of 15 m have been constructed to date.

5.13.4 Material properties

MSE wall involves the use of manufactured material (metal strips or geosynthetic) in a structure that usually is intended to be used for a long time. Tensile stress that would otherwise be experienced by the soil is transferred to the reinforcement via frictional and/or

Figure 5.35 Field construction of a geogrid reinforced retaining wall with wrap-around facing. (Courtesy of An-Bin Huang, Hsinchu, Taiwan.) (a) Placement of geogrid. (b) Backfill compaction.

Table 5.3 Gradation requirements according to AASHTO

U.S. sieve opening	Percent passing
100 mm	100
420 μm	0–60
75 μm	0–15

passive resistance at the soil–reinforcement interface. In order for this "composite" retaining structure to function properly, it is important that the various elements involved meet certain material and construction criteria and are compatible with each other.

5.13.4.1 The soil backfill

MSE walls are routinely designed for a 75-year service life; those supporting bridges are typically designed for 100 years. The primary factor affecting the service life of an MSE structure is the long-term durability of the reinforcements, which for inextensible (steel) reinforcement materials is closely related to backfill electrochemical properties. The backfill material should be free-draining and reasonably free from organic or other deleterious substances, as these materials not only enhance corrosion but also result in excessive settlements. The backfill soil particles should meet the criteria shown in Table 5.3 specified by the American Association of State Highway and Transportation Officials (AASHTO).

The soil should be placed in lifts not more than 300 mm in thickness and compacted to 95% of maximum dry unit weight (γ_d) and within ±2% of optimum water content, according to the Standard Proctor compaction test. Large soil particles can damage the geosynthetic in field compaction. For the longevity of the reinforcement, the soil should also meet the electrochemical requirements shown in Table 5.4.

5.13.4.2 The metal strip reinforcement

For steel reinforcements, the design life is achieved by reducing the cross-sectional area of the reinforcement used in design calculations by the anticipated corrosion losses over the design life period as follows:

$$E_c = E_n - E_R \tag{5.71}$$

where
E_c = thickness of the reinforcement at the end of the design life
E_n = nominal thickness of the reinforcement at construction
E_R = sacrificial thickness of metal expected to be lost due to corrosion during the service life of the structure

Table 5.4 Electrochemical requirements

Property	Criteria
Resistivity	>3000 ohm-cm
pH	5–10
Chlorides	<100 PPM
Sulfates	<200 PPM
Organic content	1% max.

The majority of MSE walls constructed to date have used zinc-galvanized steel and backfill materials with low-corrosive potential. A minimum galvanization coating of 86 μm is applied for metal strips. According to test results reported by Anderson et al. (2012), the zinc coating has a loss rate of 15 μm/yr in the first 2 years of installation, and then reduced to 4 μm/yr. After zinc depletion, the carbon steel has a loss rate of 12 μm/yr. Based on these rates, complete corrosion of galvanization with the thickness of 86 μm is estimated to occur during the first 16 years and a loss of carbon steel thickness of approximately 2 mm would be anticipated over the remaining years of a 100 year design life.

For metal strips, the allowable tensile force per unit width of the metal strip, T_a, is

$$T_a = 0.55 \frac{f_y \cdot A_s}{b} \tag{5.72}$$

Or

$$f_{all} = 0.55 f_y \tag{5.72a}$$

where

b = width of the strip
f_y = yield stress of the steel
f_{all} = allowable stress of the steel
A_s = cross-sectional area of the metal strip

The ASTM A36 structural steel that can be used as reinforcement strips has a yield stress, f_y, of 250 MPa and a mass density of 7.8 g/cm^3. The loss rates reported by Anderson et al. (2012) determine the sacrificial thickness of steel that must be added to the load-carrying cross-section to produce the design cross-section. At the end of the service life, after 75 or 100 years of metal loss, the remaining steel will have a factor of safety of 1.8 (1/0.55) against yield.

5.13.4.3 The geosynthetic reinforcement

The tensile strength is an important parameter for geosynthetics when used in the MSE wall. We need the strength parameter in the selection of vertical spacings of geosynthetics. The geosynthetic reinforcement can be damaged during construction (installation) and then degraded due to creep and mechanical and biological degradations following the construction. For the design of the MSE wall, we therefore need to project the tensile strength of the geosynthetic reinforcement at the end of its service life. The general practice in this regard is to determine the short-term ultimate tensile strength, T_{ult}, in the laboratory first. T_{ult} can be determined, for example, according to ASTM D4595, wide width strip method. According to Holtz et al. (1997), T_{ult} ranges from 12 to above 350 kN/m for woven geotextiles, 4 to 35 kN/m for nonwoven geotextiles, and 8 to 140 kN/m for geogrids. The tensile strain at T_{ult} is generally larger than 5%. The manufacturers usually provide the T_{ult} values in their product specification sheets. The long-term tensile strength, T_{al} (kN/m), is assessed by dividing T_{ult} with a series of reduction factor as follows:

$$T_{al} = \frac{T_{ult}}{RF} \tag{5.73}$$

and

$$RF = RF_{ID} \times RF_{CR} \times RF_{CD} \times RF_{BD} \times RF_{JNT} \text{ (dimensionless)} \tag{5.73a}$$

T_{ult} = ultimate geosynthetic tensile strength (kN/m) determined in laboratory tension test
RF_{ID} = reduction factor for installation damage (default value ≈ 1.4)
RF_{CR} = reduction factor for creep deformation (default value ≈ 3.0)
RF_{CD} = reduction factor for chemical degradation (default value ≈ 1.4)
RF_{BD} = reduction factor for biological degradation (default value ≈ 2.0)
RF_{JNT} = reduction factor for joints

The actual reduction factors are site specific, application specific, and product specific. The default values described above are according to Geosynthetic Research Institute (GRI) Standard Practice GT7 (2012) for retaining walls. These default values should be considered as the upper bound. Multiplying these reduction factors for a particular application is very significant in decreasing the ultimate strength. For large projects, specific procedures are used for evaluating the individual reduction factors.

For the design of the MSE walls (procedure described below), the long-term design tension load, T_a (kN/m), is determined based on the required factor of safety (FS) as

$$T_a = \frac{T_{al}}{FS} \qquad (5.74)$$

FS > 1.5 is usually required. The FS is thus in reference to the end of the service life of the MSE wall.

5.13.4.4 Pullout resistance of the reinforcement

The pullout resistance of the reinforcement is defined by the ultimate tensile load required to generate outward sliding of the reinforcement against the combined effects of frictional and passive resistance, as described previously. The following approach applies to both extensible (i.e., geosynthetic) and inextensible (i.e., metal strip) reinforcements. The pullout resistance, P_r, per unit width of the reinforcement can be estimated as

$$P_r = F^* \cdot \alpha \cdot \sigma'_v \cdot L_e \cdot 2 \qquad (5.75)$$

where
F^* = a pullout resistance factor
α = a reduction factor to account for reinforcement extensibility (1.0 for metallic, 0.8 for geogrid, and 0.6 for geotextile reinforcement)
σ'_v = effective vertical stress at the soil–reinforcement interface
L_e = embedment length in the resistant zone behind the failure surface (described later)

Ideally, F^* should be determined using pullout tests for the given reinforcement material, in the specific backfill and under design stress and density conditions (Koerner, 2012). Or, F^* can be estimated using the following approach:
For ribbed reinforcement strips, F^* is interpolated as follows:

$$F^* = 1.8 \text{ at the top of the structure} \qquad (5.76)$$

and

$$F^* = \tan\emptyset' \text{ at depth of 6 m and below} \qquad (5.77)$$

For backfills, meeting the requirements is shown in Table 5.3. For geosynthetic sheet reinforcement, F^* can be estimated as

$$F^* = 2/3 \tan \emptyset' \qquad \text{for geotextiles} \qquad (5.78)$$
$$F^* = 0.8 \tan \emptyset' \qquad \text{for geogrids} \qquad (5.78a)$$

where
$\emptyset'$ = peak-drained friction angle of the backfill soil

5.13.5 Design procedure of MSE walls

The design of an MSE wall generally involves the following steps:

1. The height of the MSE wall, H, and general geometry of the existing ground surface are usually given or should be determined prior to MSE wall design.
2. Provide a tentative sizing of the MSE wall, mainly the width of the wall, L, as shown in Figure 5.36.

    a. A preliminary length of reinforcement, L, is chosen that should be the greater of 0.7 H and 2.5 m.
3. Determine the engineering properties of the retained soil, foundation soil, and the soil to be used as backfill (i.e., their respective c', $\varnothing'$, and γ).
4. Evaluate external stability of the reinforced soil mass defined by the structure height, H, and the reinforcement length, L.

    a. The external stability evaluations for MSE walls treat the reinforced section as a composite homogeneous soil mass and evaluate the stability according to conventional failure modes for reinforced concrete retaining wall systems described previously. The reinforced soil mass is treated as a rigid "block" and subject to lateral earth pressure. The lateral earth pressure should be determined using Coulomb's active earth pressure (Section 5.3). The inclination of the soil slope behind the MSE wall (α), inclination of the back face of the MSE wall (β), and friction angle at the interface between the retained soil and the backfill soil (δ') can all be considered. The lateral stress induced by surcharge should also be included (see Section 5.7) in the calculation of lateral earth pressure.
    b. Application of vertical and horizontal forces to the block, creating eccentric loading, and determine the safety factor against sliding and overturning of the block (see Figures 5.37a and 5.37b).
    c. Treat the block as an eccentrically loaded foundation and check safety factor against bearing capacity failure (Figure 5.37c) using the bearing capacity equation provided in Chapter 4.
    d. The deep-seated shear failure (Figure 5.37d) analysis follows the same procedure typically used for slope stability analysis.
5. Evaluate internal stability of the reinforced soil mass and determine the spacings of the reinforcement.

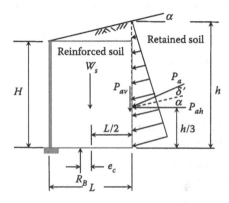

Figure 5.36 The reinforced soil mass, its lateral earth pressure and reaction force.

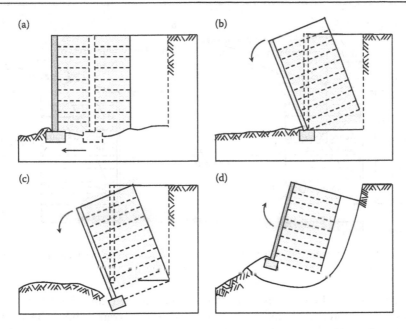

Figure 5.37 Potential external failure modes for an MSE wall. (a) Sliding, FS $\geq$ 1.5. (b) Overturning, FS $\geq$ 2. (c) Bearing capacity, FS $\geq$ 2. (d) Deep-seated stability, FS $\geq$ 1.5.

6. Design of the MSE wall facing. This includes the structural design of the facing and its connection with the reinforcement.
7. Deformation analysis of the MSE wall.

It should be noted that the flexibility of MSE walls makes the potential for overturning failure highly unlikely. However, overturning criteria (maximum permissible eccentricity) aid in controlling lateral deformation by limiting tilting, and as such should always be satisfied. The external stability analysis basically follows the same procedure for reinforced concrete retaining wall systems described previously. The procedure is demonstrated in Example 5.10.

Many methods for internal stability evaluation have been proposed in the past (Elias et al., 2001; Anderson et al., 2012). The main purpose of internal stability evaluation is to determine the reinforcement required. Differences among these methods are principally in the development of the internal lateral stress and the assumption as to the location of the most critical failure surface. In the following discussion, we concentrate on the "Simplified Method" (Elias et al., 2001) only.

5.13.5.1 Internal stability evaluation using the Simplified Method

The Simplified Method (Elias et al., 2001) to be adopted herein is applicable to MSE walls reinforced with either inextensible or extensible reinforcements. In this method, a potential failure surface that coincides with the zone of maximum tensile forces in the reinforcement layer is assumed. The characteristics of this failure surface were developed based on extensive experiments and theoretical studies. A bilinear failure surface is assumed for walls reinforced with inextensible reinforcements (see Figure 5.38a). A failure surface defined by Rankine failure plane, inclined at $45 + \varnothing'/2$ from the horizontal (see Figure 5.38b), is assumed for extensible reinforcements. The failure surface separates the active from the resistant zone. In both cases, the failure surface passes through the

(a)   (b)

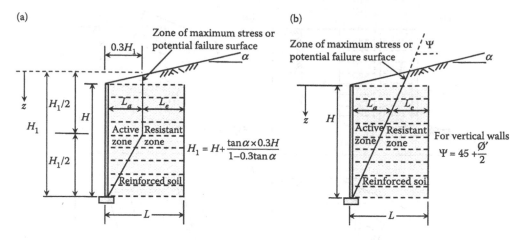

Figure 5.38 Location of potential failure surface for internal stability evaluation. (a) Inextensible reinforcement. (b) Extensible reinforcement.

toe of the MSE wall. The difference in the potential failure planes between inextensible and extensible reinforcements reflects the effects of their extensibility. The geosynthetic reinforcements are sufficiently extensible that the reinforced soil mass has enough freedom to displace and reaches active state. Therefore, the lateral earth pressure is based on K_a throughout the failure surface for geosynthetic type of reinforcements. The metal strip, with its much lower extensibility, tends to impose more restraining effects on the soil mass and thus generates higher lateral stress. The difference is consistently at about 20% from below a depth of 6 m. Such difference becomes progressively more significant toward the ground surface. From 6 m and above, the difference increases progressively from 20% to 70% at the ground surface.

The vertical stress, σ'_v, within the reinforced soil mass is assumed to be equal to the soil overburden stress due to soil self-weight for both types of reinforcements. The horizontal stress, σ'_h, along the failure surface relates to σ'_v through a lateral earth pressure coefficient, K_r as

$$\sigma'_h = K_r \sigma'_v \tag{5.79}$$

The lateral earth pressure coefficient, K_r, is determined by applying a multiplier to the active earth pressure coefficient K_a and K_a is calculated using Coulomb's method. Figure 5.39 shows the variation of K_r/K_a with depth (z) below top of the wall for geosynthetic and metal strip reinforcements. For a simple case of no wall friction, flat ground surface behind the MSE wall (i.e., $\alpha = 0$) and the back of the wall is vertical, K_a becomes the same as that of Rankine and

$$K_a = \tan^2 (45 - \varnothing'/2) \tag{5.8}$$

Note that K_a for more general ground surface profile, wall inclination, and interface friction conditions can also be determined using Coulomb's method [i.e., Equation (5.9)].

The internal stability evaluation is carried out as follows:

1. Calculate the horizontal stress, σ'_h, at each reinforcement level along the potential failure surface, due to overburden stress.

$$\sigma'_h = K_r \sigma'_v \tag{5.80}$$

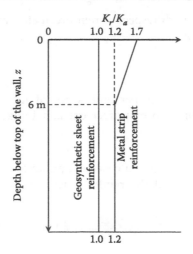

Figure 5.39 Variation of lateral stress ratio with depth.

and

$$\sigma'_v = \gamma_r z \tag{5.81}$$

where
  K_r = lateral earth pressure coefficient determined according to the type of reinforcement and Figure 5.39.
  γ_r = unit weight of the reinforced soil

Note that there can be additional vertical as well as horizontal stresses if there are surcharge loads on top of the MSE wall and/or retained soil surface. The increase of these stresses can be computed using the procedures described in Section 5.7.

2. Calculate the maximum tensile force per unit length of the wall, T_{max} based on the vertical spacing, S_v, for geosynthetic sheet reinforcements as,

$$T_{max} = \sigma'_h \cdot S_v \tag{5.82}$$

For metal strip reinforcements, T_{max} per unit length of the wall is

$$T_{max} = \frac{\sigma'_h \cdot S_v}{R_c} \tag{5.83}$$

and

$$R_c = \frac{b}{S_h} \tag{5.84}$$

where
  R_c = coverage ratio
  b = width of the metal strip reinforcing element
  S_h = center-to-center horizontal spacing between the metal strips

Note $R_c = 1$ for full coverage sheet reinforcement. In this case, Equations (5.82) and (5.83) are identical. To provide a coherent reinforced soil mass, S_v should not exceed 800 mm.

3. Check internal stability with respect to reinforcement breakage, so that

$$T_a \geq \frac{T_{max}}{R_c} \tag{5.85}$$

where

T_a = allowable tension force per unit width of the reinforcement according to Equation (5.72) for strip reinforcement and Equation (5.74) for geosynthetic sheet reinforcement

4. Check stability with respect to reinforcement pullout.

With respect to pullout of the reinforcement, T_{max} should meet the following requirement:

$$T_{max} \leq \frac{P_r}{FS_{PO}} \tag{5.86}$$

and

$$P_r = F^* \cdot \alpha_r \cdot \sigma'_v \cdot L_e \cdot 2 \tag{5.75}$$

where

FS_{PO} = factor of safety against pullout ≥ 1.5
T_{max} = maximum tensile force per unit length of the wall
P_r = pullout resistance per unit width of the reinforcement according to Equation (5.75) which is repeated above
F^* = pullout resistance factor, details of which have been discussed previously
α_r = a reduction factor to account for reinforcement extensibility (1.0 for metallic, 0.8 for geogrid, and 0.6 for geotextile reinforcement)
σ'_v = effective vertical stress at the soil–reinforcement interface
L_e = embedment length in the resistant zone behind the failure surface (see Figures 5.38a and 5.38b)

Combining Equations (5.75) and (5.86) gives

$$L_e \geq \frac{FS_{PO} \cdot T_{max}}{F^* \cdot \alpha_r \cdot \sigma'_v \cdot 2} \tag{5.87}$$

L_e should be kept at a minimum of 1 m.

If any of the safety criterion is not met for all reinforcements, the design should be adjusted and the evaluation repeated. The potential adjustments can include the use of a reinforcement with higher T_a or P_r values, increase the embedment length, and/or decrease the spacing of the reinforcements.

The total length of the reinforcement, L, is then:

$$L = L_a + L_e \tag{5.88}$$

where

L_a = length of the reinforcement in the active zone (see Figures 5.38a and 5.38b)

For walls with inextensible reinforcements (see Figure 5.38a), from the base up to $H_1/2$:

$$L_a = 0.6(H_1 - z) \tag{5.89}$$

For the upper half of the wall with inextensible reinforcements (see Figure 5.38a):

$$L_a = 0.3H_1 \tag{5.90}$$

where
H_1 = height of the MSE wall (see Figure 5.38a)
z = depth below top of the wall (see Figure 5.38a)

For walls with extensible reinforcements (see Figure 5.38b),

$$L_a = (H - z)\tan(45 - \varnothing'/2) \tag{5.91}$$

where
H = height of the wall (see Figure 5.38b)

When the backfill on top of the MSE wall is flat (i.e., $\alpha = 0$), then $H_1 = H$.

5.13.5.2 Design of the MSE wall facing and connection

The types of wall facings described in Figure 5.34 can all be used with extensible and inextensible reinforcements. When segmented concrete blocks or full-height precast panels are used, the reinforcements can be structurally connected with the wall facing. The maximum tensile force per unit length of the wall, T_{max}, for each reinforcement layer is used as the basis for the connection design. Precautions should be undertaken to avoid adverse interactions between the reinforcement and the facing material (i.e., concrete).

The geogrid or geotextile, if used as the reinforcement, can be folded to form a wrap-around facing, as shown in Figures 5.34a and 5.35. For permanent MSE walls, the exposed part of the geosynthetic should either be covered with other structural elements (such as shotcrete or concrete panels) or chemically treated for protection against ultraviolet light. The wraparound facing should maintain a minimum overlap length, L_o, as depicted in Figure 5.34a, to prevent the facing from being pulled out by the lateral earth pressure. The same procedure used for the determination of L_e for pullout resistance can be used for the estimation of L_o as follows:

$$L_o = \frac{T_{max} \cdot FS_{PO}}{F^* \cdot \alpha_r \cdot \sigma'_v \cdot 2} \tag{5.92}$$

where
T_{max} = maximum tensile force per unit length of the wall as per Equation (5.82)
FS_{PO} = factor of safety against pullout ≥ 1.5
F^* = pullout resistance factor, details of which have been discussed previously
α_r = reduction factor to account for reinforcement extensibility (0.8 for geogrid and 0.6 for geotextile reinforcement)

A maximum vertical spacing of 500 mm is typical for MSE walls with wraparound geosynthetic facings for constructability and to minimize bulging.

5.13.5.3 Deformation analysis of the MSE wall

Lateral displacement of the wall face occurs primarily during construction. The major factors that contribute to lateral displacement during construction include compaction

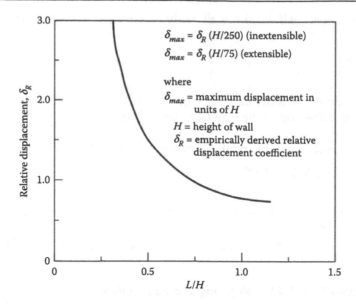

Figure 5.40 Empirical relationship for estimating the lateral displacement during construction of geosynthetically reinforced walls. (After Christopher, B.R., Gill, S.A., Giroud, J.P., Juran, I., Mitchell, J.K., Schlosser, F. and Dunnicliff, J. 1990. *Reinforced soil structures, Vol. I, Design and construction guidelines.* Report to Federal Highway Administration, No. FHWA-RD-89-043.)

intensity, reinforcement to soil stiffness ratio, reinforcement length, slack in reinforcement connections to the wall face, and deformability of the facing system. Post-construction deformation can also occur due to settlement of the structure. There is no standard method for the evaluation of overall lateral displacement of MSE walls (Christopher et al., 1990). Figure 5.40 shows an empirical relationship between reinforcement length, L (normalized with respect to wall height, H), and maximum lateral displacement, δ_{max}, for walls with granular backfill. The calculation of δ_{max} is based on a relative displacement coefficient, δ_R, as shown in Figure 5.40. The relationship between δ_R and L/W was developed based on finite element analysis and limited field evidence from 6 m high test walls.

Post-construction vertical movements can be estimated from foundation settlement analyses using the typical procedures for shallow foundations.

EXAMPLE 5.10

Given

Figure E5.10 shows an MSE wall with a height, H, of 5 m and a tentative width (reinforcement length), L, of 3.5 m. The related soil properties are included in Figure E5.10. Assume that the retained soil and foundation soil have the same unit weight and strength parameters. At the back and base of the MSE wall, assume the soil–wall interface friction, $\delta' = (2/3)\varnothing' = 20°$.

Required

Evaluate the external stability of the MSE wall against overturning, sliding, and bearing capacity failure.

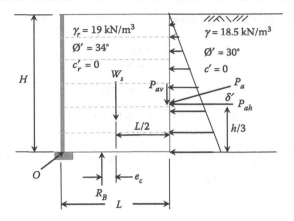

Figure E5.10 The tentative MSE wall dimensions.

Solution

The following calculations are performed for a unit width of the wall, taking advantage of the spreadsheet program provided on the publisher's website.

Step 1: For $\varnothing' = 30°$, $\alpha = 0°$, and $\beta = 0°$, K_a of Coulomb's method is

$$K_a = \frac{\cos^2(\beta - \varnothing')}{\cos^2(\beta)\cos(\beta + \delta')\left[1 + \sqrt{\dfrac{\sin(\varnothing' + \delta')\sin(\varnothing' - \alpha)}{\cos(\beta + \delta')\cos(\beta - \alpha)}}\right]^2} = 0.30 \qquad (5.9)$$

and

$$P_a = \frac{1}{2}\gamma' H^2 K_a = (0.5)(18.5)(5)^2(0.30) = 68.75\,\text{kN}$$

$$P_{ah} = P_a \cos\delta' = (68.75)(\cos 20°) = 64.61\,\text{kN}$$

$$P_{av} = P_a \sin\delta' = (68.75)(\sin 20°) = 23.52\,\text{kN}$$

Step 2:

$$W_s = (19)(3.5)(5) = 332.5\,\text{kN}$$

Take moment about point O (i.e., toe of the MSE wall),

$$FS = \frac{\sum M_R}{\sum M_O} \qquad (5.57)$$

$$\sum M_R = W_s \cdot \frac{L}{2} + L \cdot P_{av} = (332.50)(3.5/2) + (3.5)(23.52) = 664.18\,\text{kN-m}$$

$$\sum M_O = \frac{1}{3} H P_{ah} = \left(\frac{1}{3}\right)(5)(64.61) = 107.68\,\text{kN-m}$$

Thus, FS against overturning $= (664.18)/(107.68) = 6.2 > 2.0$

Step 3: For FS against sliding failure,

$$FS = \frac{(W_s + P_{av}) \cdot \tan \delta'}{P_{ah}} = \frac{(332.5 + 23.52)(\tan(20)}{(64.61)} = 2.01 > 1.5\,OK \qquad (5.60)$$

Step 4: For safety against bearing capacity failure,

$$R_B = (W_s + P_{av}) = (332.5 + 23.52) = 356.02\,kN$$

$$e_c = \frac{L}{2} - \frac{(\sum M_R - \sum M_O)}{R_B} = (3.5)(0.5) - (664.18 - 107.68)/356.02$$
$$= 0.19\,m < L/6(=0.58\,m) \qquad (5.62)$$

Following Equation (5.63),

$$q_{max} = \frac{\sum F_v}{L}\left(1 + \frac{6e_c}{L}\right) = \frac{(356.02)}{(3.5)}\left[1 + \frac{(6)(0.19)}{(3.5)}\right] = 134.30\,kN/m^2$$

Consider the wall foundation as a continuous foundation and $c' = 0$, the ultimate bearing capacity q_u of the foundation soil is

$$q_u = \bar{q}N_q F_{qs}F_{qd}F_{qi} + \frac{1}{2}\bar{\gamma}BN_\gamma F_{\gamma s}F_{\gamma d}F_{\gamma i} \qquad (4.84)$$

$$F_{qs} = F_{\gamma s} = 1$$
$$D_f = 0\,m$$

$$\bar{q} = \gamma D_f = (18.5)(0) = 0\,kN/m^2$$
$$L' = L - 2e_c = (3.5) - (2)(0.19) = 3.12\,m$$

$$F_{qd} = 1 + 2\tan\varnothing'\left(1 - \sin\varnothing'\right)^2\left(\frac{D_f}{B}\right) = 1$$

$$F_{\gamma d} = 1$$

$$\alpha = \tan^{-1}\left(\frac{P_{ah}}{\sum F_v}\right) = \tan^{-1}\left(\frac{64.61}{356.02}\right) = 10.29°$$

$$F_{\gamma i} = \left(1 - \frac{\alpha}{\varnothing'}\right)^2 = [(1) - (10.29)/(30)]^2 = 0.43$$

For $\varnothing' = 30°$,

$$N_q = \tan^2\left(45 + \frac{\varnothing'}{2}\right)e^{\pi \tan\varnothing'} = \tan^2\left(45 + \frac{(30)}{2}\right)e^{(3.1416)(\tan(30))} = 18.40 \qquad (4.23)$$

$$N_\gamma = 2(N_q + 1)\tan\varnothing' = (2)(18.40 + 1)(\tan(30)) = 22.40 \qquad (4.57)$$

$$q_u = \bar{q}N_q F_{qs}F_{qd}F_{qi} + \frac{1}{2}\bar{\gamma}L'N_\gamma F_{\gamma s}F_{\gamma d}F_{\gamma i}$$
$$= 0 + (0.5)(18.5)(3.12)(22.40)(0.43)(1) = 279.77\,kN/m^2 \qquad (4.84)$$

Thus, FS against bearing capacity failure is

$$FS = \frac{q_u}{q_{max}} = (279.77)/(134.30) = 2.08 > 2.0\,OK \qquad (5.68)$$

EXAMPLE 5.11

Given

For the same overall dimensions of the MSE wall and soil properties described in Example 5.10, use galvanized ribbed metal strips as the reinforcement. The strips are vertically spaced at 0.5 m (i.e., $S_v = 0.5$ m). The first row is located at 0.5 m from the top of the wall. Consider the metal strip has a width (b) of 50 mm and a nominal thickness (E_n) of 4.0 mm at the time of construction and it is expected to lose (E_R) 1.416 mm due to corrosion after 75 years of service life. Grade 60 steel with a yield stress (f_y) of 413.7 MPa is to be used as the metal strip.

Required

Evaluate internal stability of the metal strip reinforced wall and determine the dimensions and horizontal spacing of the metal strips. Use a service life of 75 years for the MSE wall design.

Solution

For the internal stability evaluation, the following table is prepared, taking advantage of the spreadsheet program provided on the publisher's website.

Refer to Figure E5.10 for soil properties.

$$K_a = \tan^2\left(45 - \frac{\varnothing'_r}{2}\right) = \tan^2\left(45 - \frac{34}{2}\right) = 0.28 \tag{5.8}$$

$K = \left(\dfrac{K_r}{K_a}\right) \cdot K_a$ and $\left(\dfrac{K_r}{K_a}\right)$ is interpolated according to Figure 5.39

$\sigma'_v = \gamma_r \cdot z$ where $z =$ depth in m below the top of the wall

$\sigma'_h = K \cdot \sigma'_v$

F^* is interpolated from $\tan\varnothing'_r$, at $z = 6$ m to 1.8 (at top of the wall)

L_e is calculated according to Figure 5.38a, consider a total length, $L = 3.5$ m

$\alpha_r = 1.0$ for metallic strips

$$P_r = F^* \cdot \alpha_r \cdot \sigma'_v \cdot L_e \cdot 2 \tag{5.75}$$

$$T_{max} = \frac{\sigma'_h \cdot S_v}{R_c} \tag{5.83}$$

Vertical spacing, S_v, of 0.5 m is used for the computation of the following table.

$$R_c = \frac{b}{S_h} \tag{5.84}$$

and $b = 50$ mm

Horizontal spacing, $S_h = 0.8$ and 1.0 m.

Safety factor against pullout,

$$FS_{PO} = \frac{P_r}{T_{max}} \tag{5.86}$$

$$f_{all} = 0.55 f_y = (0.55)(413.7) = 227.54 \text{ MPa} \tag{5.72a}$$

For the tensile stress in steel (f_s) after 75 years

$$f_s = T_{max}/(E_n - E_R) = T_{max}(1000)/(4.0 - 1.416)$$

z, m	σ'_v, kPa	K	σ'_h, kPa	F^*	L_e, m	P_r, kN/m	S_h, m	R_c	T_{max}, kN/m	FS_{PO}	f_s, MPa
0.5	9.5	0.47	4.45	1.71	2	64.84	0.8	0.063	35.63	1.82	13.79
1.0	19	0.46	8.68	1.61	2	122.54	0.8	0.063	69.47	1.76	26.89
1.5	28.5	0.45	12.69	1.52	2	173.12	0.8	0.063	101.52	1.71	39.29
2.0	38	0.43	16.47	1.42	2	216.58	0.8	0.063	131.78	1.64	51.00
2.5	47.5	0.42	20.03	1.33	2	252.90	0.8	0.063	160.25	1.58	62.02
3.0	57	0.41	23.37	1.24	3.26	459.81	1	0.05	233.66	1.97	90.43
3.5	66.5	0.40	26.48	1.14	3.32	504.91	1	0.05	264.77	1.91	102.47
4.0	76	0.39	29.36	1.05	3.38	539.28	1	0.05	293.65	1.84	113.64
4.5	85.5	0.37	32.03	0.96	3.44	562.29	1	0.05	320.28	1.76	123.95

According to the above table, all $FS_{PO} > 1.5$ and the tensile stress in steel (f_s) after 75 years of service life $< f_{all}$.

EXAMPLE 5.12

Given

For the same overall dimensions of the MSE wall and soil properties described in Example 5.10, use geogrid as the reinforcement. The geogrid layers are vertically spaced at 0.4 m (i.e., $S_v = 0.4$ m). The first row is located at 0.2 m from the top of the wall. Consider the combined reduction factor (RF) as 2.5. Use geogrid wraparound as the wall facing.

Required

Evaluate internal stability of the geogrid reinforced wall, choose the required ultimate tensile strength (T_{ult}) of the geogrid, and determine the overlap length for the wraparound.

Solution

For the internal stability evaluation, the following table is prepared, taking advantage of the spreadsheet program provided on the publisher's website.

$$K_a = \tan^2\left(45 - \frac{\varnothing'_r}{2}\right) = \tan^2\left(45 - \frac{34}{2}\right) = 0.28 \tag{5.8}$$

$\sigma'_v = \gamma_r \cdot z$ where $z =$ depth in m below the top of the wall

$$\sigma'_h = K_a \cdot \sigma'_v$$

$$F^* = 0.8 \tan \varnothing' = (0.8)(\tan 34) = 0.54 \tag{5.78a}$$

$$L_a = (H - z)\tan(45 - \varnothing'/2) \tag{5.91}$$

$$L_e = L - L_a = 3.5 - L_a \text{ and } L = 3.5 \text{ m}$$

$\alpha_r = 0.8$ for geogrids

$$P_r = F^* \cdot \alpha_r \cdot \sigma'_v \cdot L_e \cdot 2 \tag{5.75}$$

$$T_{max} = \sigma'_h \cdot S_v \tag{5.82}$$

Vertical spacing, S_v, of 0.4 m is used for the computation of the following table.

$$FS_{PO} = \frac{P_r}{T_{max}} \tag{5.86}$$

Combining Equations (5.73) and (5.74), the required ultimate tensile strength, T_{ult}, should be larger than $T_{max} \cdot RF \cdot FS$, where $RF =$ the combined reduction factor (given as 2.5) and FS = safety factor against reinforcement tensile stress failure. FS = 1.5 is used in the following table.

The overlap length, $L_o = \dfrac{T_{max} \cdot FS_{PO}}{F^* \cdot \alpha_r \cdot \sigma'_v \cdot 2}$ according to Equation (5.92), use $FS_{PO} = 1.5$ for the calculation of L_o.

z, m	σ'_v, kPa	σ'_h, kPa	L_e, m	P_r, kN/m	T_{max}, kN/m	FS_{PO}	L_o, m	T_{ult} req, kN/m
0.2	3.8	1.07	0.95	3.11	0.43	7.24	0.20	1.61
0.6	11.4	3.22	1.16	11.42	1.29	8.86	0.20	4.83
1.0	19.0	5.37	1.37	22.53	2.15	10.48	0.20	8.06
1.4	26.6	7.52	1.59	36.42	3.01	12.11	0.20	11.28
1.8	34.2	9.67	1.80	53.11	3.87	13.73	0.20	14.50
2.2	41.8	11.82	2.01	72.58	4.73	15.35	0.20	17.73
2.6	49.4	13.97	2.22	94.85	5.59	16.98	0.20	20.95
3.0	57.0	16.11	2.44	119.91	6.45	18.60	0.20	24.17
3.4	64.6	18.26	2.65	147.76	7.31	20.23	0.20	27.39
3.8	72.2	20.41	2.86	178.40	8.16	21.85	0.20	30.62
4.2	79.8	22.56	3.07	211.83	9.02	23.47	0.20	33.84
4.6	87.4	24.71	3.29	248.06	9.88	25.10	0.20	37.06
5.0	95.0	26.86	3.50	287.07	10.74	26.72	0.20	40.29

L_o is less than 1.0 m in all reinforcement layers, use 1.0 m as the overlap length for the wraparound.

Use a geogrid with T_{ult} of 40 kN/m.

According to the above table, the FS_{PO} is larger than 1.5 for all reinforcement layers.

5.14 SHEET PILE WALLS

Sheet pile sections can be connected via thumb-and-finger or ball-and-socket type of connections as shown in Figure 5.41, which are watertight and can enhance overall

(a)

(b)

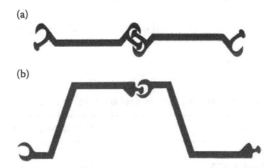

Figure 5.41 Sheet pile connections. (a) Thumb-and-finger type of sheet pile connections. (b) Ball-and-socket type of sheet pile connections.

Table 5.5 Properties of selected steel sheet-pile sections

Section	Width (w), mm	Height (h), mm	Flange (t_f), mm	Wall (t_w), mm	Section modulus, cm^3/m	Moment of inertia, cm^4/m
PZ 22	559	229	9.50	9.50	973	11,500
PZ 27	457	305	9.50	9.50	1620	25,200
PZ 35	575	378	15.21	12.67	2608	49,300
PZ 40	500	409	15.21	12.67	3263	67,000
PS 31	500	–	12.7	–	108	442

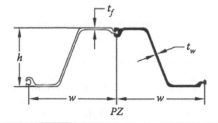

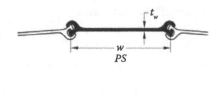

stability of the sheet pile wall after installation. There are a variety of sheet piles with Z-shaped or straight web sections. Table 5.5 shows a few steel sheet pile sections available on the market. Yield strength for some of the available steel grades typically used for steel sheet piles is shown in Table 5.6. For the design of sheet pile walls, the allowable flexural stress is usually taken as 60%–65% of the yield strength. The steel sheet piles are made of high-strength material with a relatively small cross-sectional area and very favorable moment of inertia for the amount of material used. For these reasons, steel sheet piles are versatile in their applications in civil engineering construction.

The steel sheet piles can be penetrated deep (compare to MSE and reinforced concrete retaining walls) into the ground because of their high strength and small cross-sectional areas, with the help of a vibratory hammer. When the height of the wall above the excavation line is not greater than 5 m or so, the sheet piles can be designed as a cantilever wall as shown in Figure 5.42a. Anchors can be installed to provide additional lateral support as shown in Figures 5.42b to 5.42d, where anchor rods are used to connect the sheet pile to the supporting elements placed in the soil, outside of the active zone. The supporting element can be made of concrete block (Figure 5.42b), anchor piles (Figure 5.42c), or another row of sheet piles driven parallel to each other as shown in Figure 5.42d. Tiebacks can also be used to provide the lateral support. Details of anchor blocks and tiebacks are introduced later.

Figure 5.43 shows a typical scenario and sequence for the cantilever and anchored sheet pile wall installations.

Table 5.6 Yield strength of available steel grades

Type of steel	Yield strength, MPa
ASTM A-328	270
ASTM A-572	345
ASTM A-588	345
ASTM A-690	345

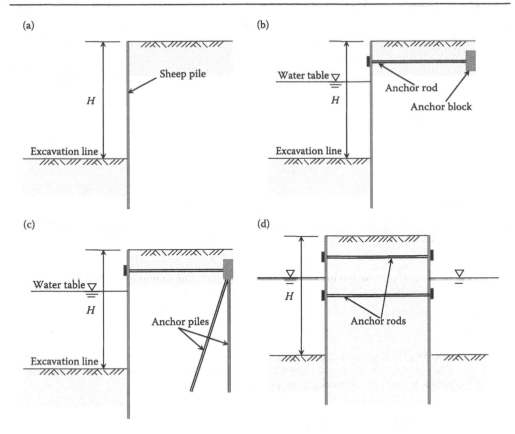

Figure 5.42 Cantilever and anchored sheet pile walls. (a) Cantilever sheet pile wall. (b) Anchored sheet pile wall supported by anchor blocks. (c) Anchored sheet pile wall supported by anchor piles. (d) Cross backfill anchor rods for mutual support of parallel sheet pile walls.

The construction sequence:

1. Drive the sheet piles.
2. Excavate the front side of the sheet pile.
3. For cantilever sheet pile wall, backfill to the top of the sheet pile wall.
4. For anchored sheet pile wall, backfill in stages to allow for the installation of anchors before finishing backfill to the top of the sheet pile wall.

Free-draining granular material is normally used as the backfill.

Available methods for sheet pile wall analysis and design can be divided into the following three categories.

5.14.1 Limit equilibrium method

The limit equilibrium method has been widely used in the design of sheet pile walls. In this approach, it is assumed that the sheet piles are rigid, and rotate with respect to a pivoting point when subject to the lateral earth load. The rotation is sufficient that the soil on both sides of the sheet pile has reached either active or passive failure (i.e., limit state). The simplification enables earth pressure, shear, and moment diagrams be established for structural analysis. This method, however, does not consider the nature of soil–structure interaction (i.e., stiffness of sheet pile and soil) and its effects.

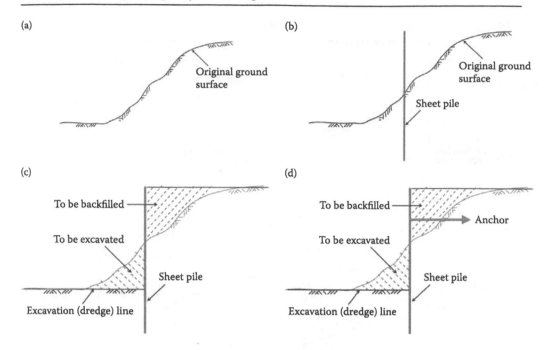

Figure 5.43 Construction sequence and types of sheet pile walls. (a) Original ground surface. (b) Sheet pile insertion. (c) Excavation and backfill for cantilever sheet pile wall. (d) Installation of lateral support for anchored sheet pile wall.

5.14.2 Finite element method

Using this powerful numerical tool, a rather comprehensive analysis for the flexible sheet pile/soil mass system can be conducted. The nonlinear soil stress–strain relationships, soil–structure interaction, and various types of lateral support systems can all be considered. The method has not been widely used in engineering practice, however, because of the complications of soil constitutive laws and other modeling efforts involved in the finite element analysis.

5.14.3 Soil–structural interaction method

In this approach, the sheet pile wall and soil mass are represented by a bending beam (sheet pile wall) resting on a series of springs (soil resistance). The springs provide resisting forces which increase with lateral deflection. The method can be traced back to its early developments in the 1950s (Rowe, 1955; Richart, 1957), where the soil resistance was represented by linear springs. With the help of numerical procedures, this method has now been extended by using a beam on nonlinear soil reaction analysis, based on realistic test results (Reese et al., 2013).

Tschebotarioff (1949) presented the result of his large-scale model tests for retaining structures. One of the most important conclusions was that the distribution of earth pressure on sheet piles is highly influenced by the wall deformations. The behavior of retaining structures is largely a matter of soil conditions and the details of the structural system. Therefore a rational method of design must include the nonlinear soil–resistance displacement relationships, pile spacing (if a row of isolated piles are used as a retaining wall), penetration depth, and structural properties of the pile.

The limit equilibrium method has been used traditionally for the analysis of sheet pile walls for both the cantilever and anchored cases. This method is based on the limit-state soil resistance instead of the mobilized soil resistance. The lateral earth pressures according to limit equilibrium method may not be realistic due to the flexibility of sheet piles coupled with constrains imposed by the anchors. Commercial software packages (e.g., Reese et al., 2013) have been developed for the design of flexible retaining structures, such as the sheet pile walls, that consider the effects of soil–structure interaction (SSI). The SSI-based method is similar in concept to that of p–y curves for laterally loaded piles (to be described in Section 7.4). Thus, the SSI-based method can also be referred to as the p–y wall method. The development of earth pressure is related to the magnitude of wall deflection. The SSI-based or p–y wall method is gaining popularity in practice because of its convenience and realistic considerations in the characteristics of flexible retaining systems.

The following sections describe the basic concepts in limit equilibrium and SSI methods in the analysis of cantilever and anchored sheet pile walls.

5.14.4 Limit equilibrium method—cantilever sheet pile walls

The goal of the analysis is to determine the depth of sheet pile penetration (D in Figure 5.44) required to maintain the overall stability of the sheet pile wall and the maximum bending moment for sizing the sheet pile section. The analysis starts with the determination of the lateral earth pressure distribution against the cantilever sheet pile wall. For this analysis, it is assumed that the sheet pile is rigid and rotates toward the excavation side against a pivoting point as shown in Figure 5.44a. It is further assumed that this movement is sufficient that the soil on both sides of the wall has reached its limit state. Thus, passive earth pressure develops in the shaded areas in Figure 5.44a where the sheet pile moves toward the soil (see Section 5.1). Active earth pressure develops in the rest of the soil mass where the sheet pile moves away from the soil. In order to facilitate computation of the earth pressure distribution, however, it is necessary to establish the active and passive pressure profiles on both sides of the wall from their respective ground surface to the base of the sheet pile. Recall that the active and passive earth pressure coefficients (K_a and K_p, respectively) are

$$K_a = \tan^2\left(\frac{\pi}{4} - \frac{\phi'}{2}\right) \tag{5.8}$$

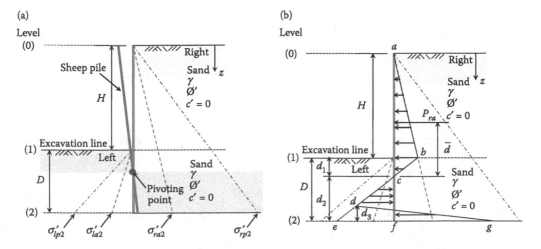

Figure 5.44 Sheet pile wall penetrating in dry sand. (a) Simplification of the cantilever sheet pile rotation. (b) Net lateral earth pressure distribution.

and

$$K_p = \tan^2\left(\frac{\pi}{4} + \frac{\phi'}{2}\right) \tag{5.12}$$

5.14.4.1 For cantilever sheet piles penetrating in sand

As a start, let's consider the groundwater table is deep and the sand is dry. For sand ($c' = 0$), the active pressure, σ'_a, can be determined as

$$\sigma'_a = \sigma'_v K_a = \gamma z K_a \tag{5.16}$$

and for passive pressure,

$$\sigma'_p = \sigma'_v K_p = \gamma z K_p \tag{5.20}$$

Therefore, on the right side of the wall, the active earth pressure changes from 0 at the ground surface (level 0) to σ'_{ra2} at the base of the wall (level 2) as

$$\sigma'_{ra2} = \gamma(H + D)K_a \tag{5.93}$$

Similarly, the passive earth pressure on the right side changes from 0 at the ground surface (level 0) to σ'_{rp2} at the base of the wall (level 2) as

$$\sigma'_{rp2} = \gamma(H + D)K_p \tag{5.94}$$

where

γ = unit weight of the sand
z = depth in reference to the ground surface on the right side of the sheet pile wall
ϕ' = drained friction angle
H = height of the wall above the excavation line
D = depth of sheet pile penetration

On the left side of the wall, the active earth pressure changes from 0 at the ground surface (level 1) to σ'_{la2} at the base of the wall (level 2) as

$$\sigma'_{la2} = \gamma D K_a \tag{5.95}$$

The passive earth pressure on the left side changes from 0 at the ground surface (level 1) to σ'_{lp2} at the base of the wall (level 2) as

$$\sigma'_{lp2} = \gamma D K_p \tag{5.96}$$

The net lateral earth pressure, σ'_n, at a given depth, z (measured from ground surface on the right side as shown in Figure 5.44a), is taken as the lateral earth pressure on the right side, σ'_r, subtracting that from the left, σ'_l, as

$$\sigma'_n = \sigma'_r - \sigma'_l \tag{5.97}$$

The lateral earth pressure is passive for shaded areas shown in Figure 5.44a, and active for the rest of the areas. Therefore
for $z = 0$ to H (from levels 0 to 1):

$$\sigma'_n = \sigma'_{ra} \tag{5.98}$$

where σ'_{ra} = active earth pressure on the right side and $\sigma'_l = 0$ in this depth range.
At level 0 (in Figure 5.44b): $\sigma'_n = 0$ (σ'_n at point a)
At level 1 (in Figure 5.44b): $\sigma'_n = \gamma H K_a$ (σ'_n at point b)
For $z = H$ to $(H + D)$ (levels 1 to 2): σ'_r changes from active to passive and σ'_l changes from passive to active, as z approaches bottom of the sheet pile. To determine the

"transition point" where the earth pressure starts to reverse its direction (i.e., point d in Figure 5.44b), we start by assuming that for $z = H$ to $(H + D)$, all σ_r' as active and all σ_l' as passive first, and

$$\sigma_n' = \sigma_{ra}' - \sigma_{lp}'$$

or

$$\sigma_n' = \gamma z K_a - \gamma(z - H)K_p \tag{5.99}$$

At level 1: $\sigma_n' = \gamma H K_a$ (σ_n' at point b)

When $d_1 = \dfrac{H K_a}{(K_p - K_a)}$, $\sigma_n' = 0$ according to Equation (5.99). (σ_n' at point c)

Then we consider that, for $z = (H + d_1)$ to $(H + D)$, all σ_r' as passive and all σ_l' as active, thus

$$\sigma_n' = \sigma_{rp}' - \sigma_{la}'$$

or

$$\sigma_n' = \gamma z K_p - \gamma(z - H)K_a \tag{5.100}$$

At level 2: $\sigma_n' = \gamma(H + D)K_p - \gamma D K_a$ (σ_n' at point g)

At this stage we do not know D (or d_2) and d_3. To determine D, we make use of the force and moment equilibrium equations; two equations for two unknowns as follows:

For stability of the wall, $\sum$ horizontal forces per unit length of the wall $= 0$, thus: area of the pressure diagram abc − area of cef + area of dge = 0, or

$$P_{ra} - \frac{1}{2}d_2[\gamma(D)K_p - \gamma(H + D)K_a]$$
$$+ \frac{1}{2}d_3[\gamma(H + D)K_p - \gamma(D)K_a - \gamma(H + D)K_a + \gamma(D)K_p] \tag{5.101}$$
$$= 0$$

where P_{ra} = area of abc $= \dfrac{1}{2}(H + d_1)\gamma H K_a$.

Rearrange Equation (5.101), we have

$$d_3 = \frac{d_2[\gamma(D)K_p - \gamma(H + D)K_a] - 2P_{ra}}{\gamma(H + 2D)(K_p - K_a)} \tag{5.102}$$

and $\sum$ moment of the forces per unit length of the wall about point $f = 0$, thus:

$$P_{ra}(\bar{d} + d_2) - \frac{1}{2}d_2[\gamma(D)K_p - \gamma(H + D)K_a]\left(\frac{d_2}{3}\right)$$
$$+ \frac{1}{2}d_3[\gamma(H + D)K_p - \gamma(D)K_a - \gamma(H + D)K_a + \gamma(D)K_p]\left(\frac{d_3}{3}\right)$$
$$= 0 \tag{5.103}$$

Combining Equations (5.101) to (5.103) yields a 4th-order equation of d_2,

$$d_2^4 + A_1 d_2^3 - A_2 d_2^2 - A_3 d_2 - A_4 = 0 \tag{5.104}$$

and

$$A_1 = \frac{HK_p + d_1(K_p - K_a)}{(K_p - K_a)} \tag{5.105}$$

$$A_2 = \frac{8P_{ra}}{\gamma(K_p - K_a)} \tag{5.106}$$

$$A_3 = \frac{6P_{ra}[2\bar{d}(K_p - K_a) + HK_p + d_1(K_p - K_a)]}{\gamma(K_p - K_a)^2} \tag{5.107}$$

$$A_4 = \frac{P_{ra}[6\bar{d}\gamma(HK_p + d_1(K_p - K_a)) + 4P_{ra}]}{\gamma^2(K_p - K_a)^2} \tag{5.108}$$

$$D = d_1 + d_2, \text{ the theoretical depth of penetration} \tag{5.109}$$

The design depth of sheet pile penetration, D_{design}, is obtained by multiplying the theoretical depth of penetration, D, from Equation (5.109) with a factor of safety, FS. The FS is usually in the range of 1.3 to 1.5 as

$$D_{design} = \text{FS} \cdot D \tag{5.110}$$

The total design length of the sheet pile is thus $H + D_{design} = H + \text{FS} \cdot D$.

The maximum bending moment should occur at a depth of d_m below point c (see Figure 5.44b), where the shear force is zero, therefore:

$$P_{ra} = \frac{1}{2}\gamma d_m^2(K_p - K_a) \tag{5.111}$$

or

$$d_m = \sqrt{\frac{2P_{ra}}{\gamma(K_p - K_a)}} \tag{5.112}$$

Taking moment at the point of zero shear, we obtain the maximum moment, M_{max}, as

$$M_{max} = P_{ra}(\bar{d} + d_m) - \left[\frac{1}{2}\gamma d_m^2(K_p - K_a)\right]\left(\frac{d_m}{3}\right) \tag{5.113}$$

The sheet pile section is then sized by computing the required sectional modulus, S_{req}, according to the allowable flexural stress, σ_{all}, of the sheet pile material as follows:

$$S_{req} = \frac{M_{max}}{\sigma_{all}} \tag{5.114}$$

5.14.4.2 For cantilever sheet piles penetrating in clay

Consider a case where the soil below the excavation line is saturated clay and groundwater table coincides with the excavation line. Assuming again that the sheet pile rotates toward the excavation side as a rigid body, the active and passive lateral earth pressure would have a distribution as shown in Figure 5.45. The lateral earth pressures in the sand above the excavation line remain the same as those shown in Figure 5.44. For saturated clay, undrained friction angle $\varnothing = 0$, undrained shear strength is s_u, saturated unit weight of clay is γ_{sat}, and $K_a = K_p = 1$.

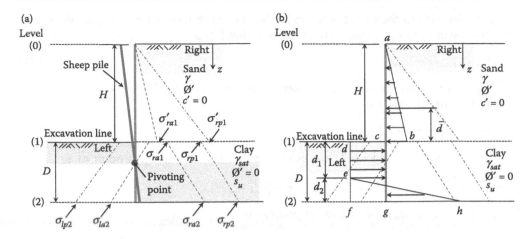

Figure 5.45 Sheet pile wall penetrating in saturated clay. (a) Simplification of the cantilever sheet pile rotation. (b) Net lateral earth pressure distribution.

On the right side of the wall, the active earth pressure changes from 0 at the ground surface (level 0) to σ'_{ra1} at level 1 just above the excavation line ($\sigma'_n = \sigma'_{ra1}$ at point b) as

$$\sigma'_{ra1} = \gamma H K_a \qquad (5.115)$$

The passive earth pressure changes from 0 at the ground surface (level 0) to σ'_{rp1} at level 1 just above the excavation line as

$$\sigma'_{rp1} = \gamma H K_p \qquad (5.116)$$

where
  γ = unit weight of the sand
  H = height of the wall above the excavation line

On the right side of the wall from just below the excavation line to the base of the wall (between levels 1 and 2), the active earth pressure changes from σ_{ra1} at level 1 to σ_{ra2} at level 2, and

$$\sigma_{ra1} = \gamma H - 2s_u \qquad (5.117)$$

and

$$\sigma_{ra2} = \gamma H + \gamma_{sat} D - 2s_u \qquad (5.118)$$

Whereas,

$$\sigma_{rp1} = \gamma H + 2s_u \qquad (5.119)$$

and

$$\sigma_{rp2} = \gamma H + \gamma_{sat} D + 2s_u \qquad (5.120)$$

where
  H = height of the wall above the excavation line
  D = depth of sheet pile penetration

On the left side of the wall, the active earth pressure changes from $-2s_u$ at the excavation line (level 1) to σ_{la2} at the base of the wall (level 2) as

$$\sigma_{la2} = \gamma_{sat}D - 2s_u \tag{5.121}$$

The passive earth pressure changes from $2s_u$ at the excavation line (level 1) to σ_{lp2} at the base of the wall (level 2) as

$$\sigma_{lp2} = \gamma_{sat}D + 2s_u \tag{5.122}$$

The net pressure σ_n at level 1, just below the excavation line (σ_n at point d), is

$$\sigma_n = \sigma_{ra1} - \sigma_{rp1} = \gamma H - 2s_u - 2s_u = -(4s_u - \gamma H) \tag{5.123}$$

The net pressure σ_n at level 2 (σ_n at point h) is

$$\sigma_n = \sigma_{rp2} - \sigma_{la2} = 4s_u + \gamma H \tag{5.124}$$

Consider force equilibrium, Σ, horizontal forces per unit length of the wall $= 0$, thus: Area of the pressure diagram abc – area of $cdfg$ + area of $ehf = 0$, or

$$P_{ra} - D(4s_u - \gamma H) + \frac{1}{2}d_2(8s_u) = 0 \tag{5.125}$$

$$P_{ra} = \text{area of } abc = \frac{1}{2}\gamma H^2 K_a \tag{5.126}$$

Combining Equations (5.125) and (5.126),

$$d_2 = \frac{D(4s_u - \gamma H) - P_{ra}}{4s_u} \tag{5.127}$$

and $\sum$ moment of the forces per unit length of the wall about point $g = 0$, thus:

$$P_{ra}(\bar{d} + D) - (4s_u - \gamma H)\left(\frac{D^2}{2}\right) + \frac{1}{2}d_2(8s_u)\left(\frac{d_2}{3}\right) = 0 \tag{5.128}$$

where
$\bar{d} =$ distance from level 1 to the centroid of area abc (see Figure 4.45b)

Combining Equations (5.127) to (5.128) yields a 2nd-order equation of D,

$$D^2(4s_u - \gamma H) - 2DP_{ra} - \frac{P_{ra}(P_{ra} + 12s_u\bar{d})}{(2s_u + \gamma H)} = 0 \tag{5.129}$$

where
$D = (d_1 + d_2)$ theoretical depth of penetration

The maximum bending moment should occur at a depth of d_m below level 1 (see Figure 5.45b), where the shear force is zero, therefore:

$$P_{ra} = d_m(4s_u - \gamma H) \tag{5.130}$$

or

$$d_m = \frac{P_{ra}}{(4s_u - \gamma H)} \tag{5.131}$$

Taking moment at the point of zero shear, we obtain the maximum moment, M_{max}, as

$$M_{max} = P_{ra}(\bar{d} + d_m) - \left[\frac{d_m}{2}(4s_u - \gamma H)\right] \qquad (5.132)$$

The following is a general procedure for the analysis of a cantilever sheet pile wall:

1. Determine the active and passive earth pressure coefficients, K_a and K_p according to the given soil properties for various soil layers.
2. Establish the net lateral earth pressure profile.
3. Calculate the areas of the various net earth pressure diagram.
4. Establish the coefficients of the equation for the determination of the depth of penetration based on force and moment equilibrium.
5. Solve the equation in Step 4 and obtain the required depth of penetration.
6. Determine the depth of zero shear force.
7. Determine the maximum moment at the point of zero shear.
8. Select the required sheet pile according to the required sectional modulus.

EXAMPLE 5.13

Given

Figure E5.13 shows the soil profile and the proposed scheme of a cantilever sheet pile wall penetrating in sand. Height of the wall, H, above excavation line is 5 m. Properties of the sand are included in the figure. Assume groundwater table is deeper than the base of the sheet pile.

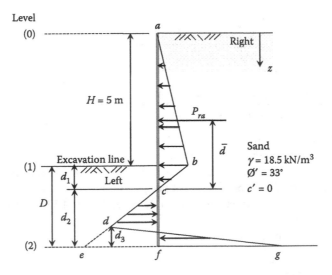

Figure E5.13 Proposed sheet pile wall penetrating in sand.

Required

Determine the total design length and required sectional modulus of the sheet pile. Use allowable flexural stress, $\sigma_{all} = 200$ MPa.

Solution

Follow the procedure described above:

Determine the active and passive earth pressure coefficients for the soil layers:

$$K_a = \tan^2\left(45° - \frac{\phi'}{2}\right) = \tan^2\left(45° - \frac{33°}{2}\right) = 0.295 \tag{5.8}$$

$$K_p = \tan^2\left(45° + \frac{\phi'}{2}\right) = \tan^2\left(45° + \frac{33°}{2}\right) = 3.392 \tag{5.12}$$

Establish the lateral earth pressure profile and determine the following parameters:

$$d_1 = \frac{HK_a}{(K_p - K_a)} = \frac{(5)(0.295)}{(3.392 - 0.295)} = 0.48 \text{ m} \tag{5.99}$$

$$P_{ra} = \frac{1}{2}\gamma H^2 K_a + \frac{1}{2}\gamma HK_a d_1 = 74.66 \text{ kN/m}$$

$$\bar{d} = \frac{\left(\frac{1}{2}\gamma H^2 K_a\right)\left(\frac{1}{3}H + d_1\right) + \left(\frac{1}{2}\gamma HK_a d_1\right)\left(\frac{2}{3}d_1\right)}{P_{ra}} = 1.98 \text{ m}$$

Determine the coefficients of Equation (5.104) and solve for d_2:

$$d_2^4 + A_1 d_2^3 - A_2 d_2^2 - A_3 d_2 - A_4 = 0 \tag{5.104}$$

$$A_1 = \frac{HK_p + d_1(K_p - K_a)}{(K_p - K_a)} = 6.00 \tag{5.105}$$

$$A_2 = \frac{8P_{ra}}{\gamma(K_p - K_a)} = 10.42 \tag{5.75}$$

$$A_3 = \frac{6P_{ra}[2\bar{d}(K_p - K_a) + HK_p + d_1(K_p - K_a)]}{\gamma(K_p - K_a)^2} = 77.50 \tag{5.107}$$

$$A_4 = \frac{P_{ra}[6\bar{d}\gamma(HK_p + d_1(K_p - K_a)) + 4P_{ra}]}{\gamma^2(K_p - K_a)^2} = 99.10 \tag{5.108}$$

By trial and error, $d_2 = 3.83$ m

Determine the total length of the sheet pile:

Theoretical depth of penetration, $D = d_1 + d_2 = (0.48) + (3.83) = 4.31$ m

Use FS $= 1.3$ for the design depth of sheet pile penetration.

Total design length of the sheet pile $= H + 1.3D = (5) + (1.3)(4.31) = 11.08$ m

Sizing the sheet pile:

Depth to zero shear below point c,

$$d_m = \sqrt{\frac{2P_{ra}}{\gamma(K_p - K_a)}} = 1.61 \text{ m} \tag{5.112}$$

Maximum bending moment,

$$M_{max} = P_{ra}(\bar{d} + d_m) - \left[\frac{1}{2}\gamma d_m^2(K_p - K_a)\right]\left(\frac{d_m}{3}\right) = 228.47 \text{ kN-m} \tag{5.113}$$

Required sectional modulus,

$$S_{req} = \frac{M_{max}}{\sigma_{all}} = 114 \times 10^{-5}\,\text{m}^3/\text{m} \tag{5.114}$$

EXAMPLE 5.14

Given

Figure E5.14 shows the soil profile and the proposed scheme of a cantilever sheet pile wall penetrating in saturated clay. The height of the sheet pile wall, H, above excavation line is 5 m. Properties of the sand and clay are included in the figure. Assume groundwater table coincides with the excavation line.

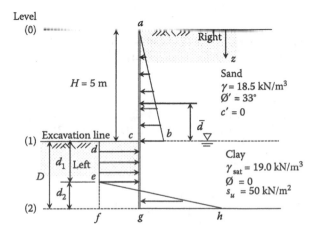

Figure E5.14 Proposed sheet pile wall penetrating in clay.

Required

Determine the design total length and required sectional modulus of the sheet pile. Use allowable flexural stress, $\sigma_{all} = 170$ MPa.

Solution

Follow the procedure described above.
   Determine the active and passive earth pressure coefficients for various soil layers. In sand:

$$K_a = \tan^2\left(45° - \frac{\phi'}{2}\right) = \tan^2\left(45° - \frac{33°}{2}\right) = 0.295 \tag{5.8}$$

$$K_p = \tan^2\left(45° + \frac{\phi'}{2}\right) = \tan^2\left(45° + \frac{33°}{2}\right) = 3.392 \tag{5.12}$$

In clay (undrained conditions, $\emptyset = 0$):
$K_a = K_p = 1.0$

Establish the lateral earth pressure profile and determine the following parameters:

$$P_{ra} = \frac{1}{2}\gamma H^2 K_a = \left(\frac{1}{2}\right)(18.5)(5)^2 = 68.17 \, \text{kN/m} \tag{5.126}$$

$$\bar{d} = \frac{H}{3} = 1.67 \, \text{m}$$

Solve Equation (5.129) for D:

$$D^2(4s_u - \gamma H) - 2DP_{ra} - \frac{P_{ra}(P_{ra} + 12s_u\bar{d})}{(2s_u + \gamma H)} = 0 \tag{5.129}$$

$$D^2[(4)(50) - (18.5)(5)] - 2D(68.17) - \frac{68.17[68.17 + (12)(50)(1.67)]}{[(2)(50) + (18.5)(5)]} = 0$$

Theoretical depth of penetration, $D = 2.61 \, \text{m}$

$$d_2 = \frac{D(4s_u - \gamma H) - P_{ra}}{4s_u} = \frac{(2.61)[(4)(50) - (18.5)(5)] - 68.17}{(4)(50)}$$
$$= 1.06 \, \text{m} \tag{5.127}$$

Determine the total length of the sheet pile:
Use an FS = 1.3 for the determination of designed depth of sheet pile penetration.
Total design length of the sheet pile = $H + 1.3D = (5) + (1.3)(2.61) = 8.40 \, \text{m}$
Sizing the sheet pile:
Depth to zero shear below point c,

$$d_m = \frac{P_{ra}}{(4s_u - \gamma H)} = \frac{(68.17)}{[(4)(50) - (18.5)(5)]} = 0.6 \, \text{m} \tag{5.131}$$

Maximum bending moment,

$$M_{max} = P_{ra}(\bar{d} + d_m) - \left[\frac{d_m}{2}(4s_u - \gamma H)\right]$$

$$= (68.17)(1.67 + 0.6) - \left\{\frac{(0.6)}{2}[(4)(50) - (18.5)(5)]\right\} = 122.77 \, \text{kN-m} \tag{5.132}$$

Required sectional modulus,

$$S_{req} = \frac{M_{max}}{\sigma_{all}} = \frac{(122.77)}{(170)} = 72 \times 10^{-5} \, \text{m}^3/\text{m} \tag{5.114}$$

5.14.4.3 Consideration of the effects of water

When sheet pile walls are used as part of a waterfront structure, part of the soil on both sides of the sheet pile can be inundated under water as shown in Figure 5.46. In this case, the water pressure on both sides of the sheet pile balances each other. For sheet piles penetrating in sand under drained conditions, the lateral earth pressure below water should be computed based on buoyant unit weight, γ'. The undrained strength parameters and saturated soil unit weight, γ_{sat}, as previously described remain the same for sheet piles penetrating in clay under water.

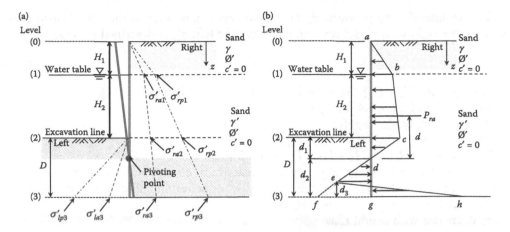

Figure 5.46 Sheet pile wall penetrating in sand and under water. (a) Simplification of the cantilever sheet pile rotation. (b) Net lateral earth pressure distribution.

On the right side of the wall (see Figure 5.46a), the active earth pressure at ground surface is 0, and increases to σ'_{ra1}, σ'_{ra2}, and σ'_{ra3}, respectively, at levels 1, 2, and 3 as

$$\sigma'_{ra1} = \gamma H_1 K_a \tag{5.133}$$

$$\sigma'_{ra2} = (\gamma H_1 + \gamma' H_2) K_a \tag{5.134}$$

and

$$\sigma'_{ra3} = (\gamma H_1 + \gamma' H_2 + \gamma' D) K_a \tag{5.135}$$

The passive earth pressure changes from 0 at the ground surface, and increases to σ'_{rp1}, σ'_{rp2}, and σ'_{rp3}, respectively, at levels 1, 2, and 3 as

$$\sigma'_{rp1} = \gamma H_1 K_p \tag{5.136}$$

$$\sigma'_{rp2} = (\gamma H_1 + \gamma' H_2) K_p \tag{5.137}$$

and

$$\sigma'_{rp3} = (\gamma H_1 + \gamma' H_2 + \gamma' D) K_p \tag{5.138}$$

On the left side of the wall, the active earth pressure changes from 0 at the ground surface (level 2) to σ'_{la3} at the base of the wall (level 3) as

$$\sigma'_{la3} = \gamma' D K_a \tag{5.139}$$

The passive earth pressure changes from 0 at the ground surface (level 2) to σ'_{lp3} at the base of the wall (level 3) as

$$\sigma'_{lp3} = \gamma' D K_p \tag{5.140}$$

where
γ = unit weight of the sand
γ' = buoyant unit weight of the sand
H_1 = distance between top of the wall and water table
H_2 = distance between water table and excavation line
D = depth of sheet pile penetration

The net lateral earth pressure, σ'_n, at a given depth, z, is taken as the lateral earth pressure on the right side, σ'_r, subtracting that from the left, σ'_l, as described previously,

$$\sigma'_n = \sigma'_r - \sigma'_l \qquad (5.97)$$

For $z = 0$ to $(H_1 + H_2)$ (from levels 0 to 2), $\sigma'_n = \sigma'_{ra}$, therefore,
At level 0 (in Figure 5.46b): $\sigma'_n = 0$
At level 1: $\sigma'_n = \sigma'_{ra1} = \gamma H_1 K_a$
At level 2: $\sigma'_n = \sigma'_{ra2} = (\gamma H_1 + \gamma' H_2) K_a$
For $z = (H_1 + H_2)$ to $(H_1 + H_2 + D)$ (levels 2 to 3):
Again, in this depth range, σ'_r changes from active to passive and σ'_l changes from passive to active, as z increases. Assuming that all σ'_r as active and all σ'_l as passive first:
At level 2: $\sigma'_n = \sigma'_{ra2} = (\gamma H_1 + \gamma' H_2) K_a$
σ'_n decreases with depth, z, as

$$\sigma'_n = (\gamma H_1 + \gamma' H_2) K_a - \gamma'(z - H_1 - H_2)(K_p - K_a) \qquad (5.141)$$

When $(z - H_1 - H_2) = d_1$, $\sigma'_n = 0$ (see Figure 5.46b), or

$$d_1 = \frac{(\gamma H_1 + \gamma' H_2) K_a}{\gamma'(K_p - K_a)} \qquad (5.142)$$

and at level 3,

$$\sigma'_n = (\gamma H_1 + \gamma' H_2) K_a - \gamma' D(K_p - K_a) \qquad (5.143)$$

Then treat all σ'_r as passive and all σ'_l as active, and σ'_n at level 3 is

$$\sigma'_n = (\gamma H_1 + \gamma' H_2 + \gamma' D) K_p - \gamma' D K_a \qquad (5.144)$$

For stability of the wall, $\sum$ horizontal forces per unit length of the wall $= 0$, thus:
Area of the pressure diagram $abcd$ – area of dfg + area of $efh = 0$, or

$$P_{ra} - \frac{1}{2} d_2 [\gamma' D K_p - (\gamma H_1 + \gamma' H_2) K_a]$$

$$+ \frac{1}{2} d_3 [(\gamma H_1 + \gamma' H_2 + \gamma' D) K_p - \gamma' D K_a - (\gamma H_1 + \gamma' H_2) K_a + \gamma' D K_p] = 0$$

$$(5.145)$$

$$P_{ra} = \left[\frac{1}{2} \gamma H_1^2 + \gamma H_1 H_2 + \frac{1}{2} \gamma' H_2^2 + \frac{1}{2} d_1 (\gamma H_1 + \gamma' H_2) \right] K_a \qquad (5.146)$$

where
$P_{ra} =$ area of the pressure diagram $abcd$ (see Figure 5.46b)
$d_2 = D - d_1$

Rearranging Equation (5.145),

$$d_3 = \frac{d_2 [\gamma' D K_p - (\gamma H_1 + \gamma' H_2) K_a] - 2 P_{ra}}{[(\gamma H_1 + \gamma' H_2 + \gamma' D)(K_p - K_a) + \gamma' D K_p]} \qquad (5.147)$$

and $\sum$ moment of the forces per unit length of the wall about point $f = 0$, thus:

$$P_{ra}(\bar{d} + d_2) - \frac{1}{2}d_2[\gamma'DK_p - (\gamma H_1 + \gamma'H_2)K_a]\left(\frac{d_2}{3}\right)$$

$$+ \frac{1}{2}d_3[(\gamma H_1 + \gamma'H_2 + \gamma'D)(K_p - K_a) + \gamma'DK_p]\left(\frac{d_3}{3}\right) = 0 \qquad (5.148)$$

Combining Equations (5.146) to (5.148) yields a 4th-order equation of d_2 as before,

$$d_2^4 + A_1 d_2^3 - A_2 d_2^2 - A_3 d_2 - A_4 = 0 \qquad (5.149)$$

and

$$A_1 = \frac{(\gamma H_1 + \gamma'H_2)(K_p + K_a)}{\gamma'(K_p - K_a)} \qquad (5.150)$$

$$A_2 = \frac{8P_{ra}}{\gamma'(K_p - K_a)} \qquad (5.151)$$

$$A_3 = \frac{6P_{ra}[2\bar{d}(K_p - K_a) + (\gamma H_1 + \gamma'H_2)(K_p + K_a)]}{\gamma'^2(K_p - K_a)^2} \qquad (5.152)$$

$$A_4 = \frac{P_{ra}[6\bar{d}[(\gamma H_1 + \gamma'H_2)(K_p + K_a)] + 4P_{ra}]}{\gamma'^2(K_p - K_a)^2} \qquad (5.153)$$

$D = d_1 + d_2$, the theoretical depth of penetration

The maximum moment should occur at a depth of d_m below point d (see Figure 5.46b), where the shear force is zero, therefore:

$$P_{ra} = \frac{1}{2}\gamma'd_m^2(K_p - K_a) \qquad (5.155)$$

or

$$d_m = \sqrt{\frac{2P_{ra}}{\gamma'(K_p - K_a)}} \qquad (5.156)$$

Taking moment at the point of zero shear, we obtain the maximum moment, M_{max}, as

$$M_{max} = P_{ra}(\bar{d} + d_m) - \left[\frac{1}{2}\gamma'd_m^2(K_p - K_a)\right]\left(\frac{d_m}{3}\right) \qquad (5.157)$$

Sizing of the sheet pile follows the same procedure as described before.

For the case of sheet pile penetrating in clay and the water table is above the excavation line as shown in Figure 5.47, the net lateral earth pressure distribution above the excavation line is similar to that described in Figure 5.46b.

On right side of the wall from just below the excavation line to the base of the wall (between levels 2 and 3), the active earth pressure changes from

$$\sigma_{ra2} = (\gamma H_1 + \gamma'H_2) - 2s_u \qquad (5.158)$$

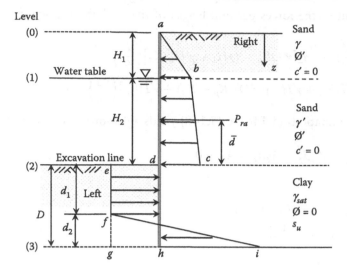

Figure 5.47 Sheet pile wall penetrating in clay with water table above the excavation line.

to

$$\sigma_{ra3} = (\gamma H_1 + \gamma' H_2 + \gamma_{sat} D) - 2s_u \tag{5.159}$$

Whereas, the passive earth pressure changes from

$$\sigma_{rp2} = (\gamma H_1 + \gamma' H_2) + 2s_u \tag{5.160}$$

to

$$\sigma_{rp3} = (\gamma H_1 + \gamma' H_2 + \gamma_{sat} D) + 2s_u \tag{5.161}$$

where
$\gamma =$ unit weight of the sand
$\gamma' =$ buoyant unit weight of the sand
$\gamma_{sat} =$ saturated unit weight of the clay
$H_1 =$ distance between top of the wall and water table
$H_2 =$ distance between water table and excavation line
$D =$ depth of sheet pile penetration

On the left side of the wall, the active earth pressure changes from $-2s_u$ at the excavation line (level 2) to σ_{la3} at the base of the wall (level 3) as

$$\sigma_{la3} = \gamma_{sat} D - 2s_u \tag{5.162}$$

The passive earth pressure changes from $+2s_u$ at the excavation line (level 2) to σ_{lp3} at the base of the wall (level 3) as

$$\sigma_{lp3} = \gamma_{sat} D + 2s_u \tag{5.163}$$

The net pressure σ_n at level 2, just below the excavation line is

$$\sigma_n = \sigma_{ra1} - \sigma_{rp1} = (\gamma H_1 + \gamma' H_2) - 2s_u - 2s_u = -[4s_u - (\gamma H_1 + \gamma' H_2)] \tag{5.164}$$

The net pressure σ_n at level 3 is

$$\sigma_n = \sigma_{rp2} - \sigma_{la2} = 4s_u + (\gamma H_1 + \gamma' H_2) \tag{5.165}$$

Consider force equilibrium, $\sum$, horizontal forces per unit length of the wall $= 0$, thus:

Area of the pressure diagram $abcd$ – area of $degh$ + area of $fgi = 0$, or

$$P_{ra} - D[4s_u - (\gamma H_1 + \gamma' H_2)] + \frac{1}{2}d_2(8s_u) = 0 \tag{5.166}$$

$$P_{ra} = \text{area of } abcd = \frac{1}{2}\gamma H_1^2 K_a + \gamma H_1 H_2 K_a + \frac{1}{2}\gamma' H_2^2 K_a$$

$$= \left[\frac{1}{2}\gamma H_1^2 + \gamma H_1 H_2 + \frac{1}{2}\gamma' H_2^2\right] K_a \tag{5.167}$$

Rearranging Equation (5.166) yields

$$d_2 = \frac{D[4s_u - (\gamma H_1 + \gamma' H_2)] - P_{ra}}{4s_u} \tag{5.168}$$

and $\sum$ moment of the forces per unit length of the wall about point $h = 0$, thus:

$$P_{ra}(\bar{d} + D) - [4s_u - (\gamma H_1 + \gamma' H_2)]\left(\frac{D^2}{2}\right) + \frac{1}{2}d_2(8s_u)\left(\frac{d_2}{3}\right) = 0 \tag{5.169}$$

where
$\bar{d}$ = distance from base of sheet pile to the centroid of area $abcd$

Combining Equations (5.166) to (5.169) results in a 2nd-order equation of D,

$$D^2[4s_u - (\gamma H_1 + \gamma' H_2)] - 2DP_{ra} - \frac{P_{ra}(P_{ra} + 12s_u\bar{d})}{[2s_u + (\gamma H_1 + \gamma' H_2)]} = 0 \tag{5.170}$$

where
$D = (d_1 + d_2)$ theoretical depth of penetration

The determination of the total design length of the sheet piles and required sectional modulus are the same as previously described.

EXAMPLE 5.15

Given

The soil profile and the proposed scheme of a cantilever sheet pile wall penetrating in sand are given in Figure E5.15. Properties of the sand are included in the figure. Water table is located at 2 m below the top of the wall, and the excavation line is at 3 m below water.

Required

Determine the design total length and required sectional modulus of the sheet pile. Use allowable stress, $\sigma_{all} = 200$ MPa.

Solution

Follow the procedure described above.

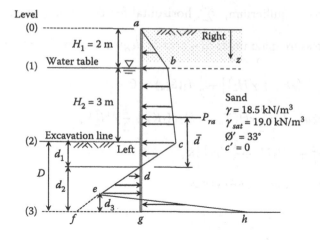

Figure E5.15 Proposed sheet pile wall penetrating in sand below a water table.

Determine the active and passive earth pressure coefficients for various soil layers:

$$K_a = \tan^2\left(45° - \frac{\phi'}{2}\right) = \tan^2\left(45° - \frac{33°}{2}\right) = 0.295 \tag{5.8}$$

$$K_p = \tan^2\left(45° + \frac{\phi'}{2}\right) = \tan^2\left(45° + \frac{33°}{2}\right) = 3.392 \tag{5.12}$$

Sand buoyant unit weight, $\gamma' = 19.0 - 9.81 = 9.19 \, \text{kN/m}^3$

$H_1 = 2 \, \text{m}$
$H_2 = 3 \, \text{m}$

Establish the lateral earth pressure profile and determine the following parameters:

$$d_1 = \frac{(\gamma H_1 + \gamma' H_2)K_a}{\gamma'(K_p - K_a)} = \frac{[(18.5)(2) + (9.19)(3)](0.295)}{(9.19)(3.392 - 0.295)} = 0.67 \, \text{m} \tag{5.142}$$

$$P_{ra} = \left[\frac{1}{2}\gamma H_1^2 + \gamma H_1 H_2 + \frac{1}{2}\gamma' H_2^2 + \frac{1}{2}d_1(\gamma H_1 + \gamma' H_2)\right]K_a$$
$$= 62.19 \, \text{kN/m} \tag{5.146}$$

$$\bar{d} = \frac{\left[\begin{array}{c}\left(\frac{1}{2}\gamma H_1^2\right)\left(\frac{1}{3}H_1 + H_2 + d_1\right) + (\gamma H_1)(H_2)\left(\frac{1}{2}H_2 + d_1\right) \\ + \frac{1}{2}\gamma' H_2^2\left(\frac{1}{3}H_2 + d_1\right) + \frac{1}{2}d_1(\gamma H_1 + \gamma' H_2)\left(\frac{2}{3}d_1\right)\end{array}\right]K_a}{P_{ra}} = 2.27 \, \text{m}$$

Determine the coefficients of Equation 5.149 and solve for d_2:

$$d_2^4 + A_1 d_2^3 - A_2 d_2^2 - A_3 d_2 - A_4 = 0 \tag{5.149}$$

$$A_1 = \frac{(\gamma H_1 + \gamma' H_2)(K_p + K_a)}{\gamma'(K_p - K_a)} = 8.40 \tag{5.150}$$

$$A_2 = \frac{8P_{ra}}{\gamma'(K_p - K_a)} = 17.48 \tag{5.151}$$

$$A_3 = \frac{6P_{ra}[2\bar{d}(K_p - K_a) + (\gamma H_1 + \gamma' H_2)(K_p + K_a)]}{\gamma'^2(K_p - K_a)^2} = 116.1 \qquad (5.152)$$

$$A_4 = \frac{P_{ra}[6\bar{d}[(\gamma H_1 + \gamma' H_2)(K_p + K_a)] + 4P_{ra}]}{\gamma'^2(K_p - K_a)^2} = 268.44 \qquad (5.153)$$

By trial and error, $d_2 = 4.45$ m.
Determine the total length of the sheet pile:

$$\text{Theoretical depth of penetration, } D = d_1 + d_2 = (0.67) + (4.45)$$
$$= 5.12 \text{ m} \qquad (5.154)$$

Use an FS = 1.3 for the designed depth of sheet pile penetration.
Total design length of the sheet pile = $H + 1.3D = (2) + (3) + (1.3)(5.12) = 12.32$ m
Sizing the sheet pile:
Depth to zero shear below point d,

$$d_m = \sqrt{\frac{2P_{ra}}{\gamma'(K_p - K_a)}} = 2.09 \text{ m} \qquad (5.156)$$

Maximum bending moment,

$$M_{max} = P_{ra}(\bar{d} + d_m) - \left[\frac{1}{2}\gamma' d_m^2(K_p - K_a)\right]\left(\frac{d_m}{3}\right) = 228.10 \text{ kN-m} \qquad (5.157)$$

Required sectional modulus,

$$S_{req} = \frac{M_{max}}{\sigma_{all}} = 114 \times 10^{-5} \text{ m}^3/\text{m} \qquad (5.114)$$

EXAMPLE 5.16

Given

For the soil profile and the proposed scheme of a cantilever sheet pile wall penetrating in clay shown in Figure E5.16. Properties of the sand are included in the figure. A water table is located at 2 m below the top of the wall, and the excavation line is at 3 m below water.

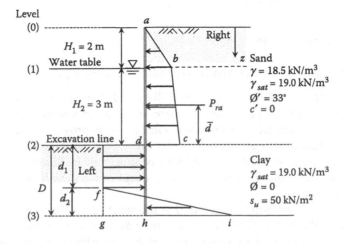

Figure E5.16 Sheet piles penetrating in clay below a water table.

Required

Determine the theoretical depth of penetration for the sheet pile.

Solution

Determine the active and passive earth pressure coefficients for various soil layers.
In sand:

$$K_a = \tan^2\left(45° - \frac{\phi'}{2}\right) = \tan^2\left(45° - \frac{33°}{2}\right) = 0.295 \tag{5.8}$$

$$K_p = \tan^2\left(45° + \frac{\phi'}{2}\right) = \tan^2\left(45° + \frac{33°}{2}\right) = 3.392 \tag{5.12}$$

Sand buoyant unit weight, $\gamma' = 19.0 - 9.81 = 9.19 \, \text{kN/m}^3$
In clay (undrained conditions, $\emptyset = 0$):

$K_a = K_p = 1.0$
$s_u = 50 \, \text{kN/m}^2$
$H_1 = 2 \, \text{m}$
$H_2 = 3 \, \text{m}$

Establish the lateral earth pressure profile and determine the following parameters:

$$P_{ra} = \left[\frac{1}{2}\gamma H_1^2 + \gamma H_1 H_2 + \frac{1}{2}\gamma' H_2^2\right] K_a = 55.82 \, \text{kN/m} \tag{5.167}$$

$$\bar{d} = \frac{\left[\left(\frac{1}{2}\gamma H_1^2\right)\left(\frac{1}{3}H_1 + H_2\right) + (\gamma H_1)(H_2)\left(\frac{1}{2}H_2\right) + \frac{1}{2}\gamma' H_2^2\left(\frac{1}{3}H_2\right)\right] K_a}{P_{ra}} = 1.81 \, \text{m}$$

Solve Equation 5.170 for D as follows:

$$D^2\left[4s_u - (\gamma H_1 + \gamma' H_2)\right] - 2DP_{ra} - \frac{P_{ra}(P_{ra} + 12s_u\bar{d})}{\left[2s_u + (\gamma H_1 + \gamma' H_2)\right]} = 0 \tag{5.170}$$

$$D^2[(4)(50) - ((18.5)(2) + (9.19)(3))] - 2D(55.82)$$
$$- \frac{55.82[55.82 + (12)(50)(1.81)]}{[(2)(50) + ((18.5)(2) + (9.19)(3))]}$$
$$= 0$$

Theoretical depth of penetration, $D = 2.35 \, \text{m}$

5.14.5 Limit equilibrium method—anchored sheet pile walls

When the height of the cantilever sheet pile wall exceeds 5 m or so (above the excavation line), the required depth of penetration and/or sectional modulus becomes excessively high and may not be practical. A common solution is to provide an additional lateral support, such as anchors near the top of the sheet pile, as conceptually described in Figure 5.48a. Simplified procedures have been proposed to facilitate the analysis under the framework of limit equilibrium. These procedures can generally be divided into two categories as follows:

Fixed earth support methods—assumes that the sheet pile is fixed at its base. The pattern of sheet pile deflection will result in active and passive earth pressure on both sides of the sheet pile, below the excavation line. A rigorous fixed earth support analysis involves the consideration of soil–structure interaction. For this reason, no further description on

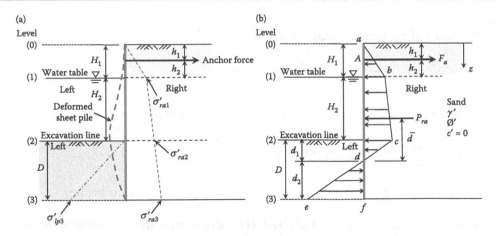

Figure 5.48 Anchored sheet pile wall penetrating in sand and under water. (a) Simplification of the sheet pile deformation. (b) Net lateral earth pressure distribution.

the fixed earth support method is presented. Instead, a section is dedicated to discuss a soil–structure interaction method that can be practically applied to cantilever and anchored sheet pile walls.

Free earth support method—assumes that the sheet pile deforms qualitatively as shown in Figure 5.48a, the base is free to rotate and therefore experiences zero moment. The sheet pile deflections are consistently toward the excavation side. Therefore the soil on the excavation side (shaded area in Figure 5.48a) will only experience passive earth pressure. The rest of the soil mass will experience active earth pressure only.

The following sections introduce the free-earth support method and then the soil–structure interaction method.

5.14.5.1 The free-earth support method

For the case shown in Figure 5.48, the active earth pressure on the right side of the wall remains the same as that of cantilever sheet pile wall with similar soil and water table conditions. The passive earth pressure on the left side changes from 0 at the excavation line (level 2) to σ'_{lp3} at the base of the wall (level 3) as

$$\sigma'_{lp3} = \gamma' D K_p \tag{5.171}$$

The net lateral earth pressure, σ'_n, above the excavation line is (level 2 in Figure 5.48b) the same as that of cantilever sheet pile wall and as shown in Figure 5.48b. If we consider sheet pile penetrating in sand, from excavation line to the base of the sheet pile:

At level 2: $\sigma'_n = \sigma'_{ra2} = (\gamma H_1 + \gamma' H_2) K_a$

σ'_n decreases with depth, z, and when $(z - H_1 - H_2) = d_1 = \dfrac{(\gamma H_1 + \gamma' H_2) K_a}{\gamma' (K_p - K_a)}$, $\sigma'_n = 0$ (see Figure 5.48b).

At level 3:

$$\sigma'_n = \gamma' d_2 (K_p - K_a) \tag{5.172}$$

Consider force equilibrium, Σ, horizontal forces per unit length of the wall $= 0$, thus: Area of the pressure diagram *abcd* – area of *dfe* – $F_a = 0$, or

$$P_{ra} - \frac{1}{2}[\gamma' d_2 (K_p - K_a)] d_2 - F_a = 0 \tag{5.173}$$

where

P_{ra} = area of the pressure diagram *abcd*
$d_2 = D - d_1$
F_a = anchor force per unit length of the wall

and Σ moment of the forces per unit length of the wall about point $A = 0$, thus:

$$P_{ra}(h_2 + H_2 + d_1 - \bar{d}) - \frac{1}{2}[\gamma' d_2(K_p - K_a)]d_2\left(\frac{2d_2}{3} + H_2 + d_1\right) = 0 \qquad (5.174)$$

Rearranging Equation (5.174) yields a 3rd-order equation of d_2 as before,

$$d_2^3 + \frac{3}{2}(h_2 + H_2 + d_1)d_2^2 - \frac{P_{ra}[(H_1 + H_2 + d_1) - (h_1 + \bar{d})]}{\gamma'(K_p - K_a)} = 0 \qquad (5.175)$$

$D = d_1 + d_2$, the theoretical depth of penetration. Again, the design depth of penetration is taken as 1.3 to 1.5D. The maximum bending moment (zero shear) is expected to be located at $H_1 < z < (H_1 + H_2)$. The anchor force per unit length of the wall, F_a, is determined using Equation (5.173) after d_2 is calculated. Execution of the free-earth support method is demonstrated in the following example.

EXAMPLE 5.17

Given

The soil profile and the proposed scheme of an anchored sheet pile wall penetrating in sand are given in Figure E5.17. Properties of the sand and water table are included in the figure. The anchor rod will be located at 1.5 m below the top of the sheet pile.

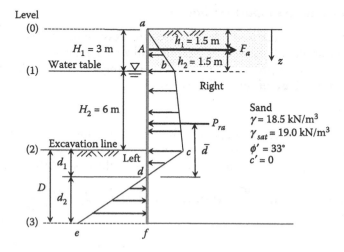

Figure E5.17 Proposed anchored sheet pile wall penetrating in sand below a water table.

Required

Determine the theoretical depth of penetration, maximum bending moment, and anchor force per unit length of the wall.

Solution

The active and passive earth pressure coefficients for the sand:

$$K_a = \tan^2\left(45° - \frac{\phi'}{2}\right) = \tan^2\left(45° - \frac{33°}{2}\right) = 0.295 \tag{5.8}$$

$$K_p = \tan^2\left(45° + \frac{\phi'}{2}\right) = \tan^2\left(45° + \frac{33°}{2}\right) = 3.392 \tag{5.12}$$

Sand buoyant unit weight, $\gamma' = 19.0 - 9.81 = 9.19\,\text{kN/m}^3$

$H_1 = 3\,\text{m}$, $H_2 = 6\,\text{m}$, and
$h_1 = h_2 = 1.5\,\text{m}$

Establish the lateral earth pressure profile and determine the following parameters:

$$d_1 = \frac{(\gamma H_1 + \gamma' H_2)K_a}{\gamma'(K_p - K_a)} = \frac{[(18.5)(3) + (9.19)(6)](0.295)}{(9.19)(3.392 - 0.295)} = 1.15\,\text{m} \tag{5.142}$$

$$\begin{aligned} P_{ra} &= \left[\frac{1}{2}\gamma H_1^2 + \gamma H_1 H_2 + \frac{1}{2}\gamma' H_2^2 + \frac{1}{2}d_1(\gamma H_1 + \gamma' H_2)\right]K_a \\ &= 190.16\,\text{kN/m} \end{aligned} \tag{5.167}$$

$$\bar{d} = \frac{\left[\begin{array}{l}\left(\frac{1}{2}\gamma H_1^2\right)\left(\frac{1}{3}H_1 + H_2 + d_1\right) + (\gamma H_1)(H_2)\left(\frac{1}{2}H_2 + d_1\right) \\ + \frac{1}{2}\gamma' H_2^2\left(\frac{1}{3}H_2 + d_1\right) + \frac{1}{2}d_1(\gamma H_1 + \gamma' H_2)\left(\frac{2}{3}d_1\right)\end{array}\right]K_a}{P_{ra}} = 4.07\,\text{m}$$

Solve for d_2:

$$d_2^3 + \frac{3}{2}(h_2 + H_2 + d_1)d_2^2 - \frac{P_{ra}[(H_1 + H_2 + d_1) - (h_1 + \bar{d})]}{\gamma'(K_p - K_a)} = 0 \tag{5.175}$$

$d_2 = 1.46\,\text{m}$

Theoretical depth of penetration, $D = d_1 + d_2 = 2.61\,\text{m}$

$$\text{Anchor force, } F_a = P_{ra} - \frac{1}{2}[\gamma' d_2(K_p - K_a)]d_2 = 159.83\,\text{kN/m} \tag{5.173}$$

Depth of zero shear, d_m, below anchor rod level:

$$F_a - \left(\frac{1}{2}\gamma H_1^2 + \gamma H_1 d_m + \frac{1}{2}\gamma' d_m^2\right)K_a = 0$$

$d_m = 5.64\,\text{m}$

Maximum bending moment, M_{max}:

$$\begin{aligned} M_{max} &= -\left(\frac{1}{2}\gamma H_1^2 K_a\right)\left(\frac{1}{3}H_1 + d_m\right) + F_a(h_2 + d_m) - \frac{1}{2}(\gamma H_1 K_a)d_m^2 - \left(\frac{1}{2}\gamma' d_m^2 K_a\right) \\ &\quad \times \left(\frac{1}{3}d_m\right) \\ &= 636.96\,\text{kN-m} \end{aligned}$$

5.14.6 Soil–structure interaction (SSI) method

The method is also known as the $p–y$ method and has been successfully used in the design of laterally loaded piles for decades. A flexible earth-retaining structure such as the

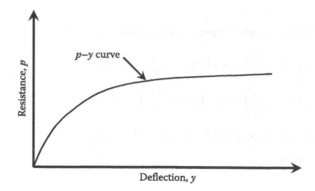

Figure 5.49 Conceptual description of a nonlinear *p–y* curve.

sheet pile wall is similar to a row of laterally loaded piles. The nonlinear soil resistance (p) – lateral deflection (y) relations that develop against the side of the sheet pile are termed *p–y* curves. Figure 5.49 shows in concept a nonlinear *p–y* curve. The solution using the SSI method allows the engineer to compute the movement in addition to shear and moment along the length of the wall under the working load. The result is then used for sizing the wall and design of the anchorage system if necessary, while ensuring that deformations throughout the system are acceptable.

The sheet pile wall is treated as a straight beam, subjected to transverse distributed loading and supported by a series of springs, as shown in Figure 5.50. In Figure 5.50, it is considered that the sheet pile wall is installed in sand and there is a water table above the excavation line. The beam deflects due to external loads and that produces distributed reaction forces from the supporting medium. The reaction forces are assumed to act perpendicular to the original axis of the beam, and opposite in direction to the deflection. The intensity of the reaction, p, at every point is a function of the magnitude of the deflection, y, at the same point, and can be expressed as

$$p = f(y) \tag{5.176}$$

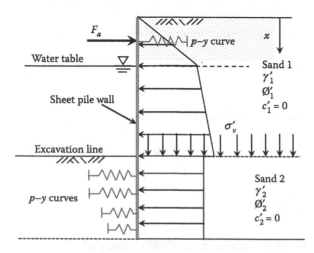

Figure 5.50 Simulation of active earth pressure and nonlinear soil reaction in the *p–y* wall method. (Adapted from Reese, L.S., Wang, S.T., Arrellaga, J.A. and Vasquez, L. 2013. *PYWALL Technical Manual. A program for the analysis of flexible retaining structures.* ENSOFT, Inc., Austin, TX, 137.)

The solution is obtained by solving the following differential equation (Hetenyi, 1946):

$$E_p I_p \frac{d^4 y}{dx^4} + P_x \frac{d^2 y}{dx^2} + E_{py} y - w = 0 \tag{5.177}$$

$E_p I_p$ = sheet pile flexural rigidity $(F\text{–}L^2)$
E_p = elastic modulus of sheet pile material (F/L^2)
I_p = moment of inertia of sheet pile (L^4)
y = sheet pile lateral deflection (L)
x = axial distance along sheet pile (L)
P_x = axial load on sheet pile (F)
E_{py} = soil modulus (F/L^2)
w = distributed load per unit length of the wall (F/L)

The soil resistance, $p = E_{py}y$. To accomplish a nonlinear $p\text{–}y$ relationship, E_{py} varies as y changes.

The numerical model used by SSI or $p\text{–}y$ wall method is based on the concept of "Winkler" foundation, where the active earth pressure is considered as external loads w in Equation 5.177 and the soil below the excavation line is modeled by a series of non-linear $p\text{–}y$ springs ($p\text{–}y$ curves) for passive resistance (see Figure 5.50). A $p\text{–}y$ spring located at the anchor level is added to represent the stiffness of the anchorage system. The formulation of soil resistance ($p\text{–}y$ curves) is simplified by considering that active earth pressure also develops on the excavation side (left side in Figure 5.50) below the excavation line. The active earth pressure on the left side thus offsets the additional active earth pressure on the right side, from below the excavation line. Therefore, only the active earth pressure (loading) created by the overburden stress, σ'_v, from the soil mass above the excavation line is considered as additional external loads below the excavation level.

The finite difference method has been one of the main numerical schemes to obtain a general solution to Equation 5.177. Readers are referred to Desai and Christian (1977) for details of the finite difference method. With the help of a spreadsheet program such as Microsoft Excel, the finite difference scheme can now be implemented in a personal computer. To apply the finite difference method, the sheet pile is discretized into segments as shown in Figure 5.51. It is assumed that the sheet pile is straight and vertical initially. All $p\text{–}y$ curves act independently at any given depth without interference from other parts of the sheet pile.

For a generic point, m in Figure 5.51, the finite difference analog for the various derivatives in the general Equation 5.177 is

$$R_p \frac{d^4 y}{dx^4} = \left(\frac{d^2 M}{dx^2}\right)_m = \frac{\left(\dfrac{dM}{dx}\right)_{m-1/2} - \left(\dfrac{dM}{dx}\right)_{m+1/2}}{h} \tag{5.178}$$

where
h = length of the segment, usually set to be uniform throughout the sheet pile
R_p = sheet pile flexural rigidity, $= E_p I_p$

$$\left(\frac{d^2 M}{dx^2}\right)_m = \frac{1}{h^4}[(R_p)_{m-1} y_{m-2} + (-2(R_p)_{m-1} - 2(R_p)_m) y_{m-1}$$

$$+ \left((R_p)_{m-1} + 4(R_p)_m + (R_p)_{m+1}\right) y_m + \left(-2(R_p)_m - 2(R_p)_{m+1}\right) y_{m+1}$$

$$+ (R_p)_{m+1} y_{m+2}] \tag{5.179}$$

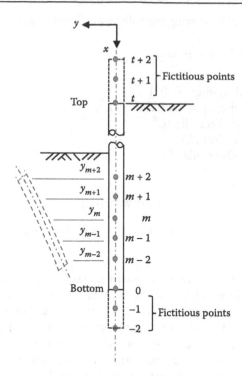

Figure 5.51 Discretization of a laterally loaded sheet pile wall.

In most cases, $P_x = 0$ for sheet piles. The combined difference equation for the entire Equation 5.177 is

$$(R_p)_{m-1} y_{m-2} + \left(-2(R_p)_{m-1} - 2(R_p)_m\right) y_{m-1}$$
$$+ \left((R_p)_{m-1} + 4(R_p)_m + (R_p)_{m+1} + (E_{py})_m h^4\right) y_m + \left(-2(R_p)_m - 2(R_p)_{m+1}\right) y_{m+1}$$
$$+ (R_p)_{m+1} y_{m+2} - w_m h^4 = 0 \qquad (5.180)$$

where y_m is the deflection and w_m is the applied lateral earth load at point m. This equation is repeated for each point 0, 1, 2, ..., t (i.e., from bottom to top) as indicated in Figure 5.51 and results in a set of simultaneous equations in y.

5.14.6.1 Boundary conditions

Boundary conditions must be given in order to solve Equation 5.177 and its equivalent finite difference equations (Equation 5.180). The boundary conditions are assigned in terms of shear, $V = EI\, d^3y/dx^3$, moment, $M = EI\, d^2y/dx^2$, or slope $S = dy/dx$, and expressed in finite difference forms with added fictitious or phantom points (points beyond the physical boundary of the pile in Figure 5.51).

5.14.6.2 Boundary conditions at the bottom

Consider a simple case where the shear and moment at the sheet pile bottom are zero. For zero moment at the bottom (point 0 in Figure 5.51),

$$y_{-1} - 2y_0 + y_1 = 0 \qquad (5.181)$$

And for zero shear at the bottom,

$$(R_p)_0 \frac{d^3y}{dx^3} + P_x \frac{dy}{dx} = 0 \tag{5.182}$$

Again, we assume that $P_x = 0$, the finite difference equation is

$$y_{-2} - 2y_{-1} + 2y_1 - y_2 = 0 \tag{5.183}$$

For approximation, it is assumed that $(R_p)_{-1} = (R_p)_0 = (R_p)_1$.

5.14.6.3 Boundary conditions at the top

For sheet piles, the load (shear), V_t, and moment, M_t, are zero at the top. The difference equations are,

For the applied V_t,

$$y_{t-2} - 2y_{t-1} + 2y_{t+1} - y_{t+2} = \frac{2V_t h^3}{(R_p)_t} = 0 \tag{5.184}$$

and for the applied M_t,

$$y_{t-1} - 2y_t + y_{t+1} = \frac{M_t h^2}{(R_p)_t} = 0 \tag{5.185}$$

By combining Equations 5.180, 5.181, and 5.183 to 5.185, a full set of simultaneous equations in y can be assembled. The solution to the simultaneous equations yields the deflections y_m. Other quantities such as slope, S_m, shear, V_m, and moment, M_m, at point m can be computed as follows:

$$S_m = \frac{(y_{m-1} - y_{m+1})}{2h} \tag{5.186}$$

$$V_m = \frac{(R_p)_m (y_{m-2} - 2y_{m-1} + 2y_{m+1} - y_{m+2})}{2h^3} + \frac{P_x(y_{m-1} - y_{m+1})}{h} \tag{5.187}$$

and

$$M_m = \frac{(R_p)_m (y_{m-1} - 2y_m + y_{m+1})}{h^2} \tag{5.188}$$

The Excel spreadsheet program has functions to solve the linear equations by computing the inverse and multiplication of the matrixes. Details of the numerical procedures are beyond the scope of this book. An Excel program to solve Equation 5.177 for a simple case of a sheet pile wall with uniform R_p and a constant E_{py} for each sheet pile segment can be downloaded from the publisher's website for registered users. Students are encouraged to download the program and acquaint themselves with the procedure and explore its capabilities.

Details of establishing p–y curves can be found in Reese (1984, 1985) and Reese et al. (2000, 2006, 2013).

EXAMPLE 5.18

Given

For the same soil profile, water table conditions and the anchored sheet pile wall penetrating in sand described in Example 5.17. The depth of sheet pile penetration is 3.5 m below the excavation line. The anchor rod is located at 1.5 m below the top of the sheet pile. The sheet pile has a moment of inertia, I_p of 1.3×10^{-4} m^4/m. The anchor system provides an equivalent resistance stiffness of 20,000 kN/m^3/m. The elastic modulus of sheet pile material, E_p, is 200 GPa. The passive resistance of the soil below the excavation line has an equivalent elastic modulus, E_{py}, of 4 MN/m^3/m, regardless of its deflection (i.e., use a linear $p-y$ relationship).

Required

Determine the deflection, shear, and moment profile of the sheet pile for the given conditions.

Solution

The active earth pressure coefficients for the sand:

$$K_a = \tan^2\left(45° - \frac{\phi'}{2}\right) = \tan^2\left(45° - \frac{33°}{2}\right) = 0.295 \qquad (5.8)$$

The sand buoyant unit weight, $\gamma' = 19.0 - 9.81 = 9.19$ kN/m^3.

The sheet pile has a total height of 12.5 m from top to bottom of the sheet pile. For the $p-y$ analysis, the sheet pile was divided into 25 equal segments, each at 0.5 m in height (h). The segments are numbered from 1 (top) through 25 (bottom). The lateral active earth pressure applied to the sheet pile is represented by a series of concentrated lateral load w_m as described in Equation 5.180. The lateral load (per unit length of the wall) for the mth segment, w_m, is computed as follows:

$$w_m = \sigma'_{ah} \cdot h = \sigma'_v \cdot K_a \cdot h$$

where

σ'_{ah} = lateral active earth pressure at the bottom of the mth segment
h = height of each segment (0.5 m)

Figure E5.18 shows the loading conditions applied for the $p-y$ wall analysis. The following table shows the key w_m values:

Level no. (see Figure E5.18)	Depth z, m	w_m, kN/m
0	0	$w_m = 0$
1	3	$w_m = \gamma \cdot 3 \cdot K_a \cdot h = (18.5)(3)(0.295)(0.5) = 8.18$
2	9	$w_m = (\gamma \cdot 3 + \gamma' \cdot 6) \cdot K_a \cdot h$
		$= [(18.5)(3) + (19.0 - 9.81)(6)](0.295)(0.5) = 16.31$
3	12.5	$w_m = (\gamma \cdot 3 + \gamma' \cdot 6) \cdot K_a \cdot h$
		$= [(18.5)(3) + (19.0 - 9.81)(6)](0.295)(0.5) = 16.31$

The matrix was assembled according to Equations 5.180, 5.181, and 5.183 to 5.185. The solution yields the lateral deflection at the bottom of each segment, y_m. The slope of the deformed sheet pile, shear, and moment at the corresponding locations are computed according to Equations 5.186, 5.187, and 5.188, respectively. The analysis by $p-y$ method is performed using the spreadsheet program mentioned previously. Figure E5.18a shows the results of the $p-y$ wall analysis. With these input parameters, the maximum deflection is 0.0076 m toward the passive side at 8 m below the top of the sheet pile, and maximum moment of 108 kN-m occurs at 6 m below the top of the sheet pile. For the assigned anchor stiffness and computed sheet pile deflection at the level of anchor rod (1.5 m from the top of the sheet pile), the anchor provides a lateral support, F_a, of 88.8 kN/m.

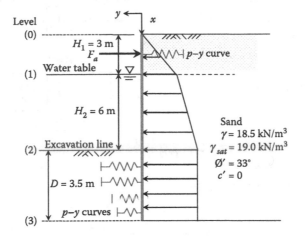

Figure E5.18 The anchored sheet pile wall penetrating in sand below a water table.

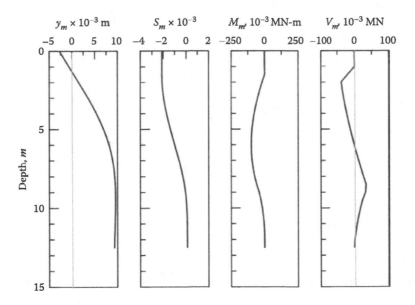

Figure E5.18a Profiles of deflection, slope, shear, and moment along the sheet pile.

5.14.7 Lateral support systems

Various types of lateral support systems have been developed and used to provide the additional lateral resistance for the anchored sheet pile walls. Figure 5.52 provides a general description of various commonly used lateral support systems. It should be noted that the anchor block depicted in Figure 5.52a can consist of isolated blocks or a long concrete wall. The concrete blocks can be precast or cast in the field. The concrete blocks can also be replaced with a row of short sheet piles. Regardless of the type of anchor block used, the

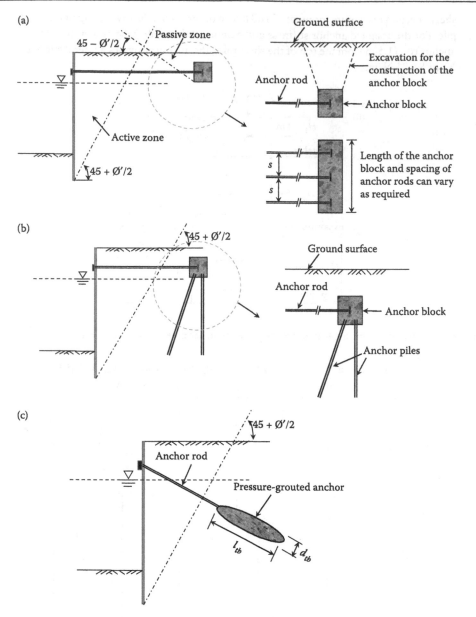

(a)

Passive zone

$45 - \emptyset'/2$

Ground surface

Active zone

$45 + \emptyset'/2$

Anchor rod

Excavation for the construction of the anchor block

Anchor block

Length of the anchor block and spacing of anchor rods can vary as required

s

s

(b)

$45 + \emptyset'/2$

Ground surface

Anchor rod

Anchor block

Anchor piles

(c)

$45 + \emptyset'/2$

Anchor rod

Pressure-grouted anchor

l_{tb}

d_{tb}

Figure 5.52 Various types of lateral support systems. (a) Anchor block. (b) Anchor piles. (c) Tieback.

passive zone in front of the anchor blocks should not overlap with the active zone behind the sheet piles as shown in Figure 5.52a. Piles either vertical or batter as shown in Figure 5.52b, installed outside of the active zone, can also be used to provide the lateral support. Tieback is a relatively new addition to the alternatives in lateral support system. To install the tieback, a hole is drilled first from the sheet pile wall to the designated depth usually at an angle from horizontal as shown in Figure 5.52c. Upon drilling, the tail end of the borehole located outside of the active zone is pressure grouted. If necessary, multiple levels of tiebacks can be installed at different depths from top of the sheet pile wall.

The following section introduces the design and analysis of anchor blocks and tiebacks. The analysis of laterally loaded piles is discussed in Chapter 7.

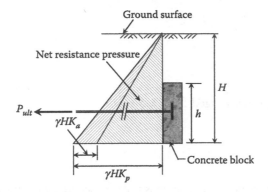

Figure 5.53 Anchorage and lateral earth resistance.

5.14.7.1 Resistance of anchor blocks

Concrete block and sheet pile anchorages similar to those shown in Figure 5.52a are designed by equating the net lateral earth force (passive minus active) to the required anchor
resistance. There can also be some favorable boundary (end) effects that enhance the lateral resistance of a block anchorage. If the height of the anchorage, h, is not less than 0.6 of
the depth from ground surface to bottom of the anchorage, H, as shown in Figure 5.53
(i.e., $h/H \geq 0.6$), the anchorage behaves as if it extends all the way to the ground surface.
There are other frictional resistances at the soil–structure interface along the surface of the
anchorage. However, full mobilization of these frictional resistances is usually not real-
ized. The lateral earth pressure coefficients, K_a and K_p, should be calculated considering
$\delta = 0$, and this assumption is on the safe side. For isolated anchorage, the soil failure
wedge is expected to extend beyond its lateral ends so that the effective width (B_e)
is about 1.6 times the actual width (B). Therefore for an isolated anchorage with its
$B \leq 1.5h$, embedded in granular soil and center-to-center spacing between the anchorages,
$S > 1.6B$, the ultimate lateral resistance, P_{ult}, per anchor can be determined as follows:

$$P_{ult} = \left(\frac{1}{2}\gamma H^2\right)(K_p - K_a)B_e = \left(\frac{1}{2}\gamma H^2\right)(K_p - K_a)1.6B \tag{5.189}$$

where
  γ = unit weight of the soil
  B = width of the anchor block
  B_e = effective width of the anchor block
  H = depth from ground surface to bottom of the anchor block

Note that the above computation should consider drained conditions, and buoyant unit
weight, γ', should be used for soil below the water table.

For isolated anchorages, the design anchor force per unit length of the wall, F_a, should
meet the following requirement:

$$F_a \leq \frac{P_{ult}}{FS \cdot S} \tag{5.190}$$

For continuous or anchorages with center-to-center spacing $S/B < 1.6$, the net ultimate
lateral resistance per unit length of the anchorage, P'_{ult}, can be calculated as

$$P'_{ult} = \left(\frac{1}{2}\gamma H^2\right)(K_p - K_a) \tag{5.191}$$

and

$$F_a \leq \frac{P'_{ult}}{FS} \tag{5.192}$$

where

FS = safety factor typically ranged from 1.5 to 2

EXAMPLE 5.19

Given

For the soil profile, water table conditions and the anchored sheet pile wall described in Example 5.18. The anchor rod is located at 1.5 m below the top of the sheet pile. The required anchor resistance, F_a, is 88.8 kN/m according to Example 5.18. It has been decided to use a continuous anchor block to provide the lateral resistance.

Required

Determine the height of the anchor block (h) so that the ultimate lateral resistance, P_{ult}, per unit length of the anchor block can offer a minimum safety factor of 1.5.

Solution

The active and passive earth pressure coefficients for the sand:

$$K_a = \tan^2\left(45° - \frac{\phi'}{2}\right) = \tan^2\left(45° - \frac{33°}{2}\right) = 0.295 \tag{5.8}$$

$$K_p = \tan^2\left(45° + \frac{\phi'}{2}\right) = \tan^2\left(45° + \frac{33°}{2}\right) = 3.392 \tag{5.12}$$

For a continuous anchor block with a height of h (see Figure 5.52a), half of its height (i.e., $0.5h$) is above the anchor rod and half below. The anchor rod is located at 1.5 m below the ground surface. The distance between the bottom of the anchor block and ground surface, H, should thus be

$H = 1.5 + 0.5h$

For continuous anchor block, the net ultimate lateral resistance per unit length of the anchorage, P'_{ult}, is

$$P'_{ult} = \left(\frac{1}{2}\gamma H^2\right)(K_p - K_a) = (0.5)(18.5)(1.5 + 0.5h)^2(3.392 - 0.295) \tag{5.191}$$

By trial and error, when $h = 1.5$ m, $P'_{ult} = 145$ kN/m and
FS $= P'_{ult}/F_a = (145)/(88.8) = 1.64 > 1.5$, OK
$h/H = 1.5/(1.5 + 0.75) = 0.67 > 0.6$, OK

Anchors embedded in cohesive soil under undrained conditions ($\emptyset = 0$) may be considered as a laterally loaded foundation-bearing capacity problem. Recall in Chapter 4 that net ultimate bearing capacity for a foundation $= N_c s_u$ where $s_u =$ undrained shear strength of the cohesive soil and N_c is the bearing capacity factor. N_c varies from slightly above 5 for shallow foundations to 9 for deep foundations (see Chapter 7). Unless the anchors are deeply buried (i.e., $H/h > 10$), the full failure mechanism analogous to that of a vertically loaded foundation may not develop (Tschebotarioff, 1973).

It is therefore recommended that, for isolated anchorages, the ultimate lateral resistance, P_{ult}, per anchor be estimated as follows:

$$P_{ult} = M_c B h \qquad (5.193)$$

For continuous or long anchorages, the per unit length ultimate lateral resistance,

$$P'_{ult} = M_c h \qquad (5.194)$$

where
$M_c \approx 3$ for $H/h = 1$
$M_c = 9$ for $H/h \geq 10$

The value of M_c may be linearly interpolated for H/h between 1 and 10.

5.14.7.2 Resistance of tiebacks

The resistance capacity of a tieback is dependent on dimensions of the tieback, soil properties, and most importantly, local experience and workmanship. In practice, the capacity of tiebacks is often verified with field load tests. In reference to Figure 5.52c, for a tieback in granular soil, the ultimate resistance, P_{ult}, per tieback may be estimated as follows:

$$P_{ult} = \pi d_{tb} l_{tb} \bar{\sigma}'_v K tan\varnothing' \qquad (5.195)$$

where
$\bar{\sigma}'_v$ = average effective vertical stress at the center of the grouted anchor
d_{tb} = diameter of the grouted anchor
l_{tb} = length of the grouted anchor
K = earth pressure coefficient
$\varnothing'$ = drained friction angle of the soil

For pressure-grouted anchor, K can be close to K_o. K can be as low as K_a, depending on the method of tieback installation and workmanship.

In cohesive soil under undrained conditions, the ultimate resistance may be approximated as

$$P_{ult} = \pi d_{tb} l_{tb} \alpha_{tb} s_u \qquad (5.196)$$

where
s_u = undrained shear strength of the soil
α_{tb} = a reduction factor may be taken as $2/3$

A safety factor in a range of 1.5 to 2 may be applied to determine the allowable resistance per tieback. More details on tiebacks are provided in Chapter 6.

5.15 REMARKS

The first part of this chapter deals with the classic theories related to lateral earth pressure. The earth behind a retaining structure is actually failed in active and passive conditions. The active condition refers to failure by reducing the lateral stress while the earth is failed by increasing the lateral stress under passive condition. Coulomb solved the problem of active and passive lateral earth pressure based on force equilibrium, while Rankine tackled the problem using the concept of stress. The earth under at-rest condition is not failed. This is why we really don't have a rigorous method to determine the at-rest lateral earth pressure other than the empirical equation proposed by Jaky (1944). It is important to recognize the differences among the various lateral earth pressure theories, and their advantages and limitations.

This chapter introduced three major types of retaining structures: concrete retaining walls, mechanically stabilized earth (MSE) retaining walls, and sheet pile walls. The analyses of these structures are primarily based on their expected limit states or failure conditions. Because of this, we were able to use the active and passive earth pressures in the analysis. While these lateral earth pressure theories are classic, the construction methods of retaining structures are evolving continuously. There is plenty of room for innovation to develop safer and more economical retaining structures. The MSE wall is a perfect example.

It should be noted that the use of active or passive lateral earth pressure does not necessarily produce conservative design. The amount of earth loading (active) or resistance (passive) is dependent on the displacement of the wall. The SSI method introduced toward the end of this chapter recognizes that the resistance (p) offered by the soil depends on the displacement (y) of the wall. The SSI method does not rely on limit state parameters such as the passive earth-pressure in the analysis. This is a more realistic way to analyze the behavior of an earth-retaining structure. The sheet pile walls introduced in this chapter are also used in braced cut, introduced in Chapter 6. Because of the differences in construction sequences, the characteristics of lateral earth pressures in braced cut are very different from those of limit state analysis. Therefore, traditionally a different set of earth pressure diagrams is often used in the design of braced cut systems as will be introduced in Chapter 6. However, because of its unique capabilities, the SSI method can also be used to analyze the behavior of a braced cut retaining structure. In fact, with the help of a computer program, the SSI method may be used to completely replace the limit state analysis methods for all types of retaining structures.

HOMEWORK

5.1. For the rigid and fixed retaining wall shown in Figure H5.1, the groundwater table is at 3 m below the ground surface. Consider soil as dry above the groundwater table with an overconsolidation ratio (OCR) of 5. The granular soil below groundwater table is normally consolidated (OCR = 1). Calculate and plot the lateral pressure distribution on the back of the retaining wall.

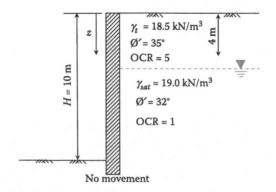

Figure H5.1 A rigid and fixed retaining wall.

5.2.   For the wall and backfill conditions described in Figure H5.2, the backfill is granular material. Consider the following soil parameters, wall, and backfill
configurations:

$\alpha = 10°$
$\beta = 10°$

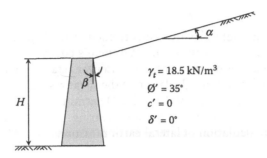

Figure H5.2 Wall and backfill conditions.

Compute K_a and K_p under these conditions using Coulomb's and Rankine's
methods.

5.3.   Figure H5.3 shows a case of vertical wall ($\beta = 0$) with flat ($\alpha = 0$) frictional cohesive
backfill. The soil parameters are included in Figure H5.3. Determine the depth of
tensile crack, z_c, Rankine active lateral earth force (P_a) before tension crack using
Equation 5.34, P_a considers the tension crack according to Equations 5.35 and 5.36.

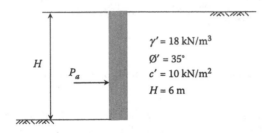

Figure H5.3 Vertical wall with a flat, frictional cohesive backfill.

5.4.   For the wall and backfill conditions described in Figure H5.2, the backfill is granular material. Consider the following soil parameters, wall, and backfill
configurations:

$\emptyset' = 35°$
$c' = 0$
$\gamma' = 18.5\ \text{kN/m}^3$
$\delta' = (2/3)\emptyset'$
$\alpha = 10°$
$\beta = 10°$
$H = $ height of the retaining wall $= 6$ m
$q = $ surcharge on the backfill $= 5\ \text{kN/m}^2$
Calculate the P_a (in kN) per unit thickness of the wall and h_c (in m) values for the
given conditions using Okabe's method [Equations 5.40 to 5.42].

5.5.   For the wall and backfill conditions described in Figure H5.2, the backfill is granular material. Consider the following soil parameters, wall, and backfill
configurations:

$\varnothing' = 33°$

$c' = 0$

$\delta' = (2/3)\varnothing'$

$\alpha = 10°$

$\beta = 10°$

$k_h = 0.2$

$k_v = 0$

Calculate the active lateral earth pressure coefficient, K_{ae}, for the given earthquake conditions and compare values with Whitman's equation.

5.6.   Consider the dimensions of a concrete cantilever retaining wall and the soil profile shown in Figure H5.6. For calculation of the shear resistance at the base of the wall foundation, assume $\delta' = (2/3)\varnothing' = 20°$. Determine the FS of the retaining wall design against overturning, sliding, and bearing capacity failure. Use Rankine's method for the calculation of lateral earth pressure.

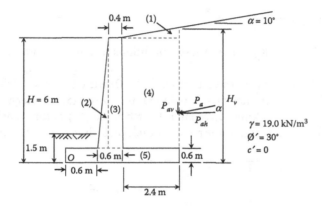

Figure H5.6 Dimensions and soil profile of the retaining wall for Rankine's method.

5.7.   Figure H5.7 shows an MSE wall with a height, H, of 6 m and a tentative width (reinforcement length), L, of 4.2 m. The related soil properties are included in Figure H5.7. Assume that the retained soil and foundation soil have the same unit weight and strength parameters. At back and base of the MSE wall, assume the soil–wall interface friction, $\delta' = (2/3)\varnothing'$. Evaluate the external stability of the MSE wall against overturning, sliding, and bearing capacity failure.

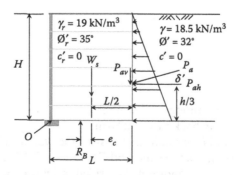

Figure H5.7 The tentative MSE wall dimensions.

5.8.   For the same overall dimensions of the MSE wall and soil properties described in H5.7, use galvanized ribbed metal strips as the reinforcement. The strips are vertically spaced at 0.5 m (i.e., $S_v = 0.5$ m). The first row is located at 0.5 m from the top of the wall. Consider the metal strip has a width (b) of 50 mm and a nominal thickness (E_n) of 4.0 mm at the time of construction and it is expected to lose (E_R) 1.416 mm due to corrosion after 75 years of service life. Grade 60 steel with a yield stress (f_y) of 413.7 MPa is to be used as the metal strip. Evaluate internal stability of the metal strip reinforced wall and determine the dimensions and horizontal spacing of the metal strips. Use a service life of 75 years for the MSE wall design.

5.9.   For the same overall dimensions of the MSE wall and soil properties described in H5.7, use geogrid as the reinforcement. The geogrid layers are vertically spaced at 0.4 m (i.e., $S_v = 0.5$ m). The first row is located at 0.25 m from the top of the wall. Consider the combined reduction factor (RF) as 2.5. Use geogrid wraparound as the wall facing. Evaluate the internal stability of the geogrid reinforced wall, choose the required ultimate tensile strength (T_{ult}) of the geogrid, and determine the overlap length for the wraparound.

5.10. Figure H5.10 shows the soil profile and the proposed scheme of a cantilever sheet pile wall penetrating in sand. Height of the wall, H, above excavation line is 4 m. Properties of the sand are included in the figure. Assume groundwater table is deeper than the base of the sheet pile. Determine the total design length and required sectional modulus of the sheet pile. Use allowable stress, $\sigma_{all} = 200$ MPa.

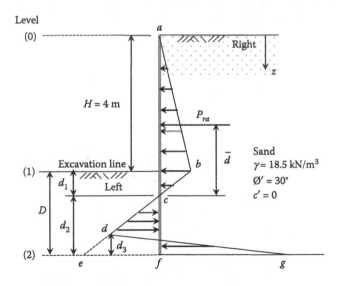

Figure H5.10 Proposed sheet pile wall penetrating in sand.

5.11. Figure H5.11 shows the soil profile and the proposed scheme of a cantilever sheet pile wall penetrating in saturated clay. The height of the sheet pile wall, H, above excavation line is 4 m. Properties of the sand and clay are included in the figure. Assume groundwater table coincides with the excavation line. Determine the design total length and required sectional modulus of the sheet pile. Use allowable stress, $\sigma_{all} = 170$ MPa.

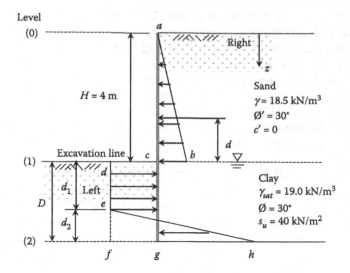

Figure H5.11 Proposed sheet pile wall penetrating in clay.

5.12. The soil profile and the proposed scheme of a cantilever sheet pile wall penetrating in sand are given in Figure H5.12. Properties of the sand are included in the figure. A water table is located at 2 m below the top of the wall, and the excavation line is at 3 m below water. Determine the design total length and required sectional modulus of the sheet pile. Use allowable stress, $\sigma_{all} = 200$ MPa.

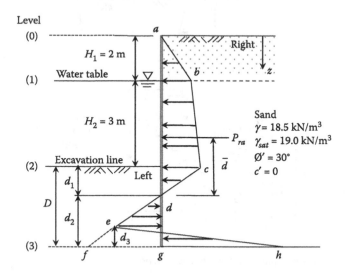

Figure H5.12 Proposed sheet pile wall penetrating in sand below a water table.

5.13. For the soil profile and the proposed scheme of a cantilever sheet pile wall penetrating in clay shown in Figure H5.13. Properties of the sand are included in the figure. A water table is located at 2 m below the top of the wall, and the excavation line is at 3 m below water. Determine the theoretical depth of penetration for the sheet pile.

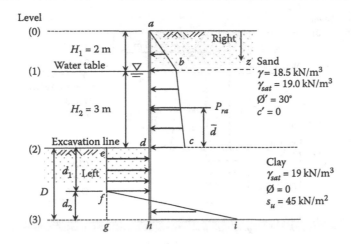

Figure H5.13 Sheet piles penetrating in clay below a water table.

5.14. The soil profile and the proposed scheme of an anchored sheet pile wall penetrating in sand are given in Figure H5.14. Properties of the sand and water table are included in the figure. The anchor rod will be located at 1.5 m below the top of the sheet pile. Determine the theoretical depth of penetration, maximum bending moment, and anchor force per unit length of the wall.

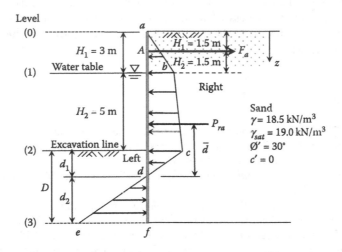

Figure H5.14 Anchored sheet pile wall penetrating in sand below a water table.

5.15. The soil profile and the proposed scheme of an anchored sheet pile wall penetrating in clay are given in Figure H5.15. Properties of the sand, clay, and water table are included in the figure. The anchor rod will be located at 1.5 m below the top of the sheet pile. Determine the theoretical depth of penetration, maximum bending moment, and anchor force per unit length of the wall (F_a) and design a continuous anchor block to resist the anchor force with a minimum safety factor of 1.5.

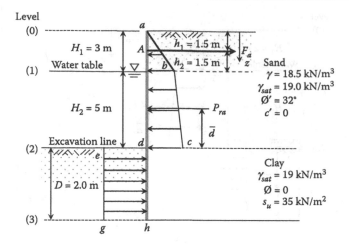

Figure H5.15 Anchored sheet pile wall penetrating in clay below a water table.

5.16. For the same soil profile, water table conditions, and the anchored sheet pile wall penetrating in clay described in H5.15. The depth of sheet pile penetration is 2 m below the excavation line. The anchor rod is located at 1.5 m below the top of the sheet pile. The sheet pile has a moment of inertia, I_p of 1.3×10^{-4} m^4/m. The anchor system provides an equivalent resistance stiffness of 20,000 kN/m^3/m. The elastic modulus of sheet pile material, E_p, is 200 GPa. The passive resistance of the soil below the excavation line has an equivalent elastic modulus, E_{py}, of 3 MN/m^3/m, regardless of its deflection (i.e., use a linear p–y relationship). Determine the deflection, shear, and moment profile of the sheet pile for the given conditions.

REFERENCES

Anderson, P.L., Gladstone, R.A., and Sankey, J.E. 2012. State of the practice of MSE wall design for highway structures. *Proceedings of the Geocongress 2012, State of the Art and Practice in Geotechnical Engineering*, Oakland, CA, 443–463.

Bowles, J.E. 1996. *Foundation Analysis and Design*. Fifth Edition. The McGraw-Hill Companies, Inc., New York, 1175.

Caquot, A. and Kerisel, J. 1948. *Tables for the Calculation of Passive Pressure, Active Pressure and Bearing Capacity of Foundations*. Gauthier-Villars, Paris (translated by M.A. Bec, London).

Chen, W.F. 1975. *Limit Analysis and Soil Plasticity*. Elsevier, Amsterdam, 638.

Christopher, B.R., Gill, S.A., Giroud, J.P., Juran, I., Mitchell, J.K., Schlosser, F., and Dunnicliff, J. 1990. *Reinforced Soil Structures, Vol. 1, Design and Construction Guidelines*. Report to Federal Highway Administration, No. FHWA-RD-89-043.

Clough, G.W. and Tsui, Y. 1974. Performance of tie-back walls in clay. *Journal of the Geotechnical Division, ASCE*, 100, 1259–1273.

Coulomb, C.A. 1776. Essai sur une Application des Régles de Maximis et Minimum à quelques roblemes de Statique Relatifs à l'Architecture. Membres Académie Royale des Sciences, Paris (Application of maximum and minimum rules to some static problems related to architecture, member of the royal academy of sciences, Paris), 3, 38.

Desai, C. and Christian, J.T. 1977. *Numerical Methods in Geotechnical Engineering*. McGraw-Hill Series in Modern Structures. McGraw-Hill, New York, 783.

DM7.2 1982. *Foundations and Earth Structures Design Manual 7.2*. Naval Facilities Engineering Command. Alexandria, VA, Department of the Navy.

Elias, V., Christopher, B.R., and Berg, R.R. 2001. *Mechanically Stabilized Earth Walls and Reinforced Soil Slopes Design and Construction Guidelines.* FHWA-NHI-00-043, National Highway Institute Federal Highway Administration U.S. Department of Transportation, Washington, DC, 418.

Geosynthetic Research Institute (GRI, http://www.geosynthetica.net) Standard Practice GT7 (2012) for the design of retaining walls.

Hetenyi, M. 1946. *Beams on Elastic Foundations.* University of Michigan Press, Ann Arbor, MI.

Holtz, R.D., Christopher, B.R., and Berg, R.D. 1997. *Geosynthetic Engineering.* BiTech, Canada, 451.

Jaky, J. 1944. The coefficient of earth pressure at rest. *Journal for the Society of Hungarian Architects and Engineers,* October, 355–358.

Janbu, N. 1957. Earth pressures and bearing capacity calculations by generalized procedure of slices. 4th ICSMFE, 2, 207–212.

Koerner, R.B. 2012. *Designing with Geosynthetics.* Sixth Edition. Xlibris, Bloomington IN, 818.

Mayne, P.W. and Kulhawy, F.H. 1982. K_o-OCR relationships in soil. *Journal of the Geotechnical Engineering Division, ASCE,* 108(GT6), 851–872.

McKittrick, D.P. 1978. Reinforced earth: Application of theory and research to practice. Keynote paper. *Symposium on Soil Reinforcing and Stabilizing Techniques.* Sydney, Australia.

Mononobe, N. 1929. On the determination of earth pressures during earthquakes. *Proceedings, World Engineering Conference,* Tokyo, 9, 177–185.

Okabe, S. 1924. General theory of earth pressure and seismic stability of retaining wall and dam. *Journal of the Japanese Society of Civil Engineering,* 10(6), 1277–1323.

Rankine, W.M.J. 1857. On stability on loose earth. *Philosophic Transactions of Royal Society,* London, Part 1, 9–27.

Reese, L.C. 1984. *Handbook on Design and Construction of Drilled Shafts under Lateral Load.* Report No. FHWA-IP-84-11, U.S. Department of Transportation, Federal Highway Administration, Washington, DC.

Reese, L.C. 1985. *Behavior of Piles and Pile Groups under Lateral Loads.* Report No. FHWA-RD-85-106, U.S. Department of Transportation, Federal Highway Administration, Washington, DC.

Reese, L.C., Wang, S.T., Isenhower, W.M., and Arrellage, J.A. 2000. *Computer Program LPLIE Plus Version 4.0 Technical Manual.* Ensoft, Austin, TX.

Reese, L.C., Isenhower, W.M., and Wang, S.T. 2006. *Analysis and Design of Shallow and Deep Foundations.* Wiley, Hoboken, NJ, 574.

Reese, L.S., Wang, S.T., Arrellaga, J.A., and Vasquez, L. 2013. *PYWALL. Technical Manual. A Program for the Analysis of Flexible Retaining Structures.* ENSOFT, Austin, TX, 137.

Richart, F.F., Jr. 1957. Analysis for sheet-pile retaining wall. *Transactions, ASCE,* 122, 1113–1132.

Rowe, P.W. 1955. A theoretical and experimental analysis of sheet pile walls. *Proceedings, Institution of Civil Engineers,* London, Paper No. 5989, 32–69.

Shields, D.H. and Tolunay, A.Z. 1973. Passive pressure coefficients by method of slices. *Journal of the Soil Mechanics and Foundations Division, ASCE,* 99(SM12), 1043–1053.

Terzaghi, K. 1943. *Theoretical Soil Mechanics.* Wiley, New York, NY, 510.

Terzaghi, K. 1954. Anchored bulkheads. *Transactions, ASCE,* 119.

Terzaghi, K. and Peck, R.B. 1967. *Soil Mechanics in Engineering Practice.* Wiley, New York, NY.

Tschebotarioff, G.P. 1949. Lateral earth pressures on flexible retaining walls: A symposium: Large-scale model earth pressure tests on flexible bulkheads. *Transactions of the American Society of Civil Engineers,* 114(1), 415–454.

Tschebotarioff, G.P. 1973. *Foundations, Retaining and Earth Structures.* Second edition. McGraw-Hill, New York, NY.

Vidal, H. 1966. *La terre Armée, Annales de l'Institut Technique du Bâtiment et des Travaux Publiques.* France, July–August, 888–938.

Whitman, R.V. 1990. Seismic design and behavior of gravity retaining walls. *Geotechnical,* SP No.25, ASCE, 817–842.

Chapter 6

Braced excavations

6.1 INTRODUCTION

Excavations are often required in congested urban areas for the construction of basements, subway stations, or sewer systems, to create underground space bounded by a system of retaining walls. It is usually not possible to build the underground retaining walls following the procedure described in Chapter 5. The excavation may have to be performed close to or sometimes immediately next to other buildings or streets. There is not enough room to excavate outside the retaining walls and backfill. Thus, the excavation is likely to be vertical, and it is important that the excavation does not cause damage to the neighboring structures. An important element of avoiding damage to other structures is to minimize ground displacement during the excavation. To overcome these constraints, we brace the vertical excavation as it progresses. For these reasons, a braced excavation has the following characteristics:

- The retaining wall (to be referred to as the excavation wall) or its major components, depending on the systems used, are installed prior to excavation.
- The bracing is installed as the excavation deepens, before the excavation reaches a critical depth and the exposed wall can have sufficient displacement and develop active failure.
- Multiple bracing levels are usually installed as the excavation continues. The bracing at all levels must maintain the required stiffness to prevent development of active failure or excessive movement.
- The bracing is often prestressed by pressing the excavation wall outward against the bracing using a jacking system. Prestressing further reduces ground movement outside the excavation wall.
- Soil below the bottom of the excavation acts as additional bracing by helping to hold the excavation open. This relieves pressure on the lowermost bracing.
- Removal of the bracing system starts after the permanent base slab and walls (depending on the system used) are completed. The various bracing levels are dismantled in sequence from bottom up as the permanent columns, beams, and floor slabs at corresponding levels are completed.
- The excavation walls are finally extracted from the ground after the bracing is completely removed, except for certain cast-in-situ walls (e.g., diaphragm walls) that will be incorporated into the permanent structure.
- The bracing and the excavation wall may be temporary structures.

It should be noted that concrete diaphragm walls (discussed later), when used as the excavation walls, can also be used as permanent walls. A reverse excavation procedure enables the permanent columns and beams to be used as the bracing system. In this

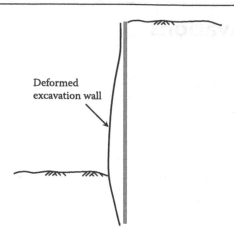

Figure 6.1 Deformation characteristics of a braced excavation wall.

case, and when coupled with the concrete diaphragm walls, all elements of the braced excavation are permanent.

Figure 6.1 shows the deformation characteristics of a typical braced excavation wall. The insertion of bracing and prestressing in the early stage of the excavation results in a reduced deflection toward the top of the wall and an appearance of bulging in the mid-level of the wall as shown in Figure 6.1. In addition to deformation, the wall may experience rotation and translation, depending on the bracing system used and workmanship of the construction. Because of this, it is often observed that the earth pressure on the upper part of the braced wall exceeds that against a conventional retaining wall (e.g., cantilever or gravity retaining wall) and the earth pressure at the middle is much larger than at the base of the wall. The resultant lateral earth force is likely to be located near the middle of the wall instead of one-third of the way from the base of the wall. In addition, the resultant lateral earth force on the braced excavation wall is about 30%–50% higher than the active earth force.

The objectives of this chapter are as follows:

- To introduce the types and major components of typical braced excavation systems.
- To describe the characteristics of lateral earth pressure involved in the design and construction of a braced excavation.
- To describe the basic design and construction procedure of a braced excavation.

6.2 COMPONENTS OF A BRACED EXCAVATION

The two major aspects of a braced excavation are the excavation wall and the bracing (or support) system. In both aspects, there are multiple choices. The type of wall and the bracing can be independently chosen to work together and serve the overall purposes. The following sections introduce commonly used wall types and bracing systems, and their basic construction procedures.

6.2.1 Wall types

The main purpose of the wall is to keep the soil, and more importantly, water, from seeping into the excavation. The demand for wall stiffness and waterproofing capability increases as the retained soil becomes softer and/or more permeable while below the groundwater table. The cost of the excavation wall is a dominant factor in the overall construction cost of the bracing system. The following introduction of wall types follows a general trend of increasing cost.

6.2.1.1 Soldier pile walls

As described in Figure 6.2, this type of wall consists of soldier piles typically placed 1 m apart, and wood plank lagging that covers the soil between the soldier piles. The soldier piles carry the full earth pressure load, while the lagging retains soil and resists relatively minor earth pressure. Soldier piles are mostly made with wide flange or H pile sections. Wood board 20 to 30 mm thick is commonly used as the lagging, as in Figure 6.2, although light steel sheeting or precast concrete board have also been used for the purpose. Timber typically used as wood lagging has an allowable flexural stress of 10 MPa. Gaps between the lagging boards are created intentionally to allow the introduction of material such as hay or geosynthetics for backpacking boards, and filtering soil to protect against ground loss from seepage.

For installation of the wall, the soldier piles are driven first, either using an impact pile-driving hammer or a vibratory hammer (see Section 8.2). The lagging boards are inserted between the soldier piles as shown in Figure 6.2a, as excavation and placement of the support system progresses. The depth of exposure below the last placed lagging board is usually less than 1.5 to 2 m; shorter for more adverse soil conditions. Figure 6.3 shows part of a completed soldier pile wall and its lateral support system. Upon completion of the excavation and construction of permanent underground structure, the soldier piles are removed. The permanent underground structure is usually built immediately next to the soldier pile wall and hence no backfill is needed. Unless otherwise specified, the lagging boards are left in place.

The soldier pile walls are often used in sand, gravel, and stiff clay (relatively competent or stable soils), and when there is no serious seepage problem. The soil and groundwater conditions allow the soldier piles to be driven and excavated soil to be exposed for a short period of time while the lagging boards are inserted and the neighboring ground remains stable. Soldier pile walls usually cost less than sheet pile or concrete diaphragm walls.

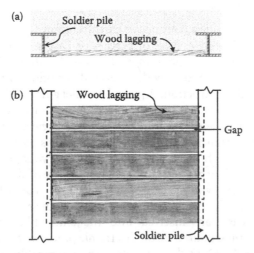

Figure 6.2 General description of a soldier pile wall. (a) Plan view. (b) Side view.

Figure 6.3 Soldier pile wall and its lateral support system. (Courtesy of An-Bin Huang, Hsinchu, Taiwan.)

6.2.1.2 Steel sheet pile walls

Characteristics and properties of steel sheet piles typically used for retaining structures are described in Section 5.14. Sheet pile walls are often used in soils that are inappropriate for soldier pile walls, such as soft clays and organic and loose soils of low plasticity below the water table (e.g., saturated silt, loose silty, or clayey sand). Individual steel sheet pile panels are interlocked via thumb-and-finger or ball-and-socket type connections (see Section 5.14). The sheet pile wall provides complete coverage of the soil within the excavation and is effective in cutting off concentrated flow through pervious layers and for protection against ground loss.

Impact or vibratory hammers (see Section 8.2) may be used to drive the steel sheet piles. The sheet piles are inserted in waves, maintaining the tips of adjoining sheet piles no more than about 2 m apart. This procedure is designed to maintain the alignment of the sheet piles and integrity of the interlocks. This is especially critical when driving in dense soil or soils with boulders. The sheet piles are extracted after the permanent underground structure is completed.

6.2.1.3 Concrete diaphragm walls

The concrete diaphragm wall is a continuous earth-retaining concrete wall built from the ground surface. The walls may consist of precast or cast-in-place concrete panels. The most common type of concrete diaphragm wall is casted within a slurry-stabilized trench using a tremie pipe. Figure 6.4a–d provides a brief description of the sequence of slurry-trenched diaphragm wall construction. A pair of parallel guide walls is built first along the line where the diaphragm wall is to be constructed. A clamshell bucket suspended from a construction crane (see Figure 6.5) is normally used to excavate the trench between the guide walls, while the trench is filled with slurry as conceptually described in Figure 6.4a. The slurry trench usually is about 3 to 6 m wide, 0.5 to 1.5 m thick, and can extend to over 50 m below ground surface. The purpose of slurry is to maintain stability of the trench during excavation. Bentonite or various types of polymer mixed with water have been used as slurry. After the individual panels are excavated, end pipes and reinforcing steel cage are inserted into the slurry-filled trench, as shown in Figure 6.4b. Concrete is then poured through a tremie pipe, and the end pipes are removed as described in Figure 6.4c. Once the concrete is set, the neighboring panel can be excavated, as depicted in Figure 6.4d.

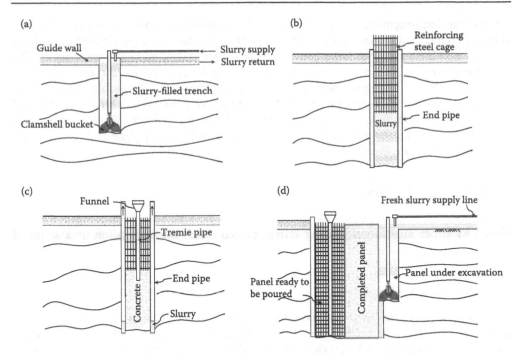

Figure 6.4 The sequence of slurry-trenched diaphragm wall construction. (a) Excavation of slurry-filled trench. (b) Placement of end pipes and reinforcing steel cage. (c) Pouring concrete while extracting the end pipes. (d) Excavation of a neighboring panel.

Figure 6.5 Construction crane with a clamshell bucket. (Courtesy of Ground Master Construction Co. Ltd., Taipei, Taiwan.)

Concrete diaphragm walls can be made with high rigidity (i.e., with increased thickness) and watertight. Techniques have been developed to construct concrete diaphragm walls in soft/loose soils or soils with large boulders, below water table, and with great depth. The equipment involved in the construction is relatively quiet and low in vibration. For these reasons, concrete diaphragm walls are often used in congested metropolitan areas where excavations are made in close proximity to other existing structures. The construction of concrete diaphragm walls in general is more costly than the sheet pile and soldier pile walls. However, the concrete diaphragm walls are usually left in place and used as part of the permanent structure.

6.2.2 Support methods

The excavation walls are rarely designed as a cantilever structure. Instead, they are supported by either an internal bracing system placed inside the excavation or a series of anchors or tiebacks.

6.2.2.1 Internal bracing

Internal bracing is often used most economically in relatively narrow excavations, where cross-lot bracing can be used without intermediate support. Figure 6.6a and b provides schematic cross-section and plan views of the wall and a cross-lot internal bracing system. In this case the excavation walls on the opposite sides are used to provide reactions against each other. The horizontal cross-lot bracing is referred to as the strut. A continuous horizontal wale is typically used to transfer loads from the excavation wall to the struts. Wale levels are normally set about 3 to 5 m apart vertically, and strut positions are set about 5 to 7 m apart longitudinally along the cut. Intermediate vertical support for the struts may

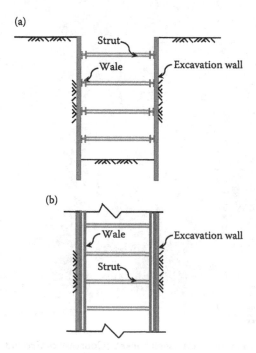

Figure 6.6 Schematic views of the cross-lot internal bracing system. (a) Section view. (b) Plan view.

Figure 6.7 Cross-lot bracing of a cut-and-cover excavation for a subway station of Kaohsiung MRT project, Taiwan. (Courtesy of An-Bin Huang, Hsinchu, Taiwan.)

become necessary as the excavation widens. Figure 6.7 shows a case of cross-lot bracing with intermediate vertical support system used in the cut-and-cover excavation for the construction of a subway station.

As the distance between the sides of the excavation increases further, internal bracing becomes less efficient, and tiebacks (described later) may be more feasible. For wide excavations where cross-lot bracing becomes impractical and suitable anchorage soil or rock layers are not available for tiebacks, inclined braces supported by a row of short wall embedded in soil within the excavated area may be used to provide the internal support. Figure 6.8 shows the schematic view of an excavation using an inclined internal bracing system. The lateral resistance came from the passive earth pressure behind the short wall. The inclined braces are referred to as the rakers. Wales are used to transfer loads from the excavation wall to the rakers as in the case of cross-lot bracing.

Displacements may occur from slack in the support system, but this can be largely eliminated by preloading. Preloading up to 50% of the design or allowable load is a common practice in areas where displacements are of concern. Extreme temperature variations can affect load experienced by the struts or rakers. To prevent overstressing, the steel members exposed to sunlight may be coated with reflective paint.

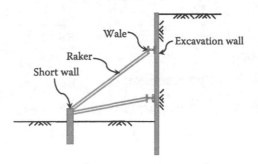

Figure 6.8 Excavation using an inclined internal bracing system.

6.2.2.2 Tiebacks

A tieback is a prestressed, grouted ground anchor installed in soil or rock that is used to transmit the applied tensile load into the ground. The basic components of a tieback include: (1) the anchorage, (2) the unbonded length, and (3) the anchor (or tendon) bond length, as schematically shown in Figure 6.9. The anchorage is the combined system of anchor head, bearing plate, and other mechanical components that facilitate transmitting the tensile force from the tendon to the excavation wall. The unbonded length is the part of the tendon that is free to elongate while transferring the tensile force from the bonded tendon to the excavation wall. The tendon bond length is that length of the bonded tendon capable of transmitting the applied tensile load into the ground. The anchor bond length should be located outside of the active failure zone.

The use of tiebacks or ground anchors to support excavation walls has increased significantly in the past few decades. The anchorage is the only exposed part of the bracing system, leaving the entire excavation area open for construction activities. Figure 6.10 shows a case of braced excavation using contiguous-bored concrete piles as the excavation wall, with tieback support.

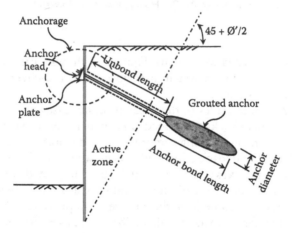

Figure 6.9 Components of a tieback.

Figure 6.10 Braced excavation at National University of Singapore campus using contiguous-bored concrete piles with tieback support. (Courtesy of An-Bin Huang, Hsinchu, Taiwan.)

The construction of a ground anchor generally follows these steps:

1. Drilling a borehole according to the planned diameter, inclination, and length.
2. Insertion of the tendon (prestressing steel element) into the borehole.
3. Grouting the tail end of the borehole (i.e., the anchor bond length in Figure 6.9).
4. Performing load test on the ground anchor and applying the lock-off load.

A variety of configurations of the grouted anchors are available and have been used successfully in braced excavations. The grouted anchor can be straight-shafted, as in Figure 6.9. The grout can be inserted into the borehole by gravity (gravity grouting) or under pressure (pressure grouting). The grouted anchor can have an enlarged toe (i.e., belled). There can be multiple bells within the anchor bond length. The pressure grouting can be applied more than once (regroutable anchor). In a given soil deposit, the actual capacity achieved in the field will depend on the equipment used, construction procedure, and workmanship. The selection of construction equipment, dimensions of the borehole, and configuration of the grouted anchor are largely based on local expertise and experience, and therefore is usually left to the discretion of the specialty anchor contractor.

The main responsibility for the designer is to define a minimum anchor capacity that can be achieved in a given ground type. Estimation of anchor capacity is therefore based on the simplest straight-shaft gravity-grouted anchor. The estimated capacity may confidently be achieved while allowing the specialty contractors to make further optimization. The design capacity of each anchor will be verified by testing before accepting the anchor. The ultimate resistance, P_{ult}, per ground anchor or tieback may be estimated as follows:

$$P_{ult} = \pi d_{tb} l_{tb} f_{bu} \tag{6.1}$$

where

d_{tb} = diameter of the grouted anchor
l_{tb} = anchor bond length
f_{bu} = average ultimate bond stress along the grouted anchor

A range of ultimate bond stress (f_{bu}) values that have been reported for gravity-grouted and pressure-grouted anchors is provided in Table 6.1. The allowable anchor design load, P_{all}, may be determined by dividing the ultimate resistance, P_{ult}, by a safety factor (FS) as follows:

$$P_{all} = \frac{P_{ult}}{FS} \tag{6.2}$$

A minimum safety factor of 2.0 is usually applied.

The purpose of prestressing steel element of the tendon is to safely transmit load in the anchor bond zone to the structure without breakage. High-strength steel wire strands, cables, and bars are commonly used for prestressing steel element. Often the choice of the type of prestressing steel element is limited by the method of installation, or by convenience. Table 6.2 shows the dimensions and strengths of typical prestressing steel elements used in ground anchors. Note that multiple wires, cables, or strands may be used in an anchor.

Every production ground anchor is load-tested by pulling the anchor usually to a maximum of 125% of the design load. If the load test results meet the acceptance criteria set for the contract (usually based on the elongation of the tendon at 125% of the design

Table 6.1 Average ultimate bond stress for ground/grout interface along straight shaft anchor bond zone

Rock		Cohesionless soil		Cohesive soil	
Rock type	f_{bu} (MPa)	Anchor type	f_{bu} (MPa)	Anchor type	f_{bu} (MPa)
Granite and basalt	1.7–3.1	Gravity-grouted	0.03–0.07	Gravity-grouted	0.07–0.14
Dolomitic limestone	1.4–2.1	Pressure-grouted		Pressure-grouted	
Soft limestone	1.0–1.4	Soft silty clay	0.03–0.07	Fine–medium sand, medium dense–dense	0.08–0.38
Slates and hard shales	0.8–1.4	Silty clay	0.03–0.07	Medium–coarse sand (w/gravel), medium dense	0.11–0.66
Soft shales	0.2–0.8	Stiff clay, medium to high plasticity	0.03–0.10	Medium–coarse sand (w/gravel), dense–very dense	0.25–0.97
Sandstones	0.8–1.7	Very stiff clay, medium to high plasticity	0.07–0.17	Silty sands	0.17–0.41
Weathered sandstones	0.7–0.8	Stiff clay, medium plasticity	0.10–0.25	Dense glacial till	0.30–0.52
Chalk	0.2–1.1	Very stiff clay, medium plasticity	0.14–0.35	Sandy gravel, medium dense–dense	0.21–1.38
Weathered marl	0.15–0.25	Very stiff sandy silt, medium plasticity	0.28–0.38	Sandy gravel, dense–very dense	0.28–1.38
Concrete	1.4–2.8				

Source: After Sabatini, P.J., Pass, D.G., and Bachus, R.C. 1999. *Geotechnical Engineering Circular No. 4. Ground Anchors and Anchored Systems*. Report No. FHWA-IF-99-015. Office of Bridge Technology, Federal Highway Administration, Washington, DC.

load), the anchor is then locked off. A lock-off load—that is, a tensile force at certain fraction of the design load—remains in the anchor after the load test. For braced excavations designed using the apparent earth pressure (described below), the lock-off load is usually set at 80%–100% of the design load.

For permanent ground anchors, a certain percentage of the anchors are tested to a much higher ratio of the design load or to failure. Selected permanent anchors may be retested at

Table 6.2 Dimensions and strengths of typical prestressing steel elements

Tendon type	Diameter (mm)	Ultimate Stress (f_{su}) (MPa)	Yield stress (f_y), fraction of f_{su}
Wire (ASTM A421)	6.35	1655	0.80
Cables or strands (ASTM A416)	6.35	1862	0.85
	12.7	1862	0.85
	15.24	1862	0.85
Bars or rods (ASTM A322)	12.7	1104	0.85
	15.88	1587	0.85
	25.4	1035	0.85
	25.4	1104	0.85
	31.75	1035	0.85
	31.75	1104	0.85
	34.93	1035	0.85
	31.75	911	0.85

a later stage after installation to verify the loading conditions and integrity of the anchor. The permanent ground anchors are usually coated with protective material against corrosion.

6.3 APPARENT EARTH PRESSURE

As described earlier, the construction sequence of a braced excavation is very different from that of conventional retaining walls introduced in Chapter 5. For braced excavation, we try to keep the excavation wall movement to a minimum to avoid damage to the neighboring structures. As a result, the lateral earth pressure distribution is rather different from that of the active condition described in Chapter 5. The following section describes the development of apparent earth pressure diagrams that we often use in the design of braced excavation.

6.3.1 Development of the apparent earth pressure diagrams

A number of empirical pressure distributions have been proposed for braced excavations. Since the actual distribution depends on several variables, including the relative stiffness of the soil and bracing and workmanship, the choice of earth pressure distribution is largely a matter of judgment. However, because of the application of bracing and prestressing at early stage of the excavation, full development of active failure is often not materialized as described previously. The lateral earth pressure is likely to be much higher than that described by the active earth pressure theory near the top of the excavation. In general, the resulting deformation pattern most closely resembles an arching active condition. Therefore a parabolic, rather than triangular, pressure distribution is most likely to act on the wall.

Apparent earth pressure diagrams are semi-empirical and were originally developed by Terzaghi and Peck (1967) and Peck (1969) to provide loadings for the design of struts in internally braced excavations. They made field measurements of the reaction forces provided by the struts for various braced excavation projects in the United States and Germany. The apparent earth pressure distribution against the excavation wall was then inferred from the field strut load measurements. Figure 6.11 shows the general procedure for developing apparent earth pressure diagrams from reaction forces (R_A, R_B, R_C,

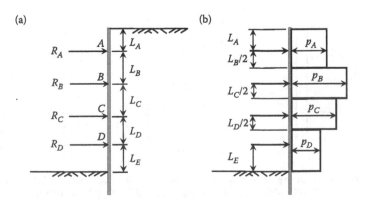

Figure 6.11 Development of apparent earth pressure diagrams. (a) Reaction forces. (b) Apparent earth pressures.

and R_D) measured at each strut level in the field. The apparent earth pressures were calculated as follows:

$$p_A = \frac{R_A}{(S_h)\left(L_A + \dfrac{L_B}{2}\right)} \tag{6.3a}$$

$$p_B = \frac{R_B}{(S_h)\left(\dfrac{L_B}{2} + \dfrac{L_C}{2}\right)} \tag{6.3b}$$

$$p_C = \frac{R_C}{(S_h)\left(\dfrac{L_C}{2} + \dfrac{L_D}{2}\right)} \tag{6.3c}$$

$$p_D = \frac{R_D}{(S_h)\left(\dfrac{L_D}{2} + L_E\right)} \tag{6.3d}$$

where
S_h = horizontal spacing between struts
p_A, p_B, p_C, and p_D = apparent earth pressure for strut level A, B, C, and D
L_A, L_B, L_C, and L_D = vertical distance from strut level A, B, C, and D to its neighboring support at higher level or ground surface

The resulting apparent earth pressure diagrams are used to develop an envelope encompassing the maximum distributed pressures. This design envelope represents the maximum strut load that can be anticipated at any stage of construction. These diagrams are likely to produce conservative design loads.

The Terzaghi and Peck apparent earth pressure envelopes are rectangular or trapezoidal in shape, as summarized in Figure 6.12. The maximum ordinate of the apparent earth pressure diagrams in Figure 6.12 is denoted as σ_{hp}. The Terzaghi and Peck envelopes were developed considering the following factors:

- The excavation is greater than 6 m deep and relatively wide. Wall movements are assumed to be large enough so that the full value of the soil shear strength may be mobilized.

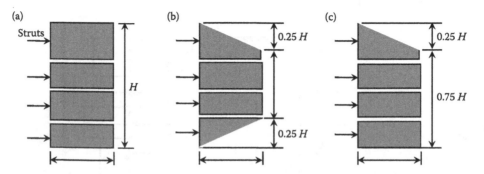

Figure 6.12 Apparent earth pressure diagrams for internally braced walls. (a) Sands: $\sigma_{hp} = 0.65K_a\gamma H$; $K_a = \tan^2(45-(\varnothing'/2))$. (b) Stiff clays: $\sigma_{hp} = 0.3\gamma H$. (c) Soft-to-medium clays: $\sigma_{hp} = 1.0K_A\gamma H$; $K_A = 1-m(4s_u/\gamma H)$. (Adapted from Peck, R.B. 1969. Deep excavation and tunneling in soft ground. State-of-the-Art Report. *Proceedings of the 7th International Conference on Soil Mechanics and Foundation Engineering*, Mexico City, Mexico, pp. 225–325.)

- Groundwater is below the base of the excavation for sands, and for clays, its position is not considered important. Loading due to water pressure is not considered in the analyses.
- The soil is assumed to be homogeneous and soil behavior during shearing is assumed to be drained for sands and undrained for clays; that is, only short-term loadings are considered.
- The loading diagrams apply only to the exposed portion of the wall and not the portion of the wall embedded below the bottom of the excavation.

Procedures to establish the apparent earth pressure diagrams are described below.

6.3.1.1 Sands

If the excavation is in a sand deposit, the earth pressure σ_{hp} is estimated as

$$\sigma_{hp} = 0.65 K_a \gamma H \tag{6.4}$$

where
K_a = active earth pressure coefficient = $\tan^2(45 - (\varnothing'/2))$
γ = total unit weight of the soil
H = depth of excavation

6.3.1.2 Clays

For clays, the apparent earth pressure is related to a stability number, N_S, which is defined as

$$N_s = \frac{\gamma H}{s_u} \tag{6.5}$$

where
s_u = average undrained shear strength of the clay soil below the base of the excavation

6.3.1.3 Soft-to-medium clay ($N_S \geq 6$)

$$\sigma_{hp} = K_A \gamma H \tag{6.6}$$

and

$$K_A = 1 - m\left(\frac{4 s_u}{\gamma H}\right) \tag{6.7}$$

In Equation 6.7, m generally equals 1, and may vary between 0.4 and 1 where open excavation is underlain by soft, normally consolidated clay.

6.3.1.4 Stiff clay ($N_S \leq 4$)

If the excavation is in a stiff clay deposit, the earth pressure σ_{hp} is estimated as

$$\sigma_{hp} = 0.3 \gamma H \tag{6.8}$$

For $4 < N_S < 6$, σ_{hp} = larger of Equations 6.6 and 6.8 and use the apparent earth pressure diagram for soft-to-medium clay.

Note that the Terzaghi and Peck apparent earth pressure diagrams were developed mainly for internally braced excavation walls, although they have also been used successfully for anchor or tieback supported walls (Goldberg et al., 1976). Modified apparent earth pressure diagrams for anchor-supported walls that are dependent on the number/location of the ground anchors have been reported by Sabatini et al. (1999). Details of the apparent earth pressure diagrams for anchor-supported walls are not described herein. Terzaghi and Peck apparent earth pressure diagrams may be used as an initial estimate in the case of anchor-supported walls.

6.3.2 Loading diagrams for stratified soil profiles

The apparent earth pressure diagrams described above were developed for reasonably homogeneous soil profiles. They are not directly applicable in cases of stratified soil deposits. For excavations in clays with significantly different undrained shear strengths, as shown in Figure 6.13a, take the average undrained shear strength $s_{u(ave)}$ and unit weight $\gamma_{(ave)}$ of the clays involved as follows:

$$\gamma_{(ave)} = \frac{\sum_{i=1}^{n} (h_i)(\gamma_i)}{H} \tag{6.9a}$$

$$s_{u(ave)} = \frac{\sum_{i=1}^{n} (h_i)(s_{ui})}{H} \tag{6.9b}$$

where
  h_i = thickness of layer i
  γ_i = unit weight of the clay in layer i
  s_{ui} = undrained shear strength of the clay in layer i

The apparent earth pressure diagram is determined according to the average undrained shear strength and unit weight.

For excavation in soils that involve both sand and clay layers, as shown in Figure 6.13b, the apparent earth pressure diagram can be established based on an equivalent undrained shear strength, $s_{u(eq)}$, proposed by Peck (1943) as follows:

$$s_{u(eq)} = \frac{1}{2H} \left[\gamma_s K_s H_s^2 \tan \emptyset_s' + 2(H - H_s) n' s_u \right] \tag{6.10a}$$

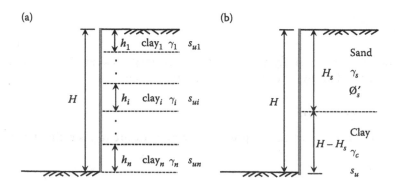

Figure 6.13 Braced excavation in layered soils. (a) Layered clays. (b) Sand and clay layers.

And the average unit weight of the layers, γ_{ave}, is calculated as

$$\gamma_{ave} = \frac{1}{H}[\gamma_s H_s + (H - H_s)\gamma_c]$$ (6.10b)

where
    H = total depth of excavation
    γ_s = unit weight of the sand
    H_s = thickness of sand layer
    K_s = lateral earth pressure coefficient for the sand layer (≈ 1.0)
    $\varnothing'_s$ = drained friction angle of sand
    s_u = undrained shear strength of the clay
    γ_c = unit weight of the clay
    n' = coefficient for progressive failure varied from 0.5 to 1.0 (use average value of 0.75)

The apparent earth pressure diagram is then determined considering the overall soil deposit as a clay, based on the equivalent undrained shear strength Equation (6.10a) and average unit weight Equation (6.10b).

6.4 ANALYSIS OF THE WALL AND BRACING SYSTEM

The braced excavation wall and many other elements of the lateral support system are continuous beams from the structural analysis point of view. Rigorous analysis of a continuous beam as an indeterminate system would require the use of a computer program. Commercial software packages are routinely used in engineering practice for the design of braced excavation systems. The following sections provide a complete but simplified, step-by-step procedure for the analysis of a braced excavation system. With these simplifications, the calculations can be executed without a computer program.

6.4.1 Determination of strut loads

The strut loads are determined by reversing the procedure used for the development of the apparent earth pressure diagrams (see Figure 6.11). A strut is designed to support a load represented by the area between the midpoints of the adjacent support levels. Consider the case shown in Figure 6.14. For a braced cut in sand, the strut load at each level (i.e., R_A, R_B, R_C, and R_D) is calculated as follows:

$$R_A = s_h \left(L_A + \frac{L_B}{2} \right) \sigma_{hp}$$ (6.11a)

$$R_B = s_h \left(\frac{L_B}{2} + \frac{L_C}{2} \right) \sigma_{hp}$$ (6.11b)

$$R_C = s_h \left(\frac{L_C}{2} + \frac{L_D}{2} \right) \sigma_{hp}$$ (6.11c)

$$R_D = s_h \left(\frac{L_D}{2} + L_E \right) \sigma_{hp}$$ (6.11d)

where
    $\sigma_{hp} = 0.65\,K_a\gamma H$ according to Equation 6.4 for braced excavation in sand
    S_h = horizontal center-to-center spacing between struts

(a) (b)

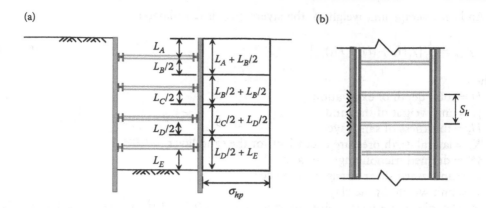

Figure 6.14 Calculation of the strut loads using the apparent earth pressure diagrams. (a) Section view and apparent earth pressure. (b) Plan view.

The struts are treated as columns hinged at both ends, carrying the loads obtained according to the above procedure. The size of the steel element to be used as struts and the need for intermediate support are then determined according to the steel design manual. For ground anchor-supported excavations, the strut load represents the horizontal component of the anchor design load at the corresponding level.

For excavations in clays, the first (top) level of strut should be placed at depth shallower than the depth of tensile crack (see Section 5.6). Regardless of the soil conditions, the first level of strut should be placed when the vertical component of the wall (i.e., soldier pile or sheet pile) is capable of supporting the lateral earth pressure as a cantilever.

6.4.2 Loading on wales

As described earlier, the main function of a wale is to transfer loads from the excavation wall to the struts. A wale is treated as a continuous beam subject to a uniform distributed load and placed at the same level as the corresponding struts. Each wale carries a tributary area of load (per unit thickness of the wall) based on the vertical spacing between adjacent struts. Figure 6.15 provides a conceptual description of the distributed load on a wale

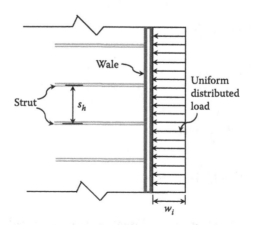

Figure 6.15 Plan view of a wale as a continuous beam subjected to a uniform distributed load.

at level i in a support system. For a continuous beam, the maximum bending moment M_{max} relates to the uniform distributed load as follows:

$$M_{max} = Cw_i s_h^2 \tag{6.12}$$

where

 C = moment coefficient
 w_i = uniform distributed load (per unit thickness of the wall) on the wale at level i
 s_h = horizontal center-to-center spacing between struts

For the braced excavation (excavation in sand) case shown in Figure 6.13, the uniform distributed load (w_A, w_B, w_C, and w_D) at the four strut levels is determined as follows:

$$w_A = \left(L_A + \frac{L_B}{2}\right)\sigma_{hp} \tag{6.13a}$$

$$w_B = \left(\frac{L_B}{2} + \frac{L_C}{2}\right)\sigma_{hp} \tag{6.13b}$$

$$w_C = \left(\frac{L_C}{2} + \frac{L_D}{2}\right)\sigma_{hp} \tag{6.13c}$$

$$w_D = \left(\frac{L_D}{2} + L_E\right)\sigma_{hp} \tag{6.13d}$$

For the determination of M_{max}, $C = 0.1$ is recommended for continuous members supporting a uniform distributed load. Selection of the wale section at each level is based on the M_{max} of that level.

6.4.3 Loading on vertical components of the wall

Depending of the system used, the load-carrying vertical component can be soldier piles, sheet piles, or concrete diaphragm walls. For a unit thickness of the wall, the vertical component can also be treated as a continuous beam subjected to a distributed load from earth pressure on one side and supported by wales on the other. The apparent earth pressure distributions as described in Figure 6.12 are not necessarily uniform in the vertical direction. For simplicity, the maximum bending moment per unit thickness (in unit of kN-m/m) in the vertical component of the wall between adjacent wales may be estimated conservatively as

$$M_{max} = C\sigma_{hp} s_v^2 \tag{6.13}$$

where

 σ_{hp} = maximum ordinate of the apparent earth pressure diagram
 s_v = vertical center-to-center spacing between the adjacent wales

Again, $C = 0.1$ can be used for the calculation of M_{max}. The vertical wall component is sized according to the largest value of M_{max} obtained from the given vertical wall component. As described earlier, the soldier piles, if used, carry the full earth pressure load while the lagging retains soil and resists relatively minor earth pressure. Each soldier pile thus carries a tributary area of load that covers the horizontal spacing between adjacent soldier piles (s_h). For a soldier pile, the maximum bending moment (in unit of kN-m) between adjacent wales should therefore be (s_h)(M_{max}), where M_{max} is obtained from Equation 6.13.

EXAMPLE 6.1

Given

A 9 m-deep braced excavation is to be conducted in a clay deposit. Figure E6.1 shows the soil profile and the proposed cross-lot bracing scheme.

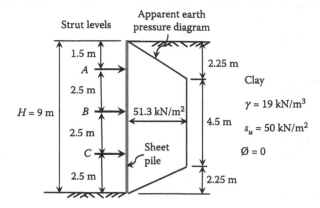

Figure E6.1 Soil profile and the proposed cross-lot bracing scheme.

Required

1. Determine and draw the apparent earth pressure diagram for the braced excavation.
2. Determine the strut load at levels A, B, and C.
3. Determine the required section modulus for the wale at levels B and C.
4. Determine the required section modulus for the sheet pile per unit thickness of the wall.

The center-to-center spacing between the struts is $s_h = 3$ m, and the allowable flexural stress of the steel elements is $\sigma_{all} = 200$ MPa.

Solution

Part 1

The stability number according to Equation 6.5, $N_s = \gamma H/s_u = (19)(9)/(50) = 3.42 < 4$, therefore use the earth pressure diagram for stiff clay.

$$\sigma_{hp} = 0.3\gamma H = (0.3)(19)(9) = 51.3 \text{ kN/m}^2$$

Part 2

Strut load at $A = (51.3)(2.25)/2 + (51.3)(0.5) = 83.4$ kN/m of the wall
  For struts horizontally spaced at 3 m, load per strut $= (83.4)(3) = 250.1$ kN
  Strut load at $B = (51.3)(2.5) = 128.3$ kN/m of the wall
  For struts horizontally spaced at 3 m, load per strut $= (128.3)(3) = 384.9$ kN
  Strut load at $C = (51.3)(2.25)/2 + (51.3)(1.5) = 134.7$ kN/m of the wall
  For struts horizontally spaced at 3 m, load per strut $= (134.7)(3) = 404.1$ kN

Part 3

At level B,

$$M_{max} = Cw_i s_h^2 \text{ and } C = 0.1$$

At level B, $w_i = (51.3)(2.5) = 128.3 \text{ kN/m}$

$$M_{max} = (0.1)(128.3)(3^2) = 115.4 \text{ kN-m}$$

Required section modulus $S = M_{max}/\sigma_{all} = (115.4)(0.001)/(200) = 0.58 \times 10^{-3} \text{ m}^3$
At level C,

$$w_i = (51.3)(2.25)/2 + (51.3)(1.5) = 134.7 \text{ kN/m}$$

$$M_{max} = (0.1)(134.7)(3^2) = 121.2 \text{ kN-m}$$

Required section modulus $S = M_{max}/\sigma_{all} = (121.2)(0.001)/(200) = 0.61 \times 10^{-3} \text{ m}^3$

Part 4

According to Equation 6.13, $M_{max} = C\sigma_{hp} s_v^2$, and s_v = vertical spacing between wales,
$C = 0.1$
$\sigma_{hp} = 51.3 \text{ kN/m}^2$ and take $s_v = 2.5 \text{ m}$ (the maximum spacing used in the design)
$M_{max} = (0.1)(51.3)(2.5^2) = 32.1 \text{ kN-m/m of the wall}$
Required section modulus $S = M_{max}/\sigma_{all} = (32.1)(0.001)/(200) = 0.16 \times 10^{-3} \text{ m}^3/\text{m}$
of the wall.

EXAMPLE 6.2

Given

A 9 m-deep braced excavation is to be conducted in a sand/clay layered deposit as
shown in Figure E6.2.

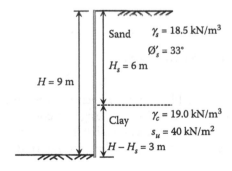

Figure E6.2 The sand/clay layered deposit.

Required

Determine the apparent earth pressure diagram using the equivalent shear strength
method.

Solution

Follow Equation 6.10a, the equivalent shear strength

$$s_{u(eq)} = \frac{1}{2H}\left[\gamma_s K_s H_s^2 \tan \varnothing_s' + 2(H - H_s)n's_u\right]$$

thus

$$s_{u(eq)} = \frac{1}{(2)(9)}\left[(18.5)(1.0)(6^2)\tan(33^0) + (2)(9-6)(0.75)(40)\right] = 34.0\ \text{kN/m}^2$$

The average unit weight $\gamma_{ave} = 1/H[\gamma_s H_s + (H - H_s)\gamma_c]$ according to Equation 6.10b

$$\gamma_{ave} = \frac{1}{9}\left[(18.5)(6) + (9-6)(19.0)\right] = 18.67\ \text{kN/m}^3$$

The stability number according to Equation 6.5, $N_s = \gamma H/s_u = (18.67)(9)/(34.0) = 4.94 > 4$

$4 < N_s < 6$, use the apparent earth pressure diagram for soft-to-medium clay and $\sigma_{hp} = $ larger of Equations 6.6 and 6.8

Following Equation 6.6, $\sigma_{hp} = K_A \gamma H$ and $K_A = 1 - m(4s_u/\gamma H)$, $m = 1.0$

$$K_A = 1 - (1.0)\frac{(4)(34.0)}{(18.67)(9)} = 0.19 \quad \text{and} \quad \sigma_{hp} = (0.19)(18.67)(9) = 31.9\ \text{kN/m}^2$$

According to Equation 6.8, $\sigma_{hp} = 0.3\gamma\,H = (0.3)(18.67)(9) = 50.4\ \text{kN/m}^2$
Establish the apparent earth pressure diagram with $\sigma_{hp} = 50.4\ \text{kN/m}^2$
The apparent earth pressure diagram is shown in Figure E6.2a.

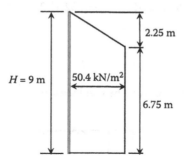

Figure E6.2a The apparent earth pressure diagram.

Numerical methods are available and have been used for the design of braced excavation systems. This is especially true for the design of concrete diaphragm walls. With the help of numerical methods, nonlinear soil behavior and effects of soil structure interaction can be considered. Interested readers are referred to Ou (2006) and Reese et al. (2013).

6.5 STABILITY AT THE BASE OF A BRACED EXCAVATION

Failure of a braced excavation can cause extensive damage to the nearby structures and potential loss of human lives. The failure can be due to overstressing of the structural elements in the wall and bracing system such as the struts, wales, or excavation walls. Stability analyses considering these types of potential failures are included in Section 6.4. This section concentrates on the stability or failure at the base of excavation. In this regard, the "push-in" failure and "basal heave" failure are the two major failure modes to be considered. Basal heave failure is limited to cohesive soils, whereas push-in failure can occur in both cohesive and cohesionless soils.

6.5.1 Basal heave

Basal heave occurs when the soils at the base of the excavation are relatively weak compared to the overburden stresses induced by the retained side of the excavation. Significant basal heave within the excavation and settlement adjacent to the excavation result when the weight of the retained soil exceeds or approaches the soil-bearing capacity at the base of the excavation. The basal heave analysis assumes a failure mechanism that is analogous to a bearing capacity failure (Bjerrum and Eide, 1956; Terzaghi et al., 1996), as shown in Figure 6.16. The excavation has a depth of H and width of B. A uniform surcharge q is placed adjacent to the excavation. The vertical pressure, q_{app}, exerted on strip cd comes from the total weight of the block of retained soil and surcharge, minus the shear resistance along plane bc. Therefore, q_{app} can be expressed as follows:

$$q_{app} = (\gamma H + q) - \frac{s_u H}{B'} \tag{6.14}$$

The factor of safety against basal heave as a bearing capacity failure is

$$FS_{bh} = \frac{N_{cb} s_u}{q_{app}} \tag{6.15}$$

where
$B' =$ width of strip cd in Figure 6.16a
$H =$ depth of excavation
$\gamma =$ total unit weight of the soil
$s_u =$ undrained shear strength of the clay
$q =$ uniform surcharge on the area adjacent to the excavation
$N_{cb} =$ bearing capacity factor from Figure 6.17

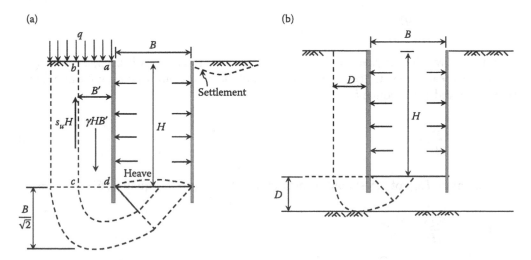

Figure 6.16 Mechanisms of a bottom heave failure. (a) In deep deposit of soft clay. (b) With stiff layer below excavation.

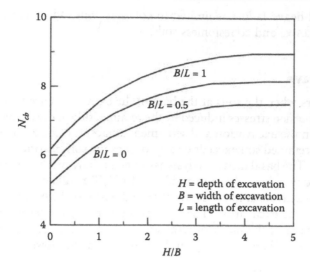

Figure 6.17 Bearing capacity factor.

Based on the geometry of the failure surface, B' cannot exceed $B/\sqrt{2}$. Thus, the minimum FS_{bh} for Equation 6.15 is

$$FS_{bh} = \frac{N_{cb}s_u}{(\gamma H + q) - \dfrac{s_u H \sqrt{2}}{B}} \tag{6.16}$$

The width B' is restricted if a stiff soil layer is located at a depth of D (and $D < B'$) below the base of the excavation, as shown in Figure 6.16b. In this case, B' is equal to D. Substituting D for B' in Equation 6.16 results in

$$FS_{bh} = \frac{N_{cb}s_u}{(\gamma H + q) - \dfrac{s_u H}{D}} \tag{6.17}$$

When the width of excavation (B) is very large and the clay layer is deep, the contribution of the shearing resistance along the exterior of the failure block is negligible and Equations 6.15 and 6.16 reduce to

$$FS_{bh} = \frac{N_{cb}s_u}{(\gamma H + q)} \tag{6.18}$$

A minimum FS_{bh} of 1.5 should be maintained. When $q = 0$, Equation 6.18 becomes

$$FS_{bh} = \frac{N_{cb}s_u}{\gamma H} = \frac{N_{cb}}{N_s} \tag{6.19}$$

where
N_s = stability number according to Equation 6.5

6.5.2 Push-in failure

Unlike basal heave, which is mostly a failure within the soil mass, push-in failure involves the design of the support system including the depth of wall penetration and level of the lowermost strut. Because of the nature of ground movement, push-in failure also results in basal heave within the excavation and settlement adjacent to the excavation, as described in Figure 6.18.

Analysis of the push-in failure follows the concept of the free earth support method (see Section 5.11) typically used for the analysis of anchored sheet pile walls. The analysis assumes active earth pressure on the retained soil side and passive earth pressure on the excavation side, as shown in Figure 6.19. The push-in failure analysis considers a free body diagram that covers part of the excavation from below the lowermost strut level, as shown in Figure 6.20. The safety factor (FS_{pi}) against push-in failure is computed as follows:

$$FS_{pi} = \frac{P_p L_p + M_s}{P_a L_a} \tag{6.20}$$

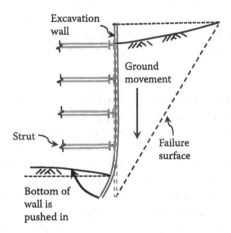

Figure 6.18 The push-in failure.

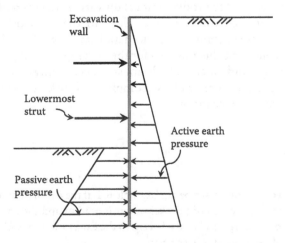

Figure 6.19 Earth pressure distribution for the push-in failure analysis.

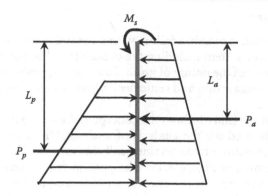

Figure 6.20 Free body diagram for push-in failure analysis.

where

P_p = passive earth force on the wall below the excavation surface

M_s = allowable bending moment of the wall at the lowest strut level

P_a = active earth force on the wall below the lowermost strut level

L_a = distance between the resultant active earth force and the lowermost strut level

L_p = distance between the resultant passive earth force and the lowermost strut level

The design usually requires a minimum FS_{pi} of 1.5. The depth of wall penetration below the excavation surface can be computed using Equation 6.20 and the required FS_{pi}. It should be noted that the active and passive earth pressure should be evaluated considering the adhesion and/or friction between the excavation wall and the surrounding soil. Ignoring this interface resistance can result in excessive wall penetration requirement because of the conservative estimates in earth pressures (i.e., active earth pressure is too high and passive earth pressure too low) (Ou, 2006). In most cases, M_s is small compared to $P_p L_p$ and thus can be assumed as zero for simplicity.

For excavation in granular soils, the soil is usually strong enough to withstand the overburden stresses induced by the retained side of the excavation, provided the excavation is properly dewatered and excessive pore water pressure does not develop. Piping can occur if there is sufficient water head to produce critical upward gradients at the base of the excavation. When pore water pressure becomes close to the total overburden stress, the strength of granular soil is substantially reduced and can flow like fluid. Seepage analysis can be conducted to determine the potential for the development of piping if excavation is to be executed below groundwater table. In most braced excavations, we either use the excavation wall (i.e., diaphragm wall) to completely block the seepage or lower the groundwater table prior to excavation.

EXAMPLE 6.3

Given

For the same soil conditions described in Example 6.1, if the depth of braced excavation remains the same (9 m), width of the excavation $B = 15$ m and the excavation is very long ($L \gg B$). No surcharge is placed adjacent to the excavation ($q = 0$). The clay layer below the base of excavation is very deep.

Required

Determine the factor of safety against basal heave, FS_{bh}.

Solution

For $q = 0$, $FS_{bh} = \dfrac{N_{cb}s_u}{\gamma H} = \dfrac{N_{cb}}{N_s}$ according to Equation 6.19.

$$N_s = N_s = \frac{\gamma H}{s_u} = (19)(9)/(50) = 3.42$$

Choose N_{cb} from Figure 6.17, consider $B/L = 0$ and $H/B = 9/15 = 0.6$, $N_{cb} \approx 5.7$

Therefore, $FS_{bh} = \dfrac{N_{cb}}{N_s} = \dfrac{5.7}{3.42} = 1.67$

6.6 PERFORMANCE OF BRACED EXCAVATIONS

Excavation inevitably induces ground displacements. The displacement is caused primarily by the unbalance of forces as a result of excavation and poor workmanship. The type and level of poor workmanship can vary significantly and are difficult to predict or quantify. The following discussion on the performance of braced excavation in terms of ground displacements excludes the factors of poor workmanship. Goldberg et al. (1976) compiled a total of 63 well-constructed, braced excavation cases and evaluated their ground displacement measurement data from outside of the excavation wall. Figure 6.21 shows the maximum horizontal displacement (δ_{hmax}) and maximum

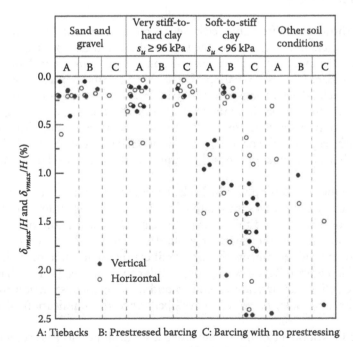

Figure 6.21 Normalized ground displacement outside of a braced excavation. (After Goldberg, D.T., Jaworski, W.E., and Gordon, M.D. 1976. *Lateral Support Systems and Underpinning, FHWA-RD-75-128.* Federal Highway Administration Offices of Research and Development, Washington, DC.)

vertical displacement (δ_{vmax}), normalized with respect to the depth of excavation H and their comparison with soil conditions. The cases include three types of support systems (tiebacks and internal bracing with and without prestressing) and four types of walls (steel sheet pile wall, soldier pile wall, diaphragm wall, and contiguous-bored pile wall).

The study by Goldberg et al. (1976) concludes the following.

For "competent soils" (granular soils, very stiff clays, etc.):

- The displacements are of insufficient magnitude to distinguish variations that may be caused by wall type or method of lateral support. The concrete diaphragm wall is likely to have less displacement than other wall types, and tiebacks appeared to perform better than internally braced walls.
- Maximum displacements are typically at 0.25%–0.35% of H. The lower range is associated with granular soils; the upper range is associated with cohesive soils.
- Maximum horizontal and vertical displacements are about equal.

For "weaker soils" (soft-to-medium clays, organic soils, etc.):

- Maximum displacements typically exceed 1% of H for flexible walls. The use of concrete diaphragm walls reduces the magnitude of displacements to about 0.25% of H.
- The maximum vertical displacements typically exceed maximum horizontal displacements.
- When the excavation is in deep deposits of weak soils, the cumulated displacements occurring below the last placed strut level amounts to about 60% of the total measured movement.

Experience shows that differential settlement is more detrimental to structures than the maximum settlement. For this reason, the profile of ground displacement or variation of settlement with distance adjacent to an excavation is more important than the maximum settlement itself. Based on his evaluation of field observations on braced sheet pile and soldier pile walls, Peck (1969) proposed a set of empirical correlation curves between vertical displacement (δ_v) and distance from the excavation wall (d) considering different soil conditions, as shown in Figure 6.22. In this figure, the displacement and distance values are normalized with respect to the depth of excavation H. These curves should be considered as envelopes that encompass the possible range of δ_v to d correlations for given soil conditions.

According to the curves proposed by Peck (1969), the excavation can cause settlement within a distance of 2 to 4 times the depth of excavation H, depending on the soil conditions. Stability number N_s and thickness of the clay layer below the base of excavation play an important role in the range of ground settlement. The curves in Figure 6.22 tend to indicate ground settlements that are significantly higher than the field observation data presented in Figure 6.21, especially for granular soils and very stiff-to-hard clays (category I). More refined analytical procedures and advanced construction methods (e.g., concrete diaphragm walls) have been developed since the publication of the chart by Peck (1969).

Clough and O'Rourke (1990) recommended a set of dimensionless settlement profiles as a basis for estimating the vertical displacement distribution adjacent to excavation in sands, stiff-to-very hard clays, and soft-to-medium clays. These dimensionless δ_v/δ_{vmax} versus d/H profiles are presented in Figure 6.23. The envelopes cover the range of settlements and distances of influence for the given types of soils according to the evaluation of a database compiled by Clough and O'Rourke (1990). To apply these settlement profiles,

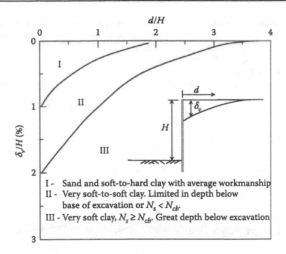

Figure 6.22 Variation of vertical displacement with distance from excavation wall. (Adapted from Peck, R.B. 1969. Deep excavation and tunneling in soft ground. State-of-the-Art Report. Proceedings of the 7th International Conference on Soil Mechanics and Foundation Engineering, Mexico City, Mexico, pp. 225–325.)

however, it is necessary to determine δ_{vmax} for the particular case of excavation. The issue of δ_{vmax} is discussed below.

Available field data have indicated that in most cases, the maximum horizontal deflection ($\delta_{hm(wall)}$) in the excavation wall is approximately the same as the maximum ground settlement δ_{vmax} adjacent to the excavation (Clough and O'Rourke, 1990; Ou, 2006).

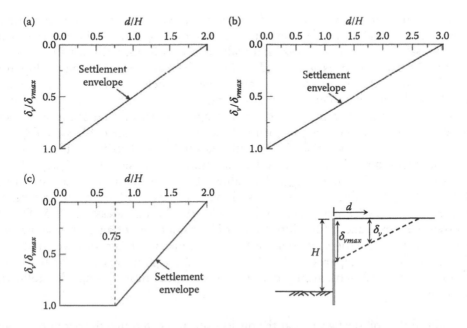

Figure 6.23 Dimensionless settlement profiles adjacent to excavation. (a) Sands. (b) Stiff-to-very hard clays. (c) Soft-to-medium clays. (Adapted from Clough, G.W. and O'Rourke, T. 1990. Construction-induced movements of in situ walls. Design and Performance of Earth Retaining Structures. ASCE Special Publication, No. 25, Figure 8, p. 448, Figure 9, p. 449 and Figure 10, p. 450. Reproduced with permission of the ASCE.)

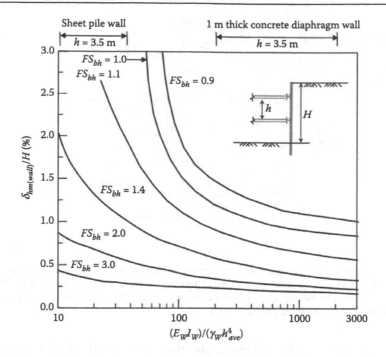

Figure 6.24 Relationship among maximum horizontal deflection of the wall, rigidity of the wall and FS_{bh}. (From Clough, G.W. et al. 1989. Movement control of excavation support systems by iterative design, Proceedings, ASCE, Foundation Engineering: Current Principles and Practices, Vol. 2, Figure 4, 874pp. Reproduced with permission of the ASCE.)

Clough et al. (1989) reported a set of curves, as shown in Figure 6.24, that relate the maximum horizontal wall deflection to the safety factor against basal heave (FS_{bh}) and wall stiffness. The maximum horizontal wall deflection is normalized with respect to the depth of excavation as $\delta_{hm(wall)}/H$. FS_{bh} is determined according to Equation 6.15. Also included in Figure 6.24 is wall stiffness, which is defined as $E_w I_w / \gamma_w h_{ave}^4$, where E_w = the elastic modulus of the wall material, I_w = moment of inertia of the wall, γ_w = unit weight of water, and h_{ave} = average support spacing of the lateral support elements (i.e., struts or ground anchors). According to Figure 6.24, when $FS_{bh} > 2$, $\delta_{hm(wall)}/H$ would be lower than 0.5% with reasonable wall stiffness. If we consider $\delta_{hm(wall)}$ as approximately the same as the maximum ground settlement δ_{vmax} adjacent to the excavation, then the $\delta_{hm(wall)}/H$ values shown in Figure 6.24 are consistent with those δ_{vmax}/H values for granular soils and very stiff-to-hard clays presented in Figure 6.21. $\delta_{hm(wall)}/H$ increases significantly as FS_{bh} becomes less than 1.5. This figure shows the importance of stability number N_s or safety factor against basal heave in affecting the ground settlement adjacent to excavation, which is also advocated in the Peck charts. Coupling Figures 6.23 and 6.24 provides a good way to estimate the distribution of ground settlement adjacent to an excavation, using the following procedure:

1. Establish the soil stratigraphy at the project site and determine the depth of excavation H and support system to be applied.
2. Determine the stability number N_s at the base of excavation and safety factor against basal heave FS_{bh}.
3. Compute the wall stiffness $E_w I_w / \gamma_w h_{ave}^4$.

4. Determine the $\delta_{hm(wall)}$ according to the soil conditions and the wall stiffness according to Figure 6.24.
5. Take δ_{max} as the same as $\delta_{hm(wall)}$ and estimate the settlement profile according to the charts presented in Figure 6.23 and soil conditions for the particular excavation.

EXAMPLE 6.4

Given

For the same soil conditions and braced excavation described in Examples 6.1 and 6.3, if the sheet piles used for the wall support have an elastic modulus (E_w) of 200 GPa, and moment of inertia (I_w) of $1.15 \times 10^{-4}\,\mathrm{m}^4/\mathrm{m}$. The safety factor against basal heave, $FS_{bh} = 1.67$ according to Example 6.3.

Required

Estimate the maximum horizontal deflection of the sheet pile wall [$\delta_{hm(wall)}$] using Figure 6.24. Consider $h_{ave} = 2.5$ m.

Solution

$$\text{The wall stiffness} = \frac{E_w I_w}{\gamma_w h_{ave}^4} = \frac{(200,000,000)(0.000115)}{(9.81)(2.5^4)} = 60.0$$

According to Figure 6.24, $\delta_{hm(wall)}/H \approx 0.7\%$ for $FS_{bh} = 1.67$ and wall stiffness of 60.0. The estimated $\delta_{hm(wall)} = (9)(0.007) = 0.063\,\mathrm{m} = 63\,\mathrm{mm}$.

6.7 REMARKS

This chapter deals with the braced excavation, a special class of earth-supporting system used to maintain the ground stability during excavation. With this system, the excavation wall is inserted first and lateral support elements are installed in stages as the excavation progresses. Throughout the process, wall movement is kept to a minimal. This includes installing lateral support at an early stage of excavation and prestressing the support system. For this reason, the lateral earth pressure experienced by the wall tends to be more concentrated toward the middle of the excavation wall.

Because of the significant differences in construction sequence and limitations in allowable wall movement, the active earth pressure described in Chapter 5 for retaining walls may not be appropriate for the design of braced excavation. The earth pressure diagram provided for the design of the braced excavation in this chapter was empirically developed based on field measurements. The apparent earth pressure diagrams described in this chapter should be treated as envelopes encompassing the maximum distributed pressures anticipated at any stage of construction for the given soil type. These diagrams are likely to produce conservative design loads.

The safety at the base of a braced excavation, from soil shear strength point of view, was discussed. The performance of a braced excavation, in terms of excavation-induced ground movement in the surrounding area, is highly dependent on the structural system used for the lateral support, and workmanship. A set of charts indicating the possible range of ground movement outside of the excavation wall was presented. These charts were developed based on the construction methods and workmanship available at the time they were prepared. It is conceivable that as we gain more experience and better construction methods become available, the ground movement associated with braced excavation will continue to improve.

HOMEWORK

6.1. Figure H6.1 shows the soil profile and the proposed cross-lot bracing scheme of a proposed 9 m-deep braced excavation. Soldier piles spaced at 2 m center-to-center will be used as the excavation wall. The struts will be spaced at 4 m center-to-center in the plan. For the steel of soldier piles, $\sigma_{all} = 170$ MPa.
   1. Determine and draw the apparent earth pressure diagram for the braced excavation.
   2. Determine the strut load at levels A, B, and C.
   3. Determine the required section modulus for the wale at levels B and C.
   4. Determine the required section modulus for the soldier pile.

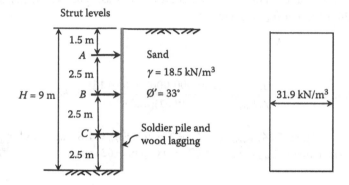

Figure H6.1 Soil profile and proposed cross-lot bracing scheme.

6.2. Refer to Figure H6.2; the proposed 7.5 m-deep braced excavation will be supported by a cross-lot bracing system. Width of the excavation $B = 10$, length of excavation $L = 20$ m. Struts will be spaced at 3 m center-to-center. Sheet piles will be used as the excavation wall. σ_{all} of the steel sheet pile $= 170$ MPa.
   1. Determine and draw the apparent earth pressure diagram for the braced excavation.
   2. Determine the strut load at levels A, B, and C.

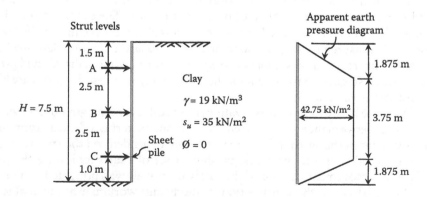

Figure H6.2 Soil profile and proposed braced excavation system.

3.   Determine the required section modulus for the wale at levels B and C.
4.   Determine the required section modulus per unit thickness of the sheet pile.
5.   Determine the safety factor against basal heave.
6.   Estimate $\delta_{hm(wall)}$; consider the sheet pile wall has an E_w of 200 MPa, $I_w = 1.3 \times 10^{-4} \, \text{m}^4/\text{m}$.

6.3.   For the soil conditions shown in Figure H6.3, determine the apparent earth pressure diagram for the layered clay.

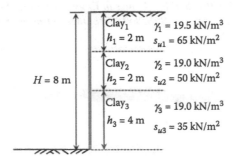

Figure H6.3 The soil profile.

6.4.   Determine the apparent earth pressure diagram for the sand/clay layer shown in Figure H6.4.

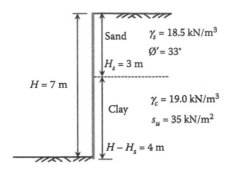

Figure H6.4 The sand/clay layered soil profile.

REFERENCES

Bjerrum, L. and Eide, O. 1956. Stability of strutted excavations in clay. *Geotechnique*, 6, 32–47.
Clough, G.W. and O'Rourke, T. 1990. Construction-induced movements of *in situ* walls. Design and Performance of Earth Retaining Structures. ASCE Special Publication, No. 25, pp. 439–470.
Clough, G.W., Smith, E.M., and Sweeney, B.P. 1989. Movement control of excavation support systems by iterative design. *Proceedings, ASCE, Foundation Engineering: Current Principles and Practices*, Evanston, IL, Vol. 2, pp. 869–884.
Goldberg, D.T., Jaworski, W.E., and Gordon, M.D. 1976. *Lateral Support Systems and Underpinning, FHWA-RD-75-128*. Federal Highway Administration Offices of Research and Development, Washington, DC.
Ou, C.Y. 2006. *Deep Excavation: Theory and Practice*. Taylor & Francis, London, 529pp.
Peck, R.B. 1943. Earth pressure measurements in open cuts, Chicago (Ill.) Subway, *Transactions*, ASCE, 108, 1008–1058.

Peck, R.B. 1969. Deep excavation and tunneling in soft ground. State-of-the-Art Report. *Proceedings of the 7th International Conference on Soil Mechanics and Foundation Engineering*, Mexico City, Mexico, pp. 225–325.

Reese, L.S., Wang, S.T., Arrellaga, J.A., and Vasquez, L. 2013. *PYWALL Technical Manual. A Program for the Analysis of Flexible Retaining Structures*. ENSOFT, Inc., Austin, TX, 137p.

Sabatini, P.J., Pass, D.G., and Bachus, R.C. 1999. *Geotechnical Engineering Circular No. 4. Ground Anchors and Anchored Systems*. Report No. FHWA-IF-99-015. Office of Bridge Technology, Federal Highway Administration, Washington, DC.

Terzaghi, K. and Peck, R.G. 1967. *Soil Mechanics in Engineering Practice*. Wiley, New York, NY.

Terzaghi, K., Peck, R.B., and Mesri, G. 1996. *Soil Mechanics in Engineering Practice*, Third Edition. Wiley, New York, NY.

Chapter 7

Static analysis and design of deep foundations

7.1 INTRODUCTION

There is no clear definition for a "deep" foundation. A shallow foundation under a deep, multiple level basement building structure may be deeper than a "deep" foundation below a structure without a basement. The foundation can be subject to axial compressive force, tensile force, lateral force, and bending moment as shown in Figure 7.1. The loadings applied to the foundation can be monotonic or cyclic and can be applied in multiple planes (e.g., in both east–west and north–south directions). Deep foundations are required when the subsurface conditions immediately below the structure or ground surface are not suitable for the support of the structure. Thus, the foundation needs to be deepened or extended to a lower soil or rock layer, usually at a higher cost. Subsurface and/or loading conditions that lead to the use of deep foundations can include:

- Inadequate strength or compressibility of the soil immediately below the structure or ground surface to support the loading conditions using a shallow foundation, as shown in Figure 7.1a.
- The soil immediately below the structure or current ground surface may be eroded by wind or water scouring during the life of the structure. These conditions can include structures to be built on sand dunes in a desert or river crossing bridges as shown in Figure 7.1b.
- Excessive movement of the soil immediately below the structure or ground surface could occur due to changes in moisture content (i.e., expansive or collapsible soil), or liquefiable under seismic loading conditions as shown in Figure 7.1c.
- For structures that impose large tensile force, lateral force, bending moment, or combinations of the above on the foundation, the deep foundation is likely to be more cost effective than a shallow foundation. Structures such as transmission towers, light poles, offshore oil rigs, wind turbines, or super high-rise buildings can involve such loading conditions.

As in the case of shallow foundations, the design or static analysis of deep foundations consider the safety and performance of the foundation. For shallow foundations, the construction usually involves excavation to the desired depth, placement of reinforcement, and pouring of concrete. For deep foundations, the construction is significantly more complicated than shallow foundations. A variety of materials can be used (i.e., timber, concrete, or steel). The construction can involve excavating a hole in the ground and pouring concrete (with or without reinforcement), or the deep foundation can be prefabricated and driven into the ground.

The construction of deep foundations is a brutal process that can alter the states of the surrounding material, making it impossible to design the deep foundation based on the projected or "correct" soil conditions with full consideration of the construction effects.

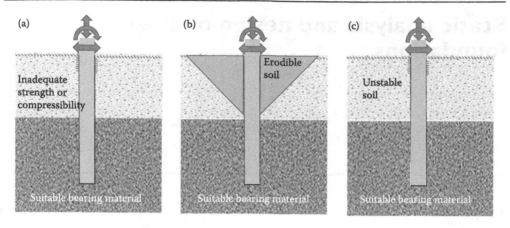

Figure 7.1 Conditions that lead to the use of deep foundations. (a) Inadequate soil for the applied load. (b) Erodible soil. (c) Unstable soil.

For these reasons, the static analysis is just the beginning. Construction issues should always be kept in mind when designing deep foundations. Instead of attempting to consider all minor details, it is preferable to concentrate on the key factors that control the behavior of a deep foundation. Assume that the foundation design has gone through the evaluation process such as that shown in Figure 4.1, and it has been decided that deep foundation is the most appropriate option.

The objectives of this chapter are to

- Introduce the commonly used types of deep foundations from the perspectives of material, geometry, load transfer mechanism, construction-induced ground displacement, and construction method (Section 7.2).
- Describe the methods for static analysis of axially loaded deep foundations and factors to be considered in the analysis (Sections 7.3 and 7.4).
- Describe the background and techniques involved in the analysis of laterally loaded piles (Section 7.5).
- Discuss the effects of installing a group of piles in close proximity in axial and lateral loading conditions (Section 7.6).
- Describe the phenomenon of negative skin friction and methods to predict the effects of negative skin friction (Section 7.7).

7.1.1 Terminology

Deep foundation or pile—these two terms will be used interchangeably in this book.
Head and toe of a pile—"head" refers to the upper end of a pile and "toe" is the lower end. This definition will be used throughout this book.
Splicing and cutting of piles—driven piles are mostly made in the factory; the pre-made pile sections are rarely exactly the length needed in the field. The connection or addition of a new pile section on top of the installed portion in the ground to increase the length is called splicing. Removing excess pile length above the ground surface is called cutting.

7.2 TYPES OF PILES

There are many types of piles and multiple ways to classify the piles. This section introduces some of the commonly used systems to classify the types of piles.

7.2.1 Classification of piles

The types or classification of pile foundations may consider the following factors:

Materials—commonly used materials include concrete, steel, timber, or composite with a combination of these materials. The concrete piles can be precast in factory, reinforced, and with or without pre-stressing, or the concrete piles can be cast in place, meaning to bore a hole in the ground and then fill it with concrete.

Geometry—straight or tapered in longitudinal direction. The pile can have a circular, square, rectangular, or polygonal cross-sectional area. Most of the steel piles are hollow, such as the steel pipe piles. The steel H-piles have an H-shaped cross-sectional area. The precast concrete piles can be hollow or solid.

Load transfer mechanism—how the load applied to the pile head is transferred to the ground. For example, a friction pile means most of the load is transferred to the surrounding soil through frictional force along the pile shaft. A toe bearing pile means the applied load is mostly supported by resistance at the pile toe. Of course, it is possible to have a pile with its applied load supported by both shaft friction and toe resistance with various proportions.

Ground displacement—the amount of ground material being pushed aside by the insertion of the pile. For example, a driven pile is expected to displace soil with the same volume as that to be occupied by the pile. A driven, solid concrete pile is a high-displacement pile. A steel H-pile with small cross-sectional area is a low-displacement pile. The amount of disturbance caused by pile driving is proportional to ground displacement. High-displacement pile installation in loose granular soil is likely to densify the surrounding soil. In dense granular soil or cohesive soil, high-displacement pile installation tends to cause ground heaving.

Method of installation—piles can be installed by driving with an impact hammer; we call these driven piles. Or the piles can be cast in place by first drilling a hole in the ground and then filling it with concrete; we called these bored piles. There are other names for bored piles; for example, drilled shafts, drilled piers, drilled caissons or caissons, and cast-in-drilled-hole piles. The term *bored pile* will be used throughout this book. There are many other pile installation methods but they are not elaborated on in this book.

A summary for some of the commonly used types of piles, grouped according to their material and construction methods, is shown in Table 7.1. It should be noted that the length and axial load capacity of the piles included in Table 7.1 are based on current experience. As the demand for mega structures increases and construction technology progresses, it is foreseeable that certain types of piles can become much bigger/longer and with significantly higher capacities. With the above background in mind, the following sections introduce a few commonly used types of piles.

7.2.1.1 Timber piles

A timber pile is a trimmed tree trunk with its branches and bark removed, driven into the ground with its narrow end (tip of tree trunk) as the pile toe, as shown in Figure 7.2. The natural taper of the tree trunk increases shaft friction of a timber pile and is recognized in the design, as will be described later. Timber is a renewable resource and plentiful in many parts of the world. Some of the timber piles installed well over 1000 years ago are still functioning in many historical cities in Europe. Because of their unique advantages, as shown in Table 7.1, timber piles remain a viable choice for deep foundations in the construction of industrial and residential buildings as well as infrastructures.

Table 7.1 Summary of various types of piles

Pile type	Timber pile	Steel H-pile	Steel pipe pile	Pre-cast pre-stressed Pile	Cast-in-place pile cased	Cast-in-place pile uncased
Typical length, m	6–23	6–30	9–36	9–15 (pre-cast) 15–36 (pre-stressed)	3–36	3–100
Typical design axial capacity, kN	156–670	450–1800	900–2200	450–1100	450–1300	450–45,000
Advantages	• Low cost • Renewable resource • Easy to handle/drive • Natural taper, higher shaft friction in granular soil than uniform piles	• Small displacement • High-load capacity • Easy to splice	• High-load capacity • Small displacement when open ended • Easy to splice	• High-load capacity • Corrosion resistance obtainable	• Can be driven with or without mandrel • Can be inspected after casing driving • Tapered section provides higher shaft friction in granular soil than uniform piles	• Significantly less noise and vibration than driven piles • Very large load can be carried by a single pile • Applicable to a wide variety of soil conditions • Changes in pile geometry possible during the progress of construction if ground conditions so dictate.
Disadvantages	• Low-axial capacity • Difficult to splice	• Vulnerable to corrosion • Not efficient as friction pile	• Open ended not efficient as friction pile	• Vulnerable to handling/driving damage	• Thin shells (mandrel driven) vulnerable to damage during driving • Displacement can be high	• Critical to construction procedure and workmanship

Source: NAVFAC, 1986. Foundations and Earth Structures, Design Manual DM 7.02. Department of the Navy, Naval Facilities Engineering Command, Alexandria, VA, 279p.

Pile head (butt of tree trunk)

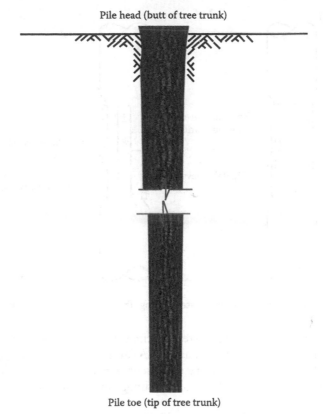

Pile toe (tip of tree trunk)

Figure 7.2 Schematic view of a timer pile.

Douglas fir and Southern yellow pine are the two major species used for timber piles in North America. The natural taper of Southern pine is approximately 0.008 mm/mm throughout the length. Douglas fir has slightly less taper than Southern pine. In some countries of Southeast Asia, small diameter timber (Bakau piles) or bamboo piles are used for support of excavations or road embankments. Table 7.2 shows the allowable pile capacity in compression for timber piles of Southern pine and Douglas fir. Specifications on the minimum requirements in toe and head circumferences for Southern pine and Douglas fir timber piles of various lengths can be found in ASTM Standard D 25.

Timber piles are susceptible to biological attack from fungi, marine borers, and insects. Pressure treatment of timber piles has proven to be an effective means of protection from biological attack. There are two broad types of wood preservatives used in pressure

Table 7.2 Allowable pile capacity in compression

| Timber species | Pile toe diameter, mm | | | | | |
	177.8	203.2	228.6	254.0	279.4	304.8
Allowable pile capacity, kN						
Southern pine	204.6	266.9	338.1	418.1	507.1	604.9
Douglas fir	213.5	280.2	355.8	435.9	529.3	627.2

Source: Collin, J.G. 2002. *Timber Pile Design and Construction Manual*. Timber Piling Council, American Wood Preservers Institute, Chicago, Illinois, USA, 145p.

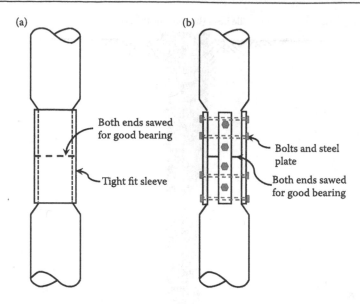

Figure 7.3 Splicing of a timber pile. (a) With steel casing. (b) With steel plates and bolts.

treatment for timber piles: oil-borne systems (primarily creosote) and waterborne preservative systems [chromated copper arsenate (CCA) and ammoniacal copper zinc arsenate (ACZA)]. The effectiveness of pressure treatment depends on the environment of the installed timber piles. Treated timber installed under favorable conditions can last over 100 years.

Timber piles are usually installed with a drop hammer. A drop hammer is the simplest form of an impact hammer for pile driving. A ram is lifted and dropped on to the pile head by using a hoisting device. Details of pile driving are described in Chapter 8.

A usual method for splicing a timber pile is by bolting two pieces of timber together with the help of steel plates or steel casing as shown in Figure 7.3. Cutting of excess timber pile length can easily be done with a saw.

7.2.1.2 Steel H-piles

The steel H-piles, with their general geometry described in Figure 7.4, are usually made with ASTM A572, A 588, or A 690, Grade 50 steel. Similar steel grades are available in Canada (CSA G40.21) and Europe (EN 10034). The yield strengths of the steel range from 300 to 500 MPa. A wide range of pile sizes is available, with different grades of steel. Table 7.3 shows the technical data for selected H-piles available in the US.

Steel H-piles typify what a low-displacement pile is. Despite their formidable capacity, steel H-piles have a relatively small cross-sectional area and are lightweight. Steel H-piles are most effective as toe bearing piles founded in hard soil or rock. Because of its low displacement, H-pile has limited compaction effects on the surrounding material if driven in loose granular soils. Thus H-piles are usually not effective as friction piles in granular soils.

A plug of soil may be formed within the flanges of H-piles. For assessment of bearing capacity, it is important to check by calculation or load testing that there is adequate shaft frictional resistance of the soil located within the flanges of the H-pile. In dense granular soil that is disturbed by pile driving, the shaft friction may not achieve the required resistance. Therefore the toe bearing of H-piles in granular soils or rocks is often calculated using only the net cross-sectional area of the steel. In granular soils, the shaft friction

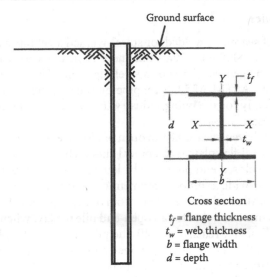

Figure 7.4 Schematic view of a steel H-pile.

on H-piles can be calculated along the entire steel surface (i.e., coating area in Table 7.3), provided that the length of soil plug after driving can be verified.

Compared to concrete piles, steel H-piles generally have better driving characteristics and can be installed to great depths. H-piles can be susceptible to deflection upon striking boulders, obstructions, or an inclined rock surface, resulting in bending on the weak axis with considerable curvature. Heavy section H-piles with appropriate toe strengthening are commonly used to penetrate and to withstand hard driving.

Steel H-piles can be easily cut off or spliced by welding. If the pile is driven into a corrosive environment, such as soil with low pH, coal-tar epoxy or cathodic protection can be applied. It is common to allow for an amount of corrosion in design by simply over-dimensioning the cross-sectional area of the steel pile.

Table 7.3 Specifications of selected steel H-piles

Section	Weight kg/m	Area cm²	d mm	b mm	t_f mm	t_w mm	Coating area m²/m	I_{xx} cm⁴	I_{yy} cm⁴
HP 200	54	684	204	207	11.3	11.3	1.19	4953	1677
HP 250	63	80.0	246	257	10.7	10.5	1.47	8741	2984
	85	108	254	259	14.4	14.4	1.50	12,237	4204
HP 310	79	100	300	305	11.0	11.0	1.77	16,358	5286
	110	141	307	310	15.5	15.4	1.80	23,683	7742
	132	167	314	313	18.3	18.3	1.84	28,700	9370
	174	222	324	327	23.6	23.6	1.91	39,400	13,800
HP 360	109	138	345	371	12.8	12.8	2.12	30,343	10,864
	152	194	356	376	17.9	17.9	2.15	43,704	15,817
HP 410	131	167	389	399	13.7	13.7	2.29	46,201	14,526
	180	231	401	404	19.1	19.1	2.32	66,180	20,978
	241	308	414	409	25.4	25.4	2.36	91,154	29,011
HP 460	201	257	445	452	19.1	19.1	2.60	91,570	29,386
	269	343	457	457	25.4	25.4	2.64	125,701	40,541

I_{xx} = moment of inertia with respect to x axis.
I_{yy} = moment of inertia with respect to y axis.

7.2.1.3 Steel pipe piles

Pipe piles are made of seamless, welded, or spiral-welded steel pipes in diameters ranging from 200 to well over 1000 mm. Wall thickness typically varies from 4.5 to 50 mm. Figure 7.5 describes the general shape of a steel pipe pile. The material should meet the grade with reference to ASTM A252. Steel pipe piles can either be made with new steel, manufactured specifically for the piling industry, or of reclaimed steel casing used previously for other purposes.

The pipe pile can be driven with either an open or closed end. For closed end pile, the pile toe is closed with a metal flat plate or a conical shoe. The closed end pipe pile is a high-displacement pile. The closed end pipe may be left open or filled with concrete from pile head after driving to provide additional moment capacity or corrosion resistance. When driven open end, soil is allowed to enter the bottom of the pipe and form a plug. Paikowsky and Whitman (1990) indicated that for an open-end pile in clay, when the ratio of penetration depth over pile diameter reached 10–20, the pile became fully plugged (no more soil can enter the interior of the pile). Plugging of open end piles in clay does not contribute significantly to the capacity of the pile. For piles in sand, the fully plugged pile behaves almost identically to closed end pile under static load. The toe bearing capacity of a fully plugged or closed end pile in sand is significantly higher than an open-end pile. The plugging of piles in sand is, however, complicated by the arching effects, and is difficult to predict.

If an empty pipe is required, a jet of water or an auger can be used to remove the soil from inside following the driving. The structural capacity of pipe piles is calculated based on steel strength and concrete strength (if filled). The strength of concrete is significantly higher when confined in steel pipe. For closed end pipe piles, it is possible to drive the pile from the bottom, by impacting the end plate. Steel pipe piles are suitable for use as toe bearing and friction piles.

Similar schemes in pile splicing, cutting, and corrosion protection as in steel H-piles can be applied to steel pipe piles. If a concrete-filled pipe pile is corroded, the load-carrying capacity of the concrete remains intact.

7.2.1.4 Precast concrete piles

Precast concrete piles are typically made with steel reinforcement and pre-stressing tendons to obtain the tensile strength required to survive handling and driving, and to

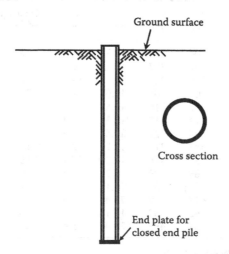

Figure 7.5 Schematic view of a steel pipe pile.

Figure 7.6 A spun-cast, pre-tensioned hollow cylinder concrete pile being tested for integrity. (Courtesy of DECL, Taipei, Taiwan.)

provide sufficient bending resistance. The piles can be made with a constant or tapered cross-section. A wide variety of sizes and geometry in cross-section and length can be accommodated. The cross-section can be circular, square, or polygonal. Long concrete piles can be difficult to handle and transport. Splicing of precast concrete pile sections requires special arrangements.

Precast concrete piles are routinely used in Asian countries for economic reasons. A popular type of precast concrete pile in Asia is made with a spun centrifugal casting process. The piles are cylindrical and hollow, with high-concrete density due to the centrifugal casting process. Typical spun-cast reinforced concrete pile sections are pre-tensioned during casting. The length of each precast pile section is usually less than 12 m. With their hollow cross-section and limited length, the pile sections can be transported without extreme measures. The ends of the precast pile section are equipped with steel rings which are also connected to the pre-stressing tendons within the pile section. Figure 7.6 shows a spun-cast concrete pile driven in the ground with its steel ring exposed. Splicing or lengthening of the spun-cast concrete piles is done by first aligning the steel rings from the two pile sections to be joined and then welding the rings. A jackhammer and torch are used to break the concrete and cut the tendons to remove the excess pile length.

Pre-stressed concrete piles are vulnerable to damage from striking through hard soil layers or obstacles. This is due to the decrease in axial compression capacity due to pre-stressing force. When driven in soft soils, care must be exercised since a great deal of tension can be generated in easy driving. The spun-cast, hollow cylinder concrete pile can be driven with its toe section open or closed. A pile shoe is attached to the toe for closed end driving. For closed end piles, the hollow space within the pile can be filled with reinforced concrete to enhance its capability to resist axial force and bending, or it can be used to install instrumentation such as telltales, inclinometer casings, or strain gages (described in Chapter 8).

7.2.1.5 Cast-in-place concrete piles

Cast-in-place piles are made by placing concrete into a borehole in the ground. The borehole can be created by either driving a casing in the ground or by drilling without the help of a casing. The steel casing can be left in place or removed after the concrete is placed.

In either case, as the concrete piles are cast in place, predetermination of the pile length is not as critical as the precast concrete piles.

7.2.1.6 Cast-in-place piles by cased method

The casing can be driven with or without a mandrel. A mandrel is a tubular steel section inserted into the casing that facilitates hard driving of relatively thin casings. After driving, the mandrel is removed. Casings driven without the mandrel have thickness in the range of 3 to 60 mm, similar to the casing thickness of steel pipe piles. The casing thickness can be reduced when using the mandrel. The mandrel-driven casings or shells are often corrugated circumferentially, as shown in Figure 7.7, to enhance their frictional characteristics and capability to prevent collapse after removal of mandrel. The sides of the casing can be straight with constant diameter throughout the pile, steadily tapered, or step tapered from head to toe.

When a mandrel is not used, driving of the casing is essentially the same as that of a steel pipe pile. The pipe pile can be driven open or closed ended. The pipe that is driven open ended may require internal cleaning to the depth where concrete is to be placed. The Monotube pile is a proprietary pile, driven without a mandrel. Monotubes are longitudinally fluted and tapered.

The cased borehole can be inspected after driving, before concrete placement. Reinforcement can be placed according to the design as per loading conditions.

7.2.1.7 Cast-in-place piles by uncased method

Bored piles are the most widely used type of cast-in-place piles by uncased method. A bored pile is constructed by drilling a hole in the ground to the desired bearing stratum and then filling the hole with concrete with or without reinforcement. The diameter and length of the bored pile can vary widely. Very large axial as well as lateral load can be resisted by a single bored pile, therefore making a pile cap unnecessary in most cases. Figure 7.8 shows the schematic view of a typical bored pile. The drilling can involve the use of casing to prevent collapse of the borehole. In this case, however, no driving is involved in the insertion of casings. The noise and vibration associated with bored piles are significantly less than with driven piles. The dimensions of the pile can readily be adjusted during the progress of construction to accommodate the ground conditions

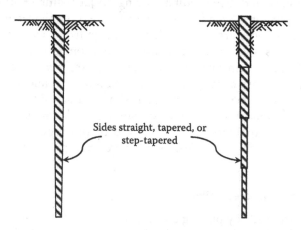

Sides straight, tapered, or step-tapered

Figure 7.7 Cast-in-place pile with mandrel-driven shells. Pile diameter 200–450 mm, corrugated shell thickness 0.5–3.3 mm.

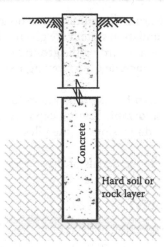

Figure 7.8 Schematic view of a bored pile.

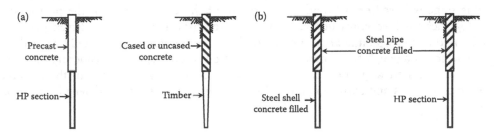

Figure 7.9 Examples of composite piles. (a) Concrete over HP section or timber. (b) Concrete filled steel pipe over concrete filled steel shell or HP section.

actually encountered in the field. Splicing and cutting of bored piles is not an issue. A properly constructed bored pile does not cause any ground displacement, thus is limited in ground heaving or settlement. For these reasons, bored piles are used almost exclusively as deep foundations in densely populated municipal areas. Bored piles represent an important class of piles. Details of their construction and testing are described in Chapter 8.

7.2.1.8 Composite piles

A review of the above descriptions of different types of piles shows that they all have advantages and disadvantages. There is no reason to limit ourselves to just one type of material or construction method in a pile. A composite pile is made up of two or more sections of different materials and/or installed by different methods. The combinations are selected to optimize the economy and performance of the piles for the given ground and loading conditions. Figure 7.9 shows a few examples of composite piles. The use of concrete for the upper section of these piles is to make the piles more resistant to corrosion and still maintain the load-carrying capacities. There is plenty of room for innovations in the design of composite piles.

7.3 GENERAL CONSIDERATIONS IN STATIC ANALYSIS

We get into the static analysis after going through the procedure outlined in Chapter 4 (i.e., Figure 4.1) and decide that a deep foundation system is the most desirable for the proposed structure and existing ground conditions. At this stage, the loading conditions to be imposed on the proposed foundation system are given. The subsurface exploration has provided the required information regarding the ground conditions. The main task in the next stage of static analysis may include the following items:

1. Select the optimal type of deep foundation system for the given conditions.
2. Estimate the ultimate and allowable bearing capacity of the piles via static analysis.
3. Determine the number and dimensions of the piles.

The following sections provide the details of these items.

7.3.1 Selection of pile type(s)

There is no clear procedure for selection of the type of deep foundation. What we have is a process of considering the pros and cons for different types of piles for the given conditions and then we narrow down the final selection in steps. Table 7.4 provides a comparison between driven and bored piles, two major groups of deep foundations. The selection at this stage is based on a broad consideration of the environment of the project site and logistics for field construction, in addition to costs. For example, driven piles are rarely used in metropolitan areas because of the noise and vibration created by pile driving. The next stage involves a more refined selection of pile type(s), considering the technical issues related to ground conditions. Table 7.5 shows a summary of the possible ground conditions and corresponding considerations in the selection of deep foundation type(s).

Table 7.4 Comparison between driven and bored piles

Item	Corresponding considerations in the selection of pile system
Characteristics of the pile system	**Driven piles** Often used in groups to resist the axial compression/uplift and lateral load. A pile cap is used to integrate the structural capacities of the piles within a group. **Bored piles** With relatively large capacities in resisting axial compression/uplift and lateral load, the pile cap can sometimes be eliminated by using bored piles as a column extension.
Noise and vibration, waste (soil) disposal – environmental constraints	**Driven piles** Excessive noise and/or vibration can be caused by pile driving. **Bored piles** Noise and vibration during bored pile construction can be significantly lower than driven piles. The excavated soil or drilling fluid from bored pile construction has to be properly disposed.
Transportation of equipment and material	**Driven Piles** Deliver pile sections through congested streets can be difficult. **Bored piles** Pre-mix concrete available locally or need to establish a pre-mix concrete plant on site.
Local expertise and equipment availability	Availability of experienced field crew and equipment can vary for both types of piles in different parts of the world.

Table 7.5 Considerations in the selection of deep foundation in response to ground conditions

Field conditions	Corresponding considerations
Boulders overlying bearing stratum	Use low-displacement piles, such as H-piles or bored piles. In the case of bored piles, costs associated with removal of boulders from borehole should be expected.
Coarse gravel deposits	The pile should be able to resist hard driving.
Loose cohesionless soil	Use tapered pile to enhance shaft friction.
Negative shaft friction	Avoid use of battered piles. Use piles with smooth surface or bituminous coating to reduce/minimize adhesion between pile shaft and surrounding soil.
Soft clay layer	Squeezing of soft clay during borehole drilling can result in settlement in the surrounding area, if bore piles are used. Should consider the use of casing or drill mud in the construction of bored piles.
Artesian pressure	Use solid or closed end pile. Avoid mandrel-driven piles with thin wall shells. The artesian pressure may collapse the shell upon removal of the mandrel.
Scour	The pile should have sufficient structural strength to act as a column through scour zone. The pile should be deep enough so that the part below the scour zone is sufficient to develop the required bearing capacity.
Corrosion	Choose the pile material or apply necessary treatment to resist the corrosion.

Source: After Cheney, R.S. and Chassie, R.G. 1993. *Soils and Foundations Workshop Manual.* Second edition, FHWA HI-88-009, Federal Highway Administration, Washington, DC.

It is possible that after going through the selection process outlined above, more than one type of pile may appear to be suitable from technical point of view. With these selections in mind, we proceed to the next stage of static analysis to determine the number and dimensions of the selected pile type(s). The final selection of the pile type, its dimensions, and number to be used for the particular project may require iterations of the selection and static analysis processes.

The construction of deep foundations can cause severe disturbance to the surrounding soil, as described earlier. Also, because of its significant embedment in soil, a typical pile foundation can develop its load bearing capacity through surface friction on the pile shaft against the surrounding soil as well as resistance at the pile toe. How the load is transferred from the pile to the surrounding soil is thus more sophisticated than with shallow foundations. For these reasons, the following two sections discuss issues related to the construction effects of piles and the background of pile load transfer mechanisms.

7.3.2 Construction effects

For driven piles, substantial soil disturbance and remolding are unavoidable. The disturbance comes mainly from vibration and displacement of soil caused by the insertion of a pile. In loose or medium dense granular soils, the pile driving can densify the soils in the vicinity of the pile. In this case, the relative density of the affected granular soil increases. Figure 7.10 shows the zone of densification and ground settlement according to studies by Broms (1966). The granular soil within a distance of up to 5.5 times the pile diameter, D_p, can be affected by the pile driving. The increase of density coupled with an increase of horizontal stress enhances the capacity of the pile. Piles with larger displacement are more effective in densifying the surrounding soil.

In dense granular soil, the pile driving causes dilation in the surrounding soil mass that loosens the soil deposit or induces negative pore water pressure if the soil is below

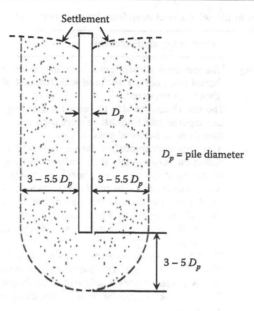

Figure 7.10 Compaction of cohesionless soils during pile driving. (Adapted from Broms, B.B. 1966. Sols-Sols, (18–19), 21–32.)

groundwater table. The increase in horizontal stress, which occurs adjacent to the pile during driving, can be lost by relaxation as the negative pre-water pressure dissipates. The end result is that the capacity of the pile remains the same or decreases.

When piles are driven into saturated cohesive materials, the soil near the pile is disturbed and displaced radially under an undrained condition. The influenced zones are generally within one pile diameter from the pile surface, as shown in Figure 7.11. In this undrained pile-driving process, saturated cohesive soil deforms under a constant

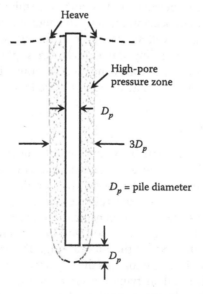

Figure 7.11 Disturbance of cohesive soils during pile driving. (Adapted from Broms, B.B. 1966. Sols-Sols, (18–19), 21–32.)

volume. Thus for high-displacement piles, the pile driving can also cause ground heaving. The disturbance and radial displacement generate positive pore pressure which temporarily weakens the soil and therefore the load capacity of the pile. Upon pile driving, the excess pre-water pressure dissipates and the cohesive soil around the pile consolidates, which leads to an increase in shear strength and thus higher pile capacity. This phenomenon is called "setup." The development of positive excess pore water pressure and the following setup are more significant in soft, normally consolidated clays. For pile driven in stiff or overconsolidated clays, the setup is less significant.

7.3.3 Load transfer mechanism

The load applied at the pile head, Q_h, is transferred to the ground through frictional resistance along the pile shaft, R_s, and toe resistance, R_t, as shown in Figure 7.12. Thus,

$$Q_h = R_t + R_s \tag{7.1}$$

The relationship between elastic deformation of the pile and Q_h applied to the pile head depends on how the load is transferred (i.e., proportions between R_s and R_t). This relationship is an important basis for the interpretation of axial pile load test described in Chapter 8. Based on Hooke's law, the elastic displacement, δ, of a pile with uniform cross-sectional area, A_p, relates to the axial stress passing internally through the pile as

$$\delta = \int_0^{L_p} \varepsilon(z) dz = \int_0^{L_p} \frac{\sigma(z)}{E_p} dz = \int_0^{L_p} \frac{Q_i(z)}{A_p E_p} dz \tag{7.2}$$

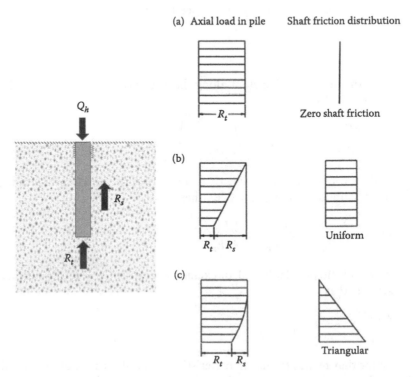

Figure 7.12 Typical load transfer profiles. (a) Zero shaft friction. (b) Uniform shaft friction. (c) Triangular distribution of shaft friction.

where

L_p = pile length
A_p = cross sectional area of the pile
$\varepsilon(z)$ = axial strain at depth z
$\sigma(z)$ = axial stress at depth z
$Q_i(z)$ = axial force passing through the pile internally at depth z
E_p = elastic modulus of the pile material

and

$$Q_i(z) = Q_b - \int_0^z \pi D_p f_{sp}(z) dz \qquad (7.3)$$

where

D_p = diameter of the pile
$f_{sp}(z)$ = shaft friction force per unit area at depth z

Replacing Equation (7.3) into Equation (7.2),

$$\delta = \frac{Q_b L_p}{A_p E_p} - \frac{1}{A_p E_p} \iint_0^{L_p} \pi D_p f_{sp}(z) dz^2 \qquad (7.4)$$

Consider the following three cases:

Case A: If shaft friction is zero and $f_{sp}(z) = 0$ (see Figure 7.12a), then

$$\delta = \frac{Q_b L_p}{A_p E_p} - 0 \qquad (7.5)$$

Case B: If the shaft friction is a constant along the pile and $f_{sp}(z) = \overline{f_{sp}}$ (i.e., uniform distribution in Figure 7.12b), then

$$\delta = \frac{Q_b L_p}{A_p E_p} - \frac{\pi D_p \overline{f_{sp}} L_p^2}{2 A_p E_p} = \frac{Q_b L_p}{A_p E_p} - \frac{R_s L_p}{2 A_p E_p} \qquad (7.6)$$

Case C: If the shaft friction increases linearly at a rate of r_s per unit depth, and $f_{sp}(z) = r_s z$ (i.e., triangular distribution in Figure 7.12c), then

$$\delta = \frac{Q_b L_p}{A_p E_p} - \frac{\pi D_p r_s L_p^3}{6 A_p E_p} = \frac{Q_b L_p}{A_p E_p} - \frac{R_s L_p}{3 A_p E_p} \qquad (7.7)$$

Consider a fully frictional pile, $R_t = 0$, or exclude R_t and set $Q_b = R_s$, Equations 7.6 and 7.7 can be combined as

$$\delta = C \frac{Q_b L_p}{A_p E_p} \qquad (7.8)$$

C is a constant that relates to the characteristics of the shaft friction distribution. For Case B, $C = 1/2$, and $C = 2/3$ for Case C.

Consider a fully end bearing pile where $R_s = 0$ (i.e., Case A), therefore $Q_b = R_t$ and $C = 1$ according to Equation 7.8.

The C parameter provides a simple but important index that relates pile elastic displacement to shaft friction distribution. The techniques of measuring the elastic displacement in a pile using telltales or strain gages and applications of their measurements in the interpretation of pile load tests are described in Chapter 8.

The objectives of static analysis are to

- Determine the ultimate axial compression or uplift capacity of a pile or pile group. The dimensions of the pile including its diameter and depth of penetration are estimated as a basis for budgeting the cost of foundation construction. The allowable capacity is then given by dividing the ultimate capacity with the factor of safety.
- If necessary, predict the performance of a pile or pile group under lateral loading conditions to ascertain the structural design of the pile or pile group can resist the applied lateral load and the induced deflections are acceptable to the structure.
- Estimate the settlement of a pile group to ensure the long-term deformation of the pile group under the sustained applied load is acceptable.

The static analysis does not deal with the construction of the selected pile type. Issues related to pile construction and quality assurance (i.e., field testing) are covered in Chapter 8. It should be emphasized that the design of a deep foundation is not complete without the consideration of construction and testing.

The remainder of this chapter introduces the procedures to fulfill the purposes of static analysis following the sequence described earlier. Pay attention to the difference between force and stress; to avoid mistakes, force will always be represented by capital letters and stress by lower case letters.

7.4 ULTIMATE BEARING CAPACITY OF SINGLE PILES UNDER AXIAL LOAD

A pile foundation fails when the applied axial load exceeds the structural capacity of the pile or exceeds the ultimate bearing capacity of the ground material surrounding the pile. The following discussion will concentrate on the ultimate bearing capacity of the ground material, or the geotechnical bearing capacity. Equation 7.1 states that any axial load applied at the pile head, Q_h, is transferred to the ground through frictional resistance along the pile shaft, Q_s, and toe resistance, Q_t. The ultimate bearing capacity of a single pile, Q_u, is the sum of frictional resistance and toe resistance when both have reached their respective maximum value, and

$$Q_u = R_{sf} + R_{tf} \tag{7.9}$$

where
R_{sf} = ultimate shaft friction resistance
R_{tf} = ultimate toe resistance

and

$$R_{tf} = A_t q_{tf} \tag{7.10}$$

$$R_{sf} = A_s f_{sf} \tag{7.11}$$

where
A_t = cross-sectional area of the pile toe
A_s = surface area of the embedded pile shaft

f_{sf} = average ultimate unit shaft friction (resistance per unit area) along the pile

q_{tf} = ultimate unit toe bearing capacity

The above equations assume that R_{sf} and R_{tf} can be determined independently and the two parameters do not interfere with each other. It is also assumed that the pile toe and shaft have moved sufficiently against the surrounding soil/rock to develop the ultimate shaft and toe resistance. In general, the relative displacement between the pile surface and adjacent soil needed to mobilize the ultimate unit shaft friction, f_{sf}, is approximately 5 to 10 mm and is independent from pile diameter (O'Neill, 2001). The R_{tf} is reached when the pile toe settles by approximately 5% of the pile diameter (O'Neill, 2001). Thus the direct addition of side and toe resistance to determine the total compressive resistance of the pile is a matter that requires engineering judgment. This is especially true when the pile diameter is large. In practice, this issue is best resolved by performing the site-specific load tests on instrumented piles to discriminate between shaft and toe resistance. Details of the pile load test and how we can distinguish shaft and toe resistance from pile load test results are described in Chapter 8.

As previously described, the construction of piles can cause severe disturbance to the surrounding soil. The process and its effects are complicated so that it is not possible or practical to design a pile with a rigorous procedure. Figure 7.13 shows a flow chart described by O'Neill (2001) regarding the development of foundation design methods that is also applicable to pile designs. The design methods have largely been developed through laboratory model tests, field pile load tests, theoretical analyses, experience, and judgment. Meyerhof (1976) also stated that "On account of the complex interaction between the soil and piles during and after construction, the behavior of single piles and groups under load can only roughly be estimated from soil tests and semi-empirical methods of analysis based on the results of pile load tests." It is not surprising that the available design methods are at least semi-empirical or entirely empirical, and these design methods are revised from time to time as we learn more from further observations and accumulate experiences.

There are basically two types of approach in estimating the ultimate unit toe bearing capacity and shaft friction. The first approach is semi-empirical that considers the field stress conditions and soil strength parameters. The second approach is entirely empirical, based on in-situ soil testing such as standard penetration or cone penetration test results.

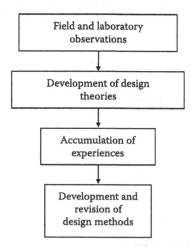

Figure 7.13 Flow chart in the development of design methods in foundation engineering.

The following sections describe how we estimate ultimate unit toe resistance, q_{tf}, and average ultimate unit shaft friction, f_{sf}.

7.4.1 Ultimate toe bearing capacity based on strength parameters

Meyerhof (1976) proposed a method that is similar to the bearing capacity equation for shallow foundations.

7.4.1.1 For piles in granular soils

The analysis should consider drained conditions and be based on effective stress. Following the bearing capacity approach for shallow foundations, the unit toe bearing capacity, q_{tf}, is expressed as

$$q_{tf} = c'N_c^* + q'N_q^* + \gamma^* D_p N_\gamma^* \qquad (7.12)$$

where
c' = drained cohesion
q' = effective overburden stress
γ^* = unit weight of soil (buoyant unit weight if below groundwater table)
D_p = diameter of the pile
N_c^*, N_q^*, and N_γ^* = modified bearing capacity factors

For piles, $\gamma^* D_p N_\gamma^*$ is relatively small and is neglected. For granular soils, $c' = 0$, thus Equation 7.12 is reduced to

$$q_{tf} = q'N_q^* \leq q_l \qquad (7.13)$$

N_q^* is estimated based on the drained friction angle, $\emptyset'$, of the bearing stratum before pile driving. q_l is the limiting value of unit toe bearing capacity when the ratio of pile depth in the bearing stratum over pile diameter (D_b/D_p) exceeds a critical value, $(D_b/D_p)_{cr}$ and

$$q_l = 0.5 \, p_a N_q^* \tan \emptyset' \qquad (7.14)$$

where
p_a = atmospheric pressure

Large-scale experiments and field observations have shown that in homogeneous sand, both q_{tf} and f_{sf} increase with depth, up to a critical depth. Below the critical depth, q_{tf} and f_{sf} become practically constant due to effects of soil compressibility, crushing, and arching. The trend of q_{tf} variation with depth of embedment, D_b, in a bearing stratum is qualitatively described in Figure 7.14. In a uniform bearing stratum, D_b starts from the ground surface as shown in Figure 7.14a. If the dense sand bearing stratum is overlain by a weak soil deposit, then D_b should start from the interface between the overlying weak deposit and the bearing stratum as shown in Figure 7.14b.

Figure 7.15 shows the correlation between N_q^*, and $\emptyset'$ with various depth ratios (D_b/D_p) according to Meyerhof (1976) for driven piles. Also included in Figure 7.15 is the correlation between $\emptyset'$ and $(D_b/D_p)_{cr}$.

Because of the significantly reduced displacement and compaction effects induced by pile installation, for bored piles, 1/3 to 2/3 of the N_q^* value shown in Figure 7.15 should be used for a given initial $\emptyset'$.

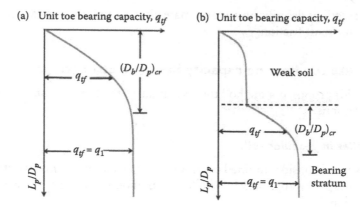

Figure 7.14 Variation of unit toe bearing capacity with embedment depth in bearing stratum. (a) Uniform bearing stratum. (b) Weak soil over dense sand.

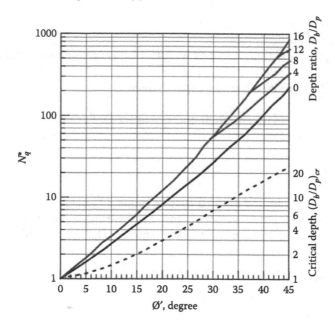

Figure 7.15 Correlation between N_q^*, critical depth and drained friction angle, $\varnothing'$. (Meyerhof, G.G. 1976. Bearing capacity and settlement of the pile foundations. *Journal of the Geotechnical Engineering Division, ASCE*, 102(GT3), Figure 1, p. 199. Reproduced with permission of the ASCE.)

EXAMPLE 7.1

Given

A closed end precast concrete pile with a diameter, D_p, of 0.5 m is driven into a uniform sand deposit. The sand has a saturated unit weight, γ_{sat}, of 17 kN/m^3 and a drained friction angle $\varnothing'$ of 35°. Groundwater table is at the ground surface.

Required

Determine q_{tf} for the pile when the toe is driven to 4 m, 5 m, and 6 m below ground surface.

Solution

$\varnothing' = 35°$, and depth ratio D_b/D_p larger than 4, $N_q^* = 143$ according to Figure 7.15.

$q_l = 0.5\, p_a N_q^* \tan \varnothing' = (0.5)(100)(143)(\tan 35°) = 5006$ kPa

$\gamma_{sat} = 17$ kN/m^3

At 4 m depth: $q_{tf} = q' N_q^* = (17 - 9.81)(4)(143) = 4113$ kPa < 5006 kPa

At 5 m depth: $q_{tf} = q' N_q^* = (17 - 9.81)(5)(143) = 5141$ kPa > 5006 kPa

At 6 m depth: $q_{tf} = q' N_q^* = (17 - 9.81)(6)(143) = 6169$ kPa > 5006 kPa

Thus $q_{tf} = 4113$ kPa at 4 m depth, and $q_{tf} = 5006$ kPa when the depth of pile toe exceeds 5 m.

7.4.1.2 For piles in cohesive soils

The analysis should consider undrained conditions ($\varnothing = 0$) and based on total stress, and

$$q_{tf} = s_u N_c^* + q N_q^* + \gamma D_p N_\gamma^* \tag{7.15}$$

where

s_u = undrained shear strength of the bearing stratum before pile driving

q = total overburden stress

γ = total unit weight of the soil

When $\varnothing = 0$, $N_q^* = 1$, $N_\gamma^* = 0$ and $N_c^* = 9$, thus Equation 7.15 becomes

$$q_{tf} = s_u 9 + q \tag{7.16}$$

The value of q is small compared to q_{tf} and thus can be neglected, or q_{tf} can be considered as the net unit toe bearing pressure (ultimate bearing pressure above the original overburden stress). The equation can be used to determine q_{tf} for both driven and bored piles.

7.4.2 Ultimate shaft friction based on strength parameters

The ultimate shaft friction is controlled by the shear strength of the soil at or near the surface of the pile shaft, which in turn is controlled by the effective normal stress, σ_h', along that surface at the time of loading as shown in Figure 7.16. The pile-driving process disturbs the soil around the pile. For piles driven in saturated clays, the soil goes through consolidation after pile driving and the driving-induced excess pore water pressure is dissipated after consolidation. This complicated process makes σ_h' rather difficult to estimate. As σ_h' is also closely related to σ_{vo}', for further discussion, the coefficient of earth pressure, $K = \sigma_h'/\sigma_{vo}'$, is used to reflect the effects of pile construction on stress states.

7.4.2.1 For piles in granular soils

The ultimate unit shaft friction along the pile in homogeneous sand may be expressed by the following equation:

$$f_{sf} = K \sigma_{vo}' \tan \delta' \le f_l \tag{7.17}$$

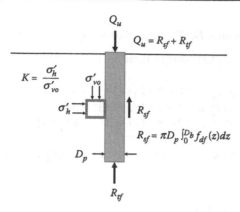

Figure 7.16 Stresses and forces acting on an axially loaded pile.

where

K = coefficient of earth pressure on pile shaft

σ'_{vo} = effective overburden stress along the pile shaft

δ' = drained friction angle between soil and the pile shaft surface

f_l = limiting value of f_{sf} when (D_b/D_p) exceeds a critical value $(D_b/D_p)_{cr}$

To apply Equation (7.17), the following procedure is recommended:

- Set the critical depth, $(D_b/D_p)_{cr}$, for shaft friction the same as that for unit toe bearing capacity.
- $f_l = f_{sf}$ when $(D_b/D_p) \geq (D_b/D_p)_{cr}$.
- Because of the disturbance caused by pile installation, $\delta' = 0.5$ to $0.8\,\varnothing'$. Lower δ' values should be used for piles with smooth shaft surface, such as steel or precast concrete piles.
- The K values can be estimated based on the type of pile and at-rest lateral earth pressure coefficient before pile construction K_o, as shown in Table 7.6.

The value of $\varnothing'$ can be estimated from cone penetration or standard penetration tests, or laboratory tests such as direct shear, or triaxial tests on sand samples retrieved from the field. Reliable K and f_l can only be deduced from pile load tests at the given site. Empirical procedures that use cone penetration or standard penetration tests without going through the estimation of $\varnothing'$ values are preferred in the estimation of f_{sf} for piles in granular soil. The penetration test-based methods will be introduced later.

Table 7.6 Estimation of K based on type of pile and K_o and $K_o = (1 - \sin\varnothing')$

Pile type	K
Non-displacement bored	$\approx K_o$
Low-displacement driven	$\approx K_o$ to $1.4\,K_o$
High-displacement driven	$\approx K_o$ to $1.8\,K_o$

Source: Das, B.M. 2015. *Principles of Foundation Engineering.* 8th Edition, Cengage Learning, Stamford, CT, p. 426. Reproduced with permission of Cengage Learning.

7.4.2.2 Effects of pile taper—the Nordlund method

Nordlund (1963, 1979) proposed a method to estimate the shaft friction resistance in cohesionless soils that considers the effects of pile taper and soil displacement caused by pile installation. The method was developed based on load tests on piles that had widths (diameters) ranging from 250 mm to 500 mm.

In the Nordlund method, the ultimate bearing capacity, Q_u, is divided into ultimate shaft resistance, R_{sf}, and toe resistance, R_{tf}, as in the case of Equation 7.9. R_{tf} is independent from pile taper and can be estimated with the method described above. The following discussion will concentrate on R_{sf} under the Nordlund method. Factors considered in the Nordlund method are illustrated in Figure 7.17, and R_{sf} is computed as follows:

$$R_{sf} = \sum_{d=0}^{d=D_b} K_\delta C_F \sigma'_{vo} \frac{\sin(\delta' + \omega)}{\cos(\omega)} D_p \Delta d \tag{7.18}$$

where
$d =$ depth
$D_b =$ pile depth in the bearing stratum
$K_\delta =$ coefficient of lateral earth pressure at depth d
$C_F =$ correction factor for K_δ when $\delta' \neq \varnothing'$
$\sigma'_{vo} =$ effective overburden stress at depth d
$\delta' =$ friction angle between pile and soil
$\omega =$ angle of pile taper from vertical
$D_p =$ pile diameter at depth d
$\Delta_d =$ length of the pile segment

Nordlund did not recommend a limiting value for the computation of R_{sf}.

The values of δ' as a ratio to ϕ', for various types of piles, according to soil displacement volume, V (in m^3/m), per meter of pile penetration are shown in Figure 7.18. Figure 7.19 provides a chart to estimate K_δ for piles when $\varnothing' = 35°$. The K_δ values for a wide range of friction angles can be found in Nordlund (1979). A chart to estimate C_F based on $\varnothing'$ and for various values of $\delta'/\varnothing'$ is shown in Figure 7.20. Example 7.2 is used to demonstrate the step-by-step procedure to use the Nordlund method for the calculation of R_{sf} for a tapered pile.

Pile taper has little effect on shaft friction in cohesive soils.

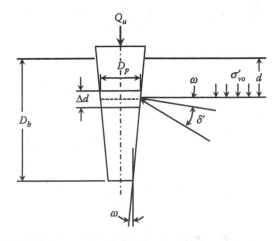

Figure 7.17 Factors considered in the Nordlund method.

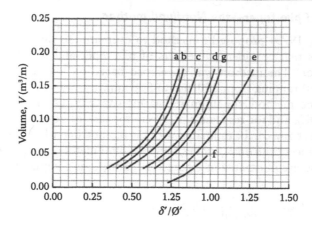

Figure 7.18 Relationship between $\delta'/\varnothing'$ and soil displacement volume, V, for various types of piles. (a) Closed end pipe and non-tapered portion of Monotube piles. (b) Timber piles. (c) Precast concrete piles. (d) Raymond step-taper piles. (e) Raymond uniform taper piles. (f) H-piles. (g) Tapered portion of Monotube piles. (After Nordlund, R.L. 1979. *Point bearing and shaft friction of piles in sand. Proceedings, 5th Annual Short Course on the Fundamentals of Deep Foundation Design*, St. Louis, Missouri, USA.)

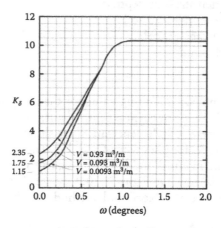

Figure 7.19 Estimation of K_δ for piles when $\varnothing' = 35°$. (After Nordlund, R.L. 1979. *Point bearing and shaft friction of piles in sand. Proceedings, 5th Annual Short Course on the Fundamentals of Deep Foundation Design*, St. Louis, Missouri, USA.)

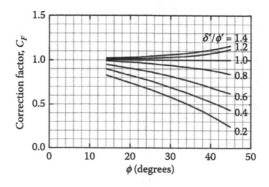

Figure 7.20 Correction factor for K_δ (After Nordlund, R.L. 1979. *Point bearing and shaft friction of piles in sand. Proceedings, 5th Annual Short Course on the Fundamentals of Deep Foundation Design*, St. Louis, Missouri, USA.)

EXAMPLE 7.2

Given

A Raymond uniform tapered pile is driven to 10 m from ground surface into a uniform sand deposit with $\gamma_{sat} = 18$ kN/m³ and drained friction angle $\varnothing'$ of 35°. The groundwater table is at the ground surface. The pile has a diameter of 0.5 m at the ground surface (pile head) and a taper (ω) of 0.7 degrees as shown in Figure E7.2.

Required

Determine the R_{sf} for the pile using the Nordlund method.

Solution

Divide the pile into five segments; Δd for each segment is 2 m.

The pile diameter at pile head is 0.5 m. With the taper (ω) of 0.7°, the pile diameter at the toe is

$$0.5 - (2)[\tan(\omega)]\, D_b = 0.5 - (2)[(0.7)](10) = 0.26 \text{ m}$$

The average diameter of the pile $(D_p)_{ave} = (0.5 + 0.26)/2 = 0.38$ m

Soil displacement caused by every meter of pile penetration, V, is approximately

$$V = \pi(D_p)_{ave}^2/4 = (3.1416)(0.38)(0.38)/4 \approx 0.11 \text{ m}^3/\text{m}$$

Follow Figure 7.18, for Raymond uniform tapered pile (curve e), $V = 0.11$ m³/m and
$\omega = 0.7$, $\delta'/\varnothing' = 1.075$
$\varnothing' = 35°$. Thus $\delta' = (35)(1.075) = 37.6°$
$K_\delta = 8$, for $\omega = 0.7$, according to Figure 7.19
$C_F = 1.05$ for $\varnothing' = 35°$ and $\delta'/\varnothing' = 1.075$, according to Figure 7.20
Divide the pile into five segments for the computation of R_{sf},

$$R_{sf} = \sum_{d=0}^{d=D_b} K_\delta C_F \sigma'_{vo} \frac{\sin(\delta' + \omega)}{\cos(\omega)} D_p \Delta d \tag{7.18}$$

The results are shown in Table E7.2.

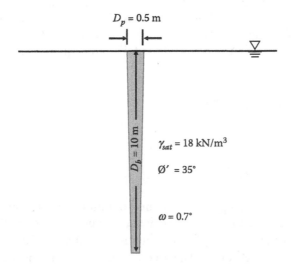

Figure E7.2 Uniform tapered Raymond pile driven into a sand deposit.

Table E7.2 Calculation of R_{sf} using the Nordlund method

Segment no.	Depth, d, to center of segment, m	$\sigma'_{vo} = (d)$ $(\gamma_{sat} - 9.81)$ kPa	D_p, m	$K_\delta C_F \sigma'_{vo} \dfrac{\sin(\delta' + \omega)}{\cos(\omega)} D_p \Delta d$ kN
1	1	8.19	0.48	40.58
2	3	24.57	0.43	109.23
3	5	40.95	0.38	161.20
4	7	57.33	0.33	196.49
5	9	73.71	0.28	215.09
			Sum	$R_{sf} = 723$ kN

7.4.2.3 For piles in cohesive soils

To apply the drained analysis similar to that of Equation 7.17, either K or σ'_h will have to be known. The construction disturbance makes it difficult to ascertain the value of either K or σ'_h. Meyerhof (1976) reported the α (total stress) and β (effective stress) methods for assessing f_{sf} for piles in saturated clays to circumvent the difficulty in estimating σ'_h. Details of these methods are described in the following sections.

7.4.2.3.1 The α method

The α method (Meyerhof, 1976) takes an undrained approach. It is assumed that

$$f_{sf} = \alpha s_u \tag{7.19}$$

where

s_u = average undrained shear strength along the length of the pile before pile driving
α = the empirical adhesion coefficient for the average s_u within embedded pile length

7.4.2.3.2 α values for driven piles

The effects on stresses along the pile during loading considering the disturbance induced by pile installation and the following consolidation are collapsed into one factor, α. This is a rather rough but practical method. Many empirical procedures have been proposed to estimate α. Research has indicated that for driven piles, α is related to s_u and stress history or overconsolidation ratio (OCR) of the clay. Based on their analysis on a rather comprehensive database of pile load tests compiled by the American Petroleum Institute (API), Randolph and Murphy (1985) developed the following equations to estimate α:

For $s_u/\sigma'_{vo} \leq 1$

$$\alpha = [(s_u/\sigma'_{vo})_{NC}]^{0.5} (s_u/\sigma'_{vo})^{-0.5} \tag{7.20}$$

For $s_u/\sigma'_{vo} > 1$

$$\alpha = [(s_u/\sigma'_{vo})_{NC}]^{0.5} (s_u/\sigma'_{vo})^{-0.25} \tag{7.20a}$$

The values of s_u and s_u/σ'_{vo} refer to the conditions before pile driving, and σ'_{vo} (= effective overburden stress before pile driving) can be estimated with reasonable accuracy. The first part (i.e., $[(s_u/\sigma'_{vo})_{NC}]^{0.5}$) of Equations 7.20 and 7.20a reflects the effects of soil plasticity, and hence $\varnothing'$ on α. The second part of Equations 7.20 and 7.20a (i.e., $(s_u/\sigma'_{vo})^{-0.5}$ and $(s_u/\sigma'_{vo})^{-0.25}$, respectively) reveals the effects of soil stress history or OCR.

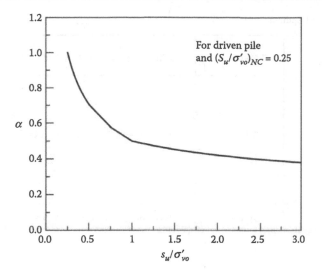

Figure 7.21 Relationship between α and s_u/σ'_{vo} for driven piles.

The value of $(s_u/\sigma'_{vo})_{NC}$ can be estimated by Skempton's empirical equation (Skempton, 1957),

$$\left(\frac{s_u}{\sigma'_{vo}}\right)_{NC} = 0.11 + 0.0037(\text{PI}) \tag{7.21}$$

where
   PI = plasticity index

For the common range of PI, Jamiolkowski et al. (1985) suggested that $(s_u/\sigma'_{vo})_{NC} = 0.23 \pm 0.04$.

For overconsolidated (OC) clays, following Ladd et al. (1977),

$$\frac{\left(\dfrac{s_u}{\sigma'_{vo}}\right)_{OC}}{\left(\dfrac{s_u}{\sigma'_{vo}}\right)_{NC}} = (\text{OCR})^{0.8} \tag{7.22}$$

Figure 7.21 shows a plot of the $\alpha - s_u/\sigma'_{vo}$ relationship for the case of $(s_u/\sigma'_{vo})_{NC} = 0.25$. The method developed by Randolph and Murphy (1985) as shown in Equations 7.20 and 7.20a has been recommended by API for pile designs in cohesive soils.

EXAMPLE 7.3

Given

A 10 m long and 0.8 m diameter precast concrete pile is driven into a saturated clay deposit as shown in Figure E7.3. The groundwater table is at the ground surface and $(s_u/\sigma'_{vo})_{NC} = 0.25$.

Required

Determine the R_{sf} for the pile using the α method.

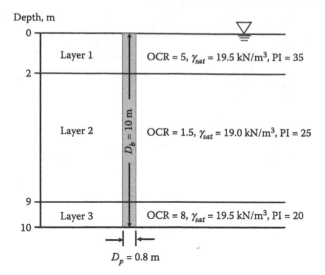

Figure E7.3 The precast concrete pile and clay soil layers.

Solution

Because all clay layers have OCR larger than 1, the computation of (s_u/σ'_{vo}) for each layer starts with the corresponding $(s_u/\sigma'_{vo})_{NC}$ values according to the given PI and

$$\left(\frac{s_u}{\sigma'_{vo}}\right)_{NC} = 0.11 + 0.037\,(\text{PI}) \tag{7.21}$$

The calculation of $(s_u/\sigma'_{vo})_{OC}$ then follows according to the given OCR for each clay layer as,

$$\left(\frac{s_u}{\sigma'_{vo}}\right)_{OC} = \left(\frac{s_u}{\sigma'_{vo}}\right)_{NC}(\text{OCR})^{0.8} \tag{7.22}$$

The computations are summarized in Table E7.3.

Table E7.3 Computations to determine $(s_u/\sigma'_{vo})_{OC}$

Layer	Layer thickness, m	PI	OCR	$\left(\dfrac{s_u}{\sigma'_{vo}}\right)_{NC}$	$\left(\dfrac{s_u}{\sigma'_{vo}}\right)_{OC}$
1	2.0	35	5.0	0.240	0.868
2	7.0	25	1.5	0.203	0.280
3	1.0	20	8.0	0.184	0.971

The average value of (s_u/σ'_{vo}) for the entire pile length, $(s_u/\sigma'_{vo})_{ave}$, is determined using a weighted average of $(s_u/\sigma'_{vo})_{OC}$ based on the thickness of each layer; thus

$$\left(\frac{s_u}{\sigma'_{vo}}\right)_{NCave} = ((0.240)(2) + (0.203)(7) + (0.971)(1))/10 = 0.21$$

$$\left(\frac{s_u}{\sigma'_{vo}}\right)_{ave} = ((0.868)(2) + (0.280)(7) + (0.971)(1))/10 = 0.47$$

The corresponding α value is determined using $(s_u/\sigma'_{vo})_{ave}$, and according to Equation (7.20),

$$\alpha = [(s_u/\sigma'_{vo})_{NC}]^{0.5}(s_u/\sigma'_{vo})^{-0.5} = [(0.21)^{\wedge}0.5][(0.47)^{\wedge} - 0.5] = 0.67$$

Use σ'_{vo} at mid-height of the pile as the average effective overburden stress:

$$\sigma'_{vo} = (19.5 - 9.81)(2) + (19 - 9.81)(3) = 46.95 \text{ kPa}$$

Average

$$s_u = \left(\frac{s_u}{\sigma'_{vo}}\right)_{ave} (\sigma'_{vo}) = (0.47)(46.95) = 21.9 \text{ kPa}$$

$$f_{sf} = \alpha\,(s_u) = (0.67)(21.9) = 14.63 \text{ kPa}$$

$$R_{sf} = \pi D_P L_p f_{sf} = (3.1416)(0.8)(10)(14.63) = 368 \text{ kN}$$

7.4.2.3.3 α value for bored piles

For bored piles, α is determined based on the average s_u along the length of the pile (O'Neill, 2001). Figure 7.22 shows a correlation between α and the average s_u normalized with respect to the atmospheric pressure, p_a. Regardless of the type of piles, the α method recognizes that the α value decreases as s_u becomes higher.

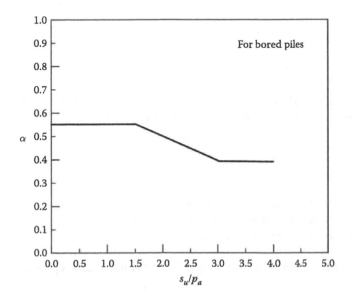

Figure 7.22 Relationship between α and s_u/p_a for bored piles. (O'Neill, M.W. 2001. Side resistance in piles and drilled shafts. *Journal of Geotechnical and Geoenvironmental Engineering*, ASCE, 127 (1), Figure 20, p. 12. Reproduced with permission of ASCE.)

EXAMPLE 7.4

Given

The same soil conditions described in Example 7.3.

Required

Determine the R_{sf} for a 0.8 m diameter and 10 m deep bored pile using the α method.

Solution

Compute s_u values using Equations 7.21 and 7.22. The results are summarized in Table E7.4.

Table E7.4 Computations to determine average s_u

Layer	Layer thickness, m	PI	σ'_{vo} at center of layer, kPa	OCR	$\left(\dfrac{s_u}{\sigma'_{vo}}\right)_{NC}$	$\left(\dfrac{s_u}{\sigma'_{vo}}\right)_{OC}$	s_u, kPa
1	2.0	35	1.0	5.0	0.240	0.868	8.40
2	7.0	25	5.5	1.5	0.203	0.280	14.40
3	1.0	20	9.5	8.0	0.971	0.971	86.00

The weighted average $s_u = ((8.40)(2) + (14.40)(7) + (86.00)(1))/10 = 20.39\ \text{kPa}$

$s_u/p_a = 20.39/100 = 0.204$

$\alpha = 0.55$ according to Figure 7.22

$f_{sf} = \alpha(s_u) = (0.55)(20.39) = 11.21\ \text{kPa}$

$R_{sf} = \pi D_P D_b f_{sf} = (3.416)(0.8)(10)(11.21) = 282\ \text{kN}$

7.4.2.3.4 The β method

In the β method, the analysis is based on effective stress (Meyerhof, 1976; O'Neill, 2001). From effective stress point of view, f_{sf} may be considered as

$$f_{sf} = \tan\delta'\sigma'_h = \tan\delta' K\sigma'_{vo} \tag{7.23}$$

where
δ' = drained pile/soil interface friction angle
σ'_h = effective horizontal stress after pile installation
K = earth pressure ratio = σ'_h/σ'_{vo}

Because of the difficulty in determining δ', σ'_h, or K individually, Equation 7.23 is rearranged as

$$f_{sf} = \beta\sigma'_{vo} \tag{7.24}$$

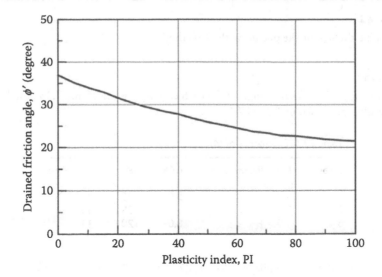

Figure 7.23 Variation of friction angle with plasticity index. (Bjerrum, L. and Simons, N.E. 1960. Comparison of shear strength characteristics of normally consolidated clays. *Proceedings, ASCE Research Conference on Shear Strength of Cohesive Soils,* Boulder, CO. Figure 4, p. 716. Reproduced with permission of ASCE.)

In essence, the effects of $\tan \delta'$ and K are lumped into a single, skin friction factor β (i.e., $\beta = \tan \delta' K$), leaving σ'_{vo} that is relatively easy to determine. β may be estimated as follows:

$$\beta = (1 - \sin \varnothing') \tan \varnothing' \quad \text{for normally consolidated clay} \tag{7.25}$$

and

$$\beta = (1.5)(1 - \sin \varnothing') \text{OCR}^{0.5} \tan \varnothing' \quad \text{for over consolidated clay} \tag{7.25a}$$

σ'_{vo} = average vertical effective stress in the soil along the pile before driving

The method recognizes that the clay surrounding the pile is thoroughly disturbed, so that the $\varnothing'$ should be taken as the remolded drained friction angle of the clay and $\delta' = \varnothing'$. For normally consolidated clay, K is expected to be close to at-rest lateral earth pressure $K_o[= (1 - \sin \varnothing')]$. Based on field load tests, O'Neill (2001) reported that $\beta = 0.16 \pm 0.06$ for normally consolidated clay, and β is independent of depth. Similar and scattered range of β values has also been reported by Meyerhof (1976). β tends to decrease as the pile length increases.

In overconsolidated clays, estimation of β relies on remolded drained friction angle $\varnothing'$ and OCR. The value of $\varnothing'$ can be obtained from laboratory tests or estimation based on plasticity index using the empirical correlation depicted in Figure 7.23, proposed by Bjerrum and Simons (1960). According to pile load tests in overconsolidated clays with K_o (before pile installation) ranging from 0.5 to 3, the corresponding β increased from approximately 0.25 to 2 (Meyerhof, 1976).

EXAMPLE 7.5

Given

A 10 m long and 0.8 m diameter pile is driven into a saturated clay deposit as shown in Figure E7.3. The groundwater table is at the ground surface.

Required

Determine the R_{sf} for the pile using the β method.

Solution

Estimate the drained friction angle $\emptyset'$ for each clay layer using Figure 7.23 based on PI. The results are shown in Table E7.5.

Table E7.5 Estimation of $\emptyset'$

Layer	Layer thickness, m	OCR	PI	$\emptyset'$, degree
1	2.0	5.0	35	28
2	7.0	1.5	25	30
3	1.0	8.0	20	33

Average OCR along the pile length $= ((5)(2) + (1.5)(7) + (8)(1))/10 = 2.85$
Average $\emptyset'$ along the pile length $= ((28)(2) + (30)(7) + (33)(1))/10 = 29.9$
Take σ'_{vo} at mid-height of the pile (5 m below ground surface) as the average value

$$= (19.5 - 9.81)(2) + (19.0 - 9.81)(3) = 46.95 \text{ kPa}$$

$$\beta = (1.5)(1 - \sin\emptyset')OCR^{0.5}\tan\emptyset' = (1.5)(1 - \sin 29.9)(2.85^{0.5})(\tan 29.9) = 0.73$$
$$(7.25a)$$

$$f_{sf} = \beta\sigma'_{vo} = (0.73)(46.95) = 34.3 \text{ kPa} \tag{7.24}$$

$$R_{sf} = \pi D_P L_p f_{sf} = (3.1416)(0.8)(10)(34.3) = 862 \text{ kN}$$

7.4.3 Ultimate bearing capacity based on field penetration tests

Small variations of the friction angle, $\emptyset'$, can considerably affect the values of K (or K_δ) and N_q^* for short piles above the critical depth or similarly influence the values of f_l and q_l for long piles. Field penetration tests such as the standard penetration test (SPT) and cone penetration test (CPT) are simple to perform and mimic the insertion of a miniature pile in many aspects. For these reasons, many empirical methods have been proposed to estimate f_{sf} and q_{tf} using these penetration test results. Details of SPT and CPT are described in Chapter 3. To take full advantage of the penetration tests, it has generally been recommended that a "direct" approach be followed. Under the direct approach, the f_{sf} and q_{tf} are determined directly from the test results (i.e., N values of SPT or cone tip resistance from CPT) without going through the estimation of soil strength parameters using the penetration test results.

It should be noted that most of the available empirical methods were proposed decades ago, and prior to some of the important testing capabilities or practices as we commonly use today in penetration tests. Necessary adjustments in the description of the penetration test-based methods may be necessary in light of the recent developments. It should also be noted that SPT is not effective in reflecting the strength of cohesive soils. Therefore, SPT is typically used only for the static analysis of piles in cohesionless soils.

7.4.3.1 Ulitimate bearing capacity in cohesionless soils based on SPT

Taking undisturbed soil samples in cohesionless soil is not practical. An in-situ test such as SPT is a viable choice and has been widely used as a means to provide profile information associated with the design of piles. As described in Chapter 3, the original N value collected in the field, N_{SPT}, is usually converted to N_{60}, or N value corresponds to an energy ratio of 60%, or E_{60} as

$$N_{60} = \frac{N_{SPT}E_{SPT}}{E_{60}} \qquad (3.8)$$

where E_{SPT} is the hammer energy delivered to the anvil during SPT. The N value in granular soil with similar relative density tends to increase with overburden stress. For engineering applications in granular soils, the N values are usually corrected to an equivalent effective overburden stress, σ'_{vo}, of 100 kPa, or N_1. The correction is expressed by a correction factor, C_N, and

$$C_N = \frac{N_1}{N_{SPT}} \qquad (3.9)$$

and it is recommended to determine C_N as follows:

$$C_N = \sqrt{\frac{100}{\sigma'_{vo}}} \ (\sigma'_{vo} \text{ in kPa}) \qquad (3.9a)$$

An N value corrected for both energy ratio to $E_r = 60\%$ and effective overburden stress σ'_{vo} of 100 kPa (1 atmosphere) is referred to as $(N_1)_{60}$.

7.4.3.1.1 Ultimate unit toe resistance, q_{tf}, for driven piles

Meyerhof (1976) recommended that the unit toe resistance, q_{tf}, in kPa for piles driven into a uniform cohesionless deposit be estimated as

$$q_{tf} = \frac{40\overline{(N_1)}_{60}D_b}{D_p} \leq 400\overline{(N_1)}_{60} \qquad (7.26)$$

The calculation of $\overline{(N_1)}_{60}$ for piles driven into a uniform cohesionless deposit is the same as the $\overline{(N_1)}_{60B}$ to be described below. If the pile toe is located near the interface of two strata with a weaker stratum overlying the bearing stratum, the unit toe resistance can be calculated as follows:

$$q_{tf} = 400\overline{(N_1)}_{60O} + \frac{(40\overline{(N_1)}_{60B} - 40\overline{(N_1)}_{60O})D_b}{D_p} \leq 400\overline{(N_1)}_{60B} \qquad (7.27)$$

where
$\overline{(N_1)}_{60B}$ = average corrected SPT $(N_1)_{60}$ value of the bearing stratum
$\overline{(N_1)}_{60O}$ = average corrected SPT $(N_1)_{60}$ value for the stratum overlying the bearing stratum
D_b = pile embedment depth into bearing stratum in meters
D_p = pile diameter in meters

$400\overline{(N_1)}_{60B}$ is the limiting value when the pile penetration reached a critical depth of 10 pile diameters into the bearing stratum. The $\overline{(N_1)}_{60B}$ should be calculated by averaging the $(N_1)_{60}$ values near the pile toe, preferably extending to three pile diameters (D_p) below the pile toe. For piles driven into non-plastic silts, the q_{tf} should be limited to $300\overline{(N_1)}_{60}$ or $300\overline{(N_1)}_{60B}$ when applying Equations 7.26 or 7.27.

7.4.3.1.2 Ultimate unit shaft resistance, f_{sf}, for driven piles

The average ultimate unit shaft resistance, f_{sf}, of driven, displacement piles, such as closed-end pipe piles or solid precast concrete piles, in kPa can be estimated as follows (Meyerhof, 1976):

$$f_{sf} = 2\overline{(N_1)}_{60} \leq 100 \text{ kPa} \tag{7.28}$$

The average ultimate unit shaft resistance, f_{sf} of driven, low-displacement piles, such as H piles, in kPa is

$$f_{sf} = \overline{(N_1)}_{60} \leq 100 \text{ kPa} \tag{7.29}$$

where

$\overline{(N_1)}_{60}$ = average corrected SPT blow count per 300 mm penetration, along the embedded length of pile

EXAMPLE 7.6

Given

A 0.8 m diameter and 10 m long precast concrete pile is driven into a sand deposit. SPT was performed at the site with one N value taken for every 1.5 m to a maximum depth of 15.0 m. Figure E7.6 shows the soil and SPT corrected blow count $(N_1)_{60}$ profiles. Ground water table is at the ground surface. The individual $(N_1)_{60}$ values are tabulated in Table E7.6.

Required

Determine the ultimate pile capacity Q_u using Meyerhof's SPT method.

Solution

This is a case of pile toe located near the interface of two strata with a weaker stratum overlying the bearing stratum. Consider the bearing stratum starts at a depth of 7.5 m.
Depth of embedment in bearing stratum, $D_b = 10 - 7.5 = 2.5$ m
Use Equation 7.27 to calculate q_{tf}.

$$q_{tf} = 400\overline{(N_1)}_{60O} + \frac{(40\overline{(N_1)}_{60B} - 40\overline{(N_1)}_{60O})D_b}{D_p} \leq 400\overline{(N_1)}_{60B} \tag{7.27}$$

Take the average of $(N_1)_{60}$ from 0 to 6 m as the $\overline{(N_1)}_{60O}$,

$$\overline{(N_1)}_{60O} = (5 + 8 + 7 + 10)/4 = 7.5$$

Take the average of $(N_1)_{60}$ from 7.5 to 13.5 m [just beyond 10 m plus three times the pile diameter $= 10 + (3)(0.8) = 12.4$ m] as the $\overline{(N_1)}_{60B}$,

$$\overline{(N_1)}_{60B} = (15 + 17 + 20 + 26 + 24)/5 = 20.4$$

$$400\overline{(N_1)}_{60B} = (400)(20.4) = 8160 \text{ kPa}$$

$$q_{tf} = (400)(7.5) + \frac{((40)(20.4) - (40)(7.5))2.5}{0.8} = 4613 \text{ kPa} < 8160$$

Thus $q_{tf} = 4613$ kPa

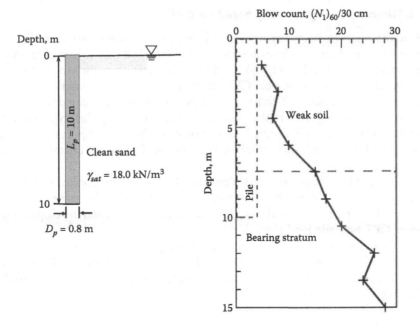

Figure E7.6 The precast concrete pile in sand and profile of $(N_1)_{60}$.

Table E7.6 The values of $(N_1)_{60}$

Depth, m	1.5	3.0	4.5	6.0	7.5	9.0	10.5	12.0	13.5	15.0
$(N_1)_{60}$	5	8	7	10	15	17	20	26	24	28

This is a high-displacement pile. For the average ultimate unit shaft resistance, f_{sf},

$$f_{sf} = 2\overline{(N_1)}_{60} \leq 100\,\text{kPa} \tag{7.28}$$

$\overline{(N_1)}_{60}$ along the entire length of the pile $= (5 + 8 + 7 + 10 + 15 + 17 + 20)/7 = 11.7$

$$f_{sf} = (2)(11.7) = 23.4\,\text{kPa} < 100\,\text{kPa} \tag{7.28}$$

Thus $f_{sf} = 23.4\,\text{kPa}$

$$Q_u = R_{sf} + R_{tf} = (\pi D_p L_p)(f_{sf}) + \left(\frac{\pi D_p^2}{4}\right)(q_{tf})$$
$$= (3.1416)(0.8)(10)(23.4) + (3.1416)(0.8)^2(4613)/4 = 589 + 2319$$
$$= 2907\,\text{kN}$$

7.4.3.1.3 Ultimate bearing capacity for bored piles

Meyerhof (1976) recommended that for bored piles, q_{tf} of 1/3 of those values from Equations 7.26 or 7.27 may be used. The f_{sf} from Equation 7.29 may be used for bored piles.

7.4.3.2 Ulitimate bearing capacity based on CPT

Because of the analogies between CPT and insertion of a miniature pile, the cone penetration resistance often correlates well with that of a full-size driven pile under static
loading conditions. Meyerhof (1976) proposed the limiting values in ultimate bearing
capacity based on observations on CPT. Details of CPT are described in Chapter 3.
Piezocone penetration test (CPTu) with the porous element located behind the face
of the cone tip (the u_2 position) is probably the most popular type of CPT used
throughout the world. As described in Chapter 3, the cone tip resistance, q_c, may
be too low for CPT in soft clay unless it is corrected for the u_2 effects. The corrected
q_c is referred to as q_t.

Unlike SPT, the CPT-based methods described below can be applied in both cohesive
and cohesionless soils. The following sections introduce two of the many available methods that are CPT based. Both methods are empirical and have been developed based on a
database of CPT and pile load tests.

It should be noted that in all of the methods by Meyerhof (1976) (i.e., the α, β, and SPT
methods) related to the estimation of pile shaft friction, f_{sf} represents the average value
along the pile length. The associated parameters used to determine f_{sf} should also be taken
as the average value along the pile length.

7.4.3.3 Eslami and Fellenius method

This method is based on the so-called "effective" cone tip resistance, q_E, and a soil classification proposed by Eslami and Fellenius (1997). q_E is defined as follows:

$$q_E = (q_t - u_2) \tag{7.30}$$

It should be noted that q_E is not the cone tip resistance under effective stress conditions. The soil classification chart based on q_E and cone sleeve friction, f_s, is shown in
Figure 7.24.

The ultimate toe capacity, q_{tf}, is estimated as follows:

$$q_{tf} = C_t q_{Eg} \tag{7.31}$$

where
   C_t = toe adjustment factor that is a function of pile diameter
   q_{Eg} = geometric average of q_E over the influence zone

For piles with diameter less than 0.4 m, $C_t = 1.0$. For piles with diameter larger than
0.4 m,

$$C_t = \frac{1}{3D_p} \tag{7.32}$$

where
   D_p = pile diameter in meters

For piles installed through a weak soil layer into a dense soil layer, consider the influence zone extends from $4 D_p$ below the pile toe to $8 D_p$ above the pile toe. When the
pile is installed through a dense soil into a weak soil, the influence zone covers a depth
range of $2 D_p$ above the pile toe to $4 D_p$ below the pile toe. The use of geometric mean
practically smoothens out the potentially extreme values in CPTu.

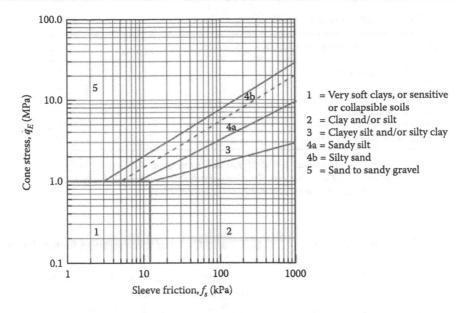

Figure 7.24 Soil classification according to Eslami and Fellenius (1997). (Eslami, A. and Fellenius, B.H. 1997. Pile capacity by direct CPT and CPTu methods applied to 102 case histories. *Canadian Geotechnical Journal*, 34(6), Figure 2, p. 891. Reproduced with permission of the NRC Research Press.)

In the Eslami and Fellenius method, the f_{sf} is calculated as

$$f_{sf} = C_s q_E \tag{7.33}$$

where

$C_s =$ shaft correlation factor, based on soil type classified according to Figure 7.24 and Table 7.7.

A step-by-step procedure to determine Q_u using the Eslami and Fellenius method is described as follows:

1. Convert all q_t values to q_E ($= q_t - u_2$), follow Equation 7.30.
2. Take a geometric average of q_E over the influence zone. For piles installed through a weak soil layer into a dense soil layer, the influence zone extends from 4 D_p below the pile toe to 8 D_p above the pile toe. When the pile is installed through a dense soil into a weak soil, the influence zone covers a depth range of 2 D_p above the pile toe to 4 D_p below the pile toe. This geometric average is called q_{Eg}.
3. Compute ultimate pile toe bearing capacity q_{tf} as follows:

$$q_{tf} = C_t q_{Eg} \tag{7.31}$$

   where $C_t = 1$ for $D_p < 0.4$ m, and $C_t = \dfrac{1}{3D_p}$ according to Equation 7.32 for $D_p \geq 0.4$ m and $D_p =$ pile diameter in meters.
4. $R_{tf} = q_{tf}(A_t)$ where $A_t =$ cross-sectional area of the pile toe ($= \pi D_p^2/4$).
5. Divide the soil surrounding the pile into major layers according to the CPTu data; perform soil classification for each soil layer according to Figure 7.23. Number the soil layers as 1,2,3..i..n.
6. Compute $\bar{q}_t$, $\bar{u}_2$, and $\bar{f}_s$, arithmetic average respective value from CPTu for each of the soil layers. For each soil layer, follow Equation 7.30, $\bar{q}_E = (\bar{q}_t - \bar{u}_2)$.

Table 7.7 Shaft correlation coefficient, C_s

Soil type according to Figure 7.24	C_s %
1. Soft-sensitive soils	8.0
2. Clay	5.0
3. Silty clay, stiff clay, and silt	2.5
4a. Sandy silt and silt	1.5
4b. Fine sand or silty sand	1.0
5. Sand to sandy gravel	0.4

Source: Eslami, A. and Fellenius, B.H. 1997. Pile capacity by direct CPT and CPTu methods applied to 102 case histories. *Canadian Geotechnical Journal* 34(6), Table 3, p. 897. Reproduced with permission of the NRC Research Press.

7. Follow Equation 7.33, where $\bar{f}_{sf} = C_s \bar{q}_E$, and C_s = shaft correlation factor, based on soil type classified according to Figure 7.23 and Table 7.7.
8. For each soil layer i, the corresponding ultimate shaft resistance $(R_{sf})_i = (\pi D_p)(\bar{f}_{sf})(L_{pi})$, where L_{pi} = length of the pile in soil layer i.
9. R_{sf} = ultimate shaft resistance = sum of all $(R_{sf})_i$ along the pile shaft.

The ultimate bearing capacity of the pile, Q_u, is

$$Q_u = R_{sf} + R_{tf} \qquad (7.9)$$

EXAMPLE 7.7

Given

A 0.8 m diameter and 15 m long precast concrete pile is driven into a sand deposit. CPTu was performed at the site to a maximum depth of 20 m. Figure E7.7 shows the soil and CPTu profiles. The groundwater table was at 0.8 m below the ground surface. The CPTu location was predrilled from 0 to 2 m below ground surface before penetration. Thus the q_t value in that depth range was close to 0. Ignore any resistance from 0 to 2 m depth in the calculation of shaft resistance.

Required

Determine the ultimate pile capacity, Q_u, using the Eslami and Fellenius method.

Solution

Consider the soil profile as piles installed through a weak soil layer into a dense soil layer; the q_{tf} is calculated with influence zone that extends from 4 D_p (15–18.2 m) below the pile toe to 8 D_p (8.6–15 m) above the pile toe.

q_{Eg} for the depth range of 8.6–18.2 m = 4.43 MPa

$$D_p > 0.4 \text{ m, thus } C_t = 1/((3)(0.8)) = 0.417 \qquad (7.32)$$

$$q_{tf} = C_t q_{Eg} = (0.417)(4.43) = 1.845 \text{ MPa} = 1845 \text{ kPa} \qquad (7.31)$$

$$R_{tf} = q_{tf} \frac{\pi D_p^2}{4} = (1845)(3.1416)(0.8)^2/4 = 927.2 \text{ kN}$$

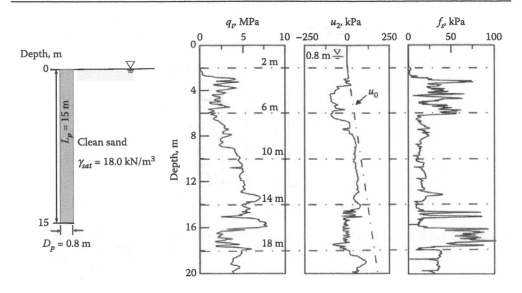

Figure E7.7 The precast concrete pile and profile of CPTu.

For the calculation of f_{sf}, divide the soil deposit into six layers according to the CPTu results. The calculations are shown in Tables E7.7.

Table E7.7 Calculation of ultimate shaft friction using Eslami and Fellenius method

Layer no.	Depth range, m	$\bar{q}_t$, MPa	$\bar{u}_2$, kPa	$\bar{q}_E$, MPa	$\bar{f}_s$, kPa	Soil type	C_s	$\bar{f}_{sf}$, kPa	$(R_{sf})_i$, kN
1	0–2	–	0.31	–	0.66	–	–	–	–
2	2–6	0.89	5.25	0.88	6.36	1	0.08	70.50	708.7
3	6–10	2.89	14.24	2.88	16.04	4b	0.01	28.76	289.1
4	10–14	5.21	53.54	5.16	17.57	5	0.004	20.63	207.4
5	14–18	4.56	1.08	4.56	41.50	4b	0.01	45.55	114.5
6	18–20	4.16	63.04	4.10	33.77	5	0.004	16.38	–
								R_{sf} = sum	1319.7

$\bar{q}_t$, $\bar{u}_2$, $\bar{f}_s$ = arithmetic average respective value from CPTu for the designated soil layer.

$$\bar{q}_E = \bar{q}_t - \bar{u}_2 \text{ for the designated soil layer} \qquad (7.30)$$

$$\bar{f}_{sf} = C_s \bar{q}_E \text{ for the designated soil layer} \qquad (7.33)$$

$(R_{sf})_i$ = shaft resistance for the designated soil layer i, and
$(R_{sf})_i = (\pi D_p)(\bar{f}_{sf})(L_{pi})$, where L_{pi} = length of the pile for the designated soil layer i
R_{sf} = ultimate shaft resistance = sum $(R_{sf})_i$ from all soil layers along the pile shaft
$Q_u = R_{tf} + R_{sf} = 927.2 + 1319.7 = 2246.9$ kN

7.4.3.4 Laboratoire central des Ponts et Chaussées (LCPC) method

The LCPC method was proposed by Bustamante and Gianeselli (1982). Similar to the Eslami and Fellenius method, the CPT sleeve friction value is indirectly used, only for the soil type identification. The LCPC method considers soil type, pile type, installation method, and q_t values from CPT or CPTu in the determination of q_{tf} and f_{sf}. In order to apply a series of empirical charts and tables involved in this method, it is necessary

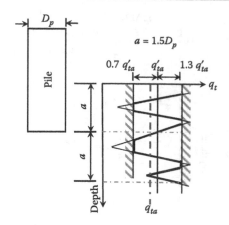

Figure 7.25 Averaging q_t for q_{tf} estimation. (Adapted from Bustamante, M. and Gianeselli, L. 1982. *Proceedings of the 2nd European Symposium on Penetration testing*, CRC Press, Amsterdam, pp. 493–500.)

to first classify the soil surrounding the pile. The soil classification can be made using the soil behavior type classification chart in Chapter 3 (e.g., Figure 3.32) according to the CPT or CPTu results.

The q_{tf} is calculated as follows:

$$q_{tf} = K_c q_{ta} \tag{7.34}$$

where
  K_c = cone bearing capacity factor
  q_{ta} = average cone tip resistance within 1.5 pile diameter above and below the pile toe

A two-phase procedure is used to determine the average cone tip resistance, q_{ta}. As described in Figure 7.25, a tentative arithmetic average value of the cone tip resistance, q'_{ta}, is taken from q_t obtained in a depth range from $1.5D_p$ above the pile toe to $1.5D_p$ below the pile toe. Extreme q_t values in the same depth range that are either larger than 1.3 q'_{ta} or lower than 0.7 q'_{ta} are eliminated. The final q_{ta} is the arithmetic average of the remaining q_t values in the same depth range.

K_c is an empirical factor selected considering the nature of soil, value of q_{ta}, and pile type as shown in Table 7.8. The piles are categorized into two groups. Group I covers mostly bored piles, and Group II relates to driven and jacked piles, and

$$R_{tf} = A_t q_{tf} \tag{7.10}$$

where
  A_t = cross sectional area of the pile toe
  q_{tf} = ultimate unit toe bearing capacity

The ultimate unit shaft resistance, f_{sf}, is estimated as follows:

$$f_{sf} = \frac{q_t}{\alpha_p} \tag{7.35}$$

where α_p is an empirical factor determined according to the nature of soil, value of q_t, and pile category as shown in Table 7.9. The f_{sf} obtained from Equation 7.35 and the selected α_p should not exceed the corresponding maximum values shown in Table 7.9.

Table 7.8 Cone bearing capacity factors for LCPC method

Nature of soil	q_{ta}, MPa	K_c Group I: bored piles	K_c Group II: driven and jacked piles
Soft clay and mud	≤ 1	0.4	0.3
Moderately compact clay	1–5	0.35	0.45
Silt and loose sand	≤ 5	0.4	0.5
Compact to stiff clay and compact silt	>5	0.45	0.55
Soft chalk	≤ 5	0.2	0.3
Moderately compact sand and gravel	5–12	0.4	0.5
Weathered to fragmented chalk	>5	0.2	0.4
Compact to very compact sand and gravel	>12	0.3	0.4

Source: Bustamante, M. and Gianeselli, L. 1982. Pile capacity prediction by means of static penetrometer CPT. *Proceedings of the 2nd European Symposium on Penetration testing*, CRC Press, Amsterdam, pp. 493–500.

Table 7.9 Selection of α_p for LCPC method

Nature of soil	q_b, MPa	α_p Pile category IA	IB	IIA	IIB	Maximum value of f_{sf} kPa IA	IB	IIA	IIB
Soft clay and mud	≤ 1	30	30	30	30	15	15	15	15
Moderately compact clay	1–5	40	80	40	80	35[a]	35[a]	35[a]	35
Silt and loose sand	≤ 5	60	150	60	120	35[a]	35[a]	35[a]	35
Compact to stiff clay and compact silt	>5	60	120	60	120	35[a]	35[a]	35[a]	35
Soft chalk	≤ 5	100	120	100	120	35[a]	35[a]	35[a]	35
Moderately compact sand and gravel	5–12	100	200	100	200	80[a]	35[a]	80[a]	80
Weathered to fragmented chalk	>5	60	80	60	780	120[a]	80[a]	120[a]	120
Compact to very compact sand and gravel	>12	130	300	150	200	120[a]	80[a]	120[a]	120

Source: After Bustamante, M. and Gianeselli, L. 1982. Pile capacity prediction by means of static penetrometer CPT. *Proceedings of the 2nd European Symposium on Penetration testing*, CRC Press, Amsterdam, pp. 493–500.

Note

[a] Higher maximum f_{sf} value can be used if construction method is well executed to assure intimate contact between pile and the surrounding material and the performance is verified with pile load test.

The bored and driven piles are categorized into four groups as shown in Table 7.10.

A step-by-step procedure to determine the ultimate shaft resistance, R_{sf}, using the LCPC method is described as follows:

1. Divide the soil surrounding the pile into major layers according to the CPT; perform soil classification for each soil layer and number the soil layers as 1,2,3..*i*..*n*.
2. Determine q_{ta} and follow the recommended averaging scheme.
3. Select K_c according to Table 7.8, consider soil types, pile category, and q_t.
4. Determine q_{tf} using Equation 7.34. The ultimate pile toe resistance, $Q_{tf} = q_{tf}$ (A_p).
5. For each soil layer, *i*, the corresponding ultimate unit shaft resistance $(f_{sf})_i$ is determined based on the arithmetic average, q_t, value for that layer and Tables 7.9

Table 7.10 Pile Categories for the selection of α_p

Pile category	Description
IA	Bored piles constructed without casing, auger piles
IB	Bored piles constructed with casing, cast-in-place pile with driven shells
IIA	Driven concrete piles, jacked concrete piles
IIB	Driven metal piles, jacked metal piles

Source: Bustamante, M. and Gianeselli, L. 1982. Pile capacity prediction by means of static penetrometer CPT. *Proceedings of the 2nd European Symposium on Penetration testing,* CRC Press, Amsterdam, pp. 493–500.

and 7.10. The ultimate shaft resistance for soil layer i, $(R_{sf})_i$, is computed as

$$(R_{sf})_i = (A_s)_i (f_{sf})_i \tag{7.36}$$

where

$(A_s)_i =$ the pile shaft surface area in soil layer i

For H-piles, the $(A_s)_i$ should be calculated considering the spaces within the flanges are fully plugged.

6. The ultimate shaft resistance, R_{sf}, for the whole pile shaft is the sum of all the $(R_{sf})_i$ values.

The ultimate bearing capacity of the pile, Q_u, is

$$Q_u = R_{sf} + R_{tf} \tag{7.9}$$

EXAMPLE 7.8

Given

For the same pile geometry, soil conditions, and CPTu profile as Example 7.7. The CPTu location was predrilled from 0 to 2 m below ground surface before penetration. Thus the q_t value in that depth range was close to 0. Ignore any resistance from 0 to 2 m depth in the calculation of shaft resistance.

Required

Determine the ultimate pile capacity Q_u using the LCPC method.

Solution

Calculation of q_{tf},

$$q_{tf} = K_c q_{ta} \tag{7.34}$$

q_{ta} is the arithmetic average of q_t value in a depth range from $1.5\,D_p$ above the pile toe to $1.5\,D_p$ (from 13.8 to 16.2 m) below the pile toe as follows.

A tentative arithmetic average value q'_{ta} is taken first in that depth range, using the q_t values shown in Figure E7.7, and $q'_{ta} = 5.72\,\text{MPa}$

Exclude the extreme values larger than 1.3 $q'_{ta} = (1.3)(5.72) = 7.44$ MPa, and exclude the extreme values less than 0.7 $q'_{ta} = (0.7)(5.72) = 4.01$ MPa

Arithmetic average of the remaining q_t values in the same depth range, $q_{ta} = 5.61$ MPa

The soil can be classified as moderately compact sand and gravel, for the range of q_{ta} and pile type, $K_c = 0.5$, according to Table 7.8.

Thus,

$$q_{tf} = (0.5)(5.61) = 2.8 \text{ MPa} = 2800 \text{ kPa} \tag{7.34}$$

$$R_{tf} = q_{tf} A_t = (2800)\left(\frac{\pi D_p^2}{4}\right) = 1410 \text{ kN}$$

For the calculation of shaft resistance, divide the soil deposit into six layers as in Example 7.7. The computation of R_{sf} is summarized in Table E7.8. The soil type in Table E7.8 is determined using the average soil behavior type index $\bar{I}_C$ for the designated soil layer and Figure 3.32. Pile category is IIA for driven concrete piles; the values of α_p are then determined according to Table 7.9, considering pile type, soil type, and q_t values. For each soil layer, the $(f_{sf})_i$ is calculated using the average $\bar{q}_t$ of that layer and the selected α_p,

$$(f_{sf})_i = \frac{\bar{q}_t}{\alpha_p} \tag{7.35}$$

$(R_{sf})_i$ = shaft resistance for the designated soil layer i, and

$$(R_{sf})_i = (A_s)_i (\bar{f}_{sf})_i \tag{7.36}$$

$$(R_{sf})_i = (\pi D_p)(\bar{f}_{sf})_i (L_{pi}),$$

where

L_{pi} = length of the pile for the designated soil layer i

R_{sf} — ultimate shaft resistance = sum of all $(R_{sf})_i$ along the pile shaft

$$Q_u = R_{tf} + R_{sf} = 1410 + 1295 = 2705 \text{ kN}$$

Table E7.8 Calculation of ultimate shaft friction using LCPC method

Layer no.	Depth range, m	$\bar{q}_t$, MPa	$\bar{f}_s$, kPa	$\bar{I}_C$	Soil behavior type	α_p	$(\bar{f}_{sf})_i$, kPa	$(R_{sf})_i$, kN
1	0–2	–	–	–	–	–	–	–
2	2–6	0.89	6.36	2.37	Sandy mixture	60	32.9	331.0
3	6–10	2.89	16.04	2.27	Sandy mixture	60	35[a]	351.9
4	10–14	5.21	17.57	2.08	Sandy mixture	100	52.1	523.8
5	14–18	4.56	41.50	1.92	Sands	60	35[a]	88.0
6	18–20	4.16	33.77	–				
							R_{sf} = sum	1295

$\bar{q}_t, \bar{f}_s$ = arithmetic average respective value from CPTu for the designated soil layer.
$\bar{I}_C$ = arithmetic average soil behavior type (SBT) value (see Chapter 3) according to Robertson (2009).

Note

[a] The original calculated value exceeded the maximum value allowed in Table 7.9.

7.4.4 Ultimate capacities for piles on rock

Piles driven to hard intact rock can carry large loads. For rocks with RQD (rock quality designation; see Chapter 3) in excess of 50%, the bearing capacity from bearing stratum can generally be higher than the structural capacity of the pile. The design of piles in these cases is thus controlled by the structural performance of the piles. Piles supported on soft weathered rock or rock with lower RQD values are generally designed based on pile load tests (described in Chapter 8) or non-penetration in-situ tests such as pressuremeter tests (PMT). Pre-bored pressuremeter (see Chapter 3) is ideal for the determination of bearing capacity and settlement analysis of pile foundations in hard soil such as glacial till or soft rock (Baguelin et al., 1978).

7.4.5 Factor of safety and allowable bearing capacity

The above static analysis provides an estimate of the ultimate pile capacity, Q_u, or soil resistance. The allowable soil resistance or design load, Q_a, is obtained by dividing the ultimate pile capacity by a factor of safety, FS. Or,

$$Q_a = \frac{Q_u}{FS} \tag{7.37}$$

As previously described, the construction of piles can cause severe disturbance to the surrounding soil. The process and its effects are complicated so that it is not possible or practical to design a pile with a rigorous procedure. The static analysis methods introduced above are semi-empirical at the best. The use of an FS is to recognize that there are uncertainties involved in the static analysis and construction of piles. These uncertainties could come from

- Uncertainties in the method of static analysis itself.
- Variability and uncertainty in soil/rock and pile material properties involved in the analysis.
- Quality or consistency in the construction of piles.
- Variability in the level of construction monitoring—static load test, dynamic formula, wave equation analysis, and dynamic testing.

The FS involved in static analysis ranges from 2 to 4, but most of the static analysis methods recommend an FS of 3. The selection of FS came from previous experience of using a specific static analysis method, consideration of the quality of construction practice involved in a given project, and how well we intend to monitor the field construction of piles. In practice, we generally allow a lower FS if more detailed testing is conducted during construction. Chapter 8 describes the details of construction and testing of piles. Of all the available pile testing methods, the static load test and the dynamic testing methods are the most effective in assuring the quality of pile construction. Thus, a relatively low FS is usually allowed when these testing methods are used during pile construction.

EXAMPLE 7.9

Given

The ultimate capacities of the pile from Examples 7.7 and 7.8 using the Eslami and Fellenius method and the LCPC method.

Required

Comparison of the Eslami and Fellenius and LCPC methods. Determine the allowable capacity based on CPT and an FS of 2.5.

Solution

Eslami and Fellenius show that:

$$Q_u = R_{tf} + R_{sf} = 927 + 1320 = 2247 \, \text{kN}$$

$$Q_a = \frac{Q_u}{FS} = (2247)/(2.5) = 899 \, \text{kN} \tag{7.37}$$

According to the LCPC method,

$$Q_u = R_{tf} + R_{sf} = 1410 + 1295 = 2705 \, \text{kN}$$

$$Q_a = \frac{Q_u}{FS} = (2705)/(2.5) = 1082 \, \text{kN} \tag{7.37}$$

For the same CPTu data, it would appear that the R_{sf} values from both methods are comparable. The R_{tf} value from LCPC is much higher than that from the Eslami and Fellenius method. Q_u from LCPC method is approximately 20% higher than that from the Eslami and Fellenius method. As each method was empirically developed based on a specific set of database, this difference is understandable and should not be taken as an evidence that one method may be consistently more conservative than the other. For large projects, it may be advisable to perform field load tests and calibrate the key parameters involved in the static analysis methods.

7.4.6 Uplift capacity of a single pile

In some cases, piles are used as an anchor to resist tensile forces, or the structure may be subject to seismic loading. In these conditions, the pile uplift capacity may control the minimum pile penetration. The ultimate uplift resistance of a pile consists of shaft friction and weight of the pile. The toe resistance is not applicable in contributing to the uplift capacity of a single pile. There have been studies on the shaft resistance in uplift conditions (e.g., Altaee et al., 1992; De Nicola and Randolph, 1993; Tomlinson, 1994). The general conclusion is that the shaft resistance of a single pile in uplift is either lower than or equal to that under compressive load. The loading rate (i.e., cyclic, sustained, or undrained static) and soil grain characteristics can all affect the shaft resistance in uplift. In practice it is advisable to use a shaft resistance that is lower than the compressive capacity or perform tensile load test to verify the uplift capacity.

7.5 STATIC ANALYSIS FOR LATERAL LOADED PILES

Piles can be subject to lateral loads in addition to axial forces. Cases where piles are subject to significant lateral loads can include:

- *High rise buildings*: The building structure is subject to significant lateral loads due to wind pressure.

- *Wind turbines*: The structure is designed to be situated in areas with strong wind for power generation. The lateral force experienced by the structure can be more significant than the axial force.
- *Bridge structures*: Traffic forces from vehicle braking, bridge bends, thrust from arch structures, flowing water against bridge piers, ship collisions, and bridge abutment.
- *Lightweight structures subject to lateral forces*: Light poles, flagpoles, power transmission towers, and overhead sign structures.
- *Slope stabilization or earth retention*: Piles are used primarily to resist lateral earth pressure.

Figure 7.26 provides a schematic description of examples where piles are used to resist lateral forces. Figure 7.27 shows a case where a row of bored piles (see Chapter 8 for construction of piles) were used for slope stabilization in Hong Kong. Under these circumstances, it is necessary to design the pile foundations to resist the expected lateral loads.

A complete solution of a laterally loaded pile for a given loading condition should provide a set of curves as shown in Figure 7.28. The most important information would be the

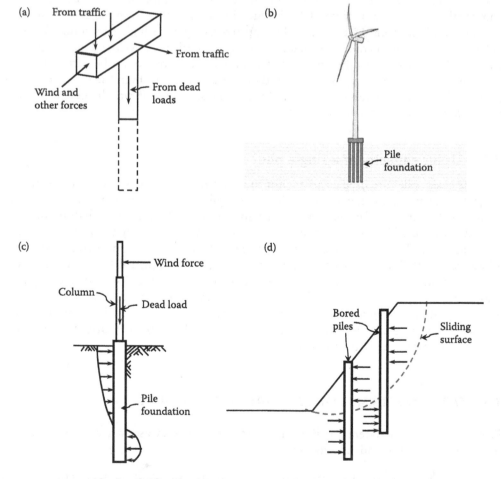

Figure 7.26 Examples of piles under lateral load. (a) Bridge structures. (b) Wind turbine. (c) Light pole structures. (d) Slope stabilization.

Figure 7.27 Bored piles for slope stabilization in Midlevel, Hong Kong. (Courtesy of An-Bin Huang, Hsinchu, Taiwan.)

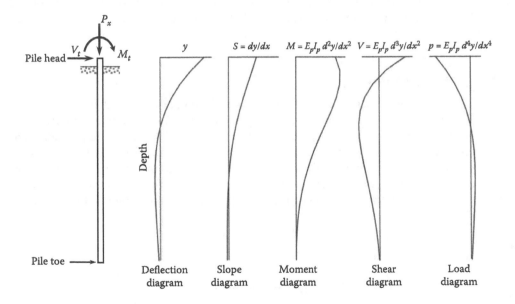

Figure 7.28 Set of curves from the complete solution of a laterally loaded pile.

deflection of the pile and bending moment in the pile, and their relationship with depth. Based on these results, the following issues can then be evaluated:

- Determining the necessary penetration of the pile to carry the lateral load at the pile head without excessive movement.
- Determining the necessary dimensions and reinforcement of the pile to resist the bending, shear, and axial thrust that will be imposed on the pile by the lateral loads in combination with the axial loads.
- Effects of pile foundation deformation on the performance of the structure.

The above solution and evaluation should cover the range of loading conditions to be applied to the pile.

The analysis of a laterally loaded pile deals with soil–pile–structure interaction. The behavior of all three elements can be nonlinear and interdependent. The solution of such problems is inherently nonlinear and usually involves iterative techniques. A rigorous approach is not practical without the help of numerical procedures. Before the advent of personal computers, most engineering design relied on simplified methods such as the equivalent cantilever method (Davisson, 1970) and limit-equilibrium solution or the Broms method (Broms, 1965). The development of the concept of p–y curve or p–y method over half a century ago (McClelland and Focht, 1958; Reese, 1984, 1985) marked the turning point for our current practice in the analysis for laterally loaded piles. The p–y method is a general method for analyzing laterally loaded structures such as piles with combined axial and lateral loads. Recall that the p–y method was also used for the analysis of sheet pile walls (see Section 5.14). With the help of a personal computer and readily available user-friendly software, most of the nonlinear analysis can be handled with reasonable accuracy and efficiency. For these reasons, the p–y method-based computer software is widely used in current engineering practice. Available software includes LPILE by ENSOFT, Inc. (Austin Texas) and FLPIER (The Florida Pier Analysis Program) developed by the University of Florida in conjunction with the Florida Department of Transportation. The following sections introduce the background of the p–y method and a simplified procedure called the characteristic load method (Duncan et al., 1994), which is an approximate method based on a parametric study of numerous p–y method solutions.

7.5.1 The p–y method

The designation "p–y" method is because the soil resistance (p)–lateral deflection (y) relations that develop against the side of the pile are termed p–y curves. For the purpose of analysis, the soil-pile system is modeled by replacing the soil with a set of p–y curves as shown in Figure 7.29. All p–y curves are assumed to act independently without interference between each another. The pile is treated as a beam-column with lateral soil support represented by the p–y curves. The p–y curves can be nonlinear to reflect the nature of soil resistance as a function of pile deflection. The problem of this laterally loaded beam-

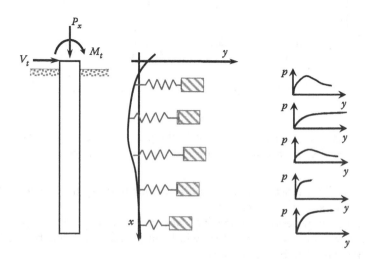

Figure 7.29 Simplification of soil resistance with a set of p–y curves.

column can be described by the following differential equation (Hetenyi, 1946):

$$E_p I_p \frac{d^4 y}{dx^4} + P_x \frac{d^2 y}{dx^2} + E_{py} y = 0 \qquad (7.38)$$

$E_p I_p$ = pile flexural rigidity
E_p = elastic modulus of pile
I_p = moment of inertia of pile
y = pile lateral deflection
x = axial distance along pile
P_x = axial load on pile
E_{py} = soil modulus

Equation 7.38 merely provides a description of how the deflection y relates to distance x for a laterally loaded structural element, in this case the pile. With simplifications, it is possible to obtain a closed-form, but limited solution to Equation 7.38. To obtain a general solution (i.e., deflection y at different axial distance x) of Equation 7.38 for given pile/soil properties (i.e., pile flexural rigidity and soil modulus) and boundary conditions (i.e., load applied at the pile head and restrictions at the pile toe) would require the use of a numerical method. The finite difference method has been used extensively to provide the numerical solution. With the help of a spreadsheet program, the finite difference method can now be easily implemented on a personal computer for that purpose. With the knowledge of y as a function of x or $y(x)$ derived from the finite difference computations, the moment M, shear force V, and soil resistance p (resistance force per unit length of the pile) along the pile (see Figure 7.28; note it is assumed that $P_x = 0$ in the figure) can be determined using the following relationships:

$$M = E_p I_p \frac{d^2 y}{dx^2} \qquad (7.39)$$

$$V = E_p I_p \frac{d^3 y}{dx^3} + P_x \frac{dy}{dx} \qquad (7.40)$$

and

$$p = \frac{d^2 M(x)}{dx^2} = E_p I_p \frac{d^4 y}{dx^4} \qquad (7.41)$$

This gives a rather powerful and comprehensive solution to the laterally loaded pile problem represented by Equation 7.38.

It is assumed that the pile is straight and vertical initially. To apply the finite difference method, the pile is first discretized into segments as shown in Figure 7.30. The ends of the discretized segments (to be called nodes) are numbered as $0, 1, 2, \ldots, t$ (from bottom to top) as indicated in Figure 7.30. The basic idea of the finite difference method is to replace the derivatives by approximations obtained by combining nearby values. For example, we approximate a first-order derivative dy/dx at a generic node m, in the finite difference form as

$$\frac{dy}{dx} = \lim_{\Delta x \to 0} \frac{y(x + \Delta x) - y(x)}{\Delta x} \approx \frac{y(x + \Delta x) - y(x - \Delta x)}{2\Delta x} = \frac{y_{m-1} - y_{m+1}}{2h} \qquad (7.42)$$

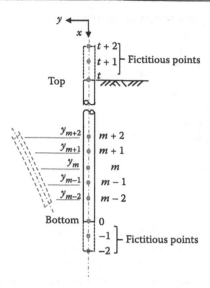

Figure 7.30 Discretization of a laterally loaded pile.

where
 h = length of the segment, usually set to be uniform throughout the pile

For the second-order derivative,

$$\frac{d^2 y}{d^2 x} \approx \frac{y_{m-1} - 2y_m + y_{m+1}}{h^2} \tag{7.43}$$

And the third-order derivative can be approximated as

$$\frac{d^3 y}{d^3 x} \approx \frac{-y_{m-2} + 2y_{m-1} - 2y_{m+1} + y_{m+2}}{2h^3} \tag{7.44}$$

Assume the pile has a uniform $E_p I_p$ and subject to a constant P_x at pile head. Follow Equations 7.39 and 7.43, at note m,

$$E_p I_p \frac{d^4 y}{dx^4} = \left(\frac{d^2 M}{dx^2}\right)_m \tag{7.45}$$

and

$$
\begin{aligned}
\left(\frac{d^2 M}{dx^2}\right)_m = \frac{1}{h^4} [& (R_p)_{m-1} y_{m-2} + (-2(R_p)_{m-1} - 2(R_p)_m) y_{m-1} + ((R_p)_{m-1} \\
& + 4(R_p)_m + (R_p)_{m+1}) y_m + (-2(R_p)_m - 2(R_p)_{m+1}) y_{m+1} \\
& + (R_p)_{m+1} y_{m+2}]
\end{aligned}
\tag{7.46}
$$

where
 $R_p = E_p I_p$ (assumed to be constant throughout the pile)

$$P_x \frac{d^2 y}{dx^2} = \frac{P_x(y_{m-1} - 2y_m + y_{m+1})}{h^2} \tag{7.47}$$

Where y_m is the deflection at node m, and axial force P_x is assumed to be constant over the length of the pile. Positive P_x is compressive. The combined finite difference equation for the entire Equation 7.38 is

$$(R_p)_{m-1}y_{m-2} + (-2(R_p)_{m-1} - 2(R_p)_m + P_xh^2)y_{m-1}$$
$$+ ((R_p)_{m-1} + 4(R_p)_m + (R_p)_{m+1} - 2P_xh^2 + (E_{py})_mh^4)y_m$$
$$+ (-2(R_p)_m - 2(R_p)_{m+1} +_x h^2)y_{m+1} + (R_p)_{m+1}y_{m+2} = 0 \qquad (7.48)$$

7.5.1.1 Boundary conditions at the pile bottom (toe)

The finite difference equation (i.e., Equation 7.48) is repeated for each point 0, 1, 2,..., t (from bottom to top or toe to head) as indicated in Figure 7.30 and results in a set of equations in y. Boundary conditions must be given in order to solve Equation 7.38 and its equivalent finite difference equations. The boundary conditions are assigned in terms of shear, $V = E_pI_p(d^3y/dx^3) + (P_x dy/dx)$, moment, $M = E_pI_p d^2y/dx^2$, or slope $S = dy/dx$ at the pile ends, and expressed in finite difference forms with added fictitious or phantom points (points beyond the physical boundary of the pile in Figure 7.30).

For a long pile, the shear and moment at the toe are small and can be assumed to be zero. For zero moment at the bottom (point 0 in Figure 7.30),

$$M = E_pI_p d^2y/dx^2 = 0 \qquad (7.49)$$

As $E_pI_p \neq 0$, thus $d^2y/dx^2 = 0$, and its finite difference form is

$$y_{-1} - 2y_0 + y_1 = 0 \qquad (7.50)$$

And for zero shear at the bottom,

$$(R_p)_0 \frac{d^3y}{dx^3} + P_x \frac{dy}{dx} = 0 \qquad (7.51)$$

Its finite difference equation is

$$y_{-2} - 2y_{-1} + 2y_1 - y_2 + \frac{P_xh^2(y_{-1} - y_1)}{(R_p)_0} = 0 \qquad (7.52)$$

It is assumed that $(R_p)_{-1} = (R_p)_0 = (R_p)_1$.

7.5.1.2 Boundary conditions at the pile top (head)

Three types of boundary conditions are possible at the pile top, depending on the interactions with the superstructure:

Case 1: Lateral load (shear), V_t and moment, M_t are applied at the top. The difference equations are:

For the applied V_t,

$$y_{t-2} - 2y_{t-1} + 2y_{t+1} - y_{t+2} + \frac{P_x h^2 (y_{t-1} - y_{t+1})}{(R_p)_t} = \frac{2 V_t h^3}{(R_p)_t} \tag{7.53}$$

and for the applied M_t,

$$y_{t-1} - 2y_t + y_{t+1} = \frac{M_t h^2}{(R_p)_t} \tag{7.54}$$

Case 2: V_t is applied and slope, S_t is fixed at the top. The difference equation for V_t is the same as Equation 7.53, and for S_t,

$$y_{t-1} - y_{t+1} = 2h S_t \tag{7.55}$$

Case 3: V_t is applied and pile top rotates at a given stiffness of M_t/S_t. The difference equation for V_t is the same as Equation 7.53, and for M_t/S_t,

$$\frac{y_{t-1} - 2y_t + y_{t+1}}{y_{t-1} - y_{t+1}} = \frac{M_t h}{2(R_p)_t S_t} \tag{7.56}$$

By combining Equations 7.48, 7.50, 7.52, and those associated with one of the three cases for the pile top, a full set of simultaneous equations in y can be assembled. The solution to the simultaneous equations yields the deflections y_m. Other quantities such as slope, S_m shear, V_m and moment, M_m at node m can be computed as follows:

$$S_m = \frac{(y_{m-1} - y_{m+1})}{2h} \tag{7.57}$$

$$V_m = \frac{(R_p)_m (y_{m-2} - 2y_{m-1} + 2y_{m+1} - y_{m+2})}{2h^3} + \frac{P_x (y_{m-1} - y_{m+1})}{h} \tag{7.58}$$

and

$$M_m = \frac{(R_p)_m (y_{m-1} - 2y_m + y_{m+1})}{h^2} \tag{7.59}$$

The assembly of simultaneous equations and solution for unknowns is called implicit scheme in numerical methods. Microsoft Excel has functions to solve the simultaneous equations by computing the inverse and multiplication of the matrixes. Details of the numerical procedures are beyond the scope of this book. An Excel program to solve Equation 7.38 for a simple case of a pile with uniform R_p and a constant E_{py} for each pile segment can be downloaded from the publisher's website by registered users. Students are encouraged to download the program and acquaint themselves with the procedure and explore its capabilities.

Ideally, the p–y curves should be derived from field lateral load tests on the pile. From field measurement of bending moment distribution along the pile $M(x)$, values of p and y

at points along the pile can be obtained by solving the following equations.

$$p = \frac{d^2 M(x)}{dx^2}$$ (7.60)

and

$$y = \int\int \frac{M(x)}{E_p I_p} dx^2$$ (7.61)

The values of moment, $M(x)$ can be derived from strain measurements on the pile (see Chapter 8 for pile load test) in a lateral load test. While y can be obtained with reasonable accuracy by double integration of $M(x)$, the corresponding p value involves double differentiation and the results can be very erratic, as in the case of taking derivatives for any field data. For this reason, the p–y curves are usually determined empirically or back-calculated from field lateral load tests through a trial-and-error process. In-situ tests such as PMT (Baguelin et al., 1978) and flat dilatometer tests (DMT) (Robertson et al., 1989; Huang et al., 2001) have been successfully used to derive p–y curves. Details of establishing p–y curves can be found in Reese (1984, 1985) and Reese et al. (2000, 2006). Readers are referred to Desai and Christian (1977) for more details on the finite difference method.

EXAMPLE 7.10

Given

Consider a 0.8 m diameter and 15 m long bored pile shown in Figure E7.10. The pile head is free to rotate. The bored pile is subject to a lateral load of 80 kN and a moment of 400 kN-m at the ground line. The bored pile is embedded in a uniform clay deposit with undrained shear strength, s_u, of 0.06 MN/m^2. The reinforced concrete pile has an overall elastic modulus of 25,000 MPa. Consider the bored pile as solid and uncracked under the applied load. Assume $E_{py} = 4$ MN/m^3 as a constant value for soil around full length of the pile.

Required

Use the p–y method to

1. Determine the ground line deflection caused by the applied loads.
2. Determine the maximum bending moment within the bored pile for the same loads.

Solution

$$D_p = 0.8 \text{ m}$$
$$I_p = \frac{\pi D_p^4}{64} = \frac{\pi 0.8^4}{64} = 0.02 \text{ m}^4$$
$$E_p I_p = R_p = (25,000)(0.02) = 500 \text{ MN-m}^2$$

$$E_{py} = 4 \text{ MN/m}^3 \text{ throughout the depth of the pile}$$

The analysis by p–y method was performed using the spreadsheet program mentioned previously.

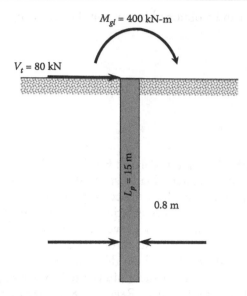

$M_{gl} = 400$ kN-m

$V_t = 80$ kN

$L_p = 15$ m

0.8 m

Figure E7.10 The bored pile and its loading conditions.

Results are plotted in Figure 7.10a, deflection at pile top or pile head, y_t was 0.017 m and maximum moment, M_{max} of 0.446 MN-m occurred at 1.0 m below the ground line as shown in Figure E7.10a.

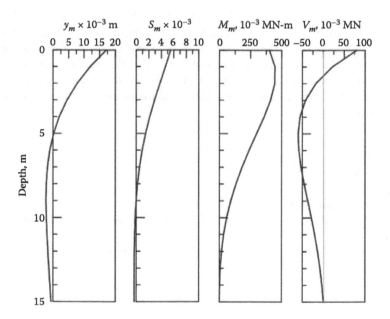

Figure E7.10a Results from spreadsheet *p–y* method analysis.

So far a constant soil modulus E_{py} was used throughout the pile and regardless of the pile deflection. In other words, the soil resistance p was assumed to have a linear relationship with pile deflection y. The numerical scheme presented above can be expanded to consider a separate non-linear *p–y* curve for various parts of the pile as described in

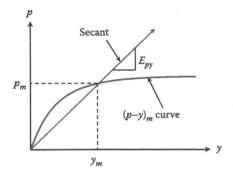

Figure 7.31 E_{py} as a secant modulus to a non-linear p–y curve.

Figure 7.29. For example, a p–y curve, $(p-y)_m$, can be assigned for each segment centered at node m. To consider a non-linear $(p-y)_m$ curve, treat E_{py} as the slope of a trial secant line stemming from the origin of $(p-y)_m$ as shown in Figure 7.31. Compare the p and y obtained at node m from the p–y method computation, with p_m and y_m at the intersection between the secant and $(p-y)_m$ curve. Adjust the slope of the secant (i.e., E_{py}) for each segment if necessary, and repeat the computation. Iterate the computation until the computed p and y are reasonably close to those at the intersection between the secant and p–y curve for all the segments. Readers are encouraged to explore this capability using the spreadsheet program provided on the website.

7.5.2 The characteristic load method

The characteristic load method (CLM) was proposed by Duncan et al. (1994). The method uses dimensional analysis to characterize the nonlinear behavior of laterally loaded piles by means of relationships among dimensional variables. The CLM was developed based on a series of p–y analyses but is simple enough that it can be used by manual calculation. The method can be used to determine (1) ground-line deflections and maximum moments under essentially all three pile top boundary conditions described above for p–y method, and (2) the location of the maximum moment in the pile.

The characteristic load and moment that form the basis for the dimensionless relationships are given as follows:

For clay

$$P_c = 7.34 D_p^2 (E_p R_I) \left(\frac{s_u}{E_p R_I} \right)^{0.68} \tag{7.62}$$

For sand

$$P_c = 1.57 D_p^2 (E_p R_I) \left(\frac{\gamma' D_p \varnothing' K_p}{E_p R_I} \right)^{0.57} \tag{7.63}$$

For clay

$$M_c = 3.68 D_p^3 (E_p R_I) \left(\frac{s_u}{E_p R_I} \right)^{0.46} \tag{7.64}$$

For sand

$$M_c = 1.33D_p^3(E_pR_I)\left(\frac{\gamma'D_p\varnothing'K_p}{E_pR_I}\right)^{0.40} \tag{7.65}$$

where
P_c = characteristic load (F)
M_c = characteristic moment (F-L)
D_p = pile width or diameter (L)
E_p = pile shaft elastic modulus (F/L^2)
R_I = moment of inertia ratio to be defined below
s_u = undrained shear strength of clay (F/L^2), in the top 8 D_p below the ground surface
γ' = effective unit weight of sand, total unit weight above the water table and buoyant unit weight below the water table (F/L^3), in the top 8 D_p below the ground surface
$\varnothing'$ = drained friction angle for sand (degrees), in the top 8 D_p below the ground surface
K_p = Rankine passive earth pressure coefficient [= $\tan^2(45 + \varnothing'/2)$, see Chapter 5]

The ratio of moment of inertia, R_I (dimensionless), is the ratio of the moment of inertia of the pile, I_p, over the moment of inertia of a solid circular cross-section, $I_{circular}$, with a diameter of D_p. Or,

$$R_I = \frac{I_p}{I_{circular}} \tag{7.66}$$

Thus $R_I = 1$ for a normal, un-cracked pile without central voids and,

$$I_{circular} = \frac{\pi D_p^4}{64} \tag{7.67}$$

7.5.2.1 Deflection due to lateral loads applied at ground line

Figure 7.32 shows the correlations between load and deflection for piles subjected to lateral load, V_{gl}, at ground line for free and fixed head conditions. The free head condition is

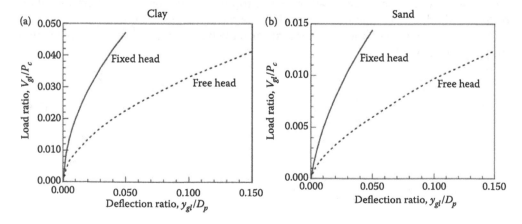

Figure 7.32 Load-deflection curves for clay and sand. (a) Load deflection curves for clay. (b) Load deflection curves for sand. (From Duncan, J.M., Evans, L.T. Jr. and Ooi, P.S.K. 1994. Lateral load analysis of single piles and drilled shafts. *Journal of Geotechnical Engineering*, ASCE, 120(6), Figure 2, p.1023. Reproduced with permission of the ASCE.)

similar to Case 1 pile top boundary condition in the p–y method described above where the pile head is free to rotate, and the fixed head condition is similar to Case 2 in p–y method where the slope of the pile head is fixed. In the free head condition, the moment applied at ground line, $M_{gl} = 0$. The pile head is located at ground line. Therefore, the ground line deflection, y_{gl}, is also the pile head deflection, y_t. The lateral load is divided or normalized by the characteristic load, P_c. The pile deflection at ground line, y_{gl}, is divided or normalized by the pile diameter, D_p. It is apparent that the pile behaves much stiffer under the fixed head condition where the pile head is fixed against rotation. For practical purposes, the stiffness of a conventional pile cap is sufficient to fix the top of a pile against rotation provided the embedment of the pile in the cap is adequate to allow full moment transfer from the pile to the cap, and vice versa. The fixed-head case is thus valid when piles are embedded in a reinforced concrete pile cap or building frame that are themselves constrained against rotation by virtue of being connected to other piles or building structure.

7.5.2.2 Deflection due to moments applied at ground line

Figure 7.33 shows the relationships between the moment applied at the ground line, M_{gl}, and deflections at the ground line, y_{gl} induced by the applied moment. The moment is divided by the characteristic moment (M_{gl}/M_c), and the deflection is divided by the pile diameter (y_{gl}/D_p).

7.5.2.3 Deflection due to lateral loads applied above ground line

Lateral loads applied above the ground line induce both lateral load and moment at the ground line, as shown in Figure 7.34. Both the load and the moment induce deflection. Because the behavior is nonlinear, it is not sufficient by just adding the deflections caused by the load and the moment. Instead, the nonlinear effects are taken into account by using

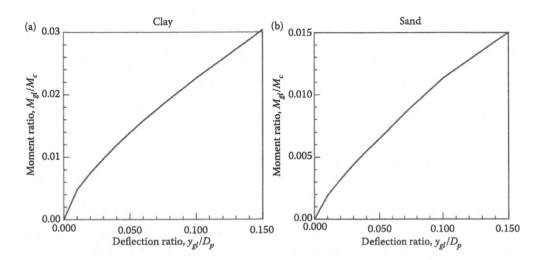

Figure 7.33 Moment-deflection curves for clay and sand. (a) Moment deflection curve for clay. (b) Moment deflection curve for sand. (From Duncan, J.M., Evans, L.T. Jr. and Ooi, P.S.K. 1994. Lateral load analysis of single piles and drilled shafts. *Journal of Geotechnical Engineering*, ASCE, 120(6), Figure 3, p. 1024. Reproduced with permission of the ASCE.)

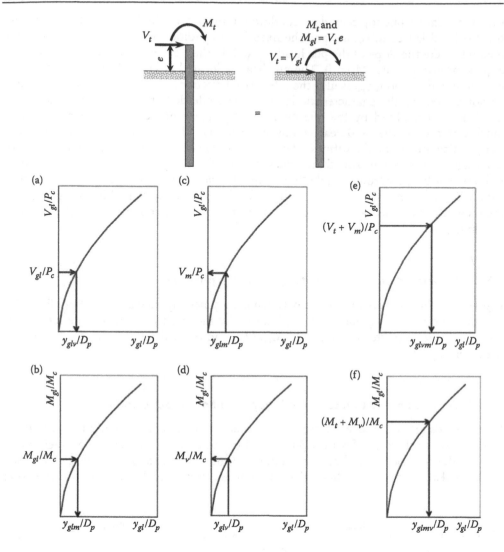

Figure 7.34 Superposition of deflections due to applied load and moment. (a) Deflection y_{glv} by lateral load V_{gl} alone. (b) Deflection y_{glm} by moment M_{gl} alone. (c) Load V_m that would cause y_{glm}. (d) Moment M_v that would cause y_{glv}. (e) Deflection by $(V_t + V_m)$. (f) Deflection by $(M_t + M_v)$.

a nonlinear superposition procedure. The induced pile deflection at the ground line is computed according to the following procedure:

Step 1: Calculate the deflections that would be caused by the load ($= V_t = V_{gl}$) acting alone (y_{glv}), and by the moment ($M_{gl} = V_t e$) acting alone (y_{glm}), as shown schematically in Figure 7.34 (a and b, respectively). These are calculated using Figures 7.32 and 7.33, as explained above.

Step 2: Determine a value of load that would cause the same deflection as the moment (this is called V_m), and a value of moment that would cause the same deflection as the load (this is called M_v). These are determined as shown schematically in Figure 7.34 (c and d, respectively).

Step 3: Determine the ground line deflections that would be caused by the sum of the real load plus the equivalent load ($V_t + V_m$), and the real moment plus the equivalent moment

$(M_t + M_v)$, as shown in Figure 7.34 (e and f, respectively). These values, called y_{glvm} and y_{glmv}, respectively, are determined using Figures 7.32 and 7.33. The estimated value of deflection due to combined effects of load and moment is the average of the two values, y_{glvm} and y_{glmv} and is calculated as follows:

$$\frac{y_{gl\,combined}}{D_p} = 0.5 \frac{(y_{glvm} + y_{glmv})}{D_p} \tag{7.68}$$

7.5.2.4 Maximum moment due to lateral load/moment applied at or above ground line

Figure 7.35 shows the correlations between normalized applied lateral load at ground line, V_{gl}/P_c and the normalized maximum moment, M_{max}/M_c for free and fixed head conditions, in clay and sand. For fixed head piles, the maximum moment, M_{max} occurs at ground line where the pile head is embedded in the pile cap and V_{gl} is applied at the ground line.

If the lateral load is applied above the ground line, it can be treated as a combination of moment, M_{gl}, and lateral load, V_{gl}, applied at the ground line, or a combination of M_{gl} and V_{gl} is applied at the ground line. In these cases, the pile is in a free head condition and the maximum moment occurs at some depth below the ground line. To determine the depth where the maximum moment occurs, we first compute $y_{gl\,combined}$ as described above (Equation 7.68), and then solve for the "characteristic length" T as follows:

$$y_{gl\,combined} = \frac{2.43 V_{gl}}{E_p I_p} T^3 + \frac{1.62 M_{gl}}{E_p I_p} T^2 \tag{7.69}$$

Once T is determined, the relation between bending moment and depth x is given by the following equation:

$$M_x = V_{gl} T A_m + M_{gl} B_m \tag{7.70}$$

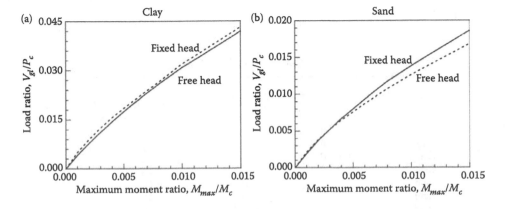

Figure 7.35 Load-maximum moment curves for clay and sand. (a) Load maximum moment curves for clay. (b) Load maximum moment curves for sand. (From Duncan, J.M., Evans, L.T. Jr. and Ooi, P.S.K. 1994. Lateral load analysis of single piles and drilled shafts. *Journal of Geotechnical Engineering*, ASCE, 120(6), Figure 5, p. 1027. Reproduced with permission of the ASCE.)

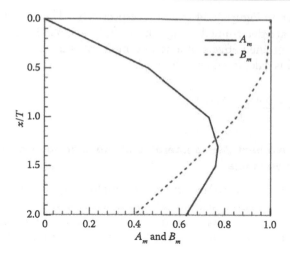

Figure 7.36 Coefficients A_m and B_m. (After Matlock, H. and Reese, L.C. 1961. *Proceedings of the Fifth International Conference on Soil Mechanics and Foundation Engineering*, Paris, France, Vol. 2, pp. 91–97.)

The relationship between A_m/B_m and normalized depth, x/T, is plotted in Figure 7.36. With the help of Equation 7.70, a moment diagram can be constructed for the portion of the pile where moment is likely to be critical. A_m relates the moment to V_{gl}. The maximum moment due to the lateral load at ground line can be seen to occur at a depth of 1.3 T (depth corresponds to the maximum value of A_m). B_m relates the moment to M_{gl} and the maximum moment due to the moment load at ground line ($x/T = 0$). The depth where the maximum moment occurs depends upon the combination of the applied lateral load and moment at the ground line. The procedure to execute this method is demonstrated in Example 7.11.

For the characteristic load method to be valid, the pile should have a minimum depth of penetration as shown in Table 7.11. These minimum penetrations are based on the principle that the base of the pile should not deflect when the head is loaded. If the pile penetration is less than those specified in Table 7.11, the ground line deflection will be underestimated and the maximum moment will be overestimated by the characteristic load method.

Table 7.11 Minimum pile penetration depth for CLM

Soil type	$\dfrac{E_p R_I}{s_u}$	$\dfrac{E_p R_I}{(\gamma' D_p \varnothing' K_p)}$	Minimum penetration
Clay	1×10^5		$6\ D_p$
Clay	3×10^5		$10\ D_p$
Clay	1×10^6		$14\ D_p$
Clay	3×10^6		$18\ D_p$
Sand		1×10^4	$8\ D_p$
Sand		4×10^4	$11\ D_p$
Sand		2×10^6	$14\ D_p$

Source: Duncan, J.M., Evans, L.T., Jr. and Ooi, P.S.K. 1994. Lateral load analysis of single piles and drilled shafts. *Journal of Geotechnical Engineering*, ASCE 120(6), Table 5, p.1028. Reproduced with permission of the ASCE.

Limitations of the characteristic load method for piles embedded in soils include:

- It considers soil conditions in the top 8 D_p to be uniform.
- It does not consider the effects of axial loads, P_x, on bending moments.
- It does not consider changes in flexural rigidity if cracking occurs due to excessive bending moment, for example.

EXAMPLE 7.11

Given

For the same pile dimensions described in Example 7.10 and shown in Figure E7.10. The pile head is free to rotate. The bored pile is subject to a lateral load of 80 kN and a moment of 400 kN-m at the ground line. The bored pile is embedded in a uniform clay deposit with undrained shear strength, s_u, of 0.06 MPa. The reinforced concrete pile has an overall elastic modulus of 25,000 MPa. Consider the bored pile as solid and uncracked under the applied load.

Required

Use the characteristic method to

1. Determine the ground line deflection caused by the applied loads.
2. Determine the maximum bending moment within the bored pile for the same loads.

Solution

$$D_p = 0.8 \text{ m} \qquad V_t = 80 \text{ kN} = 0.08 \text{ MN}$$

$$I_p = \frac{\pi D_p^4}{64} = \frac{\pi 0.8^4}{64} = 0.02011 \text{ m}^4$$

$$E_p I_p = R_p = (25,000)(0.02) = 500 \text{ MN-m}^2$$

By characteristic load method:
  For uncracked pile, $R_I = 1$

$$\frac{E_p R_I}{s_u} = (25,000)(1)/0.06 = 4.2 \times 10^5, \text{ therefore for the CLM to be valid, the pile}$$
length has to be $>14\, D_p$, or $(14)(0.8) = 11.2$ m. (Table 7.11)
  The given pile length is 15 m, thus is acceptable.
  For clay, compute P_c using Equation 7.62

$$P_c = 7.34 D_p^2 (E_p R_I) \left(\frac{s_u}{E_p R_I} \right)^{0.68} = (7.36)(0.8)^2 (25,000 \times 1) \left(\frac{0.06}{25,000 \times 1} \right)^{0.68}$$
$$= 17.76 \text{ MN}$$

Compute M_c using Equation 7.64,

$$M_c = 3.86 D_p^3 (E_p R_I) \left(\frac{s_u}{E_p R_I} \right)^{0.46} = (3.86)(0.8)^3 (25,000 \times 1) \left(\frac{0.06}{25,000 \times 1} \right)^{0.48}$$
$$= 128.44 \text{ MN-m}$$

Step 1: Ground line deflection due to lateral load alone,
$V_{gl}/P_c = 0.08/17.76 = 0.0045$, $y_{glv}/D_p = 0.0038$ according to Figure 7.32a
(free head), $y_{glv} = (0.0038)(0.8) = 0.0030$ m
Ground line deflection due to moment only,
$M_{gl}/M_c = 0.4/128.44 = 0.0031$, $y_{glm}/D_p = 0.0068$ according to
Figure 7.33a, $y_{glm} = (0.0068)(0.8) = 0.0054$ m

Step 2: The corresponding lateral load, V_m/P_c for y_{glm}/D_p of 0.0068
$V_m/P_c = 0.0075$, according to Figure 7.32a (free head),
The corresponding moment, M_v/M_c for y_{glv}/D_p of 0.0038
$M_v/M_c = 0.002$, according to Figure 7.33a

Step 3: Ground line deflection due to $(V_{gl} + V_m)/P_c = 0.0045 + 0.0075 = 0.012$,
$y_{glvm}/D_p = 0.016$, according to Figure 7.32a (free head),
Ground line deflection due to $(M_{gl} + M_v)/M_c = 0.0031 + 0.002 = 0.0051$,
$y_{glmv}/D_p = 0.011$, according to Figure 7.33a

According to Equation 7.68,

$$\frac{y_{gl\ combined}}{D_p} = 0.5\frac{(y_{glvm} + y_{glmv})}{D_p} = 0.5(0.016 + 0.011) = 0.0135$$

Thus $y_{gl\ combined} = (0.0135)(0.8) = 0.0108$ m
To obtain M_{max}, determine T following Equation 7.69 using $y_{gl\ combined}$

$$y_{gl\ combined} = \frac{2.43V_{gl}}{E_pI_p}T^3 + \frac{1.62M_{gl}}{E_pI_p}T^2$$

$T = 2.23$ m from trial and error

$0.0108 = (2.43)(0.08)(2.23)^3/500 + (1.62)(0.4)(2.23)^2/500$

Use Equation 7.70 to determine the maximum bending moment,

$M_x(MN-m) = V_{gl}TA_m + M_{gl}B_m = (0.08)(2.23)\,A_m + (0.4)B_m = 0.1784A_m + 0.4B_m$

Using Figure 7.36 to determine M_x for a series of x/T as follows,

x/T	M_x (MN-m) $= 0.1784A_m + 0.4B_m$
0.4	$(0.178)(0.382) + (0.4)(0.981) = 0.461$
0.5	$(0.178)(0.460) + (0.4)(0.970) = 0.470$
0.6	$(0.178)(0.512) + (0.4)(0.950) = 0.471$
0.7	$(0.178)(0.565) + (0.4)(0.925) = 0.470$
0.8	$(0.178)(0.620) + (0.4)(0.899) = 0.470$

The maximum moment occurs at $x/T = 0.6$, so $x = (0.6)(2.23) = 1.34$ m
Maximum moment $= 0.471$ MN-m

7.6 GROUP EFFECTS

The static analysis discussed so far has been limited to that of a single pile. In practice, however, the piles are often applied in groups especially for driven piles. Multiple piles are structurally connected to a pile cap, as schematically described in Figure 7.37. The pile cap adds significant rigidity and axial/lateral resistance to the pile group, especially when the cap is buried in soil. It should be mentioned that for bored piles, as their dimensions can be adjusted to meet the requirement, a single pile is often used to resist the designed loading conditions.

When piles are installed in close proximity, the disturbance caused by pile construction can alter the states of the surrounding soil and thus the performance of the adjacent piles. A conceptual description of pile group effect is shown in Figure 7.38. The pile shaft friction drags the surrounding soil downward, and thus the downward ground surface movement as a pile is loaded in compression. This downward movement dissipates away from the pile. When piles are closely spaced, the influence zones overlap, causing the downward movement in group (Figure 7.38b) to be more significant and more widespread than the single pile (Figure 7.38a). As a result, the capacity of the pile group is not necessarily the same as the sum of the individual piles in the group.

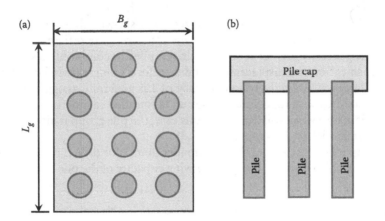

Figure 7.37 Schematic views of a pile group. (a) Plan view. (b) Side view.

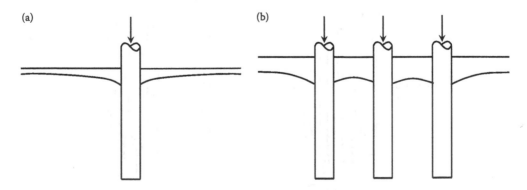

Figure 7.38 Group effects between closely spaced piles. (a) Single pile. (b) Group behavior.

7.6.1 Pile group effects in axial compression

The efficiency of a pile group in supporting the foundation load in axial compression is defined as the ratio of the ultimate capacity of the group to the sum of the ultimate capacities of the individual piles comprising the group. This may be expressed as follows:

$$n_g = \frac{Q_{ug}}{nQ_u} \qquad (7.71)$$

where

n_g = pile group efficiency
Q_{ug} = ultimate capacity of the pile group
n = number of piles in the pile group
Q_u = ultimate capacity of each individual pile in the pile group

To describe the group effects, a block failure model as shown in Figure 7.39 is often used as a benchmark (Reese et al., 2006). It is assumed that piles and soil mass within the block fails simultaneously as one rigid body. The load carried by the block is the sum of resistances from the toe and friction on the perimeter of the block. The ultimate capacity of the block is compared with the sum of the individual single piles in the group. The smaller of these two is selected as the capacity of the group.

Studies (Vesić, 1969; O'Neill, 1983) have generally showed that the group effects for axially loaded piles in cohesive and cohesionless soils are quite different. The pile group efficiency, n_g, is generally greater than 1 for driven piles in cohesionless soil. Pile driving causes soil densification and displacement, and therefore an increase in lateral stress along the pile shaft. In practice, a pile group efficiency of 1 is generally assigned for group piles driven in cohesionless soils unless the pile group is founded in a dense sand with limited thickness overlain by a weak soil deposit. In this case, the pile group is also evaluated using the block failure model as shown in Figure 7.39.

The pile group efficiency, n_g, is generally smaller than 1 for driven piles in cohesive soil. For driven piles in cohesive soils, the load-carrying capacity of the pile group is often taken

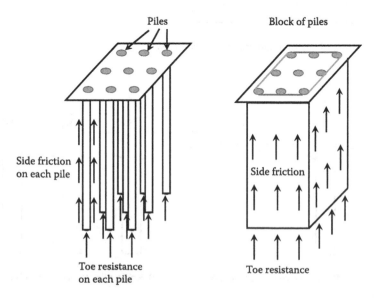

Figure 7.39 Block failure model for block of closely spaced piles.

as the smaller of the sum of the individual single piles in the group and the ultimate bearing capacity of the block failure model as shown in Figure 7.39. In saturated clays, the pile driving induces pore water pressure increase. The affected zone of pore water pressure increase is significantly larger for pile group than a single pile. The following dissipation of pore water pressure and therefore the process of pile setup (i.e., increase of pile load, carrying capacity with time as the surrounding soil consolidates and pore water pressure dissipates) is much slower for pile group.

Pile group effects can be different between driven and bored piles (Huang et al., 2001). For pile groups, especially bored pile groups, the axial compression capacity can be very large, and full-scale load tests are not practical and rarely performed. For important projects, the pile group efficiency is usually evaluated using 3D numerical analysis and physical model tests such as centrifuge tests.

EXAMPLE 7.12

Given

A 3 × 3 pile group of 0.8 m diameter and 15 m long precast concrete piles is to be installed in a saturated clay deposit as shown in Figure E7.12. The piles are spaced at 2 m center-to-center. The groundwater table is on the ground surface. Consider the pile has an ultimate unit shaft resistance, f_{sf}, of 18 kPa. The bearing stratum has an undrained shear strength, s_u, of 80 kPa.

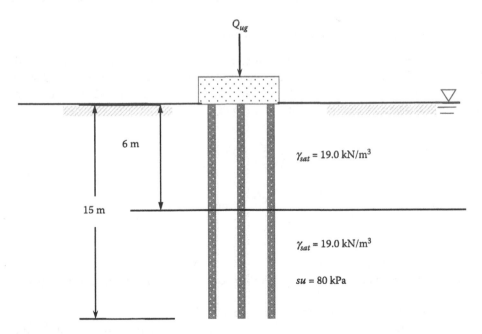

Figure E7.12 Clay soil profile for the group pile.

Required

Determine the ultimate bearing capacity of the pile group, Q_{ug}.

Solution

If consider Q_{ug} as the sum of the Q_u of the 9 single piles, Q_u of a single pile is calculated as follows:

$$R_{sf} = f_{sf}\pi D_p D_b = (18)(3.1416)(0.8)(15) = 678.6\,\text{kN}$$

$$R_{tf} = s_u N_c \frac{\pi D_p^2}{4} = (80)(9)\left(\frac{(3.1416)(0.8)(0.8)}{4}\right) = 361.9\,\text{kN}$$

The single pile $Q_u = R_{sf} + R_{tf} = (678.6) + (361.9) = 1040.5\,\text{kN}$
The sum of 9 piles $= (1040.5)(9) = 9364\,\text{kN}$

If Q_{ug} is considered as the ultimate capacity of a block failure model, the block should have sides of 4.8 m by 4.8 m (i.e., $B_g = L_g = 4.8$ m) and a depth (D_b) of 15 m.

$$R_{sf} = f_{sf}(B_g D_b + L_g D_b)2 = (18)((4.8)(15) + (4.8)(15))(2) = 5184\,\text{kN}$$

$$R_{tf} = s_u N_c B_g L_g = (80)(9)(4.8)(4.8) = 16{,}588.8\,\text{kN}$$

For the failure block, $Q_u = R_{sf} + R_{tf} = 5148 + 16{,}588.8 = 21{,}772.8\,\text{kN} > 9364\,\text{kN}$
Thus $Q_{ug} = 9364\,\text{kN}$

7.6.2 Settlement of pile groups under axial compression load

The settlement of a pile group is likely to be much greater than the settlement of an individual pile carrying the same load per pile in the group. The soil influence zone under a pile group due to vertical stress increase is considerably larger than that under a single pile.

7.6.2.1 Settlement of a pile group in cohesionless soils

For pile groups in cohesionless soils, the settlement occurs immediately after loading. Meyerhof (1976) proposed to use the same empirical method for spread foundations to estimate the settlement for a pile group in sand and gravel not underlain by more compressible soil, using SPT $(N_1)_{60}$ values as follows:

$$\text{Sand and gravel} \qquad s = \frac{0.96 q_a \sqrt{B_g} I_f}{(N_1)_{60}} \quad \text{(in mm)} \tag{7.72}$$

$$\text{For silty sand} \qquad s = \frac{1.92 q_a \sqrt{B_g} I_f}{(N_1)_{60}} \quad \text{(in mm)} \tag{7.73}$$

where
$s =$ settlement of the pile group
$q_a =$ design group bearing pressure in kPa, group design load divided by group area $L_g \times B_g$ as shown in Figure 7.37
$B_g =$ width of the pile group in m (see Figure 7.37)
$(N_1)_{60} =$ average N value corrected for both energy ratio to $E_r = 60\%$ and effective overburden stress σ'_{vo} of 100 kPa, within a depth of B_g below the pile group
$I_f =$ influence factor for group embedment $= [1 - (z_p/B_g)] \geq 0.5$
$z_p =$ pile embedment depth in m

Alternatively, the pile group settlement can be estimated using q_t from CPT as follows:

$$s = \frac{42 q_a B_g I_f}{q_t} \tag{7.74}$$

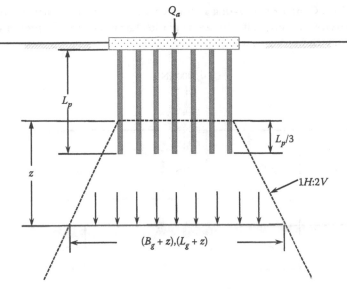

Figure 7.40 Pile group as an equivalent footing $\Delta\sigma_v = \dfrac{Q_a}{(B_g + z)(L_g + z)}$.

where

> s, q_a, B_g, and I_f are defined above, s is in the same unit as B_g
>
> $\overline{q_t}$ = average cone tip resistance in kPa (same unit as q_a) within a depth of B_g below the pile group

7.6.2.2 Settlement of a pile group in cohesive soils

Terzaghi and Peck (1967) proposed that pile group settlements be evaluated using an equivalent footing situated at 1/3 of the pile length above the pile toe. This concept is illustrated in Figure 7.40. The equivalent footing has a plan area B_g by L_g that corresponds to the perimeter dimensions of the pile group as shown in Figure 7.37. The pile group load over this plan area is then the bearing pressure transferred to the underlying soil through the equivalent footing. The load can be assumed to spread within a pyramid of sides sloped at $1H{:}2V$ and to cause uniform additional vertical stress in the underlain soil layers. The vertical stress increase, $\Delta\sigma_v$, at any level is equal to the load carried by the group divided by the plan area at that level. Consolidation settlement within the underlying cohesive soil layer is calculated based on the vertical stress increase in that layer, following the procedure described in Chapter 4. It should be noted that a main reason for using piles is to extend the foundation through the compressible layer. For properly designed pile foundations, highly compressible, normally consolidated clay layer under the pile toe should not be allowed.

A better approach to analyze pile group settlement in cohesive soil may be to use the concept of neutral plane, which is introduced along with the phenomenon of negative skin friction in Section 7.7.

EXAMPLE 7.13

Given

The 3 × 3 pile group of 0.8 m diameter and 15 m long pile group and the soil profile in Example 7.12 is subject to a sustained design load, Q_a, of 3000 kN. The soil profile along with additional properties needed for the settlement analysis is given in

Figure E7.13. Consider the bearing stratum consists of overconsolidated clay and the effective overburden stress after the group pile loading is less than the preconsolidation stress within the bearing stratum.

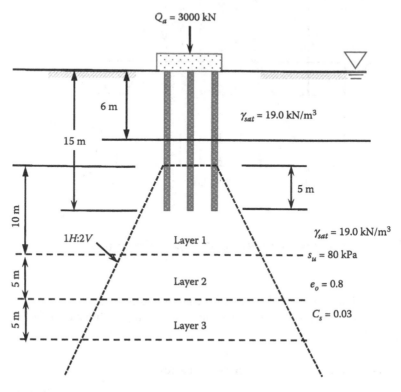

Figure E7.13 The clay soil layers for pile group settlement analysis.

Required

Estimate the settlement of the pile group using the equivalent footing method.

Solution

Divide the bearing stratum below the pile toe into three 5 m thick layers. Use the stress values at the center of each layer for the settlement analysis.

For the original effective overburden stress at the center of each layer,

$$\sigma'_{vo} = (\gamma_{sat} - \gamma_w)z$$

where γ_w is the unit weight of water and z is depth to the center of soil layer from ground surface.

The stress increase, $\Delta\sigma_v$, at the center of each layer due to the loading at equivalent footing, stemming from 1/3 pile length above the pile toe, is computed as

$$\Delta\sigma_v = \frac{Q_a}{(B_g + z)(L_g + z)}$$

where z is the depth to the center of each layer from 1/3 pile length above the pile toe.

The settlement, S, is computed considering that the effective overburden stress after the group pile loading is less than the pre-consolidation stress within the bearing stratum. Thus

$$S = \frac{C_s H_c}{1 + e_o} \log \frac{\sigma'_{vo} + \Delta \sigma_v}{\sigma'_{vo}}$$

where C_s is the recompression index and H_c is the thickness of the clay layer ($H_c = 5$ m or 5000 mm), $e_o = 0.8$ and $C_s = 0.03$ as shown in Figure E7.13. Details on the foundation settlement analysis are described in Chapter 4.

Related computations for the settlement analysis are summarized in Table E7.13. As indicated in Table E7.13, at the center of layer 3, the stress increase, $\Delta \sigma_v$, is 6.0 kPa, significantly less than 10% of the σ'_{vo} of that layer.

The settlement due to consolidation of the underlain clay layer is 6.8 mm.

Table E7.13 Computations to determine consolidation settlement

Layer	Depth below ground surface at center of layer, m	σ'_{vo} at center of layer, kPa	Depth below 1/3 pile length, m	$\Delta\sigma_v$ kPa	Settlement, mm
1	17.5	161	7.5	19.8	4.2
2	22.5	207	12.5	10.0	1.7
3	27.5	253	17.5	6.0	0.9
Sum					6.8

7.6.3 Pile group effects in uplift loads

When piles with uplift loads are driven to a relatively shallow bearing stratum, uplift capacity may control the foundation design. The procedure proposed by Tomlinson (1994) is often used in evaluating the ultimate uplift capacity of a pile group. The ultimate uplift capacity of a pile group in cohesionless soils is taken as the effective weight of the block of soil extending upward from the pile toe level at a slope of 1H:4V, as shown in Figure 7.41. The weight of the piles within the soil block is considered equal to the weight

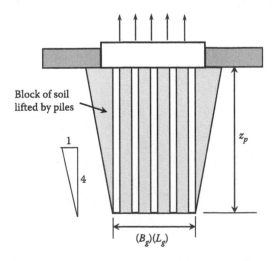

Figure 7.41 Pile group under uplift load in cohesionless soil. (Adapted from Tomlinson, M.J. 1994. *Pile Design and Construction Practice*. E & FN Spon, London, 411p.)

of the soil. The shear resistance around the perimeter of the soil block is ignored in the calculation.

For pile groups in cohesive soils, the same block failure model as shown in Figure 7.39 is used to estimate the group uplift capacity (Tomlinson, 1994). The calculation is based on the undrained shear resistance of the soil within the block and the effective weight of the pile cap and pile–soil block. The ultimate group capacity against block failure in uplift, Q_{ug} in kN, is calculated as follows:

$$Q_{ug} = 2z_p(B_g + L_g)s_{u1} + W_g \tag{7.75}$$

where
 z_p = embedded length of piles in m
 B_g = width of pile group in m
 L_g = length of pile group in m
 s_{u1} = undrained shear strength of the soil along the pile group perimeter in kPa
 W_g = effective weight the pile/soil block including the pile cap weight in kN

Regardless of the soil type, the ultimate group uplift capacity determined from the block failure method should not exceed the sum of the ultimate uplift capacities of the individual piles in the group. A factor of safety of 2 to 3 is generally applied to determine the allowable uplift capacity of the pile group.

7.6.4 Lateral capacity of pile groups

The ability of a pile group to resist lateral loads such as those from vessel impact, earth pressure, wind or wave loading, seismic events, and other sources is an important aspect in foundation design. The group effects for a pile group under lateral load are much more obvious than those under axial loading conditions. If we apply the pile group efficiency here using Equation 7.71, n_g is consistently less than 1. The deflection of a pile group under a lateral load is usually much larger than the deflection of an isolated single pile under the same amount of lateral load shared by that pile. Figure 7.42 illustrates the nature of pile–soil–pile interaction for a laterally loaded pile group. The shades in Figure 7.42 represent the stress bulb (zone influenced by the stress variation induced by

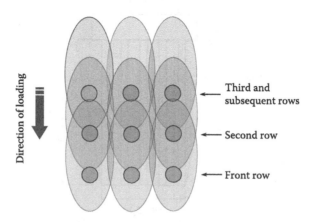

Figure 7.42 Pile–soil–pile interaction of piles in a laterally loaded group. (Adapted from Wang, S.T. 2014. *Lecture note on analysis of drilled shafts and piles using p-y and t-z curves.* Department of Civil Engineering, National Chiao Tung University, Hsin Chu, Taiwan.)

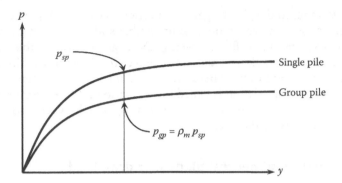

Figure 7.43 Application of p-multiplier to modify the $p-y$ curve of a single pile.

pile deflections) around the deflected piles. For closely placed piles in a group, these stress bulbs are overlapped for the piles in the trailing rows and thus weaken the soil in these regions. This is the main reason that piles in trailing rows of a pile group have significantly less resistance to a lateral load than piles in the lead or front row.

Bogard and Matlock (1983) and Brown et al. (1988) reported the use of a p-multiplier, ρ_m, to modify the $p-y$ curve of a single pile. An illustration of the p-multiplier concept is presented in Figures 7.42 and 7.43. For a pile in a given location in the group, the same ρ_m is applied to all $p-y$ curves along the length of that pile. For a given deflection y, the corresponding lateral resistance of a single pile, p_{sp}, is multiplied by ρ_m to obtain the equivalent lateral resistance under the pile group conditions, p_{gp}. The available data suggest that the value of ρ_m is affected by center-to-center pile spacing (s) and relative position with the neighboring piles. The multiplier consists of three elements in the reduction factors: line-by-line reduction factor, ρ_l, for leading or trailing piles, side-by-side reduction factor, ρ_s, and reduction for skewed piles, ρ_{sk}. The value of ρ_m is the multiplication of ρ_l, ρ_s, and ρ_{sk}. Depending on the spacing between piles and its relationship with pile diameter, the value of these individual reduction factors can range from 0.5 to 1. The multiplication of ρ_l, ρ_s, and ρ_{sk} (i.e., ρ_m) can often be less than 0.5 (Reese et al., 2006a).

7.7 NEGATIVE SKIN FRICTION

So far, in the static analysis of piles, we have considered that both shaft friction and toe bearing contribute to the support of the pile. Both of these resistance components point upward and we consider them positive. For this statement to be valid, it is necessary that the pile moves downward in reference to the surrounding soil. Recall in Section 7.3, in the discussion of load transfer mechanisms for axially loaded piles, the elastic deformation, δ, of the pile, shaft friction, f_{sf}, distribution along the pile surface, and axial force passing through the pile internally, Q_i, are related to one another. As shown in Figure 7.12, unless the pile shaft is frictionless (Figure 7.12a), Q_i decreases with depth (Figures 7.12b and 7.12c). At the toe of the pile, Q_i reduces to R_t, the toe resistance.

The shaft friction reverses its direction and becomes negative when piles are installed through a compressible soil layer that undergoes consolidation and the downward soil movement exceeds that of the pile. This reversed shaft friction is called the "negative skin friction" and it points downward. In this case, the consolidating soil moves downward in reference to the pile. Negative skin friction can develop whenever the effective overburden stress is increased on a compressible soil layer through which a pile is driven, due to placement of new fill or lowering of the groundwater table. Instead of supporting the pile, the negative skin friction adds load to the Q_i values, making Q_i increases with depth within the depth range where negative skin friction is present. The additional Q_i caused by negative skin friction is called the "drag load." In extreme cases, this drag load can cause structural damage to the pile. The drag load, however, does not affect the ultimate bearing capacity of the pile.

7.7.1 Determination of neutral plane and drag load

Consider a case shown in Figure 7.44, where the pile is installed through a compressible soil and the shaft friction is evaluated using the effective stress method (i.e., the β approach). A key element in the evaluation of drag load is the determination of the depth of "neutral plane." Note that in an earlier discussion of β method, a representative ultimate unit shaft friction f_{sf} [$f_{sf} = \beta\sigma'_{vo}$ according to Equation 7.24] taken at the mid-height of the pile was used for the determination of ultimate shaft friction resistance, R_{sf}. The representative f_{sf} was used as a constant throughout the pile length L_p to compute R_{sf} (i.e., $R_{sf} = \pi D_P L_p f_{sf}$). For the analysis of negative skin friction, f_{sf} will still be computed according to Equation 7.24 but will be considered as the ultimate unit shaft friction at a given depth z. Since σ'_{vo} increases with depth, f_{sf} should also increase with depth. For a uniform soil layer with constant soil unit weight, σ'_{vo} increases linearly with depth. Consider a constant β, f_{sf} should therefore increase linearly with depth as well. For a linearly

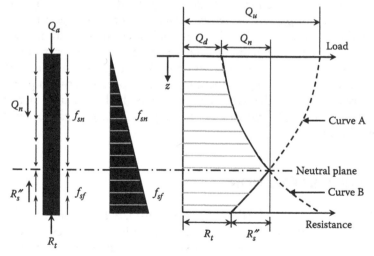

Q_d = sustained dead load	f_{sf} = ultimate unit shaft friction
Q_u = ultimate capacity	f_{sn} = negative unit shaft friction
Q_n = drag load	Q_i = axial force passing through the pile internally
R_s'' = positive shaft resistance	R_t = toe resistance

Figure 7.44 Shaft friction distribution, axial force passing through pile internally and neutral plane when there is negative skin friction. (Adapted from Fellenius, B.H. 1988. *Unified design of piles and pile groups.* Transportation Research Record 1169, Transportation Research Board, Washington, DC, pp. 75–82.)

increasing unit shaft friction, Q_i should be a second-order curve according to Equation 7.3 (also see Figure 7.12c) as described in Section 7.3. The meaning of neutral plane and procedure to find the depth of neutral plane are described as follows.

Establish Curve A in Figure 7.44—Curve A represents the variation of Q_i with depth under ultimate loading conditions when shaft friction is positive. We assume that the soil along the pile length is uniform with a constant β and soil unit weight γ_{sat}, and groundwater table is at the ground surface. In this case, f_{sf} increases linearly with depth as,

$$f_{sf} = \beta\sigma'_{vo} = \beta\gamma' z \tag{7.76}$$

where
$\gamma' =$ soil buoyant unit weight $= \gamma_{sat} - \gamma_w$
$z =$ depth

Q_i starts as Q_u at the pile head (i.e., $Q_h = Q_u$) in ultimate loading condition, decreases with depth, and reduces to R_t at the pile toe. In ultimate loading condition, the unit shaft friction $f_{sp}(z) =$ ultimate unit shaft friction $f_{sf}(z)$, following Equation 7.3 and combine Equation 7.76,

$$Q_i(z) = Q_h - \int_0^z \pi D_p f_{sp}(z)dz = Q_u - \int_0^z \pi D_p f_{sf}(z)dz = Q_u - \frac{\pi D_p \beta\gamma'}{2}z^2 \tag{7.77}$$

Equation 7.77 describes Curve A that starts with a value of Q_u and decreases non-linearly with depth z as a second-order equation, depicted in Figure 7.44 and as discussed in Section 7.3. Curve A is the same as that of Q_i in ultimate loading condition, without the effects of negative skin friction.

Establish Curve B in Figure 7.44—Curve B reflects the variation of Q_i with depth when the pile head is subject to a sustained dead load Q_d (i.e., $Q_i = Q_d$ at pile head). The design or allowable load of a single pile, $Q_a = Q_u/FS$, according to Equation 7.37. Q_a is used to support the sustained dead load Q_d and the transient life load Q_l. Drag load is a long-term behavior and therefore only relates to Q_d. This is why for Curve B, Q_i starts with Q_d.

When the compressible soil moves downward against the pile, the shaft friction reverses its direction and is denoted as f_{sn} in Figure 7.44. As it takes very little movement to develop ultimate shaft friction regardless of its direction (see Section 7.4), we can reasonably assume f_{sn} and f_{sf} at a given depth are equal in magnitude but in opposite directions. This assumption is especially reasonable when the pile is surrounded by compressible soils when the required soil movement to reach ultimate shaft friction (f_{sn} or f_{sf}) is easily accomplished. The distribution of Q_i with depth z now becomes

$$Q_i(z) = Q_a + \int_0^z \pi D_p f_{sn}(z)dz = Q_a + \frac{\pi D_p \beta\gamma'}{2}z^2 \tag{7.78}$$

Q_i increases with depth and the change of Q_i also has a non-linear relationship with depth z.

The neutral plane depth—The depth where Curve A intersects Curve B, or Q_i from Equations 7.77 and 7.78 is equal—is the depth where the soil ceases to move downward against the pile and the shaft friction changes its direction from negative to positive. The horizontal plane passing through this intersection is called the "neutral plane." At the depth of neutral plane z_{np}, there is no relative displacement between soil and pile, and

Q_i reaches its maximum value. By equating Equations 7.77 and 7.78, we can calculate the depth of neutral plane z_{np} as

$$z_{np} = \sqrt{\frac{(Q_u - Q_d)}{\pi D_p \beta \gamma'}} \qquad (7.79)$$

As stated earlier, f_{sn} and f_{sf} are practically equal in magnitude but in opposite directions, f_{sn} can be estimated following the same procedures for f_{sf} as described earlier (i.e., the β method). The value of Q_n can be calculated as follows:

$$Q_n = \int_0^{z_{np}} \pi D_p f_{sn}(z) dz \qquad (7.80)$$

When shaft friction is positive, Q_i has its maximum value at pile head and becomes smaller with depth as the positive shaft friction shares the axial compressive load, or $Q_i < Q_d$ below pile head. However, with negative skin friction, Q_i increases with depth until the depth of neutral plane. The drag load, Q_n, causes higher structural force within the pile. Bituminous coating on the surface of piles (Fellenius, 1988) has been used effectively as a lubricant to lower the shaft friction and minimize the drag load.

From Equation 7.78, we can see that, for a toe bearing pile (e.g., a pile penetrates through compressible soil and penetrates its toe into a hard material) with a large R_t and thus Q_u significantly larger than Q_d, it is conceivable that z_{np} from Equation 7.79 can be larger than the depth of the compressible soil layer as conceptually described in Figure 7.45a. In this case, the neutral plane is at the surface of the hard material. On the other hand, if the pile is fully frictional and practically floats in the compressible soil with very little R_t, then it can be demonstrated that the neutral plane is at $L_p/\sqrt{2}$ from the ground surface, as shown in Figure 7.45b where $L_p =$ length of the pile. Interestingly, $L_p/\sqrt{2} \approx 2L_p/3$, location of the equivalent footing according to Terzaghi and Peck (1967) shown in Figure 7.40. Students are encouraged to consider different scenarios of load transfer characteristics and find out how they affect the neutral plane depth.

Example 7.14 provides a description of how to use the above procedure to estimate the depth of neutral plane and drag load.

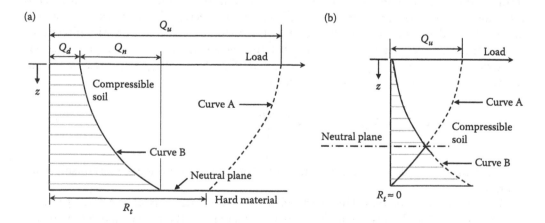

Figure 7.45 Characteristics of the Q_i distributions for toe bearing and fully frictional pile. (a) Distribution of Q_i with a hard bearing stratum. (b) Distribution of Q_i for a fully frictional pile.

EXAMPLE 7.14

Given

A 0.8 m diameter (D_p) and 15 m long bored pile ($L_p = 15$ m) was installed in a saturated clay deposit, as shown in Figure E7.14. Water table is at the ground surface. The clay around the pile shaft is normally consolidated. The clay changed to overconsolidated and stiff from below the pile toe level. A new fill is to be placed at the ground surface. The ultimate bearing capacity of the pile $Q_u = 840$ kN. A β value of 0.15 was used to calculate the ultimate unit shaft friction. A sustained dead load, Q_d, of 340 kN is applied at the pile head.

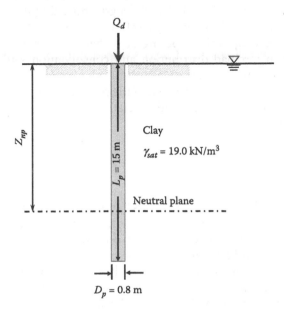

Figure E7.14 Soil profile and the bored pile subject to negative skin friction.

Required

Determine the depth of neutral plane z_{np} and drag load Q_n on the bored pile due to negative skin friction.

Solution

$$\beta = 0.15$$

$$\gamma' = \gamma_{sat} - \gamma_w = (19.0) - (9.81) = 9.19 \text{ kN/m}^3$$

$$D_p = 0.8 \text{ m}$$

$$Q_u = 840 \text{ kN}$$

$$Q_d = 340 \text{ kN}$$

$$z_{np} = \sqrt{\frac{(Q_u - Q_d)}{\pi D_p \beta \gamma'}} = \sqrt{\frac{(840 - 340)}{(3.1416)(0.8)(0.15)(9.19)}} = 12.0 \text{ m} \qquad (7.79)$$

$$Q_n = \int_0^{z_{np}} \pi D_p f_{sn}(z) dz = \frac{\pi D_p \beta \gamma'}{2} (z_{np})^2$$

$$= \frac{(3.1416)(0.8)(0.15)(9.19)}{2}(12.0)^2 = 250\,\text{kN} \tag{7.80}$$

7.7.2 Settlement of pile groups using the neutral plane

In their pile group consolidation settlement analysis, Terzaghi and Peck (1967) considered an equivalent footing situated at 1/3 of the pile length above the pile toe. The group load is distributed below this equivalent footing level at a slope of $1H{:}2V$. A more rigorous approach is to place the equivalent footing at or below the neutral plane depth. The rest of the calculation for load distribution and consolidation settlement remains the same. This method is preferred because at neutral plane, the pile has an equal amount of settlement as the soil. The total settlement of the pile then equals to the elastic deformation of the pile above the neutral plane and the consolidation settlement below the neutral plane (Fellenius, 1988).

7.8 REMARKS

In this chapter we introduced the "office work" in the design of deep foundations, after describing the various types of deep foundations. The title of this chapter emphasizes that it deals with the static analysis because certain types of deep foundations are installed by driving with a hammer, a dynamic procedure. The pile installation causes severe disturbance to the surrounding soil. For this reason, the static analysis is mostly approximate, based on soil parameters obtained from laboratory tests on field samples or in-situ test results. It was emphasized in the beginning of this chapter that it is impossible to design a deep foundation based on the projected or "correct" soil conditions and have a full consideration of the construction effects. Instead of attempting to consider all minor details, it was suggested that it is preferable to concentrate on the key factors that control the behavior of a deep foundation. It should also become obvious that there are many "key factors" that we need to pay attention to in a typical deep foundation design.

An axially loaded deep foundation may develop its bearing resistance from the friction between the pile shaft surface and the surrounding soil. The load transfer mechanism, or how the load applied to the pile head is distributed between shaft friction and toe bearing, is an important part of the characteristics of an axially loaded pile. This load transfer mechanism was not considered for shallow foundations. A deep foundation can also develop significant lateral load resistance. Because of this, the static analysis of laterally loaded piles is also an important part of this chapter.

The procedures we use in our analysis are mostly empirical. A good understanding in the local engineering practice and calibration of the empirical rules for the local geotechnical conditions are essential for the success in our design. It is also important to understand that static analysis is only the beginning and the results are used as the basis to estimate the amount of material and the budget required for the construction. The outcome of our analysis will have to be verified during construction. The construction and testing of piles, or the "field work," are described in Chapter 8.

HOMEWORK

7.1 A closed end precast concrete pile with a diameter, D_p, of 0.8 m is driven into a uniform sand deposit. The sand has a saturated unit weight, γ_{sat}, of 18 kN/m³ and a drained friction angle $\varnothing'$ of 34°. Groundwater table is at the ground surface. Determine the q_{tf} for the pile when the pile toe is driven to 5, 6, and 7 m below ground surface.

7.2 A Raymond step tapered pile is driven to 10 m into a uniform sand deposit from ground surface. The pile has a diameter of 0.8 m at the ground surface (pile head) and a taper (ω) of 0.5 degrees as shown in Figure H7.2. The sand has a saturated unit weight, γ_{sat}, of 18 kN/m³ and a drained friction angle $\varnothing'$ of 34°. The groundwater table is at the ground surface. Determine the R_{sf} for the pile using the Nordlund method.

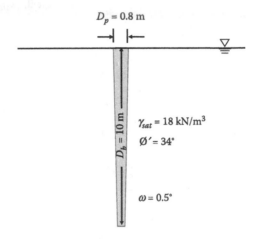

$D_p = 0.8$ m

$\gamma_{sat} = 18$ kN/m³
$\varnothing' = 34°$

$\omega = 0.5°$

$D_b = 10$ m

Figure H7.2 Step tapered Raymond pile driven into a sand deposit.

7.3 A 15 m long and 0.8 m diameter precast concrete pile is driven into a saturated clay deposit as shown in Figure H7.3. The groundwater table is at the ground surface. Determine the R_{sf} for the pile using the α method.

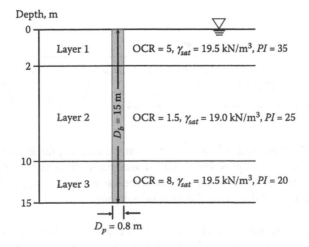

Depth, m

Layer 1 — OCR = 5, $\gamma_{sat} = 19.5$ kN/m³, $PI = 35$

Layer 2 — OCR = 1.5, $\gamma_{sat} = 19.0$ kN/m³, $PI = 25$

Layer 3 — OCR = 8, $\gamma_{sat} = 19.5$ kN/m³, $PI = 20$

$D_b = 15$ m

$D_p = 0.8$ m

Figure H7.3 The precast concrete pile and clay soil layers.

7.4 A 15 m long and 0.8 m diameter precast concrete pile is driven into a saturated clay deposit as shown in Figure H7.2. The groundwater table is at the ground surface. Determine the R_{sf} for the pile using the β method.

7.5 For the same soil conditions described in HW7.3, determine the R_{sf} for a 0.8 m diameter and 15 m deep bored pile using the α method.

7.6 A 0.8 m diameter and 15 m long precast concrete pile is driven into a sand deposit. SPT was performed at the site with one N value taken for every 1.5 m to a maximum depth of 15.0 m. Figure H7.6 shows the soil and SPT corrected blow count, $(N_1)_{60}$ profiles. Groundwater table is at the ground surface. The individual $(N_1)_{60}$ values are tabulated in Table H7.6. Determine the ultimate pile capacity Q_u using the Meyerhof SPT method.

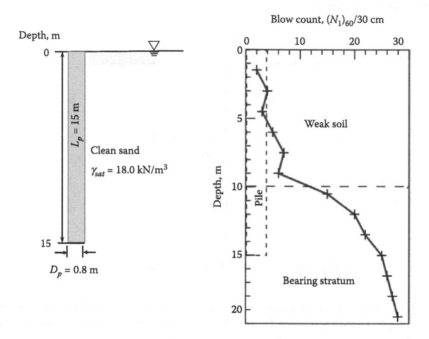

Figure H7.6 The precast concrete pile in sand and profile of $(N_1)_{60}$.

Table H7.6 The values of $(N_1)_{60}$

Depth, m	1.5	3.0	4.5	6.0	7.5	9.0	10.5	12.0	13.5	15.0	16.5	18.0	19.5
$(N_1)_{60}$	2	4	3	5	7	6	15	20	22	23	25	26	28

7.7 A 0.8 m diameter and 15 m long precast concrete pile is driven into a sand deposit as shown in Figure H7.7. CPTu was performed at the site to a maximum depth of 24 m. Table H7.7 shows the average values $\bar{q}_t$, $\bar{u}_2$, and $\bar{f}_s$ from CPTu. The groundwater table is at the ground surface.

- Determine the ultimate pile capacity Q_u using the Eslami and Fellenius method.
- Determine the ultimate pile capacity Q_u using the LCPC method.

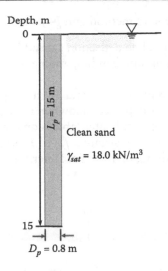

Depth, m

0

$L_p = 15$ m

Clean sand

$\gamma_{sat} = 18.0$ kN/m^3

15

$D_p = 0.8$ m

Figure H7.7 The precast concrete pile and profile of CPTu

Table H7.7 Average values from CPTu

Depth range, m	0–4	4–8	8–12	12–16	16–20	20–24
$\bar{q}_t$, MPa	0.32	0.70	3.35	5.45	4.12	5.82
$\bar{u}_2$, kPa	0.12	5.00	13.80	50.35	40.26	61.23
$\bar{f}_s$, kPa	5.12	5.90	25.05	27.35	35.75	40.75

7.8   Consider a 0.8 m diameter and 15 m long bored pile shown in Figure H7.8. The pile head is free to rotate. The bored pile is subject to a lateral load of 100 kN and a moment of 500 kN-m at the ground line. The bored pile is embedded in a uniform clay deposit with undrained shear strength, s_u, of 0.08 MN/m^2. Assume $E_{py} = 4$ MN/m^3 as a constant value for soil around the full length of the pile. The reinforced concrete pile has an overall elastic modulus of 25,000 MPa. Consider the bored pile as solid and uncracked under the applied load.

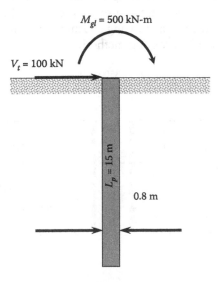

$M_{gl} = 500$ kN-m

$V_t = 100$ kN

$L_p = 15$ m

0.8 m

Figure H7.8 The bored pile and its loading conditions.

Use the characteristic load method and p–y method to:

1. Determine the ground line deflection caused by the applied loads.
2. Determine the maximum bending moment within the bored pile for the same loads.

7.9 A 3 × 3 pile group of 0.8 m diameter and 18 m long precast concrete piles is to be installed in a saturated clay deposit as shown in Figure H7.9. The piles are spaced at 2 m center to center. The groundwater table is on the ground surface. Consider the pile has an ultimate unit shaft resistance, f_{sf}, of 15 kPa. The bearing stratum has an undrained shear strength, s_u, of 100 kPa. Determine the ultimate bearing capacity of the pile group, Q_{ug}.

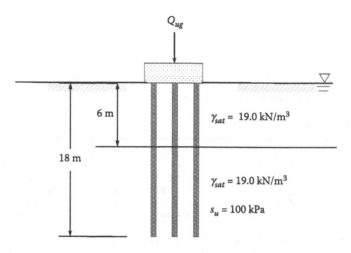

Figure H7.9 Clay soil profile for the group pile.

7.10 The 3 × 3 pile group of 0.8 m diameter and 18 m long pile group and the soil profile in HW7.9 is subject to a sustained design load, Q_a, of 3500 kN. The soil profile along with additional properties needed for the settlement analysis are given in Figure H7.10. Consider the bearing stratum consists of overconsolidated clay and the effective overburden stress after the group pile loading is less than the pre-consolidation stress within the bearing stratum. Estimate the settlement of the pile group using the equivalent footing method.

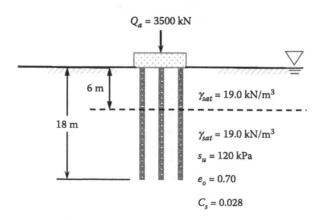

Figure H7.10 The clay soil layers for pile group settlement analysis.

7.11 A 0.8 m diameter and 20 m long bored pile is installed in a saturated clay deposit as shown in Figure H7.11. A β value of 0.18 was used to calculate the ultimate unit shaft friction. A new fill is to be placed at the ground surface. The neutral plane is estimated to be at 15 m below the ground surface. Determine the drag load on the bored pile due to negative skin friction.

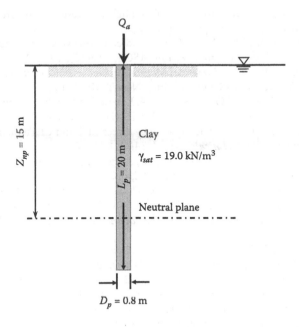

Q_a

$Z_{np} = 15$ m

$L_p = 20$ m

Clay

$\gamma_{sat} = 19.0$ kN/m^3

Neutral plane

$D_p = 0.8$ m

Figure H7.11 Soil profile and the bored pile subject to negative skin friction.

7.12 Prove that for a fully frictional and floating pile in compressible soil with very little R_t, the neutral plane is at $L_p/\sqrt{2}$ from the ground surface where L_p = length of the pile.

REFERENCES

Altaee, A., Evgin, E., and Fellenius, B.H. 1992. Axial load transfer for piles in sand. I: Tests on an instrumented precast pile. *Canadian Geotechnical Journal*, 29(1), 11–20.

Baguelin, F., Jézéquel, J.F., and Shields, D.H. 1978. *The Pressuremeter and Foundation Engineering*. Trans Tech Publications, Clausthal, Germany, 617p.

Bjerrum, L. and Simons, N.E. 1960. Comparison of shear strength characteristics of normally consolidated clays. *Proceedings, ASCE Research Conference on Shear Strength of Cohesive soils*, Boulder, CO, pp. 711–726.

Bogard, D. and Matlock, H.M. 1983. Procedures for analysis of laterally loaded pile groups in soft clay. *Proceedings, Geotechnical Practice in Offshore Engineering*, ASCE, pp. 499–535.

Broms, B.B. 1965. Design of laterally loaded piles. *Journal of the Soil Mechanics and Foundations Division*, ASCE, 91(SM3), 79–99.

Broms, B.B. 1966. Methods of calculating the ultimate bearing capacity of piles—A summary. *Sols-Sols*, (18–19), 21–32.

Brown, D.A., Morrison, C., and Reese, L.C. 1988. Lateral load behavior of pile group in sand. *American Society of Civil Engineers, Journal of Geotechnical Engineering*, 114(11), 1261–1276.

Bustamante, M. and Gianeselli, L. 1982. Pile capacity prediction by means of static penetrometer CPT. *Proceedings of the 2nd European Symposium on Penetration Testing*, CRC Press, Amsterdam, pp. 493–500.

Cheney, R.S. and Chassie, R.G. 1993. *Soils and Foundations Workshop Manual*. Second edition, FHWA HI-88-009, Federal Highway Administration, Washington, DC.

Collin, J.G. 2002. *Timber Pile Design and Construction Manual*. Timber Piling Council, American Wood Preservers Institute, Chicago, Illinois, USA, 145p.

Das, B.M. 2015. *Principles of Foundation Engineering*. 8th Edition, Cengage Learning, Stamford, CT, 944p.

Davisson, M.T. 1970. Lateral load capacity of piles. Highway Research Record No. 333, Highway Research Board, Washington, DC, pp. 104–112.

De Nicola, A. and Randolph, M.F. 1993. Tensile and compressive shaft capacity of piles in sand. *Journal of Geotechnical Engineering Division, ASCE*, 119(12), 1952–1973.

Desai, C. and Christian, J.T. 1977. *Numerical Methods in Geotechnical Engineering*. McGraw-Hill Series in Modern Structures, McGraw-Hill Inc., New York, 783 p.

Duncan, J.M., Evans, L.T. Jr., and Ooi, P.S.K. 1994. Lateral load analysis of single piles and drilled shafts. *Journal of Geotechnical Engineering, ASCE*, 120(6), 1018–1033.

Eslami, A. and Fellenius, B. H. 1997. Pile capacity by direct CPT and CPTu methods applied to 102 case histories. *Canadian Geotechnical Journal*, 34(6), 886–904.

Fellenius, B.H. 1988. Unified design of piles and pile groups. Transportation Research Record 1169, Transportation Research Board, Washington, DC, pp. 75–82.

Hetenyi, M. 1946. *Beams on Elastic Foundations*. University of Michigan Press, Ann Arbor, MI.

Huang, A.B., Hsueh, C.K., O'Neill, M.W., Chern, S., and Chen, C. 2001. Effects of construction on laterally loaded pile groups. *Journal of Geotechnical and Geoenvironmental Engineering, ASCE*, 125, 385–397.

Jamiolkowski, M., Ladd, C.C., Germaine, J.T., and Lancelotta, R. 1985. New development in field and laboratory testing of soils. *11th International Conference on Soil Mechanics and Foundation Engineering*, San Francisco, CA, Vol. 1, pp. 57–153.

Ladd, C.C., Foote, R., Ishihara, K., Schlosser, F., and Poulos, H.G. 1977. Stress deformation and strength characteristics. *Proceedings, 9th International Conference on Soil Mechanics and Foundation Engineering*, Tokyo, Japan, Vol. 2, pp. 421–494.

Matlock, H. and Reese, L.C. 1961. Foundation analysis of offshore pile-supported structures. *Proceedings of the Fifth International Conference on Soil Mechanics and Foundation Engineering*, Paris, France, Vol. 2, pp. 91–97.

McClelland, B. and Focht, J.A. Jr. 1958. Soil modulus for laterally loaded piles. *Transactions of the ASCE*, 123(1), 1049–1063.

Meyerhof, G.G. 1976. Bearing capacity and settlement of the pile foundations. *Journal of the Geotechnical Engineering Division, ASCE*, 102(GT3), 195–228.

NAVFAC, 1986. *Foundations and Earth Structures, Design Manual DM 7.02*. Department of the Navy, Naval Facilities Engineering Command, Alexandria, VA, 279p.

Nordlund, R.L. 1963. Bearing capacity of piles in cohesionless soils. *Journal of the Soil Mechanics and Foundation Engineering Division, ASCE*, 89(SM3), 1–35.

Nordlund, R.L. 1979. Point bearing and shaft friction of piles in sand. *Proceedings, 5th Annual Short Course on the Fundamentals of Deep Foundation Design*, St. Louis, Missouri, USA.

O'Neill, M.W. 1983. Group action in offshore piles. *Proceedings, Conference on Geotechnical Practice in Offshore Engineering, ASCE*, University of Texas, Austin, TX, pp. 25–64.

O'Neill, M.W. 2001. Side resistance in piles and drilled shafts. *Journal of Geotechnical and Geoenvironmental Engineering, ASCE*, 127(1), 1–16.

Paikowsky, S.G. and Whitman, R.V. 1990. The effects of plugging on pile and design. *Canadian Geotechnical Journal*, 27(4), 429–440.

Randolph, M.F. and Murphy, B.S. 1985. Shaft capacity of driven piles in clay. *Proceedings. Offshore Technology Conference*, Houston, TX, Vol. 1, pp. 371–378.

Reese, L.C. 1984. Handbook on design and construction of drilled shafts under lateral load. Report No. FHWA-IP-84-11, U.S. Department of Transportation, Federal Highway Administration, Washington, DC.

Reese, L.C. 1985. Behavior of piles and pile groups under lateral loads. Report No. FHWA-RD-85-106, U.S. Department of Transportation, Federal Highway Administration, Washington, DC.

Reese, L.C., Wang, S.T., Isenhower, W.M., and Arrellage, J.A. 2000. *Computer Program LPLIE plus Version 4.0 Technical Manual*. Ensoft, Inc., Austin, TX.

Reese, L.C., Isenhower, W.M., and Wang, S.T. 2006. *Analysis and Design of Shallow and Deep Foundations*. John Wiley and Sons, Hoboken, NJ, 574p.

Reese, L.C., Wang, S.T., and Vasquez, L. 2006a. Analysis of a group of piles subjected to axial and lateral loading, GROUP Version 7.0 technical manual. Ensoft, Inc., Austin, TX.

Robertson, P.K. 2009. Interpretation of cone penetration tests—A unified approach. *Canadian Geotechnical Journal*, 49(11), 1337–1355.

Robertson, P.K., Davies, M.P., and Campanella, R.G. 1989. Design of laterally loaded driven piles using the flat dilatometer. *ASTM Geotechnical Testing Journal*, 12(1), 30–38.

Skempton, A.W. 1957. Discussion: The planning and design of the new Hong Kong Airport. *Proceedings, Institute of Civil Engineers*, 7, 305–307.

Terzaghi, K. and Peck, R.B. 1967. Soil mechanics in engineering practice. John Wiley and Sons, New York, NY.

Tomlinson, M.J. 1994. *Pile Design and Construction Practice*. E & FN Spon, London, 411p.

Vesić, A.S. 1969. Experiments with instrumented pile groups in sand. Performance of Deep Foundations, ASTM, Special Technical Publication, West Conshohocken, PA, USA, No. 444, pp. 172–222.

Wang, S.T. 2014. Lecture note on analysis of drilled shafts and piles using p-y and t-z curves. Department of Civil Engineering, National Chiao Tung University, Hsin Chu, Taiwan.

Chapter 8

Construction and testing of deep foundations

8.1 INTRODUCTION

For driven piles, the pile is inserted into the ground by repeated impact forces or vibration. Driven piles are fabricated in the factory. The material and dimensions of the driven piles are determined in the static analysis. For the construction of driven piles, the following items will have to be carried out prior to the field operation:

- Select the type and capacity of the driving equipment for the installation of piles. A wide variety of hammers by different manufacturers with different energy output are available. A good hammer or driving system selection will provide adequate driving energy and be cost effective.
- Establish a driving criteria as a basis for accepting the piles during construction. The driving criteria usually consist of a simple and quick indicator or instrument reading that can be used to decide if a pile is to be accepted while the pile is being driven in the field.
- Determine a test program to verify the integrity and performance of the piles—the type and number of tests to be performed to provide more detailed information as to the safety and/or performance of the piles. This type of test is usually elaborate and time consuming.

Bored piles are made completely in the ground. A borehole with its dimensions determined in static analysis is drilled in the ground first and then the borehole is filled with concrete. For bored piles, engineers responsible for the project will have to deal with the following issues prior to construction:

- The procedure to be used for the construction of bored piles—selection of the construction method that is compatible to the subsurface conditions and the job site.
- Selection of the equipment to be used for the bored pile construction—the equipment should be capable of performing the procedure to be used for the bored pile.
- Establishment of a quality assurance scheme for borehole preparation and concrete placement—measurements and testing to be undertaken during construction to document the dimensions of the borehole, procedure, and quantity and quality of concrete placement.
- Determination of a test program to verify the integrity and performance of the piles—similar to the case of driven piles, except the capacity of bored piles can be substantially greater than driven piles.

The main objectives of this chapter are to

- Provide a general description to the construction of driven piles, the equipment used, and procedures applied (Section 8.2).
- Describe the principles of dynamic formulas and wave equation analysis, and introduce their applications in preconstruction planning for driven pile installations.
- Introduce the techniques of pile driving analysis and their applications as a tool for quality assurance in pile driving.
- Describe the available methods and key issues related to the construction of bored piles (Section 8.3).
- Discuss the necessary quality assurance procedures involved during the construction of bored piles.
- Introduce various types of axial and lateral pile load test methods, their principles, and interpretation of test results (Section 8.4).

The applications of driven and bored piles constructed by machines can be traced back to at least a century ago. The construction methods have been subject to intensive research. Related construction equipment and engineering expertise are readily available in most parts of the world. This chapter is divided into three parts. The first part deals with the construction and quality assurance of driven piles, the second concentrates on the construction and quality assurance of bored piles, and the third describes the technique of pile load test and its interpretation. The pile load test and interpretation methods can generally be used for both driven and bored piles.

8.2 CONSTRUCTION AND QUALITY ASSURANCE OF DRIVEN PILES

The driven piles are inserted into the ground by repeated and violent impact forces powerful enough to overcome the soil resistance, but not so powerful as to damage the pile in the process. A well-executed pile driving demands a delicate balance among cost, efficiency, and integrity of the installed pile. The following sections introduce the types of available equipment and tools to help in selecting the equipment and setting up construction specifications.

8.2.1 Equipment and procedure of pile driving

Numerous types of hammers are available to drive piles. Table 8.1 shows a list of the possible range of available pile hammer types and their characteristics. These pile hammers can generally be divided into five categories according to their operation mechanisms, as shown in Figure 8.1. Table 8.2 shows the specifications of a few selected hammers under the categories of steam or air (external combustion hammer), diesel (internal combustion), and vibratory hammers. More complete lists of pile hammers can be found in the literature or handbooks (e.g., Hannigan et al., 2006), and manufactures' websites.

The simplest form of a pile hammer is a drop hammer (Figure 8.1a). A ram is lifted and dropped by using a hoisting device. The driving energy created by the hammer is controlled by the weight of the ram (W_r) and stroke (h) of the hammer drop. For a single-acting hammer (Figure 8.1b), the ram is lifted by a piston. Air, steam, or hydraulic pressure is injected into the lower chamber of the piston bore to push the piston and the hammer upward. During lifting, the exhaust pressure is vented from the upper chamber. The

Table 8.1 Typical pile hammer characteristics

| Hammer type | Drop | Steam or air | | | Diesel | | Hydraulic | | Vibratory |
		Single acting	Double acting	Differential	Single acting	Double acting	Single acting	Double acting	
Rated energy, kJ per blow	9–81	10–2440	1–29	20–86	12–667	11–98	35–2932	35–2945	–
Impact velocity, m/sec	7–10	2.5–5	4.5–6	4–4.5	3–5	2.5–5	1.5–5.5	1.5–7	–
Blows/minute	4–8	35–60	95–300	98–303	40–60	80–105	30–50	40–90	750–2000 pulses/minute

Source: After Hannigan, P.J. et al., 2006. Design and construction of driven pile foundations. Volume II. Report No. FHWA-NHI-05-043, National Highway Institute, Federal Highway Administration, U.S. Department of Transportation, Washington, DC, 486pp.

hammer is dropped by releasing the pressure from the lower chamber. In a double-acting hammer (Figure 8.1c), the downward movement of the hammer is assisted by injecting pressure in the upper chamber while exhausting the pressure from the lower chamber. The double-acting hammer is differential, if the pressure applied to the upper chamber is different from that in the lower chamber. The diesel hammer (Figure 8.1d) is similar to a single-cylinder diesel combustion engine. The cylinder in this case is the ram. Combustion of the diesel fuel injected into the lower chamber lifts the ram. Dropping of the diesel hammer ram can also be single-acting or double/differential-acting. A vibratory driver (Figure 8.1c) consists of a weight mounted on an oscillator. The oscillator typically consists of one or more pairs of eccentric masses. The eccentric masses rotate in opposing directions but with the same frequency; they generate vertical vibration but nullify horizontal centrifugal forces. The generated vertical force equals the sum of the vertical forces of all the eccentric masses.

Figure 8.2 shows the field setup for driving a precast concrete pile using a diesel hammer. Figure 8.3 demonstrates the installation of a steel H pile using a vibratory driver. Vibratory drivers are not routinely used for pile installation. The following discussion on construction of driven piles concentrates on impact hammers.

Once a decision is made to use driven piles, the material and dimensions of the piles are chosen and a series of questions related to the installation of the piles will have to be answered. These questions include:

- What kind of pile hammer should be used to drive the piles? (The hammer should be powerful enough to drive the pile to the required capacity but not so powerful as to damage the pile in the process.)
- What driving criteria is to be used for accepting the driven pile in the field during construction?

8.2.2 The dynamic formulas

The dynamic formula was used before the advent of computers as a main tool to answer the above questions. Pile is assumed to be rigid (no elastic deformation). The energy from dropping a ram with a weight W_r and a drop height h is $W_r h$. This energy is absorbed by pushing the pile into the ground with a permanent settlement S, under an ultimate resistance, R_u. This permanent settlement, S, is called "set" for a given hammer blow. The energy consumed in pushing of the pile is thus $R_u S$. However, because of the use of cushion and imperfection in the pile hammer system, there is inevitable energy loss in the hammer impact. The Engineering News Record (ENR) formula was proposed based on these

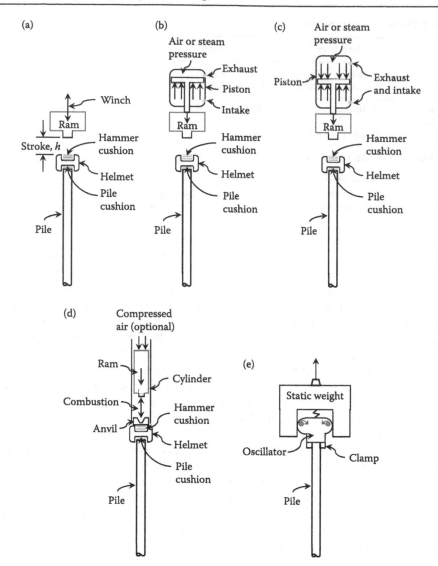

Figure 8.1 Major types of pile-driving hammers. (a) Drop hammer. (b) Single-acting hammers. (c) Differential and double-acting hammers. (d) Diesel hammers. (e) Vibratory driver.

premises. In ENR formula, the effects of energy losses in pile driving are lumped into an empirical factor C. The energy generated by hammer $W_r h$ equals to the energy consumed by the pile penetration $R_u(S + C)$, or

$$R_u = \frac{W_r h}{S + C} \tag{8.1}$$

where $C = 25.4$ mm for drop hammer, 2.54 mm for steam hammer, if S and h are in mm. R_u and W_r have the same unit. According to Equation 8.1, a pile driven by a hammer with rated energy of $W_r h$ should reach R_u when the last hammer blow causes the pile to settle by the amount of S. With this system, the chosen pile hammer should have its rated energy larger than $W_r h$. The field pile-driving criteria is established based on S that would yield

Table 8.2 Specifications of a few selected hammers

Manufacturer	Model	Rated energy, kJ	Ram weight, kN	Stroke, m	Hammer action
Steam, air, or hydraulic					
BSP	HH 3	35.31	29.41	1.20	Single acting
BSP	HH 11–1.5	161.78	107.91	1.50	Single acting
BSP	HH 16–1.2	188.43	156.95	1.20	Single acting
Conmaco	C 50	20.34	22.25	0.91	Single acting
Conmaco	C 100	44.07	44.50	0.99	Single acting
Conmaco	C 200	81.36	89.00	0.91	Single acting
McKiernan-Terry	5	1.36	0.89	1.52	Double acting
McKiernan-Terry	11B3	25.97	22.25	1.17	Double acting
McKiernan-Terry	S-20	81.36	89.00	0.91	Single acting
Raymond International	65C	26.44	28.93	0.91	Double acting/differential
Raymond International	150C	66.11	66.75	0.99	Double acting/differential
Raymond International	R 60X	203.40	267.00	0.78	Single acting
Vulcan	08	35.26	35.60	0.99	Single acting
Vulcan	200C	68.07	89.00	0.77	Double acting/differential
Vulcan	040	162.72	178.00	0.91	Single acting
Diesel					
Berminghammer	B-200	24.41	8.90	2.74	Single acting
Berminghammer	B-300	54.66	16.69	3.28	Single acting
Berminghammer	B-500 5	124.81	34.71	3.60	Single acting
Delmag	D5	14.24	4.90	2.93	Single acting
Delmag	D55	169.51	52.78	3.40	Single acting
Delmag	D100-13	360.32	98.21	4.11	Single acting
ICE	180	11.03	7.70	1.43	Double acting
ICE	520	41.18	22.56	1.83	Double acting
ICE	660	70.01	33.69	2.08	Double acting
Kobe	K25	69.86	24.52	2.85	Single acting
Kobe	K35	51.5	5.51	2.85	Single acting
Kobe	K60	176.53	58.87	3.00	Single acting

Vibratory

Manufacturer	Model	Power, kW	Ram weight, kN	Frequency, Hz
ICE	223	242	2.05	38.3
ICE	812	375.00	8.10	26.70
ICE	66–80	597.00	8.68	26.70
American Pile-driving Equipment (APE)	15	59.67	0.49	30.00
APE	50	194.00	1.02	30.00
APE	200	466.00	1.29	30.00

the required R_u according to Equation 8.1 and the rated energy for the selected hammer. The maximum stress experienced by the pile is estimated as R_u/A_p, where A_p is the cross-sectional area of the pile. This maximum stress is used to evaluate if the pile can be damaged during driving using the selected hammer and driven to set S. R_u from dynamic formula makes no distinction between dynamic (resistance to high-speed penetration) and static resistance.

Figure 8.2 Driving of a precast concrete pile with a single-acting Delmag D-100-13 diesel hammer in Jiayi, Taiwan. (Courtesy of An-Bin Huang, Hsinchu, Taiwan.)

Figure 8.3 Driving of a steel H pile with a vibratory driver in Houston, Texas, USA. (Courtesy of Dr. Kenneth Viking, Stockholm, Sweden.)

EXAMPLE 8.1

Given

A single-acting Vulcan 08 steam hammer with a ram weight, W_r, of 35.6 kN and drop height (stroke), h, of 991 mm was used to drive a precast concrete pile.

Required

Use ENR formula to calculate the ultimate resistance, R_u, when S reached 5 mm.

Solution

$W_r = 35.6\,\text{kN}$

$h = 991\,\text{mm}$

$S = 5\,\text{mm}$

For steam hammer $C = 2.54\,\text{mm}$

According to Equation 8.1

$$R_u = \frac{W_r h}{S + C} = \frac{(35.6)(991)}{(5 + 2.54)} = 4679\,\text{kN}$$

ENR was developed in the late nineteenth century primarily for evaluating the timber piles driven by a drop hammer in sands. Studies have indicated that the ENR formula performed poorly in predicting the static capacities of modern pile foundations. Various forms of dynamic formulas have been proposed in the past to improve its performance, but they were all based on a similar framework as the ENR formula and considered the pile as a rigid material. Pile stiffness and length effects were ignored. Eventually, the dynamic formulas were replaced by the numerical wave equation analysis.

8.2.3 Dynamic analysis by wave equation

One-dimensional wave equation that describes wave propagation through an elastic rod—in this case the pile—is more realistic in describing the dynamics of pile driving. It is a second-order partial differential equation derived by applying Newton's second law that relates stress wave propagation with time and distance as

$$\frac{\partial^2 d}{\partial t^2} = \left(\frac{E_p}{\rho_p}\right)\left(\frac{\partial^2 d}{\partial z^2}\right) \pm R \tag{8.2}$$

where

d = displacement of an element of the pile

E_p = elastic modulus of the pile material

ρ_p = mass density of the pile material

R = soil resistance against the pile at given depth, z, and time, t

Other than describing a physical phenomenon, Equation 8.2 serves little purpose in engineering applications. Smith (1960) is believed to have been the first to solve Equation 8.2 by using a finite difference method to obtain a quantitative solution for engineering analysis. The concept of using a finite difference method to solve partial differential equation was described in Chapter 7 (Section 7.5). The finite difference method was also used by Smith (1960) to solve Equation 8.2, but with the explicit scheme. The explicit scheme calculates the state of the system at a later time based on the state of the system at the current time without the need to solve algebraic equations as in the case of implicit scheme. A time step, Δt, is chosen for the computation. The magnitude of Δt determines the accuracy of the approximate solutions as well as the number of computations. Typically, we set $\Delta t = \Delta L_p / c_w$ in wave equation analysis where ΔL_p = length of the discretized pile segment and c_w = velocity of wave propagation in the pile = $\sqrt{E_p/\rho_p}$. Time is the controlling factor in the wave equation, and the event of pile driving starts with the hammer hitting the head of the pile. The explicit scheme is therefore ideal for solving the wave equation.

In this numerical scheme, the hammer and the pile are first discretized into a series of segments with concentrated weights (denoted as WAM) connected by weightless internal springs (denoted as XKAM), as shown in Figure 8.4. The springs represent the stiffness of the pile, hammer cushion, and pile cushion. The discretization is similar to that in the implicit finite difference method described in Section 7.5.

The main purpose of the cushion between hammer and pile is to limit impact stresses in both the pile and the ram. The cushion is ideally a spring with load-deformation characteristics and a coefficient of restitution, e, that is compatible with the cushion material. This is accomplished by simulating the cushion load deformation curve by two straight lines with different stiffness, as shown in Figure 8.5. The stiffness of the loading line is $A_c E_c / L_c$, where A_c is the cross-sectional area of the cushion, E_c is Young's modulus of the cushion, and L_c is the thickness of the cushion. The stiffness of the unloading line is $A_c E_c / e^2 L_c$ where e is the coefficient of restitution. The cushion inevitably absorbs much of the impact energy, thus reducing the efficiency of driving. Properties for some of the commonly used cushion materials are shown in Table 8.3.

The soil medium is assumed to be weightless. The soil resistance is simulated by a spring and a damper (dashpot) (external spring denoted as XKIM) on each pile segment, as shown in Figure 8.4. The soil spring can deform linearly to a limiting value, Q, after which no additional load is required to continue deformation (Figure 8.6a). The static soil resistance corresponding to a deformation Q is denoted as R_u. Q is called "quake." The dashpot simulates the viscosity or dynamic effects in soil resistance (Figure 8.6b). Dynamic resistance is the additional soil resistance that is linearly proportional to the velocity of soil spring deformation, as shown in Figure 8.7. The slope of this relationship is called the damping coefficient. For soil along the side of the pile, the damping coefficient is called J', and J for soil at the pile toe. It is often assumed that $J' = J/3$.

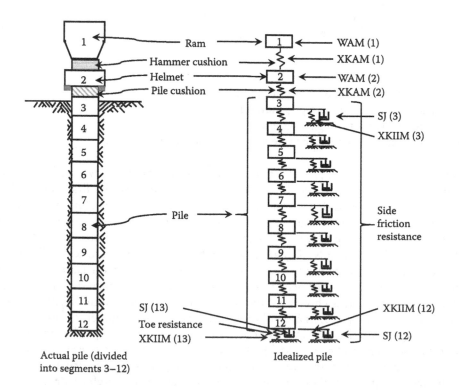

Figure 8.4 Discretization of the pile and driving system.

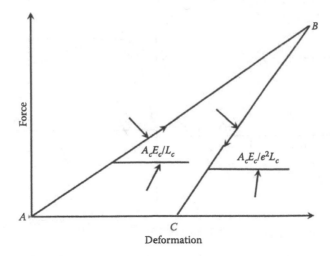

Figure 8.5 Stiffness of cushion in loading and unloading.

The combined (static and dynamic soil resistance) load deformation path is depicted by the thick curve $OABCDEFG$ in Figure 8.6b for soil along the side of the pile. The soil resistance acts in the opposite direction of the pile movement. At points O, B, and E, the relative displacement between the pile segment against the surrounding soil is reversing its direction and the soil spring has zero velocity against the pile shaft, thus there is no dynamic resistance. For the soil spring at the pile toe, the load deformation curve is $OABC$, as the pile is free to rebound at the toe.

Each segment of the pile system has a corresponding soil spring. The distribution of soil resistance along the length of the pile can be specified by the proper choice of the spring stiffness and damping coefficient, via the choice of R_u, Q, J', and J for each individual soil spring. Table 8.4 shows some empirical values of Q and J used in wave equation analysis.

The ram [represented by WAM (1) in Figure 8.4] impacting its neighboring spring [represented by XKAM (1)] with a known velocity (velocity of the ram at impact or the impact velocity, v_i) starts the chain action of wave propagation simulation from a single hammer blow. The rated energy per hammer blow, EN_r, shown in Tables 8.1 and 8.2, represents the theoretical energy generated by a single hammer blow under ideal conditions. Due to inevitable friction, misalignment, and other inaccuracies involved in hammer

Table 8.3 Properties of commonly used cushion materials

Material	Elastic modulus, E_c MPa	Coefficient of restitution, e
Asbestos	276	0.5
Mixture of 25.4 mm Micarta disks and 12.7 mm aluminum disks	4827	0.8
Micarta	3103	0.8
Oak, load perpendicular to grain	310	0.5
Fir plywood, load perpendicular to grain	241	0.4
Pine plywood, load perpendicular to grain	172	0.3

Source: Vesić, A.S. 1977. Design of Pile Foundations, National Cooperative Highway Research Program Synthesis of Highway Practice 42, Transportation Research Board, Washington, DC, 68pp.

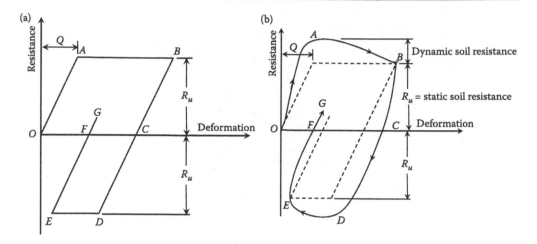

Figure 8.6 Representation of soil resistance for each pile segment. (a) Static. (b) Dynamic.

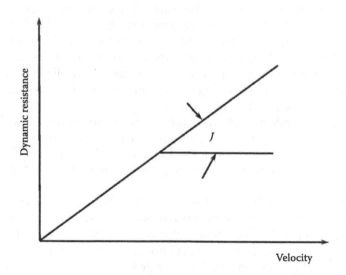

Figure 8.7 Relationship between dynamic resistance and velocity of soil spring deformation.

Table 8.4 Empirical values for Q and J

Soil	Q, mm	J (10^{-4}), s/mm
Coarse sand	2.54	4.92
Sand gravel mixed	2.54	4.92
Fine sand	3.81	6.56
Sand and clay or loam, at least 50% of pile in sand	5.08	6.56
Silt and fine sand underlain by hard strata	5.08	4.92
Sand and gravel underlain by hard strata	3.81	4.92

operation, the efficiency of the hammer, eff_r (ratio of the actual energy delivered in an impact over EN_r), is usually less than 100%. The energy generated by a hammer blow just prior to the impact is converted to kinetic energy of the ram. Thus the ram velocity, v_i, at impact can be calculated using the following equation:

$$\frac{W_r v_i^2}{2g} = (EN_r)(eff_r) \tag{8.3}$$

where
  W_r = weight of ram
  g = gravity

Calculation of v_i based on hammer energy and efficiency is demonstrated in Example 8.2.

EXAMPLE 8.2

Given

For a single-acting hammer:

Hammer-rated energy $EN_r = 35.3$ kN-m
Ram weight (WAM(1)) $W_r = 35.6$ kN
Hammer efficiency $eff_r = 0.66$

Required

Determine-impact velocity, v_i, of the ram in m/sec.

Solution

Using Equation 8.3,

$$\frac{W_r v_i^2}{2g} = \frac{(35.6)v_i^2}{(2)(9.81)} = (EN_r)(eff_r) = (35.3)(0.66)$$

Thus,

$$v_i^2 = \frac{(35.3)(0.66)(2)(9.81)}{35.6} = 12.84$$

$$v_i = \sqrt{12.84} = 3.58 \, \text{m/sec}$$

The hammer impact produces displacements in the individual pile segments as time progresses, with time intervals Δt. The displacements of the two adjacent segments (represented by WAM in Figure 8.4) produce a compression or extension in the spring (XKAM and XKIM in Figure 8.4) between them. The spring compression or extension produces a force in the spring. The forces of the two springs on an individual segment along with the resistance from ground produce a net force on the segment which either accelerates or decelerates it. This results in a new velocity and a new displacement in the succeeding time interval. The following finite difference equations represent the numerical scheme for wave equation analysis of pile (WEAP) driving developed by Smith (1960).

$$u(m, t) = u(m, t - \Delta t) + \Delta t\, v(m, t - \Delta t) \tag{8.4}$$

$$C(m, t) = u(m, t) - u(m + 1, t) \tag{8.5}$$

$$F(m, t) = C(m, t)\, K(m) \tag{8.6}$$

$$R(m, t) = [u(m, t) - u_p(m, t)]\, K_s(m)[1 + J(m)\, v(m, t - \Delta t)] \tag{8.7}$$

$$v(m, t) = v(m, t - \Delta t) + [F(m - 1, t) + M(m)g - F(m, t) - R(m, t)]\frac{\Delta t}{M(m)} \tag{8.8}$$

where
  m = element number
  t = time interval number
  M = mass of the segment element
  u = displacement of segment element
  v = velocity of segment element
  C = compression of the internal spring
  K = stiffness of the internal spring
  F = force generated by the internal spring
  R = dynamic soil resistance
  K_s = stiffness of soil (external) spring
  J = damping coefficient
  u_p = irrecoverable deformation of the surrounding soil (see Example 8.3)
  Q = quake
  Δt = time interval = $\Delta L_p/c_w$

The above equations describe the propagation of displacement and forces from the pile head as well as the development of soil resistance induced by a single hammer strike. This hammer strike can be in the early stage when only part of the pile is inserted in the ground, or it can be toward the final stage of construction, when the pile is fully embedded in the ground. The distribution of soil resistance between shaft friction and toe bearing can be assumed. The shaft friction can be assumed as uniform, linearly increasing with depth or other shapes of distribution, considering the nature of the bearing stratum. There is no limitation to the number of time intervals that can elapse in the WEAP computation. The following equation can be used to estimate the number of maximum time intervals (NSTOP) required to obtain reasonable results for a given hammer blow:

$$\text{NSTOP} = 30\, L_p/L_{\min} \tag{8.9}$$

where
  L_p = length of pile
  $L_{\min}$ = length of shortest pile segment used in the analysis

Experience shows that reasonable results can be obtained using the above-recommended time steps.

Summation of the R_u values from all pile segments is the static ultimate resistance of the pile RUT. The net permanent displacement of the pile toe (u_p at the pile toe) per hammer blow is called the "set," S. The inverse of S is the penetration resistance, expressed in blows for a given pile penetration distance (usually expressed in blow counts per 300 mm of penetration), for the corresponding ultimate resistance, RUT.

The Smith (1960) procedure (Equations 8.4 through 8.8) can be executed with the help of spreadsheet computer software which is commonly available. The spreadsheet program for a simple wave equation analysis is available on the publisher's website for free download by registered users. The following example shows the execution of WEAP using this spreadsheet program.

EXAMPLE 8.3

Given

A pre-stressed square concrete pile driven in clay with a Vulcan 08 hammer, the 18.3 m long pile was 9.15 m embedded in clay, as shown in Figure E8.3. Assume uniform shaft friction distribution that accounts for 95% of the static resistance; 5% of the static resistance is taken by toe resistance.

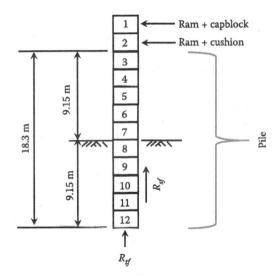

Figure E8.3 The pile configuration.

Required

1. Determine the bearing graph.
2. Determine the maximum compressive and tensile stresses induced by the hammer blow when RUT = 1600 kN.

Solution

Summary of the input parameters:

Driving system

Hammer:
Volcan 08, ram weight $W_r = 35.6$ kN, stroke $h = 990.6$ mm, hammer efficiency $eff_r = 0.66$

Capblock:

Material: oak, diameter (circular disk) = 355.6 mm, thickness L_c = 152.4 mm, modulus of elasticity E_c = 310.3 MPa, coefficient of restitution e = 0.5

Helmet:

Weight: 4.45 kN

Cushion:

Material: oak, width (square) = 304.8 mm, thickness L_c = 152.4 mm, modulus of elasticity E_c = 310.3 MPa, coefficient of restitution e = 0.5

Pile

Pre-stressed concrete pile:

Size (square): 304.8 mm square, length L_p = 18.3 m, modulus of elasticity E_p = 20,685 MPa, unit weight = 23.6 kN/m³, number of pile segments = 10, length per pile element ΔL_p = 1.83 m, weight per pile element WAM = $g\rho_p A_p \Delta L_p$ = 4.0 kN, where g = gravity and ρ_p = mass density of the pile material (concrete).

Soil

Quake Q = 2.54 mm, J (toe) = 0.0007, J' (shaft) = 0.0003, shaft soil spring stiffness $K_s = R_u/Q$ = 11.2 kN/mm, where R_u = ultimate static resistance along the pile shaft per pile segment. Toe soil spring stiffness $K_s = R_u/Q$ = 2.96 kN/mm, where R_u = ultimate static resistance at the pile toe.

For wave equation computation

Wave propagation velocity $c_w = \sqrt{E_p/\rho_p}$ = 2934.5 m/sec, where E_p = elastic modulus of the pile material (concrete). Time interval $\Delta t = \Delta L_p/c_w$ = 0.00006 sec.

Total static soil ultimate resistances RUT analyzed

250, 500, 600, 750, 1000, 1250, 1375, 1500, 1600, and 1635 kN

Figure E8.3a shows the bearing graph of driving resistance (blow count per 30 cm of pile penetration) versus the total static soil resistance, according to the computations.

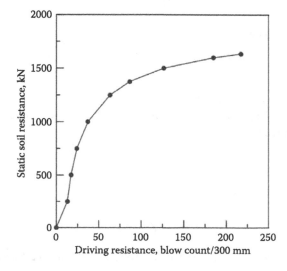

Figure E8.3a Bearing graph from WEAP computations.

Figure E8.3b shows the displacement of the pile, u, and the permanent displacement of the adjacent soil, u_p, with time, when $RUT = 1600$ kN. u_p lags behind u during wave propagation as the pile penetrates relative to the surrounding soil. The relative displacement between the pile shaft and neighboring soil $(u - u_p)$ is the deformation (horizontal axis) of Figure 8.6. The value of $(u - u_p)$ in reference to the quake Q determines the shearing resistance R following Figure 8.6 and Equation 8.7.

Figure E8.3c shows the displacement at selected pile segments after the hammer impact. The displacement is more significant toward the pile head. The displacement at the pile toe represents the set, S, of the pile induced by the hammer impact.

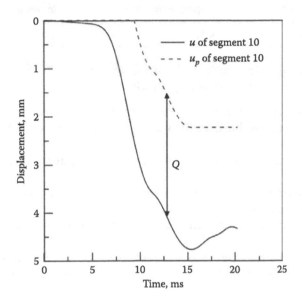

Figure E8.3b Displacement of pile segment 10 and its surrounding soil versus time.

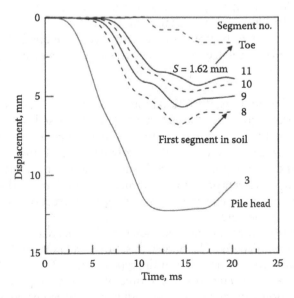

Figure E8.3c Displacement at selected pile segments after the hammer impact.

The maximum compressive and tensile force (computed according to Equation 8.7) passing through the pile segments (represented by XKIM in Figure 8.4), divided by the cross-sectional area of the pile A_p, is the maximum compressive and tensile stress, respectively induced by the hammer blow. Results for the case of RUT = 1600 kN, according to the computation, are presented in Table E8.3. No significant tensile stress was obtained in the computation.

Table E8.3 Maximum stresses in pile during driving

RUT, kN	In compression, MPa	In tension, MPa
1600	4.42	≈ 0

8.2.3.1 Applications and limitations of WEAP

For a given set of input parameters of the pile (material and dimensions) and driving system (hammer, cushion, and helmet), the wave equation analysis can be used to

- Establish a bearing graph that relates pile static ultimate resistance to hammer blow counts
- Analyze the stresses (maximum compressive and tensile stress) to be experienced by the pile during driving
- Analyze the effects of hammer energy and cushion properties on pile driving

In design and for preparation of pile installation, the input parameters of piles and hammer system can be readily changed in WEAP. With the above results, we can determine:

- If the selected pile can provide the required static ultimate capacity.
- If the selected hammer is cost effective, efficient, and suitable for the pile installation. The required ultimate capacity can be achieved with reasonable blow counts, usually set at less than 98 blows/0.25 m (Hannigan et al., 2006).
- The driving resistance to be used as a basis to accept the pile in the field.
- If the maximum compressive and tensile stresses are excessive to damage the pile during construction.

In the wave equation analysis, the ultimate soil resistance and its distribution along the pile are assumed. Although these assumptions are usually based on static analysis using the soil conditions from subsurface exploration, there is no guarantee that these assumed soil parameters are correct. The analyses also involve specific parameters related to WEAP such as quake Q and damping coefficients J' and J. The wave equation analysis is very effective in analyzing the effects of various parameters related to pile driving. The analysis itself, however, does not guarantee these parameters are correct or if the pile can develop its required capacity in the field.

The soil ultimate resistance can increase (soil setup) or decrease (relaxation) with time following the initial pile driving. The nature of setup or relaxation depends on the characteristics of the soils around the pile and the type of piles. The wave equation analysis reflects the pile-driving characteristics at the time of hammer impact. The magnitude of soil setup or relaxation can be assessed by adjusting the expected soil parameters at various times following the initial pile driving and comparing the differences in the ultimate resistance.

Wave equation analysis programs are commercially available, and some can be accessed free of charge on the internet. An Excel-based wave equation analysis program as used in Example 8.3 is available for registered readers of this book.

8.2.4 Pile driving analysis and dynamic load testing

The development of dynamic testing and analysis for driven piles started in the late 1950s at Case Institute of Technology (now Case Western Reserve University) (Goble and Rausche, 1970; Rausche et al., 1972; Goble et al., 1975). In contrast to the wave equation analysis, which is strictly numerical computation, dynamic testing and analysis involves physical field measurements. These measurements typically consist of two strain transducers and two accelerometers bolted to diametrically opposite sides of the pile, as shown in Figure 8.8. The purpose of these sensors is to offset the non-uniform impacts or bending during pile driving. The transducers are reusable and generally are attached near the pile head. The current field data acquisition and analysis system, often referred to as the pile-driving analyzer (PDA), that handles signal conditioning, analog/digital conversion, data logging, and processing, is typically controlled by a notebook computer, as shown in Figure 8.9. The system is compact and powerful. After more than half a century of research and development, the techniques of pile driving analysis (PDA) and dynamic load testing (DLT) are now widely used throughout the world.

In analyzing the data for each hammer blow, the strain, $\varepsilon(t)$, recorded at a given time, t, is converted to force, $F(t)$, as

$$F(t) = E_p A_p \varepsilon(t) \qquad (8.10)$$

where
E_p = pile elastic modulus
A_p = pile cross-sectional area

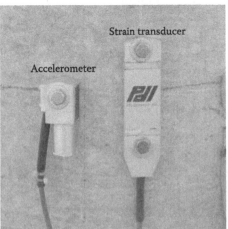

Figure 8.8 Strain transducers and accelerometers mounted near the top of a pile. (Courtesy of DECL, Taipei, Taiwan.)

Figure 8.9 Notebook-controlled field pile dynamic testing data logging and analysis system. (Courtesy of An-Bin Huang, Hsinchu, Taiwan.)

Integrating the acceleration readings, $a(t)$, with time yields velocity, $V(t)$,

$$V(t) = \int a(t)dt \qquad (8.11)$$

To understand how we can interpret the $F(t)$ and $V(t)$ data, let's start with simple cases of wave propagation through a pile with uniform cross-section and material (i.e., treated as a uniform, elastic rod). Figure 8.10 shows a case of impact wave

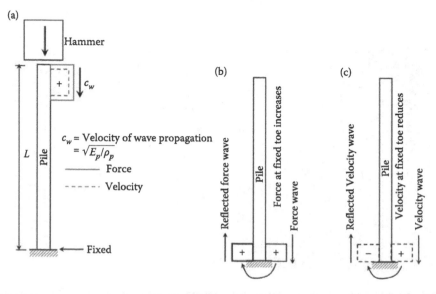

Figure 8.10 Wave propagation through a pile with fixed toe and no shaft resistance. (a) Hammer impact and initiation of force and velocity waves. (b) Reflection of force wave at the fixed toe. (c) Reflection of velocity wave at the fixed toe.

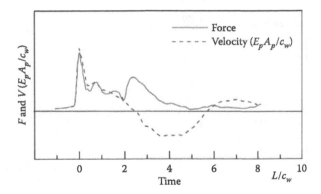

Figure 8.11 Force and velocity versus time for a mostly toe bearing pile.

propagating through the pile with a fixed toe (i.e., a toe bearing pile), after it is struck by an impact force on the pile head. The pile has no resistance on the shaft. The impact on the pile head generates force and velocity waves in the pile, as shown in Figure 8.10a. The waves travel through the pile with a velocity of c_w. The amount of time required for the wave to travel from the pile head, the measurement point, to the toe and back to the pile head is $2L/c_w$. For wave propagating through an elastic pile, its impedance, $E_p A_p/c_w$, is the ratio of force within the pile over wave velocity, c_w, where E_p = elastic modulus of the pile material and A_p = cross-sectional area of the pile. As there is no resistance on the pile shaft, the force measurement is identical to velocity measurement V multiplied by the impedance ($E_p A_p/c_w$), until $t = 2L/c_w$. Figure 8.11 shows a case of pile driving through soft soil with limited shaft friction but relatively high toe resistance (i.e., a mostly toe bearing pile). The force measurement F increases and the velocity decreases when $t > 2L/c_w$, as shown in Figure 8.11. In an ideal condition of zero shaft friction and completely fixed pile toe, the force measurement doubles and velocity becomes zero when $t = 2L/c_w$.

To extend the above discussion to a more general condition, the force and velocity measurements versus time are proportional (with a ratio of impedance) at impact and remain so until affected by soil resistance or cross-sectional variations. Reflections from the location where such variation occurs reach the measurement point at time $2D/c_w$ where D is the distance from transducers (i.e., measurement point) to the location of variation. Increase of soil resistance or pile cross-section will cause an increase in the force record and decrease in velocity. Conversely, cross-sectional reduction such as that due to pile damage will cause a decrease in the force record and an increase in velocity.

Figure 8.12 presents a pile with minimal shaft resistance except at depths A and B. Resistance at B is more significant than at A. The force record then shows an increase at time $2A/c_w$ and $2B/c_w$. The velocity record decreases correspondingly. The force becomes zero and velocity record increases significantly when $t = 2L/c_w$ because the toe is free and the reflections of force and velocity are opposite those shown in Figure 8.10b and c.

A uniform driven pile that develops its resistance mostly through shaft friction, as conceptually demonstrated in Figure 8.13, can be viewed as a pile with numerous shaft resistances, shown in Figure 8.12, distributed throughout the pile shaft. In this case, upon hammer impact, the force record remains high or increases when reaching the shaft friction, and the velocity record decreases correspondingly. The velocity curve

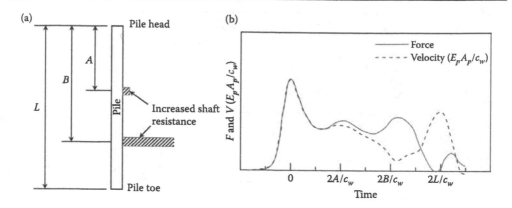

Figure 8.12 Force and velocity records for a pile with increased shaft resistance at depths A and B. (a) Pile with increased shaft resistance at depths A and B. (b) The force and velocity record. (Adapted from Hannigan, P.J. et al., 2006. Design and construction of driven pile foundations. Volume II. Report No. FHWA-NHI-05-043, National Highway Institute, Federal Highway Administration, U.S. Department of Transportation, Washington, DC, 486pp.)

separates and falls below the force curve when $t < 2L/c_w$, as qualitatively described in Figure 8.14.

It should be noted that PDA readings are usually taken toward the end of a pile driving. In the field, it is unlikely that a pile develops zero shaft friction and low toe resistance toward the end of pile driving. Useful, qualitative description about the nature of soil resistance along the pile shaft and toe or damage to the pile can be made based on visual inspection of the force and velocity records.

With rigorous interpretations, much more useful information regarding the pile and hammer system can be drawn from the PDA record. The following sections demonstrate a case of dynamic pile load test applied in the field. Figure 8.15 shows a set of field force and velocity records for a 40 m long closed-end steel pipe pile. The pipe pile had an outside diameter of 800 mm and inside diameter of 762 mm. The pile was driven into

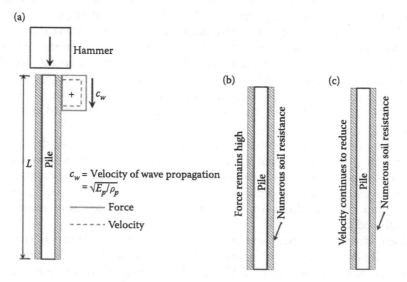

Figure 8.13 Wave propagation through a friction pile. (a) Hammer impact and initiation of force and velocity waves. (b) Reflection of force wave. (c) Reflection of velocity wave.

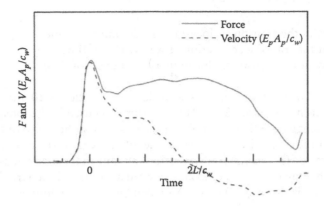

Figure 8.14 Force and velocity versus time for a friction pile.

a deep deposit of medium-dense silty fine sand, using a BSP HH 16-1.2 hydraulic single-acting hammer (see Table 8.2 for specifications). The pile developed its capacity mostly through frictional resistance along the pile shaft. The strain and acceleration transducers were mounted at 1.2 m from the pile head. In this case, $E_p = 2.1 \times 10^5$ MPa and $c_w = 5123$ m/sec for steel, $A_p = 0.0466$ m^2 and $L = 38.8$ m (distance between the transducers to the pile toe). The readings reflect the driving of mostly a friction pile with limited resistance at the toe. This is demonstrated by the significant deviation of F and V ($E_p A_p / c_w$) curves immediately after reaching their respective initial peaks, similar to those shown in Figure 8.14. These peaks correspond to the time when hammer impact stress wave passes through the transducer location for the first time or time of initial impact, marked as t_1 in Figure 8.15. The F values remain above V ($E_p A_p / c_w$) until $L / 2 c_w$ is larger than 4.

8.2.5 Determination of energy transfer

Energy transferred to the pile and received at the transducer location can be calculated from the integral of force and velocity records as

$$E_h(t) = \int_0^t F(t) V(t) dt \qquad (8.12)$$

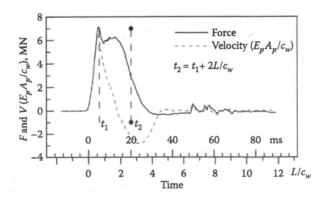

Figure 8.15 Force and velocity records from field measurement of a steel pipe pile installation.

where

$E_h(t)$ = energy at the transducer location as a function of time
$F(t)$ = force at the transducer location as a function of time
$V(t)$ = velocity at the transducer location as a function of time

$F(t)$ is calculated from strain readings according to Equation 8.10 and $V(t)$ from acceleration readings using Equation 8.11. The time duration used in the integration of Equation 8.12, equivalent to 2–3 times that of $2L/c_w$, is usually sufficient to obtain a maximum $E_h(t)$ value. The ratio of maximum $E_h(t)$ over the rated energy is referred to as the hammer energy transfer ratio. The energy transfer ratio is an important index in assessing the performance of the hammer system. For example, what appears to be a hard driving situation with high-blow counts for a given penetration can be caused by low-energy transfer ratio instead of high-soil resistance. Note that a similar procedure is used to measure Standard Penetration Test (SPT) hammer energy efficiency (see Chapter 3).

8.2.6 Analysis by CASE method

Two methods have been developed for more rigorous or quantitative applications of the dynamic load test data: the CASE method and the case pile wave analysis program (CAPWAP) method. The CASE method was derived from a closed-form solution to the one-dimensional wave propagation theory. For a pile with linear elastic material that has a constant cross-section, the total static and dynamic resistance of a pile during driving, RTL (see Figure 8.15), is:

$$\text{RTL} = \frac{1}{2}[F(t_1) + F(t_2)] + \frac{1}{2}[V(t_1) - V(t_2)]\frac{E_p A_p}{c_w} \qquad (8.13)$$

where

E_p = elastic modulus of the pile material
A_p = cross-sectional area of the pile
c_w = wave speed in pile material
L = pile length below the transducer location
F = force measurement at the transducer location
V = velocity measurement at the transducer location
t_1 = time of initial impact
t_2 = time when waves reflected back from the toe, $t_1 + 2L/c_w$

To obtain the static capacity RSP, the dynamic resistance due to damping is subtracted from RTL. Goble et al. (1975) proposed a method that approximates the dynamic resistance as a linear function of a damping factor times the pile toe velocity. RSP is then determined as

$$\text{RSP} = \text{RTL} - J_t\left[V(t_1)\frac{E_p A_p}{c_w} + F(t_1) - \text{RTL}\right] \qquad (8.14)$$

where

J_t = dimensionless damping factor based on soil type near the pile toe

Typical values of J_t for various types of soils are shown in Table 8.5. It should be noted that J_t, referred to as the CASE damping, is nondimensional and is not the same as the damping coefficients shown in Table 8.4 for Smith type of wave equation analysis.

Table 8.5 Damping factors for CASE RSP equation

Soil type at pile toe	CASE damping
Clean sand	0.05–0.20
Silty sand, sandy silt	0.15–0.30
Silt	0.20–0.45
Silty clay, clayey silt	0.40–0.70
Clay	0.60–1.10

Source: After Goble, G.G. et al., 1975. Bearing capacity of piles from dynamic measurements. Final Report, Department of Civil Engineering, Case Western Reserve University, Cleveland, OH.

The computation is very simple and fast; usually the RSP value is provided by the PDA, such as the one shown in Figure 8.9, in real time during pile driving. A CASE computation can be conducted for each hammer blow. The results, shown in Figure 8.15, are the direct output from PDA from one hammer blow.

EXAMPLE 8.4

Give

The force and velocity curves shown in Figure 8.15 from driving of a closed end steel pipe pile.

For steel, $E_p = 2.1 \times 10^5$ MPa and $c_w = 5123$ m/sec

Distance between the transducers to the pile toe, $L = 38.8$ m

Cross-sectional area of the pipe pile, $A_p = 0.0466$ m^2 for the pile of outside diameter of 800 mm and inside diameter of 762 mm

Required

Determine RTL and RSP for the case shown in Figure 8.15.

Use $J_t = 0.3$

Solution

$t_1 = 4.95$ ms, $t_2 = t_1 + 2L/c_w = (4.95) + (2000)[(38.8)/(5123)] = 20.1$ ms

$F(t_1) = 7.13$ MN and $F(t_2) = 2.81$ MN are taken from the force curve of Figure 8.15.

$V(t_1) E_p A_p / c_w = 6.25$ MN and $V(t_2) E_p A_p / c_w = -2.21$ MN are taken from the velocity curve of Figure 8.15.

$$\text{RTL} = \frac{1}{2}[F(t_1) + F(t_2)] + \frac{1}{2}[V(t_1) - V(t_2)]\frac{E_p A_p}{c_w} \tag{8.13}$$

$$= (0.5)[(7.13) + (2.81)] + (0.5)[(6.25) - (-2.21)] = 9.20\,\text{MN}$$

$J_t = 0.3$

$$\text{RSP} = \text{RTL} - J_t\left[V(t_1)\frac{E_p A_p}{c_w} + F(t_1) - \text{RTL}\right] \tag{8.14}$$

$$= (9.20) - (0.3)[(6.25) + (7.13) - (9.20)] = 7.95\,\text{MN}$$

The PDA readings provide the maximum compressive stress passing the sensor location using the measured strain and pile elastic modulus. It should be noted that the maximum compressive stress in the pile may be higher than the maximum compressive stress measured at the sensor location. A semi-empirical procedure is also available in PDA to determine the maximum tensile stress along the pile by superposition of the upward and downward traveling waves. These stress readings are important parameters in evaluating if there is a mismatch between the pile and hammer system. For the record shown in Figure 8.15, the maximum compressive stress at the transducer location was 154.05 MPa and maximum tensile stress was 29.41 MPa along the pile. A maximum energy of 122.11 kN-m was transmitted to the pile that represents a hammer energy transfer ratio of 0.65 (see Table 8.2 for rated energy of BSP HH 16-1.2). The maximum displacement in the vertical direction measured at the transducer location, by twice integration of acceleration with time, was 21.46 mm.

8.2.7 Analysis by the CAPWAP method

The CAPWAP method takes a more rigorous approach than the CASE method. A propriety software based on the CAPWAP method is available on the market. Other commercial software such as TNOWAVE performs similar analysis to CAPWAP, using similar techniques (Reiding et al., 1988). In this book, CAPWAP is used as a generic term to describe the interpretation method of pile dynamic testing data that involves matching the measured and computed records. The CAPWAP analysis is usually performed on the record from an individual hammer blow toward the end of pile driving. The wave equation analysis is used in CAPWAP, but in a different way. Recall that, in wave equation analysis, the computation initiates with the hammer impacting on the hammer cushion with an impact velocity computed based on the rated hammer energy and assumed efficiency. It is not certain if this impact velocity is correct, nor it is certain if all the soil parameters (ultimate resistance, quake, and damping coefficients) used in the wave equation analysis are appropriate.

CAPWAP analysis is much more time consuming than the CASE method. The analysis is typically done in the office using the measured velocity and force records from PDA. In CAPWAP, the pile is also assumed as a series of segments, and the soil resistance is modeled as a series of springs and dashpots, as depicted in Figure 8.4. A series of iterative wave equation analysis is conducted. The soil resistance distribution along the pile, the quake, and damping coefficients are assumed first. The measured acceleration near the pile head is used to initiate the wave equation analysis. Curve matching is then conducted where the computed and measured force/velocity versus time records near the pile head are compared. Adjustments are made to the soil model assumptions and the computation process is repeated until no further agreement between the computed and measured forces near the pile head can be obtained. The static capacity computed from CAPWAP is therefore based on wave equation analysis but calibrated according to field measurements.

Figure 8.16 shows the comparisons between the time histories of the measured and computed force and velocity values from a CAPWAP analysis using the same PDA record shown in Figure 8.15. Figure 8.17 shows the static load movement curves at pile head and toe, according to CAPWAP. Figure 8.18 depicts the internal load transfer and shaft friction force along the pile induced by the hammer blow according to CAPWAP. The total static capacity of the pile was 7.17 MN, frictional resistance from pile shaft was 5.90 MN, and the toe bearing was 1.27 MN. The selected hammer blow generated a maximum movement of 27.4 mm in x direction (vertical direction).

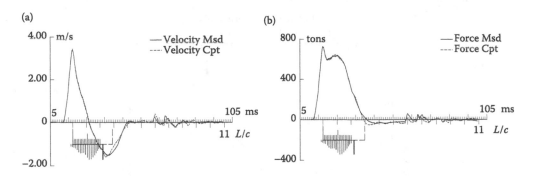

Figure 8.16 CAPWAP from field measurement for a steel pipe pile installation. (a) Computed and measured velocity. (b) Computed and measured force. (Courtesy of DECL, Taipei, Taiwan.)

A comparison of the advantages and disadvantages between the CASE and CAPWAP methods is shown in Table 8.6.

The static capacities from dynamic testing represent the capacity at the time of testing. Pile capacity is known to increase or decrease with time due to setup or relaxation. To evaluate time-dependent pile capacity requires taking PDA readings by re-striking the pile after a waiting period. This may involve additional mobilization of the pile-driving equipment and PDA.

8.3 CONSTRUCTION AND QUALITY ASSURANCE OF BORED PILES

In principle, the construction of a bored pile is straightforward: drill a hole and fill the void with concrete. In reality, however, the borehole can be very deep and very large. The drilling may have to go through unstable material that can collapse without a protective measure. The reinforced concrete may be cast under muddy conditions. The finished product

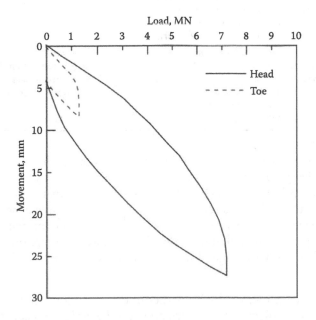

Figure 8.17 Load movement curve from CAPWAP.

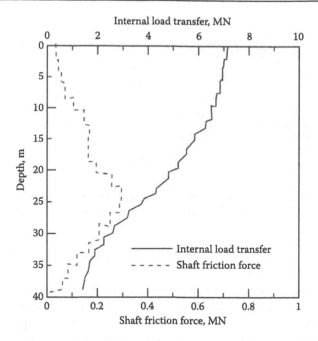

Figure 8.18 Load transfer mechanism according to CAPWAP.

Table 8.6 Comparison between CAPWAP and CASE methods

Item	Advantages	Disadvantages
CASE	• Real-time computation, can be applied for *every* hammer blow throughout the pile-driving process • Provide useful information regarding the capacity and integrity of the pile throughout the driving process • A useful tool for field quality assurance	• Involve semi-empirical parameters or procedures
CAPWAP	• Computations are based on wave equation analysis • More refined and more accurate than CASE method • Calibrates wave equation analysis • Refines CASE method computations	• Time consuming • Performed for *one* hammer blow usually selected from the end of driving or beginning of restrike

should be a competent structural element capable of sustaining the loading conditions for which the pile is designed. The success of bored piles hinges mostly on the quality of construction. How we drill the borehole, with what kind of equipment and quality assurance plan for keeping track of the construction process, including the placement of concrete are all important aspects to be considered. Knowledge of construction methods, exercise of sound engineering judgment, attention to detail, and thorough preparation are imperative to the success of bored piles. A good practice in the construction of bored piles should include:

Thorough subsurface exploration—Properties of the bearing stratum, soil, and groundwater conditions are all important information for the design and construction of bored piles. Saturated granular material under groundwater or soft cohesive soils are considered "caving" soil. The existence of caving soil is a key element in the selection of construction method.

Practical design in favor of constructability—The design should consider the availability of local expertise, supply of equipment, and practicality for the subsurface conditions. These elements assure quality construction with reasonable cost.

Reasonable and sound quality assurance plan—The specifications for field work should be compatible with the performance of the pile and doable by the contractor. As the bored pile is entirely built in the field, thorough inspection and record keeping are imperative.

The following sections provide details for the available bored pile construction methods. The requirements in ground conditions for the success of these methods and their construction procedures are described.

8.3.1 Methods of bored pile construction

Depending on the ground conditions, there are generally three methods available for the construction of bored piles: the dry method, the slurry method, and the casing method. In many cases, the construction of bored piles may involve combinations of these three methods. This section describes the details of these construction methods.

8.3.1.1 Dry method

This is the most economical way of constructing a bored pile. The procedure is illustrated in Figure 8.19. The borehole is advanced with a rotary auger such as the one shown in Figure 8.19a. The auger is attached to a telescoping drive shaft (kelly) and rotated by a powerful engine. The dry method can be applied in the following ground conditions:

*   Soils with sufficient cohesion to keep the borehole open and stable without the protection of a casing.
*   Minimal seepage into the borehole can be maintained during drilling and placement of reinforcement cage and concrete.

Therefore an ideal soil condition would be medium to stiff clays or moist sand with some fine sand or clay content. Because there is no need for inserting a casing or for the

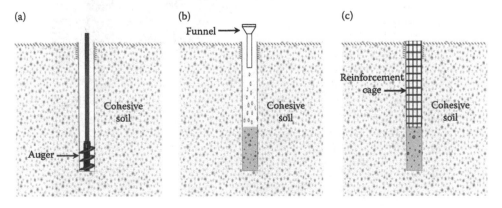

Figure 8.19 Bored pile construction by dry method. (a) Drilling. (b) Concrete placement. (c) Cage placement.

use of slurry for protection of the borehole, the drilling process is very efficient, especially when using a powerful drill rig. Concrete is usually poured by free fall, as shown in Figure 8.19b. A device such as a funnel or chute is used to direct the concrete flow into the center of the shaft and avoid hitting the reinforcement cage or sides of the borehole. A reinforcement cage (Figure 8.19c) can be placed in the upper part or full length of the borehole, depending on the structural design of the foundation system. If a reinforcement cage is used, it is placed before pouring concrete (Figure 8.19c).

8.3.1.2 Slurry method

The slurry method is used when drilling and pouring concrete under the dry conditions are not feasible, such as the following conditions:

- Drilling toward the toe of the pile is in caving soil. Soft clays or granular soils below the groundwater are caving soils that can collapse or slump into the borehole without protection.
- Drilling toward the toe of the pile is in permeable soil or rock stratum where seepage is excessive.

Depending on the nature of the caving soil, the slurry can be as simple as the mixture of water and soil resulting from drilling of the borehole (drilling fluid), or the slurry can be made of a mixture of Bentonite (i.e., Montmorillonite) and water (Bentonite slurry). Bentonite consists of very small, plate-like clay particles capable of absorbing the large amounts of water molecules on their surface. When mixed properly, the Bentonite slurry has a unit weight slightly heavier than water, and a viscosity favorable for maintaining the stability of the borehole but not so excessive as to adversely affect the concrete placement. As shown in Figure 8.21a and b, drilling and placement of a reinforcement cage proceeds in a slurry-filled borehole. The concrete is placed in the slurry using a tremie pipe, as shown in Figure 8.21c. A tremie pipe is a funnel with a long neck that allows the concrete to be placed from the base of the slurry-filled

Figure 8.20 Typical power auger used to advance the borehole. (Courtesy of An-Bin Huang, Hsinchu, Taiwan.)

(a)   (b)   (c)

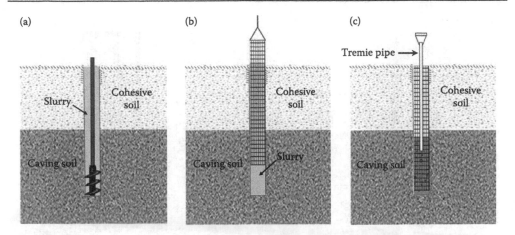

Figure 8.21 Bored pile construction by slurry method. (a) Drill with slurry. (b) Cage placement. (c) Tremie concrete.

borehole. The tremie pipe is extended to the bottom of the borehole initially. The tremie pipe is lifted by keeping the bottom of the pipe submerged at a certain distance below the rising surface of the freshly placed concrete so that the concrete that comes out of the tremie pipe does not mix with slurry.

Disposing of Bentonite or other types of mineral slurry can be an environmental hazard. Biodegradable polymer slurry is becoming popular and is required by law in some parts of the world. The polymer slurry can serve similar functions as the mineral slurry but with much less problems of waste disposal.

Reverse circulation is a preferred technique in places where the slurry method is used. A schematic illustration of the reverse circulation and photographs of its field operation are shown in Figures 8.22 and 8.23. The drill rig advances the full face of the borehole

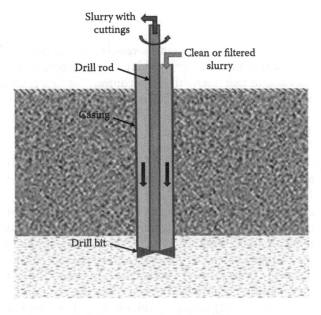

Figure 8.22 Schematic illustration of the reverse circulation drilling.

(a)   (b)

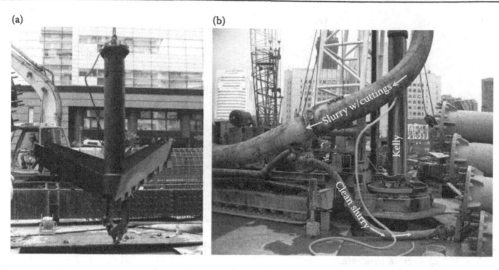

Figure 8.23 Reverse circulation field operation. (a) Tri-wing drill bit. (b) Field operation of reverse circulation. (Courtesy of Sino-Geotechnics Research and Development Foundation, Taipei, Taiwan.)

using a large cutting bit, with the help of a protection casing. Clean slurry enters the borehole via the annular space between the casing and the drill rod. Cuttings from the drilling are conveyed upward through the center of the drill rod. The term *reverse circulation* comes from the fact that this process is reverse to rotary drilling in soil borings where the drilling fluid is conveyed down through the drill rod and up through the annular space between the drill rod and the borehole. The significantly reduced cross-sectional area within the drill rod forces a discharge velocity that is favorable in lifting the cuttings from the bottom of the borehole, which would otherwise not be possible through the much larger space between the drill rod and the casing.

8.3.1.3 Casing method

The casing method is used when drilling of the borehole must go through water or permeable or caving soil before reaching competent and non-caving bearing stratum. The casing method can be conducted in two alternative procedures. The first is to drill the borehole before inserting the casing. This procedure is described in Figure 8.24. Initial drilling can be in dry or under-slurry, as shown in Figure 8.24a, until the slightly oversized borehole is extended beyond the caving soil layer. A casing is inserted and sealed into the stable soil and the interior slurry is removed using a slurry bailer (Figure 8.24b). The drilling then continues in dry condition to the bearing stratum (Figure 8.24c). The reinforcement cage and concrete are placed under dry conditions similar to those in dry method (Figure 8.19b and c).

The second alternative is to insert the casing through the caving soil before borehole drilling, as shown in Figure 8.25. The casing may be driven by impact or vibratory hammers, or using a casing oscillator (Figure 8.25a). Significant torque and downward force is required to insert the casing. The casing is inserted and sealed into the underlying cohesive soil. Upon insertion of the casing, soil is excavated from the inside of the casing with a clamp bucket or hammer grab (Figure 8.25b) to the bearing stratum. The reinforcement cage and concrete are placed in dry condition (Figure 8.25c). Figure 8.26 shows the field operation of an oscillator and a hammer grab.

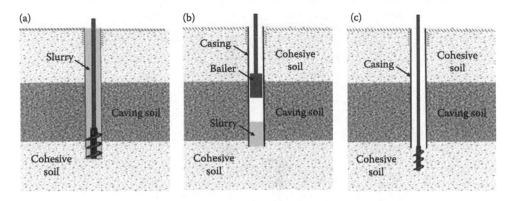

Figure 8.24 Bored pile construction by casing method—borehole drilling before casing insertion. (a) Drill with slurry. (b) Insert casing and bailout slurry. (c) Drill in dry.

0.3.2 Use of casing in bored piles

Long and heavy casings are used often in the construction of bored piles. They can be part of the drilling equipment, such as in Figure 8.26, for advancing the borehole, or they can be part of the required protection when a worker is lowered to the bottom of the borehole for excavation, cleaning, or inspection. In any case, the temporary casing usually is removed and reused because of its high cost. Often there is ground water or slurry on the outside of the casing, as these are usually the reasons for the use of casing. After the borehole is drilled, the temporary casing is lifted in a controlled manner after a sufficient amount of fresh concrete has been placed inside the casing to offset the slurry pressure from outside of the casing, as shown in Figure 8.27. In reference to Figure 8.27,

$$\frac{h_c \gamma_c}{h_s \gamma_s} > 1 \tag{8.15}$$

where
h_c = head of concrete inside the casing
h_s = head of ground water/slurry outside the casing
γ_c = unit weight of concrete
γ_s = unit weight of slurry or water

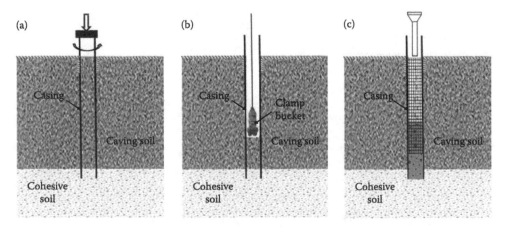

Figure 8.25 Bored pile construction by casing method—casing insertion before excavation. (a) Casing insertion. (b) Excavation. (c) Reinforcement cage and concrete placement.

Figure 8.26 Field operation of an oscillator rig used to insert casing and a hammer grab. (Courtesy of An-Bin Huang, Hsinchu, Taiwan.)

The heads, h_c and h_s, are in reference to the bottom of the casing. The ratio of $h_c\gamma_c/h_s\gamma_s$ should be larger than 1.2. Premature removal of the casing may cause the slurry to enter from the bottom of the lifted casing and mix with fresh concrete. On the other hand, the concrete can develop adhesion with the casing due to initial hardening of the concrete if

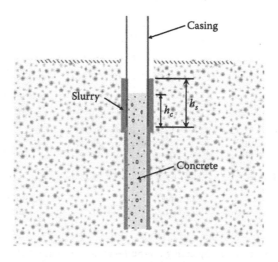

Figure 8.27 Slurry and concrete heads during casing removal.

Figure 8.28 Use of a permanent casing in bored pile construction. (Courtesy of An-Bin Huang, Hsinchu, Taiwan.)

the casing is lifted too slowly. Lifting in this condition may cause breakage in the hardened concrete. The bored pile can be seriously damaged in either case.

Where possible, permanent casing made of relatively low-cost corrugated metal pipe, as shown in Figure 8.28, is used. In this case, the casing is not removed from the borehole.

8.3.3 Underreams (Bells)

As most of the bored piles involve relatively high toe bearing capacities, an enlarged pile toe can be very beneficial because the toe bearing load can be increased significantly with limited cost increase. This process is called underreaming, as schematically shown in Figure 8.29. The underream typically has the shape of a bell. The diameter of the bell should not exceed three times the diameter of the pile shaft. The pile shaft is bored first to the bearing depth; the auger is then removed and a belling bucket, as shown in

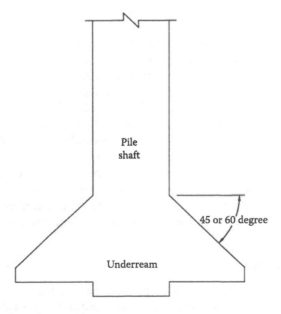

Figure 8.29 Schematic view of an underream.

Figure 8.30 A belling bucket with unfolded arms. (Courtesy of An-Bin Huang, Hsinchu, Taiwan.)

Figure 8.30, is attached to the kelly and lowered to the bottom of the borehole. The belling bucket has two hinged arms that are folded and stored inside the bucket when the tool is lowered to the bottom of the borehole. The arms are forced open by a downward-pushing force from the drive shaft while the belling bucket is rotated. The rotation cuts a bell-shaped hole as the arms unfold. The soil cutting is swept inside the bucket when the belling tool is lifted and the arms are retracted.

The method is feasible only if the underream can be constructed while the surrounding material remains dry and stable and can be cut by the belling bucket. Materials suitable for underreaming include glacial till and intact soft rock. Because of the potential risk of excessive seepage or collapse when belling in fractured soil or rock formations, underreams are not as popular as they were a few decades ago.

8.3.4 Barrettes

There is a trend to use the same grab-bucket (or clamshell bucket) typically used to excavate diaphragm walls (described in Chapter 6) for the construction of bored piles. This is especially cost effective when diaphragm walls are used at the same project site, as the same tools are used for both the piles and diaphragm walls. This type of bored pile is called barrette. The simplest barrettes are made with one stroke of a standard sized grab-bucket, with a rectangular cross-section. The dimensions of this simple rectangular cross-section can vary depending on the size of the grab-bucket, as shown in Figure 8.31a. Starting from these basic rectangular dimensions, bigger piles can be formed. Possible configurations can include strips, crosses, *H*, or *T*, as shown in Figure 8.31b. The configurations are chosen to optimize the pile performance for the loading conditions.

The construction of barrettes follows the slurry method described previously. This is also the typical procedure used for the construction of diaphragm walls (Chapter 6).

8.3.5 Quality assurance

As a bored pile is made in the field, it is important that the piles are made as planned or specified in the contract. Quality assurance is an important part of the construction and a legally correct term for "field inspection," to be carried out by "inspectors," a job

(a)

(b)

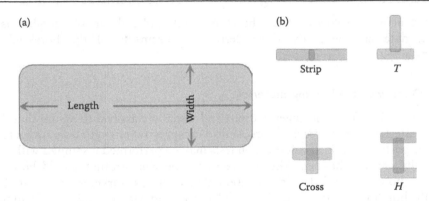

Strip

T

Cross

H

Length

Width

Figure 8.31 Basic element of barrettes and their configurations. (a) Basic rectangular element. (b) Possible configurations.

usually assigned to junior engineers, such as students who have taken a foundation engineering course and are fresh out of school. It is important for the students to understand at this stage that the purpose of quality assurance is to assure that the field construction meets the requirements specified in the contract. Errors according to field measurements are within the acceptable tolerances. Whether the finished product is capable of fulfilling its function is not the main purpose of quality assurance. Major items to be verified and/or documented in the field include the following.

8.3.5.1 Dimensions and verticality of the borehole

When drilling is conducted in dry condition, these parameters can be made with direct measurements. For boreholes filled with slurry, the borehole diameter profile can be measured using acoustic pulse-echo sensors, as shown in Figure 8.32. The sensors, suspended in the slurry with a hoisting device shown in Figure 8.32a, measure the transit time of acoustic waves between the sensor and the borehole wall and transmit the results to the surface in real time. The surface computer uses this information to derive the corresponding standoff distance and borehole diameter, as shown in Figure 8.32b, based on the

(a)

(b)

Figure 8.32 Diameter measurement of slurry-filled borehole using acoustic pulse-echo sensors. (a) Hoisting device. (b) Readout unit. (Courtesy of DECL, Taipei, Taiwan.)

calculated acoustic velocity of the borehole fluid under downhole conditions. The recorded data can be analyzed to derive measurements of the borehole shape and verticality.

8.3.5.2 Quality of the bearing material

Samples are taken from the auger cuttings and visually classified to assure that desired bearing material has been reached. Simple field compressive strength tests such as pocket penetrometer or unconfined compression tests may be performed if samples with reasonable quality can be obtained. The surface of the bearing stratum should be properly cleaned for boreholes drilled in dry condition. If necessary, a worker is lowered to the bottom of the borehole, while the borehole is fully protected with a casing, for cleaning and/or taking measurements. Concrete is poured after the observation and necessary measurements are completed.

8.3.5.3 Quality and quantity of concrete

The quantity and time are recorded when the concrete is poured. The quantity of poured concrete should be compatible with the dimensions of the borehole. After mixture, the premix concrete should be delivered to the job site within the time limit specified in the contract. Concrete slump tests and cylinder samples are taken as required by the specifications.

8.4 PILE LOAD TEST AND ITS INTERPRETATION

The basic idea of a pile load test is to apply a known loading condition on the pile and measure its response. The load can be applied in axial compression, axial tension, or lateral direction, usually from the head of the pile. The magnitude of the applied load and the kind of measurements taken in a load test depend on its purpose. The purpose of pile load testing usually includes the following.

Proof load test—to make sure the test pile can sustain the design load with adequate safety factor with a tolerable displacement. The maximum applied load in this case equals to the design load multiplied by a required safety factor (usually 2). Unless otherwise specified, the measurements are limited to the load applied and the corresponding movement at the pile head. If the requirements are met in the load test, then the pile will be used in the future to serve its structural purposes. Usually a portion of the production piles are involved in this type of load test.

Load test to ascertain the ultimate static capacity and load transfer mechanism of the pile—the pile is usually tested to failure and discarded after the load test. This kind of measurement can be much more elaborate and thus more expensive than that involved in proof load test. In addition to the load and movement at the pile head, the measurement often includes strains and/or movement at different parts of the pile.

Static load test is the most ideal way to determine the load capacity of a pile. However, dynamic and pseudodynamic load test methods have been developed in the past few decades for this purpose. The dynamic testing method described in Section 8.2 can be used to determine the static capacity of a pile. The statnamic method (described later) may be considered as a pseudodynamic method, or the load (at least in axial compression load) can be applied from the bottom of the pile, such as the Osterberg cell (O-cell) method, described later. The same load test methods can be applied to both bored and driven piles. The difference usually lies in the capacity of the piles (thus the magnitude of the applied load) and the need to mobilize the equipment specifically for the load test (such as the use of dynamic testing on a bored pile by re-striking).

The following sections introduce the conventional static load tests in axial compression, the test setup, its procedure, and the interpretation of test data. Lateral load tests and some of the recent developments in pile load test techniques are also described.

8.4.1 Conventional static axial compression pile load test

8.4.1.1 Principles of static axial compression load test

Figure 8.33 describes the basic concept of the conventional static pile load test in axial compression. This type of pile load test generally involves the following steps:

- The load is applied at the pile head incrementally, or the pile can be loaded at a constant rate of penetration.
- The load Q, concurrent movement δ at the pile head, and time are recorded.
- The result of the load test is presented with a plot of Q versus δ curve (the load movement curve).
- The static load and movement at failure of the test pile are determined according to a chosen interpretation method.

For most of the proof load tests, this type of Q versus δ plot is sufficient. The elastic deformation of the pile and nature of the soil friction distribution along the pile shaft (load transfer mechanism) can contribute to the variations of pile movement at different depths. Thus if the load test involves the analysis of load transfer mechanism, it would be highly desirable to measure movements or strains at different depths within the pile. A telltale (see Figure 8.33) consists of a solid rod protected by a tube. The rod, loosely placed inside the tube and extending from pile head to the bottom of the tube, is used to measure the movement at the bottom of the telltale during load test. Multiple telltales extending to different depths can be installed depending on the need and dimensions of the pile.

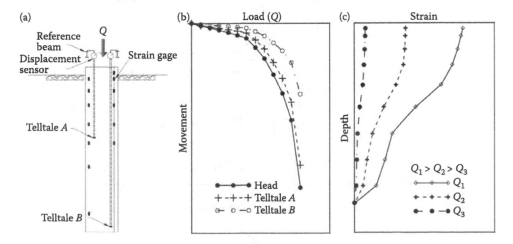

Figure 8.33 Basic concept of static axial compression pile load test. (a) Internal measurements. (b) Load movement curves. (c) Strain measurements.

All movements are measured against a steady reference beam installed on the ground surface near the pile head, as indicated in Figure 8.33. Attaching strain gages at different depths of the pile can also be used to determine the load transfer mechanism. For steel piles, the strain gages can be welded or epoxied directly to the surface of the pile. For reinforced concrete piles, the strain gages are usually attached to a short piece of reinforcement steel, called sister bar. The sister bars are then overlapped or welded to the pile reinforcement steel cage before casting concrete. Details of interpreting the telltale and strain gage readings from a load test are described later.

8.4.1.2 Field setup of conventional static axial compression load test

Depending on the capacity of the pile, the load test can involve imposing tens to thousands of metric tons of force at the pile head and measuring the related movements. Figure 8.34 shows a schematic view of a typical pile load test setup. In this case, a stiffened steel beam is used as a reaction frame. The reaction frame is anchored by a group of tension piles or ground anchors. The axial load is applied by a hydraulic jack against the reaction frame. The spherical bearing minimizes potential bending movement created by the axial force due to eccentricity. It is often required that the axial force be measured independently with a load cell.

The movements of the pile head and telltales are measured with displacement sensors such as dial gages or linear variable differential transformers (LVDTs). These displacement sensors are supported by a reference beam and pointed to their respective target. The reference beam is extended and supported by footings away from the test pile so that the reference beam remains stationary and is not affected by the test pile movement. The target can be a bracket extended from pile head or top end of the telltales. A reaction plate with a smooth surface (usually made of glass) is epoxied to the target to receive the dial gage or LVDT. The smooth surface minimizes potential reading errors resulting from friction between the tip of the dial gage or LVDT and target surface of movement measurement. The sensor readings are usually connected to an automated data logging system.

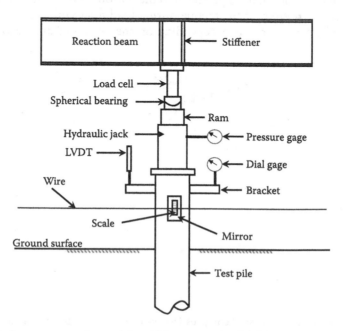

Figure 8.34 Typical arrangement for applying axial compression load test.

For a pile load test, as in most of the field geotechnical monitoring, it is important to have redundancy in the measurements in case of system error or failure of automated data logging system. The hydraulic jack pressure gage readings can be used as a redundancy for axial force readings. The wire shown in Figure 8.34 is used to provide redundant movement readings manually. A telescope is used to take the readings off the scale according to the position of the wire image on the mirror. The height of the telescope is adjusted until the wire and its reflected image from the mirror are aligned before recording the scale reading. Figure 8.35 shows a field setup for a pile load test with a maximum axial load capacity of 75,000 kN. Multiple hydraulic jacks and load cells were used to apply and measure the axial force. Instead of using tension piles, dead weights stacked on top of a platform (i.e., the Kentledge method), as shown in Figure 8.36, have also been used to provide reaction force in axial compression pile load tests.

8.4.1.3 The load test procedure

Several loading procedures are allowed, and described in the ASTM standard D-1143. The quick load test method is a popular procedure and is described as follows:

1. The load is applied in increments of 5% of the anticipated failure load.
2. Each load increment is maintained with a constant time interval from 4 to 15 minutes, the same time interval is used for all load increments.
3. The procedure is continued until failure load is reached where continuous jacking is required to maintain the test load, or until the capacity of the loading system is reached, whichever occurs first.
4. Upon reaching the maximum load, the pile is unloaded in five to ten equal decrements for 4–15 minutes each, using the same time interval for all unloading decrements.
5. Longer time intervals are used for the failure load to assess creep behavior and for the final zero load to assess rebound behavior.
6. Readings of load, movement, and time are recorded immediately during and after each load increment and decrement.

Figure 8.35 Pile load test using a reaction frame with 75,000 kN capacity. (Courtesy of DECL, Taipei, Taiwan.)

Figure 8.36 Pile load test using the Kentledge method. (Courtesy of DECL, Taipei, Taiwan.)

8.4.1.4 Presentation and interpretation of the load test results

Figure 8.37 shows a load movement curve from an axial compression load test on a bored pile. It is important to present the load movement curve in conformance with the pattern of Figure 8.37. Movement at the pile head is presented in the vertical coordinate, and load in horizontal coordinate. An important parameter in the interpretation of the test result is the elastic deformation of the pile, Δ, which is computed from

$$\Delta = \frac{QL_p}{A_p E_p} \qquad\qquad (8.16)$$

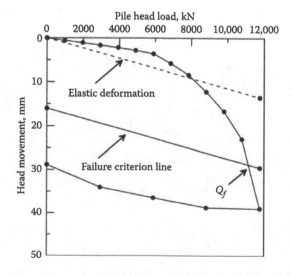

Figure 8.37 Load–movement curve from static compression test on a bored pile.

where
   Q = test load applied at the pile head
   L_p = pile length
   A_p = cross-sectional area of the pile
   E_p = elastic modulus of the pile

The scales of the load and movement should be selected so that the elastic deformation Δ versus load inclines at approximately 20° from the horizontal coordinate, as shown in Figure 8.37.

Numerous methods have been proposed for the interpretation of the pile load test results in the past few decades. Recent studies have concluded that a good interpretation method should consider the elastic deformation of the pile. Ignoring elastic deformation could lead to overestimation of the failure capacity for short piles and underestimation for long piles. It is also important to recognize that unlike the shaft friction, the ultimate resistance at pile toe is proportional to pile diameter. The method proposed by Davisson (1972) is widely accepted for the interpretation of pile load test results, as it considers these two important factors. The failure load Q_f is the load that causes a movement s_f at pile head, according to an empirical equation:

$$s_f = \Delta + (4.0 + 0.008D_p) \tag{8.17}$$

where
   s_f = pile head movement at failure in mm
   Δ = elastic deformation of the pile in mm
   D_p = diameter of the pile in mm

To apply this method, the following procedure is recommended:

1. Plot a failure criterion line that corresponds to Equation 8.17, and parallel to the elastic deformation line as shown in Figure 8.37. The intercept of the failure criterion line with the vertical coordinate is $(4.0 + 0.008D_p)$.
2. The load at which this straight failure criterion line intersects the load movement curve is the failure load Q_f.

If the failure criterion line does not intersect the load movement curve, then the failure load is larger than the maximum applied load in the load test.

The allowable load is determined by dividing the failure load Q_f by a factor of safety. A factor of safety of 2.0 is often used.

EXAMPLE 8.5

Given

The load–movement-curve shown in Figure 8.37 came from a compression load test on a 1.5 m diameter and 34.5 m long bored pile.

Elastic modulus of the pile, $E_p = 17\,\text{GPa} = 17,000,000\,\text{kN/m}^2$
Pile diameter, $D_p = 1.5\,\text{m} = 1500\,\text{mm}$
Pile length, $L_p = 34.5\,\text{m} = 34,500\,\text{mm}$

Required

Determine the failure load by Davisson's method.

Cross-sectional area of the pile,

$$A_p = \frac{\pi D_p^2}{4} = \frac{(3.1416)(1.5)^2}{4} = 1.767 \, \text{m}^2$$

$$\frac{\Delta}{Q} = \frac{L_p}{A_p E_p} = \frac{(34,500)}{(1.767)(17,000,000)} = 0.00115 \, \text{mm/kN}$$

$$(4.0 + 0.008 D_p) = 16 \, \text{mm}$$

$$s_f = \Delta + (4.0 + 0.008 D_p) = 0.00115 Q + 16$$

The failure criterion line starts at point (0, 16) with a slope of 0.00115.

The failure criterion line intersects the load–movement curve shown in Figure 8.37 at 11,130 kN; thus $Q_f = 11,130 \, \text{kN}$ and,

$$s_f = \Delta + (4.0 + 0.008 D_p) = (0.00115)(11,130) + 16 = 28.8 \, \text{mm}$$

It should be noted that for piles with diameter or width greater than 610 mm, s_f from Equation 8.17 may be too small and thus can lead to underestimation of Q_f. This is especially true for large toe bearing piles where the pile takes more movement to develop its full capacity. Kyfor et al. (1992) have suggested that for piles with diameter or width greater than 610 mm, $s_f = \Delta + D_p/30$ should be used.

8.4.1.5 Load transfer mechanism analysis

Information about the force distribution within a pile in a given loading condition can be determined from displacement or strain measurements within the pile, and elastic properties of the pile. This analysis can lead to important understanding of how the applied load is transferred from pile to the surrounding soil via side friction and toe bearing, or the load transfer mechanism. As it is not possible to measure side friction (a shear force) directly on the pile surface, we rely on axial deformation measurements within the pile to infer load transfer mechanism. Consider a test pile shown in Figure 8.38 with two telltales installed. The average internally transferred load between any two telltale measurement points 1 and 2, Q_{iavg}, can be calculated based on Hooke's law, and the difference in movements between the two measurement points as

$$Q_{iavg} = A_p E_p \frac{s_1 - s_2}{d_{12}} \tag{8.18}$$

where

s_1 = movement readings from upper measurement point
s_2 = movement readings from lower measurement point
d_{12} = distance between the two measurement points
A_p = cross sectional area of the pile
E_p = elastic modulus of the pile

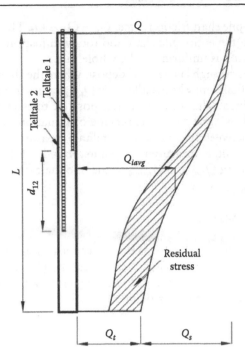

Figure 8.38 Load distribution from telltales.

If the shaft friction between the two telltales is uniform, Q_i decreases linearly within the pile, as part of the load is taken by the surrounding soil via friction/adhesion force (see Section 7.3). Q_{iavg} represents the average axial force passing internally between telltale locations 1 and 2.

$$Q_{iavg} = A_p E_p \frac{(s_1 - s_2)}{d_{12}} = (Q_{i(1)} + Q_{i(2)})/2 \qquad (8.19a)$$

If Q_i at the upper measurement point ($Q_{i(1)}$) is known, then

$$Q_{i(2)} = 2Q_{iavg} - Q_{i(1)} \qquad (8.19b)$$

If the upper measurement point is the pile head (no need of a telltale at pile head), and the lower measurement point is the pile toe, d_{12} = pile length L_p, then Q_{iavg} represents the average axial force passing internally through the entire pile, or

$$Q_{iavg} = A_p E_p \frac{(s_1 - s_2)}{L_p} = \frac{(Q_h + Q_t)}{2} \qquad (8.20a)$$

and

$$Q_t = 2Q_{iavg} - Q_h \qquad (8.20b)$$

where
   Q_h = load applied at the pile head
   Q_t = load transmitted to the pile toe

Load resisted by the pile shaft friction force, $Q_s = Q_b - Q_t$. This leads to an estimate of load distribution along the entire pile shaft and toe with just one telltale at the pile toe, provided the shaft friction is uniform for the whole pile.

For a pile penetrating through layered soil deposit where the shaft friction is not likely to be uniform, multiple telltales may be installed and Q_{iavg} values from different depth ranges of the pile can be obtained. With these data, it is possible to estimate how the pile load is transferred based on the simple averaging scheme of Equation 8.19 and the assumption that the shaft friction between two consecutive telltale measurement points is uniform.

For two consecutive telltale measurement points, n and $n + 1$ in a pile, the difference between the corresponding $Q_{i(n)}$ and $Q_{i(n+1)}$ is the pile shaft frictional resistance between points n and $n + 1$, $Q_{s(n,n+1)}$, or

$$Q_{s(n,n+1)} = (Q_{i(n)} - Q_{i(n+1)})$$
(8.21)

In this case, n is the upper and $n + 1$ is the lower measurement point. As the distance between the telltale measurement points often varies, the $Q_{s(n,n+1)}$ value can be misleading. It is more desirable to present the shaft friction in terms of frictional resistance per unit area, $q_{s(n,n+1)}$. For a round pile with a diameter D_p,

$$q_{s(n,n+1)} = \frac{Q_{s(n,n+1)}}{\pi D_p d_{n,n+1}}$$
(8.22)

where
$d_{n,n+1} = $ distance between the two telltale measurement points

However, it is inevitable that some residual stresses remain between the pile shaft and the surrounding soil due to locked-in frictional forces from pile driving (Fellenius, 1990). The Q_{iavg} values from telltales therefore include the effects of residual stress, as shown in Figure 8.38. The effects of residual stress cannot be isolated from the use of telltales alone and thus could result in errors in the determination of load distribution.

If strain gages are installed, the strain reading, $\varepsilon_{(n)}$, taken from measurement point n in the pile, the applied load transmitted internally at this point, $Q_{i(n)}$, is:

$$Q_{i(n)} = A_p E_p \varepsilon_{(n)}$$
(8.23)

It is usually easier to install a series of strain gages in the test pile than the same number of telltales, and strain gage readings can be taken with an automated data logger. For these reasons, the use of strain readings is gaining popularity in pile load tests. With $Q_{i(n)}$ values known, the interpretations for $Q_{s(n)}$ and $q_{s(n)}$ are the same as those for telltale readings. Again, it is assumed that the shaft friction between two consecutive strain reading locations is uniform. Details of using these measurement and interpretation methods are demonstrated in the following examples.

EXAMPLE 8.6

Given

For a compression load test on a 1.5 m diameter and 34.5 m long bored pile, when the applied load at pile head $Q_h = 9810$ kN, the movement measurements at pile head and toe are 16.67 mm (s_1) and 10.10 mm (s_2), respectively.

Elastic modulus of the pile, $E_p = 17$ GPa $= 17,000,000$ kN/m^2

The shaft friction (q_s) is expected to be uniformly distributed along the pile, as shown in Figure E8.6a.

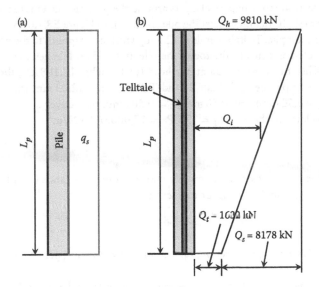

Figure E8.6 Load transfer for uniformly distributed shaft friction. (a) Uniform shaft friction along the pile. (b) Load transfer according to telltale reading.

Required

Determine the resistance at pile toe (Q_t) and frictional resistance from the pile shaft (Q_s).

Solution

For 1.5 m diameter pile, its cross-sectional area, $A_p = 1.767\ \text{m}^2$

Distance between the pile head and tip of the telltale (pile toe), $d_{12} = L_p = 34.5\ \text{m} = 34,500\ \text{mm}$

According to Equation 8.19a,

$$Q_{iavg} = A_p E_p \frac{(s_1 - s_2)}{d_{12}} = A_p E_p \frac{s_1 - s_2}{L_p}$$

$$= (1.767)(17,000,000)\frac{[(16.67) - (10.10)]}{(34,500)} = 5721\ \text{kN}$$

$$Q_t = 2Q_{iavg} - Q_h = (2)(5721) - (9810) = 1632\ \text{kN}$$

$$Q_s = Q_h - Q_t = (9810) - (1632) = 8178\ \text{kN}$$

A plot of load transferred internally through the pile, Q_i, versus depth is shown in Figure E8.6b. The results show that 17% $(1632/9810 = 17\%)$ of the applied load at pile head is taken by toe bearing and the rest is resisted by shaft friction.

EXAMPLE 8.7

Given

For a test barrette pile with cross-sectional dimensions of 0.8 m by 2.8 m and a pile length of 40.55 m, there are two major soil layers around the pile; each layer is expected to provide a uniform shaft friction against the pile, as shown in Figure E8.7a. Two telltales are installed in the test pile. Telltale 1 ends at 12.5 m, which corresponds to the bottom of soil layer 1. Telltale 2 extends to the toe of the pile at 40.55 m. Pile head is considered as Telltale 0. When the applied load at pile head (Q_h) reaches 17,168 kN, the movement measurements at pile head, Telltale 1, and Telltale 2 are 12.03 mm ($s_0 = 12.03$ mm), 6.50 mm ($s_1 = 6.50$ mm), and 0.50 mm ($s_1 = 0.50$ mm), respectively.

Elastic modulus of the pile, $E_p = 17$ GPa $= 17,000,000$ kN/m^2

Required

Determine the load transferred internally and frictional resistance from the pile shaft (Q_s) at Telltale 1 and Telltale 2 (or toe) levels.

Solution

For 0.8 m by 2.8 m pile cross-section, $A_p = (0.8)(2.8) = 2.24$ m^2, perimeter, $p_p = 2(0.8 + 2.8) = 7.2$ m.

Distance between the pile head (measurement point 0) and Telltale 1:

$$d_{0,1} = 12.5 \text{ m} = 12,500 \text{ mm}$$

Load transfer between pile head and Telltale 1:
According to Equation 8.19a,

$$Q_{iavg} = A_p E_p \frac{s_0 - s_1}{d_{0,1}} = (2.24)(17,000,000)\frac{[(12.03) - (6.50)]}{(12,500)} = 16,237 \text{ kN}$$

$$Q_{i(0)} = Q_h = 17,168 \text{ kN (given)}$$

$$Q_{i(1)} = 2Q_{iavg} - Q_h = (2)(16,237) - (17,168) = 15,307 \text{ kN}$$

$$Q_{s(0,1)} = Q_h - Q_{i(1)} = 17,168 - 15,307 = 1861 \text{ kN}$$

$$q_{s(0,1)} = \frac{Q_{s(0,1)}}{p_p d_{0,1}} = \frac{(1861)}{(7.2)(12.5)} = 20.7 \text{ kPa}$$

Load transfer between Telltale 1 and Telltale 2 (pile toe):
Distance between the Telltale 1 and Telltale 2,

$$d_{1,2} = 40.55 - 12.5 \text{ m} = 28.05 \text{ m} = 28,050 \text{ mm}$$

According to Equation 8.19a,

$$Q_{iavg} = A_p E_p \frac{s_1 - s_2}{d_{1,2}} = (2.24)(17,000,000)\frac{[(6.70) - (0.50)]}{(28,050)} = 8417 \text{ kN}$$

$$Q_{i(2)} = Q_t = 2Q_{iavg} - Q_{i(1)} = (2)(8417) - (15,307) = 1527 \text{ kN}$$

$$Q_{s(1,2)} = Q_{i(1)} - Q_{i(2)} = 15{,}307 - 1527 = 13{,}779\,\text{kN}$$

$$q_{s(1,2)} = \frac{Q_{s(1,2)}}{p_p d_{1,2}} = \frac{(13{,}779)}{(7.2)(28.05)} = 68.2\,\text{kPa}$$

A plot of load transferred internally through the pile, Q_i, versus depth is shown in Figure E8.7b. The results show that this barrette pile is almost completely frictional under the test load conditions.

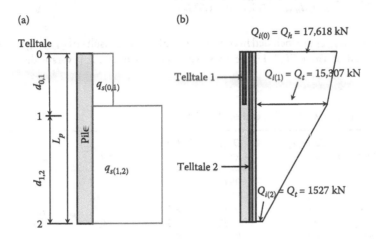

Figure E8.7 Load transfer from pile load test in two-layered soil. (a) Shaft friction along the pile. (b) Load transfer according to telltale readings.

EXAMPLE 8.8

Given

A set of strain readings obtained from a static compression load test on a 1.5 m diameter and 34.5 m long bored pile is shown in Table E8.8. The applied load at pile head (Q_h) was 9810 kN.
Elastic modulus of the pile, $E_p = 17\,\text{GPa} = 17{,}000{,}000\,\text{kN/m}^2$
For 1.5 m diameter pile, $A_p = 1.767\,\text{m}^2$

Required

Compute and plot the distribution of internally transmitted load $Q_{i(n)}$ and frictional resistance per unit area, $q_{s(n,n+1)}$, versus depth.

Solution

The given data are shown in the first three columns of Table E8.8.
Load transmitted internally at a given strain gage, $Q_{i(n)}$, following Equation 8.23,

$$Q_{i(n)} = A_p E_p \varepsilon_{(n)} = (1.767)(17{,}000{,}000)\varepsilon_{(n)} = 30{,}039{,}000\,\varepsilon_{(n)}$$

Results of $Q_{i(n)}$ in kN are shown in column 4 of Table E8.8. A plot of $Q_{i(n)}$ versus depth is shown in Figure E8.8.

The calculation of $q_{s(n,n+1)}$ starts with $Q_{s(n,n+1)}$ according to Equation 8.21,

$$Q_{s(n,n+1)} = (Q_{i(n)} - Q_{i(n+1)}) = A_p E_p (\varepsilon_{(n)} - \varepsilon_{(n+1)}) = (1.767)(17,000,000)$$
$$(\varepsilon_{(n)} - \varepsilon_{(n+1)}) = 30,039,000 \, (\varepsilon_{(n)} - \varepsilon_{(n+1)})$$

Results of $Q_{s(n,n+1)}$ are shown in column 6 of Table E8.8, and following Equation 8.22,

$$q_{s(n,n+1)} = \frac{Q_{s(n,n+1)}}{\pi D_p d_{n,n+1}} = \frac{Q_{s(n,n+1)}}{(3.1416)(1.5)d_{n,n+1}}$$

Values of $d_{n,n+1}$ are included in column 7 of Table E8.8; results of $q_{s(n,n+1)}$ are shown in column 8 of Table E8.8. Figure E8.8a shows a plot of $q_{s(n,n+1)}$ versus depth.

Table E8.8 Strain readings when Q_h reached 9810 kN

Number, n	Depth, m	Strain $\varepsilon_{(n)} \times 10^{-6}$	$Q_{i(n)}$, kN	Depth range, m	$Q_{s(n,n+1)}$, kN	$d_{n,n+1}$, m	$q_{s(n,n+1)}$, kPa
0	0	–	9810	–	–	–	–
1	1.0	310	9313	0–1	497	1	105
2	2.5	305	9163	1–2.5	150	1.5	21
3	5.0	295	8862	2.5–5	300	2.5	25
4	8.0	280	8412	5–8	451	3	32
5	11.8	247	7420	8–11.8	991	3.8	55
6	16.0	176	5287	11.8–16	2133	4.2	108
7	21.0	110	3305	16–21	1983	5	84
8	26.0	90	2704	21–26	601	5	26
9	31.0	64	1923	26–31	781	5	33
10	34.3	13	391	31–34.3	1532	3.3	99

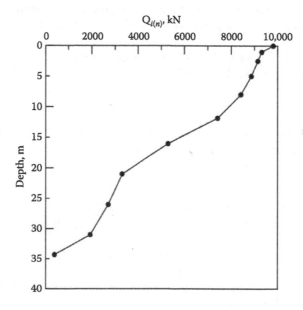

Figure E8.8 $Q_{i(n)}$ versus depth.

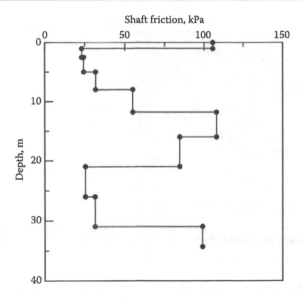

Shaft friction, kPa

Depth, m

Figure E8.8a $q_{s(n,n+1)}$ versus depth.

8.4.2 Static lateral pile load test

Piles for structures such as light poles, roller coasters, on-land or offshore wind turbines, offshore oil drilling rigs, and power line towers are usually subject to significant lateral loads. ASTM D-3966 describes the standard procedure for performing lateral load test on piles. Figure 8.39 shows a typical setup for a lateral load test on piles. Similar to the case of axial load test, a hydraulic jack mounted against a reaction beam, or another pile is used to push the pile head laterally. The lateral load is measured with a load cell. The lateral load is applied in increments until a maximum of 200% of the design load. Duration of each load increment varies from 10 to 60 minutes. Readings of time, load, and lateral movement are recorded in each load increment. In addition to displacement

Figure 8.39 Typical setup for lateral pile load test. (Courtesy of DECL, Taipei, Taiwan.)

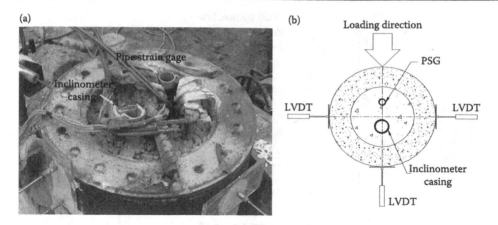

Figure 8.40 Measurement of lateral deflection at different depths using inclinometer casing and a pipe strain gage. (a) Picture. (b) Schematic view.

sensors such as dial gages or LVDTs mounted on a reference beam around the pile head, lateral deflection at different depths (Kyfor et al., 1992) is often measured to facilitate calibration of *p–y* curves.

Figure 8.40 shows the layout of LVDTs, inclinometer casing, and a pipe strain gage (PSG) for the measurement of a 25 m long, 500 mm outside diameter (OD), and 90 mm wall thickness precast concrete pile in a lateral load test. The PSG was made of an *s* series of strain sensors attached to the surface of a 28 mm OD, 20 m long PVC pipe. The PSG and the inclinometer casing were grouted inside the hollow concrete pile upon pile installation. The strain sensors measure the flexural strains, ε, experienced by the pipe as it is forced to deflect with the pile. Double-integrating ε with depth yields the deflection, y, as

$$y = \frac{1}{r} \int \left(\int \varepsilon dx \right) dx \qquad (8.24)$$

where
r = outside radius of the pipe
x = depth

The strain readings taken from PSG during a lateral load test are shown in Figure 8.41a. The corresponding lateral deflections and their comparison with the manual inclinometer readings under three loading conditions are demonstrated in Figure 8.41b.

8.4.3 Static axial compression pile load test with Osterberg cell

The Osterberg load test method was developed and patented by Prof. Jorj Osterberg in the 1980s (Osterberg, 1995). An Osterberg cell, or O-cell, is basically a hydraulic jack placed at the pile toe. For concrete piles, the O-cell and the associated plumbing system are cast in the pile. There are techniques to attach the O-cell to the pile toe for open- or closed-end steel pipe piles. As schematically shown in Figure 8.42, during the load test, the O-cell expands vertically from pile toe against the underlying ground. In doing so, the shaft friction force (Q_s) acts as a reaction against toe bearing (Q_t), and vice versa. Therefore the axial force imposed by the O-cell, Q_o, always equals Q_t and Q_s in quantity. The load test ends when either Q_s or Q_t reaches its limit.

Upon installation of the test pile, a manual or electric pump is used to control the hydraulic jack. The applied load is determined based on the hydraulic pressure readings.

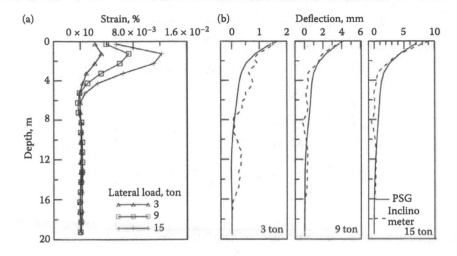

Figure 8.41 Strain and deflection profile measurements from a lateral load test. (a) Strain readings. (b) Deflection of the pile. (After Ho, Y.T. et al., 2005. Ground movement monitoring using an optic fiber Bragg grating sensored system. *Proceedings, 17th International Conference on Optical Fiber Sensors*, SPIE, Bruges, Belgium, Vol. 5855, pp. 1020–1023.)

Telltales can be used to monitor displacement of the pile toe and other parts of the pile. Pile head movement is measured using the conventional displacement sensor and reference beam as typically applied in conventional static load tests. The tests using O-cell can follow the quick loading procedure described in ASTM D-1143.

Figure 8.43 shows the results from an O-cell load test on a 15 m long and 1.2 m diameter bored pile with higher toe resistance than shaft friction. The symbols marked in Figure 8.43 represent the respective movement of the pile shaft (δ_s, upward movement

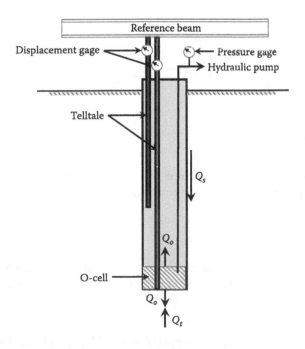

Figure 8.42 The concept of pile load test using an O-cell.

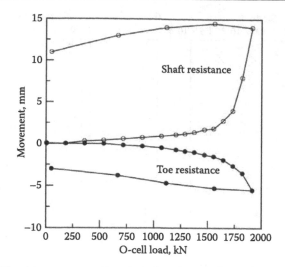

Figure 8.43 Results from an O-cell load test.

is positive) and pile toe (δ_t, upward movement is positive) for each load increment. Throughout the test, the shaft and toe resistance (Q_s and Q_t, respectively) values are always identical. To convert the test result into a conventional pile head load movement curve, set

$$Q_h = Q_s + Q_t \tag{8.25}$$

and

$$\delta_h = \delta_s - \delta_t \tag{8.26}$$

where

Q_h = total resistance at pile head
δ_h = total displacement at pile head

As $Q_s = Q_t$, so $Q_h = 2Q_s$ or $2Q_t$. Details of the data conversion and their interpretation are described in Example 8.9.

EXAMPLE 8.9

Given

Table E8.9 shows the recorded movement and resistance values from a load test on a 15 m long, 1.2 m diameter bore pile using an O-cell. The original test data are plotted in Figure 8.43.

Required

Compute the equivalent load and movement at pile head for the loading part, plot the pile head load movement curve, and determine the failure load by Davisson's method.

Solution

The equivalent load at pile head is computed by doubling the individual load values in Table E8.9. The equivalent movement at pile head is calculated by adding the upward

Table E8.9 Values from the load test with O-cell

Load Q_o, kN	Upward movement δ_s, mm	Downward movement δ_t, mm	Load Q_o, kN	Upward movement δ_s, mm	Downward movement δ_t, mm
5.3	0.1	0.0	1468.9	1.7	−1.3
182.3	0.0	0.0	1556.2	1.8	−1.5
359.3	0.3	0.0	1643.2	2.8	−1.9
536.4	0.4	0.0	1733.2	4.0	−2.6
716.0	0.6	−0.2	1823.0	8.0	−3.5
893.0	0.7	−0.3	1915.5	14.0	−5.5
1070.0	0.9	−0.5	1561.4	14.5	−5.3
1204.7	1.1	−0.7	1114.9	14.0	−4.7
1289.3	1.2	−0.9	668.5	13.0	−3.8
1376.4	1.4	−1.0	50.2	11.0	−3.0

and downward movement (multiply by −1) for a given load. The equivalent load move-ment values at pile head are shown in Table E8.9a.

The load movement curve is shown in Figure E8.9.

Elastic modulus of the pile, $E_p = 17$ GPa $= 17,000,000$ kN/m^2

$D_p = 1.2$ m $= 1200$ mm and $L_p = 15$ m $= 15,000$ mm

$$A_p = \frac{(3.1416)(1.2^2)}{4} = 1.31 \text{ m}^2$$

Elastic deformation at 4000 kN,

$$\Delta = \frac{Q_b L_p}{A_p E_p} = \frac{(4000)(15,000)}{(1.31)(17,000,000)} = 3.12 \text{ mm}$$

$s_f = \Delta + (4.0 + 0.008 D_p) = 3.12 + [4 + (0.008)(1200)] = 3.12 + 13.6 = 16.72$ at $Q_b = 4000$ kN.

The failure criterion line connects points (0, 13.6) and (4000, 16.72).

The failure criterion line intersects the load–movement curve at 3752 kN, thus $Q_f = 3752$ kN.

Advantages of O-cell:

- No need for a reaction frame or dead weight to apply the axial compression load.
- The field setup can be simple, economical, and safe.

Table E8.9a Equivalent load and movement values at pile head

Q_b, kN	δ_b, mm	Q_b, kN	δ_b, mm	Q_b, kN	δ_b, mm
10.6	0.09	2140.0	1.38	3286.5	4.58
364.6	0.03	2409.4	1.84	3466.4	6.57
718.7	0.32	2578.7	2.11	3646.0	11.51
1072.7	0.38	2752.9	2.48	3831.0	19.52
1431.9	0.84	2937.9	3.01		
1785.9	1.02	3112.3	3.32		

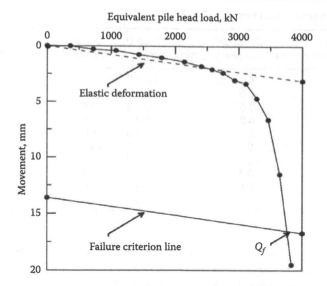

Figure E8.9 Equivalent pile head load movement curve for the result shown in Figure 8.43 and its interpretation to determine Q_f.

Disadvantages of O-cell:

- Test load is limited by the lesser of Q_s and Q_t; the test may be of little value when the pile is mostly frictional or toe bearing.
- The O-cell load is based on hydraulic pressure reading only. The readings can be subject to error due to hydraulic piston friction and there is a lack of redundancy.

8.4.4 The Statnamic load test method

The Statnamic method was developed by Berminghammer Foundation Equipment and TNO of the Netherlands (Berminghammer and Janes, 1989). It is a recognized load test method by ASTM D7383. The test method is based on Newton's laws of motion. A reaction mass is placed on top of the test pile as shown in Figure 8.44. By igniting solid fuel (propellant) in a gas chamber, the generated gas pressure pushes the reaction mass upward while an equal and opposite force pushes downward on the pile. The gas chamber is subsequently vented and gas pressure released. Gravel is placed around the reaction mass and surrounded by a retention bin. After launching, the gravel slumps into the void created by upward movement of the reaction mass and thus catches the reaction mass as it falls down. Other procedures that involve the use of hydraulic or mechanical devices have also been used to catch the reaction mass. A load cell, accelerometers, and/or other optical sensors are used to measure load, movement, and velocity of the pile head during Statnamic load test. A typical Statnamic load test lasts no more than 0.5 second. The magnitude and duration of the applied load are controlled by the selection of piston and cylinder size, the amount and type of propellant, amount of reaction mass, and the gas ventilation technique. Figure 8.45 shows the field setup of a Statnamic load test.

Figure 8.46 shows an original load movement curve (solid curve) from a Statnamic pile load test. The resistance from Statnamic load test is likely to overpredict the static resistance due to the high–loading rate and viscosity effects from the surrounding soil, except at the point of maximum movement shown in Figure 8.46. Assuming the pile as a rigid body, the pile changes from downward to upward movement at the point of maximum movement, and pile velocity at this point is zero. The static and Statnamic load should thus be

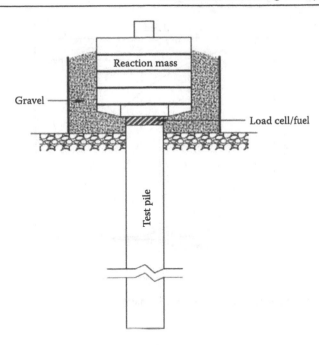

Figure 8.44 Schematic illustration of a Statnamic load test.

the same when the pile reaches the point of maximum movement. Methods (Middendorp et al., 1992; Brown, 1994; Paikowski, 2002) evolved from the point of maximum movement have been proposed to determine the equivalent static load movement curve (dashed curve in Figure 8.46) by estimating the extra dynamic resistance based on the acceleration measurement and characteristics of the original Statnamic load movement curve.

There is no need for a reaction frame in Statnamic load test. The method can be used to apply inclined or lateral load on the pile. The savings in cost and time can be significant over the conventional static load tests. In order to have valid interpretation, it is important that the applied Statnamic force is larger than the ultimate pile capacity. The interpretation of Statnamic load test result is relatively complex. Additional sensors are required to provide the desirable redundant measurements.

Figure 8.45 Field setup of the Statnamic load test. (Courtesy of DECL, Taipei, Taiwan.)

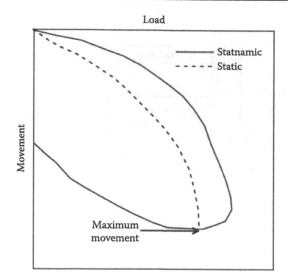

Figure 8.46 Typical load movement curves from a Statnamic load test.

8.5 REMARKS

The static analysis for deep foundations began in Chapter 7 and continued in Chapter 8 regarding the construction and testing of deep foundations. Because of the disturbance caused by construction activities, it is understandable that the static analysis provides, at best, an estimate of the type and dimensions of deep foundations. For driven piles, the static analysis is just the beginning. To put the piles in the ground, we need to determine what type of driving equipment (i.e., hammer and cushion, etc.) to use and driving criteria for the specific driving system and pile involved. A series of analytical and monitoring tools are available and can be used to assure proper installation of driven piles. The use of these tools involves many empirical rules or parameters.

For bored piles, the static analysis is also a beginning. The selection of construction method (i.e., dry, casing, or slurry method) and workmanship can have significant effects on the quality and behavior of bored piles. Local experience and knowledge on the specific construction method/equipment for the given geology and ground conditions are imperative to the success of the bored piles.

It appears that the static load test is the ultimate tool to ascertain the failure capacity of a finished pile. Unfortunately, the determination of failure capacity is again subject to interpretation of the pile load test result. This is true for both driven and bored piles. For the same set of test results, the selection of interpretation method and/or involved empirical parameters can affect the outcome of the interpretation. Significant progress has been made in the past few decades in improving the efficiency and consistency in pile construction and testing; however, some of the inherent uncertainties remain.

HOMEWORK

8.1.   A single-acting Raymond International 150C steam hammer with a ram weight, W_r, of 66.75 kN and drop height, h, of 990 mm, was used to drive a precast concrete pile. Use the ENR formula to calculate the ultimate resistance, R_u, in kN when S reached 5 mm.

8.2. Repeat Problem 8.1 with the same hammer and pile. Use ENR formula to calculate the ultimate resistance, R_u when S reached 6, 8, 10, 12, and 14 mm. Present the results in a table.

8.3. For a single-acting hammer, given:
   Hammer-rated energy $(EN_r) = 44.07$ kN-m
   Ram weight (WAM (1)) $W_r = 44.50$ kN
   Hammer efficiency $(eff_r) = 0.7$
   Determine the impact velocity of the ram in m/sec

8.4. Figure H8.4 shows a set of field force and velocity records for an 18 m long steel pipe pile. The pipe pile has an outside diameter of 232 mm and inside diameter of 152 mm. The pile was driven into loose silty sand with some gravel. The strain and acceleration transducers were mounted at 450 mm from the pile head. In this case, $E_p = 2.1 \times 10^5$ MPa, $c_w = 6000$ m/sec for steel, $A_p = 0.024$ m^2 and $L_p = 17.55$ m (distance between the transducers to the pile toe). Determine RTL and RSP. Use $J_t = 0.45$.

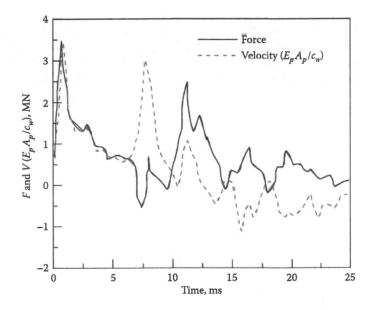

Figure H8.4 Force and velocity record from field measurement from a steel pipe pile.

8.5. The load movement record from a compression load test on a 0.8 m diameter and 20 m long driven precast concrete pile is shown in Table H8.5. Plot the load–movement curve and determine the failure load by Davisson's method. Elastic modulus of the pile, $E_p = 17$ GPa.

Table H8.5 Load movement record from a compression load test on a 0.8 m diameter and 20 m long driven pile

Load, kN	0	491	981	1472	1962	2453	2943
Movement, mm	0	0.29	0.86	1.57	2.28	3.15	4.13
Load, kN	3434	3924	4415	4905	5396	4905	3434
Movement, mm	5.21	8.00	12.00	20.00	33.00	32.50	29.00
Load, kN	1962	981					
Movement, mm	25.00	20.00					

8.6. The load movement record from a compression load test on a barrette pile with cross-sectional dimensions of 0.8 m by 2.8 m and a pile length of 40 m is shown in Table H8.6. Plot the load–movement curve and determine the failure load by Davisson's method. Elastic modulus of the pile, $E_p = 17$ GPa. For D_p value in the computation of s_f, use the diameter of a circular area that equals to the cross-section area of 0.8 m by 2.8 m.

Table H8.6 Load movement record from a compression load test on the 0.8 m by 2.8 and 40 m long barrette pile

Load, kN	0	1717	3434	5150	6867	8584	0.301
Movement, mm	0	0.38	0.73	1.50	2.08	3.19	4.30
Load, kN	12,017	13,734	15,451	17,168	18,884	20,601	22,318
Movement, mm	5.64	7.60	9.59	12.03	15.30	19.98	27.51
Load, kN	24,035	26,751	20,650	15,548	10,448	5346	0
Movement, mm	41.76	93.69	93.73	93.00	88.66	85.35	79.98

8.7. For the compression load test on the 0.8 m diameter and 20 m long pile described in Problem 8.5, when the applied load at pile head (Q_h) reached 3924 kN, the movement measurements at pile head and toe were 8 mm (s_1) and 2 mm (s_2), respectively. The shaft friction (q_s) is expected to be uniformly distributed along the pile, as shown in Figure H8.7a. Determine the resistance at pile toe (Q_t) and frictional resistance from the pile shaft (Q_s). Elastic modulus of the pile, $E_p = 17$ GPa.

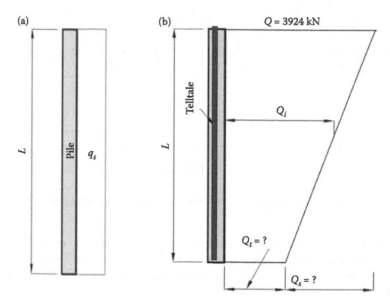

Figure H8.7 Load transfer for uniformly distributed shaft friction. (a) Uniform shaft friction along the pile. (b) Load transfer according to telltale reading.

8.8. For a test pile with the same material and dimensions as in Problem 8.6, there are two major soil layers around the pile. Each layer is expected to provide a uniform shaft friction against the pile as shown in Figure H8.8a. Two telltales are installed in the pile load test. Telltale 1 ends at 8 m below ground surface, which corresponds to the bottom of soil layer 1. Telltale 2 extends to the toe of the pile at 20 m deep. Pile head is considered as Telltale 0. When the applied load at pile head (Q_h)

reached 3924 kN, the movement measurements at pile head, Telltale 1, and Telltale 2 are 8 mm ($s_0 = 8$ mm), 5 mm ($s_1 = 5$ mm), and 1 mm ($s_2 = 1$ mm), respectively. Determine the load transferred internally and frictional resistance from the pile shaft (Q_s) at Telltale 1 and Telltale 2 (or toe) levels. Elastic modulus of the pile, $E_p = 17$ GPa.

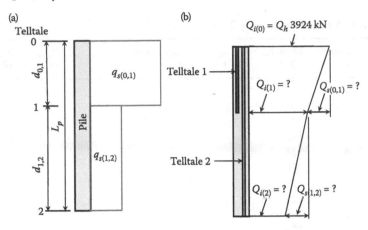

Figure H8.8 Load transfer for pile load tests in two layered soil. (a) Shaft friction along the pile. (b) Load transfer according to telltale readings.

8.9.   For a compression load test on a 1.5 m diameter and 34.5 m long bored pile, there are two major soil layers around the pile. Each layer is expected to provide a uniform shaft friction against the pile, as shown in Figure H8.9. Two telltales are installed in the pile load test. Telltale 1 ends at the mid-height of the pile, which corresponds to the bottom of soil layer 1. Telltale 2 extends to the toe of the pile. Pile head is considered as Telltale 0. When the applied load at pile head (Q_h) reached 9810 kN, the movement measurements at pile head, Telltale 1, and Telltale 2 are 16.67 mm ($s_0 = 16.67$ mm), 11.20 mm ($s_1 = 11.10$ mm), and 8.40 mm ($s_2 = 8.40$ mm), respectively. Determine the load transferred internally and frictional resistance from the pile shaft (Q_s) at Telltale 1 and Telltale 2 (or toe) levels. Elastic modulus of the pile, $E_p = 17$ GPa.

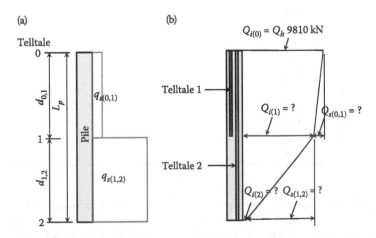

Figure H8.9 Load transfer for pile load tests in two-layered soil. (a) Shaft friction along the pile. (b) Load transfer according to telltale readings.

8.10. Given a set of strain readings as shown in Table H8.10, from a static compression load test on a 0.8 m diameter and 20 m long bored pile, when the applied load at pile head (Q_h) reached 3000 kN. Compute and plot the distribution of internally transmitted load $Q_{i(n)}$ and frictional resistance per unit area, $q_{s(n,n+1)}$ versus depth. Is this an end bearing or friction pile?

Table H8.10 Strain readings when Q_h reached 3000 kN

Number, n	Depth, m	Strain $\varepsilon_{(n)} \times 10^{-6}$	$Q_{i(n)}$, kN	Depth range, m	$Q_{s(n,n+1)}$, kN	$d_{n,n+1}$, m	$q_{s(n,n+1)}$, kPa
0	0	–					
1	1.0	351					
2	2	348					
3	3	341					
4	4	336					
5	5	332					
6	6	329					
7	7	324					
8	8	315					
9	9	306					
10	10	294					

8.11. Given a set of strain readings as shown in Table H8.11, from a static compression load test on a 0.8 m diameter and 20 m long bored pile, when the applied load at pile head (Q_h) reached 3000 kN. Compute and plot the distribution of internally transmitted load $Q_{i(n)}$ and frictional resistance per unit area, $q_{s(n,n+1)}$, versus depth. Is this an end bearing or friction pile?

Table H8.11 Strain readings when Q reaches 3000 kN

Number, n	Depth, m	Strain $\varepsilon_{(n)} \times 10^{-6}$	$Q_{i(n)}$, kN	Depth range, m	$Q_{s(n,n+1)}$, kN	$d_{n,n+1}$, m	$q_{s(n,n+1)}$, kPa
0	0	–					
1	1.0	347					
2	2	339					
3	3	329					
4	4	318					
5	5	276					
6	6	259					
7	7	212					
8	8	165					
9	9	106					
10	10	24					

8.12. Table H8.12 shows the recorded movement and resistance values from a load test on an 18 m long, 0.8 m diameter bored pile using an O-cell. The original test data are plotted in Figure H8.12. Compute the equivalent load and movement at pile head, plot the pile head load movement curve, and determine the failure load by Davisson's method.

Table H8.12 Values from the load test with O-cell

Load, kN	Upward movement, mm	Downward movement, mm	Load, kN	Upward movement, mm	Downward movement, mm
5.0	0.1	0.0	950.0	6.0	−0.9
50.0	0.2	0.0	1150.0	10.0	−1.4
125.0	0.3	0.0	1350.0	18.0	−2.5
225.0	0.8	0.0	1100.0	17.9	−2.5
350.0	1.2	−0.2	800.0	17.3	−2.3
500.0	1.7	−0.3	400.0	15.8	−1.8
650.0	2.8	−0.5	50.0	14.0	−1.2
800.0	4.0	−0.7	–	–	–

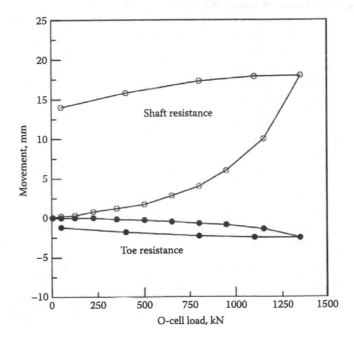

Figure H8.12 Original O-cell load movement curve.

REFERENCES

American Society for Testing and Materials, ASTM D7383. Standard Test Methods for Axial Compressive Force Pulse (Rapid) Testing of Deep Foundations.

American Society for Testing and Materials, ASTM D1143. Standard Test Methods for Deep Foundations under Static Axial Compressive Load.

American Society for Testing and Materials, ASTM D3966. Standard Test Methods for Deep Foundations under Lateral Load.

Berminghammer, P. and Janes, M. 1989. An innovative approach to load testing of high capacity piles. In Burland, J.B. and Mitchell, J.M., eds., *Proceedings, International Conference on Piling and Deep Foundations*, Vol. 1, Balkema, Rotterdam, pp. 409–413.

Brown, D.A. 1994. Evaluation of static capacity of deep foundations from Statnamic testing. *ASTM Geotechnical Testing Journal, GTJODJ*, 17(4), 403–414.

Davisson, M.T. 1972. High capacity piles. *Proceedings, Soil Mechanics Lecture Series on Innovations in Foundation Construction*, ASCE, Illinois Section, Chicago, pp. 81–112.

Fellenius, B.H. 1990. *Guidelines for the Interpretation of the Static Loading Test*. Deep Foundation Institute Short Course, First Edition, Vancouver, British Columbia, Canada, 44pp.

Goble, G.G., Linkins, G.E., and Rausche, F. 1975. Bearing capacity of piles from dynamic measurements. Final Report, Department of Civil Engineering, Case Western Reserve University, Cleveland, OH.

Goble, G.G. and Rausche, F. 1970. Pile load test by impact driving. Highway Research Board, No. 333, Washington, DC.

Hannigan, P.J., Goble, C.C., Linkins, G.E., and Rausche, F. 2006. Design and construction of driven pile foundations. Volume II. Report No. FHWA-NHI-05-043, National Highway Institute, Federal Highway Administration, U.S. Department of Transportation, Washington, DC, 486pp.

Ho, Y.T., Huang, A.B., Ma, J., Zhang, B., and Cao, J. 2005. Ground movement monitoring using an optic fiber Bragg grating sensored system. *Proceedings, 17th International Conference on Optical Fiber Sensors*, SPIE, Bruges, Belgium, Vol. 5855, pp. 1020–1023.

Kyfor, Z.G., Schnore, A.S., Carlo, T.A., and Bailey, P.F. 1992. Static testing of deep foundations. Report No. FHWA-SA-91-042, U.S. Department of Transportation, Federal Highway Administration, Office of Technology Applications, Washington, DC, 174pp.

Middendorp, P., Berminghammer, P., and Kuiper, B. 1992. Statnamic load testing of foundation piles. *Proceedings, 4th International Conference on the Application of Stress-Wave Theory to Piles*, Balkema, Rotterdam, pp. 581–588.

Osterberg, J. 1995. The Osterberg cell for load testing drilled shafts and driven piles. Report No. FHWA-SA-94-035, U.S. Department of Transportation, Federal Highway Administration, Office of Technology Applications, Washington, DC, 92pp.

Paikowski, S. 2002. Innovative load testing systems. Report NCHRP 21-08, National Cooperative Highway Research Program, Transportation Research Program, Washington, DC.

Rausche, F., Moses, F., and Goble, G.G. 1972. Soil resistance predictions from pile dynamics. *Journal of the Soil Mechanics and Foundation Engineering Division*, ASCE, 98(SM9), 917–937.

Reiding, F.J., Middendorp, P., Schoenmakere, R.P., Middendorp, F.M., and Bielefeld, M.W. 1988. A new generation of foundation diagnostic equipment. In Fellenius, B.H., ed., *Proceedings, 3rd International Conference on the Application of Stress Wave Theory to Piles*, BiTech, Ottawa, Canada, pp. 123–134.

Smith, E.A.L. 1960. Pile-driving analysis by the wave equation. *Journal of the Engineering Mechanics Division, Proceedings of the American Society of Civil Engineers*, 86(EM 4), 35–61.

Vesić, A.S. 1977. *Design of Pile Foundations, National Cooperative Highway Research Program Synthesis of Highway Practice 42*, Transportation Research Board, Washington, DC, 68pp.

Chapter 9

Slope stability analysis

9.1 INTRODUCTION

Slopes were created or made less stable by nature or due to human activities long before we had the knowledge to make them safe. We built levees by piling up locally available material with very little controlled compaction. This was the case for the Yellow River levees in China, and many other cases in the world. Historically, the Yellow River levee failures usually involved severe loss of lives and property while changing the course of the river. Natural forces such as earthquakes, waves, current, and water infiltration can also cause slope failure. The material composing a slope has a tendency to slide under the influence of driving forces that include gravity and other natural activities such as tectonic movement and earthquakes. This tendency to slide is resisted by shear strength of the material within the slope. Instability occurs when the shear strength is not sufficient to withstand the driving forces along any surface within a slope. This instability can be caused by the increase of driving forces and/or decrease of the material strength. The increase of driving forces can be due to

- Construction activities such as excavation or filling on or adjacent to a slope that alter either the geometry and/or the loading/stress conditions of the slope or ground mass.
- Sudden lowering of water level adjacent to a slope, a phenomenon called "sudden drawdown."
- Seismic activities such as earthquakes or other forms of earth tremor.
- Strong or prolonged current that can create erosion to the slope.

The slope material can be weakened for the following reasons:

- Progressive decrease in shear strength due to weathering of the slope material or stress release due to excavation.
- Increase of pore water pressure or decrease of suction (negative pore water pressure existing in unsaturated soils) due to infiltration of precipitation (rainfall, snowmelt) into the ground.
- Seepage force associated with leakage around a pond or reservoir.

Natural or manmade slopes that could be stable for many years may suddenly fail due to one or more of the above causes. Schuster et al. (2002) compiled 23 catastrophic landslides of South America in the period 1941–1994. Among them, 10 were triggered by heavy or prolonged rainfall, 5 by earthquake, 3 by valley down-cutting due to long-term erosion, 2 by failure of natural dam, 1 by volcanic activity, and 1 by leakage from manmade pond.

Figure 9.1 Su-Hua Highway carved into steep slopes along the east coast of Taiwan. (Courtesy of Prof. Ming-Lang Lin, National Taiwan University, Taipei, Taiwan.)

Figure 9.1 shows the scenic Su-Hua Highway along the east coast of Taiwan on a bright sunny day. The highway was built by cutting into a steep slope composed of mostly metamorphic rock formations covered by a thin layer of weathered or colluvial material. Figure 9.2 shows the scar created by a fatal landslide along the same highway during typhoon Megi of 2010 that brought in heavy rainfall. A bus along with its 26 passengers was pushed by the debris into the cliff and sank in the Pacific Ocean.

Varnes (1978) and Cruden and Varnes (1996) proposed a system to classify landslides based on the type of movement and material involved in the landslide as shown in Table 9.1. A conceptual description of the five types of landslide movement is shown in Figure 9.3. The system is commonly used in describing landslides. This chapter will concentrate on the "slide" type of landslides. It should be noted that with a sufficient amount of water and under other favorable ground conditions, a slide can turn into a "flow," where the failed ground mass flows like liquid.

Figure 9.4 shows a landslide triggered by the 2016 Kumamoto earthquake (magnitude 7.3) in Kyushu, Japan. The slide moved along a 35° steep slope, traveled 700 meters, crushed a bridge, and killed one driver. Figure 9.5 shows the result of a debris flow occurred during the 2001 Typhoon Toraji in Central Taiwan. Large rock pieces as shown in the picture along with muddy water gushed into the lobby of a nearby hotel and nearly destroyed it.

We continue to build slopes such as levees, embankments, mine tailing dams, and earth dams, or cut natural slopes to build infrastructures today. A comprehensive evaluation of slope stability requires a multidisciplinary approach. We often need geologists to provide

Figure 9.2 Scar created by a fatal landslide along Su-Hua Highway after typhoon Megi. (Courtesy of An-Bin Huang, Hsinchu, Taiwan.)

information regarding the rock formations, fault activities, and other geological characteristics. The information may be qualitative, but that information usually covers the region of interest. For example, where we should select the route of a highway to avoid areas where construction activities may destabilize the neighboring slopes or be too costly to maintain stability. Civil or geotechnical engineers usually deal with a specific location and are responsible for the quantitative analysis to assure the safety of a given slope. This chapter concentrates on the mechanics and available tools related to the site-specific quantitative analysis of a given slope.

Table 9.1 Classification of landslides

Type of movement		Bedrock	Type of material	
			Engineering soils	
			Coarse grained	Fine grained
Fall		Rock fall	Debris fall	Earth fall
Topple		Rock topples	Debris topple	Earth topple
Slides	Rotational	Rock slump	Debris slump	Earth slump
	Translational	Rock block slide, rock slide	Debris block slide, debris slide	Earth block slide, earth slide
Lateral spreads		Rock spread	Debris spread	Earth spread
Flows		Rock flow (deep creep)	Debris flow	Earth flow
Complex		Combines more than one type of movement		

Source: Varnes, D.J. 1978. Slope movement types and processes. Chapter 2, Landslides: Analysis and control. *Special Report 176*, TRB, National Academy of Sciences, Washington, DC, 234p.

(a)   (b)   (c)

(d)

(e)

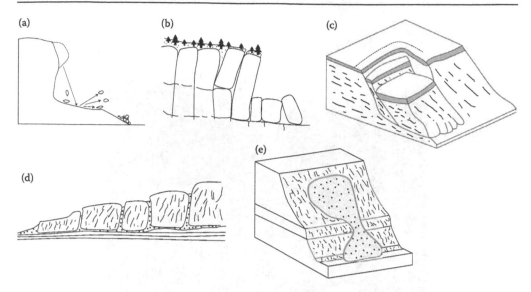

Figure 9.3 Description of the major types of slope movement. (a) Falls. (b) Topples. (c) Slides. (d) Spreads. (e) Flows. (Adapted from Varnes, D.J. 1978. Slope movement types and processes. Chapter 2, Landslides: Analysis and Control. *Special Report 176*, TRB, National Academy of Sciences, Washington, DC, 234p.)

Extensive developments have been made in slope stability analysis-related software and computer packages in the past few decades. Sophisticated stress and deformation (i.e., stress and strain) analyses of slopes, considering elasto-plastic characteristics of the slope materials, have been made. For engineering practice, however, we are usually more concerned with the safety of a given slope. Limit equilibrium methods based on forces and strength of the slope materials are routinely used for slope stability analysis in engineering practice. The following sections concentrate on the principles of limit equilibrium methods and their applications. Recent developments in the use of limit analysis for slope stability analysis are also briefly introduced.

Figure 9.4 Landslide triggered by the 2016 Kumamoto earthquake in Kyushu, Japan. (Courtesy of Prof. Takaji Kokusho, Tokyo, Japan.)

Figure 9.5 Debris flow during Typhoon Toraji in 2001 nearly destroyed a hotel in Nan-Tou, Taiwan. (Courtesy of An-Bin Huang, Hsinchiu, Taiwan.)

The objectives of this chapter are to

- Introduce the concept of the limit equilibrium method in slope stability analysis (Section 9.2). Under rather restricted material and geometrical conditions of the assumed slip surface, the analysis in this category may be conducted by hand calculations. The effects of sudden draw-down and groundwater seepage on slope stability are evaluated using a simplified close form solution considering an infinitely long slope with uniform material and a planar failure surface.
- Describe the principles of the method of slices that can consider non-homogeneous soil conditions and irregular slip surfaces in the slope stability analysis (Section 9.3). Except for extremely simplified conditions, a numerical procedure is usually required to execute the method of slices. A few commonly used methods of slices are introduced, their unique capabilities and limitations are discussed.
- Make a brief introduction to other numerical methods that have been used or demonstrated great potential for the purpose of slope stability analysis (Section 9.4).
- Describe the possible loading conditions and nature of the slope that we often consider in the slope stability analysis (Section 9.5).
- Introduce some of the commonly used techniques to monitor the stability of a slope and engineering measures to improve the stability of a slope (Section 9.6).

9.2 LIMIT EQUILIBRIUM METHODS

Limit equilibrium methods are generally not concerned with stress distribution within the slope. An important aspect of the limit equilibrium analysis is to assume that there exists a slip surface in the slope. Materials above and below the slip surface form two rigid bodies. The lower block remains stationary and the upper block is referred to as the sliding block, that is, the block that has the potential to slide. There is no guarantee that the assumed slip surface is the most critical one in the slope. A limit equilibrium analysis typically involves consideration of many assumed slip surfaces (often in the hundreds). Each of these slip surfaces will have a different factor of safety (FS). The minimum of these values is taken as the representative factor of safety for the slope. The slip surface, with which this minimum value is associated, is called a "critical slip surface." This is basically the same

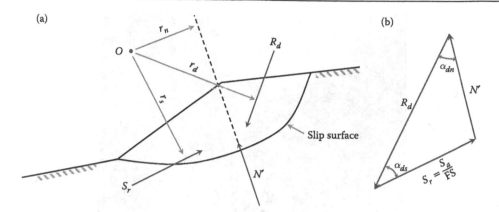

Figure 9.6 Forces acting on the free body with a general shape slip surface. (a) Slip surface of any permissible shape and the free body diagram. (b) Force polygon.

procedure that Coulomb used to develop the active earth force imposed on a retaining wall. It is no surprise that Coulomb also used the limit equilibrium method. For slope stability analysis, the search for the critical slip surface is much more complicated than the case of retaining walls. A wide variety of optimization methods in search of the critical slip surface have been implemented in commercial slope stability analysis programs. The following introduction concentrates on the mechanics involved in the stability analysis of a given slip surface. Methods in the search of the critical slip surface are not discussed.

Intuitively, a stress-based analysis is more desirable for slopes, as the strength of slope material is, after all, a quantity of stress. The available stress-based methods are mostly in the research stage. As a result, we routinely use the limit equilibrium methods in engineering practice. In any case, comparisons made by Chen (1975) showed close agreement between limit equilibrium and stress-based analyses. Ideally, a slope failure is a three-dimensional (3D) problem. Methods are available for 3D slope stability analysis (Chen and Chameau, 1982; Duncan, 1992). For simplicity, we evaluate the stability of a slope usually as a two-dimensional (2D), plane strain problem. It is assumed that the slope is infinitely long in the direction perpendicular to the paper. In most cases, this assumption leads to conservative results (i.e., lower factor of safety) in the slope stability analysis. The following discussion begins with limit equilibrium-based 2D slope stability analysis.

9.2.1 Introduction to limit equilibrium method

The limit equilibrium method (LEM) was also used for foundation bearing capacity, and lateral earth pressure analysis as described in Chapters 4 and 5, respectively. The application of LEM for slope stability analysis basically involves the following procedure. In reference to Figure 9.6 and for illustration purposes, we consider the case of drained conditions (effective stress analysis):

- Assume a slip surface. The slip surface can be a plane or a curved surface. Figure 9.6a shows a more general form of a curved slip surface of any permissible shape.
- To facilitate limit equilibrium analysis, we use a force-based Mohr–Coulomb yield criterion as follows,

$$S_a = C'_a + N' \tan \emptyset'_a \qquad (9.1)$$

where

S_a = available resistance force along the slip surface
C'_a = sum of drained cohesion forces along the slip surface
N' = sum of effective normal contact forces along the slip surface
$\emptyset'_a$ = peak drained friction angle for the material along the slip surface

• Using the concept of free body diagram, the sliding block is subject to three forces: the driving force, R_d, that tends to destabilize the sliding block, the required resistance force, S_r, to maintain stability of the stability block, and N' that represents the sum of all normal forces acting on the slip surface, excluding pore water pressure effects. R_d is associated with gravity and passes through the center of gravity of the sliding block. The magnitude and line of action of R_d are the important known quantities in LEM. The required resistance force, S_r, is the portion of S_a required to maintain stability, sometimes referred to as the mobilized resistance, and

$$S_r = \frac{S_a}{\text{FS}} = \frac{C'_a + N' \tan \emptyset'_a}{\text{FS}} \tag{9.2}$$

where

FS = factor of safety for the sliding block

The magnitude and direction of N' are unknown as it relates to S_r. There are four equations to be satisfied to maintain stability of the sliding block:

1. Force-based Mohr–Coulomb yield criterion,

$$S_a = C'_a + N' \tan \emptyset'_a \tag{9.1}$$

2. Moment equilibrium at an arbitrary point O,

$$R_d r_d - N' r_n - S_r r_s = 0 \tag{9.3}$$

3. Equilibrium of forces parallel to R_d,

$$R_d - S_r \cos \alpha_{ds} - N' \cos \alpha_{dn} = 0 \tag{9.4}$$

4. Equilibrium of forces perpendicular to R_d,

$$S_r \sin \alpha_{ds} - N' \sin \alpha_{dn} = 0 \tag{9.5}$$

Refer to Figure 9.6a and b. There are seven unknowns involved in the equilibrium analysis for the sliding block: FS, S_a, N', r_n, r_s, α_{ds}, and α_{dn}. There are four equations to be satisfied. The system is therefore indeterminate, meaning there is no unique solution.

Some simplifications or restrictions will have to be applied in order to obtain a unique solution in the above limit equilibrium analysis. The following sections show some cases of simplifications.

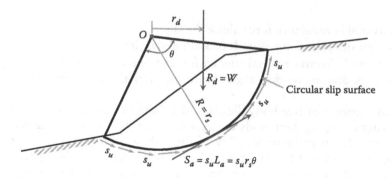

Figure 9.7 Sliding block with a circular slip surface.

9.2.2 The circular arc method

For a circular slip surface in a uniform cohesive material under undrained conditions ($\varnothing = 0$), a unique solution can be obtained. Refer to Figure 9.7, where R_d is the weight of the sliding block, W, and r_d is the distance between the center of the circular arc (O in Figure 9.7) to the line of action of R_d. Resistance force consists of the undrained shear strength, s_u, along the slip surface. Radius of the arc, R, is also the moment arm of the resistance force or r_s.

The arc length of the circular slip surface, L_a is,

$$L_a = r_s\theta \tag{9.6}$$

The factor of safety against sliding can be computed based on moment equilibrium as follows:

$$FS = \frac{s_u L_a r_s}{W r_d} = \frac{s_u r_s^2 \theta}{W r_d} \tag{9.7}$$

where
θ = angle of the circular slip surface, in radian

The use of an Excel spreadsheet program to compute arc length, L_a, W, and the location of the line of action of W, based on the geometry of the slope and slip surface, is introduced later in the method of slices. These parameters will be given at this stage for the purpose of introducing the circular arc and friction circle methods and Examples 9.1 and 9.2.

EXAMPLE 9.1

Given

Consider 1 m thick sliding block in the direction perpendicular to the paper. For the circular slip surface shown in Figure E9.1,

$R = r_s = 20.1$ m
Arc length, $L_a = 40.3$ m
$\gamma = 19$ kN/m^3
$s_u = 60$ kPa ($\varnothing = 0$)
Weight of the sliding block, $W = 5180$ kN
Distance from center of circle to W, $r_d = 6.1$ m

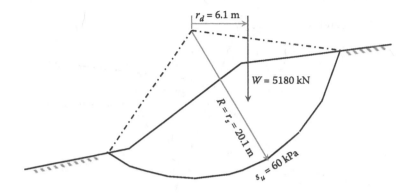

Figure E9.1 Slope of cohesive soil with a circular slip surface.

Required

Determine the FS of the circular slip surface using the circular arc method.

Solution

Following Equation 9.7,

$$FS = \frac{s_u L_a r_s}{W r_d} = \frac{(60)(40.3)(20.1)}{(5180)(6.1)} = 1.54$$

9.2.3 The friction circle method

This is another simplified method applicable for circular slip surface where a unique solution can be obtained. The method was proposed before computers were readily available. The friction circle was a graphical procedure to replace the cumbersome trigonometry computations without a computer. The method is applicable for a slope with homogeneous, cohesive frictional soil.

Assume the resultant of all normal stresses acting on the slip surface can be combined into a single normal force. Refer to Figure 9.8a: the cohesion along arc *ab* has a resultant C_m that acts in parallel to the direction of the chord *ab*. Taking moment about the center of the circle, the line of action of C_m can be located at

$$R_c = \frac{L_a}{L_c} R \qquad\qquad (9.8)$$

where
   L_a = length of arc *ab* of the circular slip surface
   L_c = length of the chord *ab* in Figure 9.8a

The point of application, A, is located at the intersection of the weight force W and C_m. U (if pore water pressure is present) points toward the center of the circle. Resultant of all normal and shear forces, P, is inclined. In the force polygons shown in Figure 9.8b and c, the magnitude and direction of W and U (computation of U is based on the geometry of the slope, phreatic surface, and slip surface) are known. α is the slope of the chord or C_m,

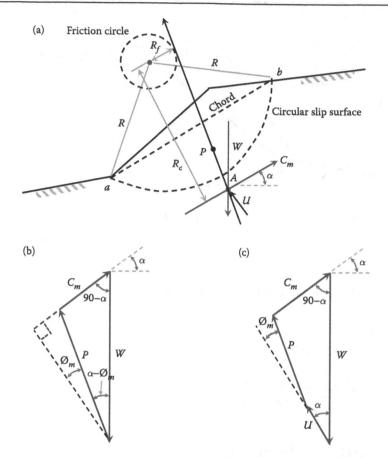

Figure 9.8 The friction circle method. (a) Slip surface and the friction circle. (b) The force polygon. (c) The force polygon with pore pressure.

a known quantity. Therefore, the directions of C_m and P are known but their respective magnitudes need to be determined.

Procedure to apply the friction circle method is as follows:

1. Calculate W, U, R_c, and location of the intersection point A.
2. Assume a factor of safety for $\varnothing$ (or $F_\varnothing$), then the required or mobilized friction angle, $\varnothing_m$, is

$$\varnothing_m = \tan^{-1}\left(\frac{\tan \varnothing}{F_\varnothing}\right) \tag{9.9}$$

3. Draw the friction circle with its radius R_f as

$$R_f = R \sin \varnothing_m \tag{9.10}$$

4. Plot a tangent line to the friction circle that starts from A as shown in Figure 9.8a. This is also the direction of P, determined graphically with the help of friction circle.

5. Find C_m using the force polygon shown in Figure 9.8b (or Figure 9.8c if U is present). Then derive the safety factor for cohesion, c (or F_c) as

$$F_c = \frac{cL_c}{C_m} \tag{9.11}$$

6. Iterate steps 2–4 by varying $F_\emptyset$ until

$$F_\emptyset \approx F_c \tag{9.12}$$

Which is also the factor of safety (FS) against failure of the sliding block.

In the original friction circle method, the values of P and C_m of the force polygon were determined graphically by measuring the length of all sides of the force polygon using the known magnitude of W as a scale. Nowadays, with the availability of computers, it may be easier to use a spreadsheet program such as Excel and apply the law of sines to solve the force polygon.

EXAMPLE 9.2

Given

For the same slope and circular slip surface as shown in Example 9.1, consider 1 m thick sliding block in the direction perpendicular to the paper.

For the circular slip surface shown in Figure E9.1,

$R = 20.1$ m
Arc length, $L_a = 40.3$ m
Length of chord, $L_c = 33.61$ m
Slope of the chord, $\alpha = 22.92°$
$\gamma = 19$ kN/m^3
$c = 20$ kPa
$\emptyset = 25°$
Weight of the sliding block, $W = 5180$ kN
Ignore the effects of U

Required

Determine the FS using the friction circle method.

Solution

Follow Equation 9.8,

$$R_c = \frac{L_a}{L_c} R = \frac{(40.3)}{(33.61)}(20.1) = 24.1 \text{ m}$$

An Excel spreadsheet program was used to compute the related quantities. The program is available to all registered users. Results are summarized in the following table. A range of $F_\emptyset$ was used in the trials to determine the corresponding F_c value. As shown in the table, $F_c \approx F_\emptyset$ when $F_\emptyset = 1.436$, and thus is chosen as the safety factor for the slip surface.

$F_{\varnothing}$	$\varnothing_m$	C_m, kN	F_c
1.450	17.83	483.00	1.391
1.440	17.94	472.35	1.423
1.436	17.99	468.05	1.436
1.420	18.18	450.59	1.491
1.410	18.30	439.48	1.529
1.400	18.42	428.21	1.569

9.2.4 Limit equilibrium analysis with a planar slip surface

If the slip surface is a plane, then the limit equilibrium analysis can be uniquely solved with two equations. Refer to Figure 9.9; the direction of N' is normal to the slip surface, or α_{dn} is known. The direction of S_r is parallel to the slip surface, or α_{ds} is known. Only two unknowns, N' and FS, remain, and they can be determined uniquely with the following two equations:

$$S_r - R_d \sin \alpha_{dn} = 0 \tag{9.13}$$

and

$$N' - R_d \cos \alpha_{dn} = 0 \tag{9.14}$$

where

$$S_r = \frac{S_a}{FS} = \frac{C'_a + N' \tan \varnothing'_a}{FS} \tag{9.2}$$

Slope failure with a planar slip surface may be applied to the analysis of rock slopes with a clearly defined planar discontinuity such as joint or bedding plane with an inclination angle, β, less than the slope angle, i, as shown in Figure 9.9a.

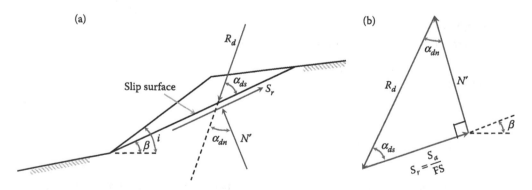

Figure 9.9 Forces acting on the free body with a planar slip surface. (a) Planar slip surface and free body diagram. (b) Force polygon.

EXAMPLE 9.3

Given

Consider a slope of inclination i, height H, and a potential failure plane of inclination β as shown in Figure E9.3a. The upper ground surface is horizontal.

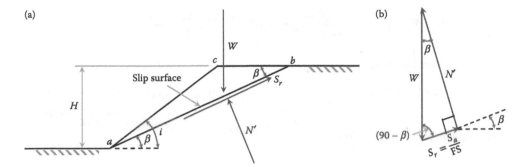

(a)

(b)

Figure E9.3 Slope with a plane slip surface and horizontal ground surface. (a) The slope with a planar slip surface. (b) Force polygon.

Required

Show that that slope has a factor of safety,

$$FS = \frac{2c' \sin i}{\gamma H \sin(i - \beta) \sin \beta} + \frac{\tan \varnothing'}{\tan \beta} \tag{9.15}$$

Solution

Follow Equations 9.13 and 9.14 and consider the force polygon in Figure E9.3b,

$$S_r - R_d \sin \alpha_{dn} = 0 \tag{9.13}$$

$$N' - R_d \cos \alpha_{dn} = 0 \tag{9.14}$$

$R_d = W$, and $\alpha_{dn} = \beta$, thus
$S_r = W \sin \beta$, and $N' = W \cos \beta$

Consider a unit thickness of the slope.
Weight of the sliding block, $W = (\gamma)(A_{abc})$ (1)
Where $\gamma =$ total soil unit weight, $A_{abc} =$ area of triangle abc, and

$$A_{abc} = \frac{1}{2}(H)(bc)$$

Refer to the triangle abc in Figure E9.3a, $ab = H/(\sin \beta)$, and invoke the law of sines,

$$\frac{ab}{\sin(180 - i)} = \frac{ab}{\sin i} = \frac{bc}{\sin(i - \beta)}$$

thus

$$bc = \frac{(ab)(\sin(i - \beta))}{\sin i} = \frac{(H)(\sin(i - \beta))}{(\sin i)(\sin \beta)}$$

and

$$A_{abc} = \frac{1}{2}(H)(bc) = \frac{(H^2)[\sin(i-\beta)]}{2(\sin i)(\sin \beta)}$$

$$W = (\gamma)(A_{abc})(1) = \frac{(\gamma)(H^2)[\sin(i-\beta)]}{2(\sin i)(\sin \beta)}$$

$$S_r = W \sin \beta = \frac{(\gamma)(H^2)(\sin(i-\beta))}{2(\sin i)(\sin \beta)} \sin \beta = \frac{(\gamma)(H^2)(\sin(i-\beta))}{2(\sin i)}$$

Following Equation 9.2,

$$FS = \frac{C_a' + N' \tan \varnothing_a'}{S_r} = \frac{C_a' + N' \tan \varnothing_a'}{W \sin \beta} = \frac{C_a' + W \cos \beta \tan \varnothing_a'}{W \sin \beta} + \frac{C_a'}{W \sin \beta} + \frac{\tan \varnothing_a'}{\tan \beta}$$

$$C_a' = (c')(ab)(1) = \frac{c'H}{\sin \beta}$$

replacing W and C_a' into the first term,

$$\begin{aligned} FS &= \frac{C_a'}{W \sin \beta} + \frac{\tan \varnothing_a'}{\tan \beta} = \frac{2c'H(\sin i)(\sin \beta)}{(\gamma)(H^2)(\sin(i-\beta))\sin \beta} + \frac{\tan \varnothing_a'}{\tan \beta} \\ &= \frac{2c' \sin i}{\gamma H \sin(i-\beta) \sin \beta} + \frac{\tan \varnothing'}{\tan \beta} \end{aligned} \tag{9.15}$$

With further simplification, consider an infinitely long slope with a planar slip surface; many important mechanisms related to slope stability analysis can be demonstrated and evaluated. Most importantly, the analysis can be readily conducted based on stress. These simplified cases are presented in the following section.

9.2.5 Analysis of an infinitely long slope

Consider an infinitely long slope with uniform soil properties and a planar slip surface that is parallel to the slope surface. Let's start with a case where the slope is inundated under water and we consider a drained (effective stress) analysis as shown in Figure 9.10.

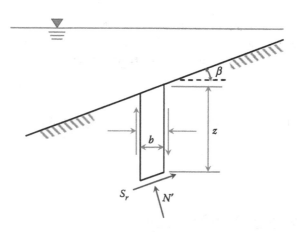

Figure 9.10 Infinite slope with a planar slip surface parallel to the slope surface.

Consider a soil element with depth z and width b, in the slope as shown in Figure 9.10. For an infinitely long slope, the normal and shear forces acting on the vertical surface on both sides of the element cancel each other. Considering a unit thickness of the slope (unit length perpendicular to the paper), the buoyant weight of the element, W' is,

$$W' = \gamma' bz(1) = \gamma' bz \tag{9.16}$$

where
γ' = buoyant unit weight of the soil

The equilibrium of forces requires that,

$$S_r - W' \sin\beta = 0 \tag{9.17}$$

and

$$N' - W' \cos\beta = 0 \tag{9.18}$$

Combining Equations 9.17 and 9.18 yields,

$$S_r = N' \tan\beta \tag{9.19}$$

Dividing both sides of Equation 9.19 by the length of the base of the element, l, and $l = b \sec\beta$, yields a stress-based relationship as follows:

$$\tau_r = \sigma' \tan\beta \tag{9.20}$$

where
τ_r = required or mobilized shear strength to maintain stability
σ' = effective normal stress on the base of the element

and

$$\sigma' = \gamma' z \cos^2\beta \quad \text{and} \quad \tau_r = \gamma' z \sin\beta\cos\beta \tag{9.21}$$

According to the Mohr–Coulomb failure criterion, the shear strength, τ_f, along the base of the element is

$$\tau_f = c' + \sigma' \tan\varnothing' \tag{9.22}$$

Thus,

$$FS = \frac{\tau_f}{\tau_r} = \frac{c' + \sigma' \tan\varnothing'}{\gamma' z \sin\beta\cos\beta} = \frac{c' + \gamma' z \cos^2\beta \tan\varnothing'}{\gamma' z \sin\beta\cos\beta} \tag{9.23}$$

For cohesionless soil, $c' = 0$, Equation 9.23 reduces to

$$FS = \frac{\tan\varnothing'}{\tan\beta} \tag{9.24}$$

The equation simply says that the FS is larger than 1, if the angle of the slope, β, is smaller than the drained friction angle of the soil, $\varnothing'$.

9.2.6 Factor of safety under a rapid drawdown

For a slope above water table, the buoyant unit weight, γ', in Equation 9.23 is replaced with the total unit weight, γ, in both numerator and denominator while Equation 9.24 remains the same. In an extreme condition, if the water level as shown in Figure 9.10 drops rapidly from its original level to the surface of the slope, before the water within the slope has a chance to seep out, the soil remains saturated but without the full benefit of buoyancy. A "rapid or sudden drawdown" can occur, for example, in a reservoir when the pool level drops quickly due to breakage of the dam or other forms of significant leakage. We take a conservative approach by computing the driving force W based on saturated soil unit weight. Equation 9.16 now becomes

$$W = \gamma_{sat} b \cdot z \tag{9.25}$$

where
γ_{sat} = saturated unit weight of the soil

and calculation of the required or mobilized shear strength, or Equation 9.21, becomes

$$\sigma' = \gamma_{sat} z \cos^2 \beta \quad \text{and} \quad \tau_r = \gamma_{sat} z \sin \beta \cos \beta \tag{9.26}$$

The shear strength is still computed using the buoyant unit weight, and Equation 9.23 becomes

$$FS = \frac{\tau_f}{\tau_r} = \frac{c' + \sigma' \tan \varnothing'}{\gamma_{sat} z \sin \beta \cos \beta} = \frac{c' + \gamma' z \cos^2 \beta \tan \varnothing'}{\gamma_{sat} z \sin \beta \cos \beta} \tag{9.27}$$

If the soil is cohesionless, or $c' = 0$, Equation 9.27 reduces to

$$FS = \frac{\gamma' \tan \varnothing'}{\gamma_{sat} \tan \beta} \tag{9.28}$$

For most soils, γ'/γ_{sat} is approximately $1/2$. Thus for a slope of cohesionless soil, the rapid drawdown can reduce the FS by approximately 50%. In cohesive slope, with the help of c', the reduction of FS is not as significant in a rapid drawdown condition.

9.2.7 Factor of safety under transient seepage conditions

In addition to the simplified and/or extreme conditions described in the previous sections, we have also used the infinitely long slope case to evaluate some of the more realistic slope stability problems. For example, in the case of artesian water or infiltration of surface water due to heavy rainfall, the pore water pressure within a slope is not necessarily hydrostatic, or the pore water pressure does not necessarily increase linearly with depth. Also, the pore water pressure distribution is transient and varies with time. To accommodate this situation, we modify Equation 9.23 as

$$FS = \frac{c' + (\sigma - u) \tan \varnothing'}{\gamma z \sin \beta \cos \beta} = \frac{c' + (\gamma z \cos^2 \beta - u) \tan \varnothing'}{\gamma z \sin \beta \cos \beta} \tag{9.29}$$

A slope failure occurs when a combination of depth z and pore pressure u results in FS = 1. This critical depth z is referred to as d_{cr}, and d_{cr} is derived by setting FS = 1 and reorganizing Equation 9.29 as

$$d_{cr} = \frac{c' - u \cdot \tan \varnothing'}{\gamma \cos^2 \beta (\tan \beta - \tan \varnothing')} \qquad (9.30)$$

Figure 9.11 shows the results from a transient numerical seepage analysis of an infinitely long slope reported by Collins and Znidarcic (2004). The groundwater table was initially at 4 m below ground surface. The soil above the groundwater table was unsaturated and the initial pore water pressure was negative, as indicated in the figure. The infinitely long slope had a slope angle $\beta = 28°$. The slope material had strength parameters of $c' = 3$ kPa and $\varnothing' = 30°$, the total unit weight, $\gamma = 20$ kN/m³. A plot of d_{cr} versus u according to Equation 9.30 and the above soil parameters are included in Figure 9.11 as a stability envelope. A software package called SEEP/W (GEO-SLOPE International Ltd., 1994) was used to simulate water infiltration from ground surface. The infiltration was assumed to be in vertical direction only. Figure 9.11 shows the pore water pressure profile at 0.3, 0.8, 1.5, and 1.8 hours after the infiltration started. As the rainfall continues, soil becomes saturated (i.e., $u = 0$ as shown in Figure 9.11) at ground surface first and this saturation front progresses downward into the unsaturated zone as the rainfall continues. Within the saturated zone, the pore water pressure increased linearly with depth, as in the case of hydrostatic conditions below a groundwater table. The negative pore pressure remains, however, in the unsaturated zone below the saturation front. At 1.8 hour, the pore water pressure profile touches the stability envelope at 2.1 m below ground surface

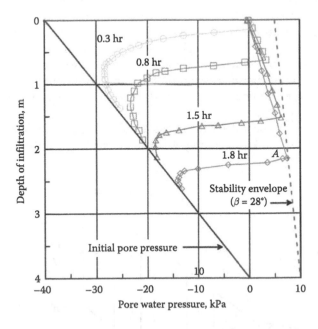

Figure 9.11 Stability of an infinitely long slope with transient seepage. (Collins, B.D. and Znidarcic, D. 2004. Stability analyses of rainfall induced landslides. *Journal of Geotechnical and Geoenvironmental Engineering*, 130, 4. Figure 4, p. 364. Reproduced with permission of the ASCE.)

(point A in Figure 9.11), meaning the FS at that point is 1 and reached a failure condition. The depth of point A corresponds to d_{cr} of Equation 9.30.

9.3 METHOD OF SLICES

The method of slices is a versatile and powerful tool under the category of limit equilibrium analysis method for dealing with slopes with irregular slip surface and in non-homogeneous soils in which the values of c and $\varnothing$ are not necessarily constant. The method was pioneered by Fellenius (1927, 1936) and Taylor (1937, 1948) and has since been improved extensively in its capabilities. With the help of computer software, the method of slices is rather effective and thus has been widely used in engineering practice. The method of slices generally involves the following procedure (Abramson et al., 1996):

- Divide the sliding block above an assumed slip surface into vertical slices as shown in Figure 9.12. For the case shown in Figure 9.12, there are 10 slices. The width of the slices is not necessarily uniform and the number of slices can vary.
- The bottom of each slice is simplified as a straight chord.
- The properties and the shear strength of the slices can vary, thus allowing non-uniform soil conditions to be considered.
- The analysis is based on forces. The stresses acting on the sides of a slice are replaced with equivalent concentrated forces. Figure 9.13 shows the possible forces that can be involved for a given slice. The line that connects the location of all the interslice forces (Z_R and Z_L in Figure 9.13) is referred to as the thrust line.

Note that $\bar{C}$ in Figure 9.13 represents the effects, in terms of force, of c or c', and u along the base of a slice, and $\bar{C} = c\Delta l$ or $c'\Delta l$ and $U_\alpha = u\Delta l$ where $\Delta l =$ length of the base of the slice along the assumed slip surface. $\bar{\varnothing}$ is used to represent the total stress friction angle ($\varnothing$) or drained, effective stress friction angle ($\varnothing'$). In an undrained, total stress slope stability analysis, U_α is ignored, c should be replaced with the undrained shear strength, s_u, and $\bar{\varnothing} = 0$. In drained or effective stress analysis, c' and $\varnothing'$ are used to determine the strength parameters $\bar{C}$ and $\bar{\varnothing}$. Further discussion on the selection of soil strength parameters as they relate to the types (drained or undrained) of slope stability analysis is presented in the later sections.

The system remains to be indeterminate, as demonstrated in Table 9.2, for a case of n slices, there are $6n-2$ unknowns and $4n$ equations. The number of equations and

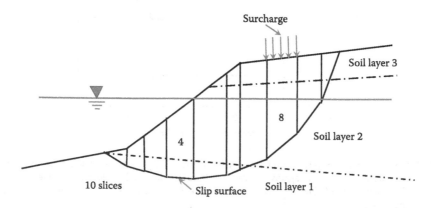

Figure 9.12 Simplification of a sliding block above an assumed slip surface as a series of vertical slices.

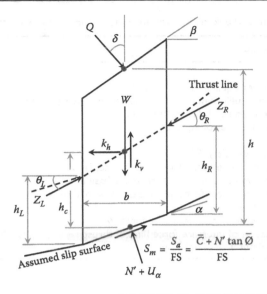

Figure 9.13 Forces acting on a slice in general conditions.

Known quantities	Unknown quantities
W = weight of the slice	Z_R, Z_L = interslice force on the right and left side of the slice
h, b = height and width of the slice	θ_R, θ_L = angle of the right and left interslice force
h_c = height to centroid of slice	N' = effective normal force at the base of the slice
$\bar{C}, \bar{\varnothing}$ = strength parameters	S_m = mobilized shear force
k_h, k_v = horizontal and vertical seismic coefficient	FS = safety factor
Q = surcharge load	
U_α = normal force from pore water pressure	
α, β = inclination of base and top of the slice	

unknowns are equal only when $n = 1$, which then is identical to the case of a planar slip surface described above. An important part of the earlier developments in the method of slices was to simplify the parameters involved so that the numbers of equations and unknowns are equal and a unique solution can be obtained. A few of the commonly used methods of slices are introduced in the following sections. Examples that do not involve pore water pressure and seismic loading conditions are presented first. Considerations of pore water pressure (i.e., U_α) and seismic loadings (i.e., k_h, k_v) in the methods of slices are described in subsequent sections.

9.3.1 Ordinary method of slices

It is difficult to evaluate the interslice forces (Z and θ in Figure 9.13), which depend on factors including the stress–strain and deformation characteristics of the material of the slope. The inclusion of interslice forces can also complicate the computations, especially without the help of computer. A simple early approach assumed the interslice forces to be zero and determined the total normal stress on the base of each slice by resolving all forces perpendicular and tangential to the base of a slice. The ordinary method of slices (OMS), also known as the Swedish circle or Fellenius method (Fellenius, 1927, 1936),

Table 9.2 Equations and unknowns involved in the method of slices (number of slices = n)

	Description
Equations	
n	Moment equilibrium for each slice
2n	Force equilibrium in two mutually perpendicular directions for each slice
n	Mohr–Coulomb failure criterion
4n	Total number of equations
Unknowns	
1	Factor of safety, FS
n	Normal force at the base of each slice, N'
n	Location of normal force, N' at the base of each slice
n	Mobilized shear force at the base of each slice, S_m
n−1	Interslice force, Z
n−1	Inclination of each interslice force, θ
n−1	Location of interslice force (line of thrust)
6n−2	Total number of unknowns

follows this approach. The OMS involves the following assumptions and/or simplifications:

1. A circular slip surface is assumed.
2. All interslice forces (Z and θ in Figure 9.13 and Table 9.2) are ignored as shown in Figure 9.14.
3. Consider force equilibrium in the N' direction (perpendicular to the base) for each slice, then

$$N' = -U_\alpha - k_h W \sin\alpha + (1 - k_v)\cos\alpha + Q\cos(\delta - \alpha) \tag{9.31}$$

4. Consider moment equilibrium of the entire sliding block about the center of the circular slip surface. The factor of safety is defined as a ratio of resisting over

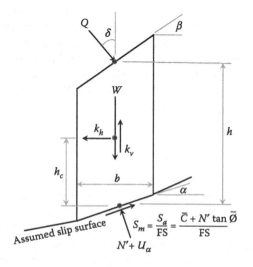

Figure 9.14 Simplification of forces on a slice.

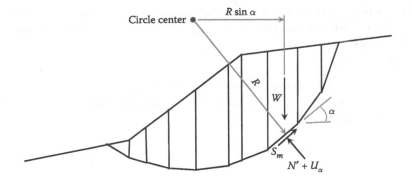

Figure 9.15 Summing moment with respect to the center of the circular slip surface.

disturbing/driving moments. The location of N' at the base of each slice is irrelevant, as all N' points toward the center of the circle as shown in Figure 9.15.

$$FS = \frac{\sum_{i=1}^{n}(\bar{C} + N' \tan \bar{\emptyset})}{\sum_{i=1}^{n} A_1 - \sum_{i=1}^{n} A_2 + \sum_{i=1}^{n} A_3} \tag{9.32}$$

where

$$A_1 = (W(1 - k_v) + Q \cos \delta) \sin \alpha \tag{9.33}$$

$$A_2 = Q \sin \delta \left(\cos \alpha - \frac{h}{R} \right) \tag{9.34}$$

$$A_3 = k_h W \left(\cos \alpha - \frac{h_c}{R} \right) \tag{9.35}$$

5. The analysis is determinate with one equilibrium equation ($\sum$ moments about the center of the circular slip surface) and one unknown (safety factor), as represented by Equation 9.32.

An Excel spreadsheet program for the execution of method of slices is introduced in the following examples. The spreadsheet program is available on the publisher's website for registered users. The ordinary method of slices is known to be conservative and tends to underestimate the factor of safety. The error, albeit on the safe side, becomes greater for deep critical circles with large variations in α.

EXAMPLE 9.4

Given

For the same slip surface and soil conditions in Example 9.1.

Required

Repeat Example 9.1 using the ordinary method of slices.

Solution

The sliding block is divided into 10 slices as shown in Figure E9.4.

This example does not involve any surcharge load ($Q = 0$), pore water pressure ($U_\alpha = 0$), and there is no seismic load ($k_h = k_v = 0$). Under these conditions, Equation 9.31 becomes

$$N' = W \cos \alpha$$

The Excel spreadsheet program is used to setup the coordinates of the slope profile and compute the related quantities for the slices. Table E9.4 summarizes the results of the computations.

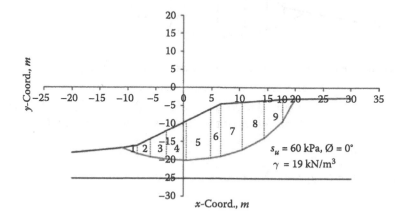

Figure E9.4 Profile of the slope and slices.

Table E9.4 Summary of OMS computations

Slice no.	b (m)	h (m)	W (kN)	α, radian	W sin α (kN)	W cos α (kN)	Δl (m)	$\bar{C} = s_u \Delta l$ (kN)
1	2.94	1.16	64.73	−0.508	−31.50	56.55	3.37	201.91
2	2.32	3.67	161.87	−0.364	−57.58	151.28	2.48	148.94
3	2.88	6.48	354.71	−0.228	−80.30	345.51	2.96	177.41
4	3.66	9.34	649.17	−0.064	−41.34	647.85	3.67	220.05
5	4.34	12.16	1003.95	0.136	135.96	994.70	4.38	263.06
6	1.84	13.99	488.20	0.293	141.18	467.34	1.92	115.14
7	3.73	13.77	975.62	0.442	417.24	881.90	4.13	247.58
8	3.98	11.74	887.57	0.668	549.76	696.81	5.08	304.02
9	3.24	8.34	513.10	0.925	409.92	308.61	5.43	323.02
10	2.02	3.44	132.36	1.194	123.07	48.73	5.68	329.72
Sum					1566.41			2330.86

b = width of slice.
h = average height of slice.
Δl = base length of slice.
W = weight height of slice.

Equation 9.32 with the consideration of the given conditions becomes

$$FS = \frac{\sum_{i=1}^{n} \bar{C}}{\sum_{i=1}^{n} A_1}$$

and $A_1 = W \sin \alpha$. Using the summation of $\bar{C}$ and $W \sin \alpha$ values from Table E9.4,

$$FS = \frac{2330.86}{1566.41} = 1.49$$

9.3.2 Simplified Bishop method

Bishop (1955) proposed a rigorous scheme by including the interslice forces in the equations of equilibrium of a typical slice. An iterative procedure was used to calculate the interslice forces. He also proposed a simplified procedure which gives fairly accurate results even though the interslice forces are ignored. As the rigorous Bishop method is more involved and has been included in more recent and comprehensive procedures, only the simplified Bishop method is presented below. A later section is devoted to discussion of the more recent and comprehensive slope stability analysis methods.

In the simplified Bishop method, a factor of safety is derived for an assumed slip surface by combining the vertical force equilibrium for each slice and moment equilibrium for the entire sliding block. This arrangement allows the interslice shear forces ($Z \sin \theta$ in Figure 9.13) to be ignored. By assuming a circular slip surface and taking moment about the center of the circular slip surface, the position and magnitude of the interslice horizontal forces ($Z \cos \theta$ in Figure 9.13) become irrelevant, as they are canceled in the computation of the moment equilibrium. A drawback to this approach is that the method is applicable only for the analysis of a circular slip surface. The simplified Bishop method involves the following assumptions and/or simplifications:

1. A circular slip surface is assumed.
2. All interslice shear forces ($Z \sin \theta$ in Figure 9.13) are ignored.
3. Consider force equilibrium in the vertical direction for each slice (see Figure 9.13), then

$$N' = \frac{1}{m_a}\left[W(1 - k_v) - \frac{\bar{C}\sin\alpha}{\mathrm{FS}} - U_\alpha \cos\alpha + Q\cos\delta\right]\tag{9.36}$$

where

$$m_a = \cos\alpha\left[1 + \frac{\tan\alpha\,\tan\bar{\varnothing}}{\mathrm{FS}}\right]\tag{9.37}$$

and FS is an assumed factor of safety. This arrangement practically neglected the interslice horizontal forces ($Z \cos \theta$ in Figure 9.13) as well.

4. Consider moment equilibrium of the entire sliding block about the center of the circular slip surface. The location of N' at the base of each slice is irrelevant, as all N' points toward the center of the circle as shown in Figure 9.15. The factor of safety for the assumed slip surface is

$$\mathrm{FS} = \frac{\sum_{i=1}^{n}(\bar{C} + N'\tan\bar{\varnothing})}{\sum_{i=1}^{n}A_4 - \sum_{i=1}^{n}A_5 + \sum_{i=1}^{n}A_6}\tag{9.38}$$

$$A_4 = (W(1 - k_v) + Q\cos\delta)\sin\alpha\tag{9.39}$$

$$A_5 = Q\sin\delta\left(\cos\alpha - \frac{h}{R}\right)\tag{9.40}$$

$$A_6 = k_h W\left(\cos\alpha - \frac{h_c}{R}\right)\tag{9.41}$$

5. The procedure is determinate with $n + 1$ equilibrium equations (n equations for $\sum$ in vertical direction for each slice and 1 equation for $\sum$ moments about the center of circular slip surface) and $n + 1$ unknowns (factor of safety and n normal forces on the base of slices) where n is the number of slices.

The FS is determined using an iteration procedure. An FS is assumed first in Equations 9.36 and 9.37 of step 3 for the determination of N'. FS is calculated again using Equation 9.38 in step 4. The procedure is repeated until the assumed FS is reasonably close to the calculated FS. Convergence is usually very rapid.

EXAMPLE 9.5

Given

For the slope geometry and circular slip surface as shown in Example 9.4. The cross-section of the slope and its slip surface are shown in Figure E9.4, but with different soil parameters, as follows:

$\gamma = 19\,\text{kN/m}^3$
$c' = 20\,\text{kPa}$
$\emptyset' = 25°$

Required

Determine the factor of safety for the slip surface using the simplified Bishop method.

Solution

The sliding block is divided into 10 slices as shown in Figure E9.4.

This example does not involve any surcharge load ($Q = 0$), pore water pressure ($U_\alpha = 0$) and there is no seismic load ($k_b = k_v = 0$). Under these conditions, Equation 9.36 becomes

$$N' = \frac{1}{m_a}\left[W - \frac{\bar{C}\sin\alpha}{\text{FS}}\right]$$

Equation 9.37 remains the same, Equation 9.38 becomes

$$\text{FS} = \frac{\sum_{i=1}^{n}(\bar{C} + N'\tan\emptyset)}{\sum_{i=1}^{n} A_4}$$

and A_4 (Equation 9.39) is simplified as

$$A_4 = W\sin\alpha$$

The Excel spreadsheet program was used to set up the coordinates of the slope profile and compute the related values. A series of trial FS values was used and compared with the computed FS values. Table E9.5 summarizes the results of the computations when assumed FS = 2.12, after many iterations.

$$\sum_{i=1}^{n}(\bar{C} + N'\tan\emptyset) = 782.20 + 2541.84 = 3324.04\,\text{kN}$$

and

$$\sum_{i=1}^{n} A_4 = 1566.41\,\text{kN}$$

$$\text{FS} = \frac{\sum_{i=1}^{n}(\bar{C} + N'\tan\emptyset)}{\sum_{i=1}^{n} A_4} = \frac{(3324.04)}{(1566.41)} = 2.12$$

The computed FS after iterations is essentially the same as the assumed value.

Table E9.5 Summary of the simplified Bishop method computations

Slice no.	b (m)	h (m)	W (kN)	α, radian	$W \sin \alpha$ (kN)	Δl (m)	$\bar{C} = c' \Delta l$ (kN)	m_a	N' (kN)	$N' \tan \bar{\varnothing}$ (kN)
1	2.94	1.16	64.73	−0.51	−31.50	3.37	67.41	0.77	104.62	48.79
2	2.32	3.67	161.87	−0.36	−57.58	2.48	49.68	0.86	198.76	92.68
3	2.88	6.48	354.71	−0.23	−80.30	2.96	59.15	0.92	390.62	182.15
4	3.66	9.34	649.17	−0.06	−41.34	3.67	73.35	0.98	661.99	308.69
5	4.34	12.16	1003.95	0.14	135.96	4.38	87.70	1.02	978.22	456.15
6	1.84	13.99	488.20	0.29	141.18	1.92	38.40	1.02	473.08	220.60
7	3.73	13.77	975.62	0.44	417.24	4.13	82.62	1.00	960.87	448.06
8	3.98	11.74	887.57	0.67	549.76	5.08	101.66	0.92	931.13	434.20
9	3.24	8.34	513.10	0.93	409.92	5.43	108.64	0.78	607.54	283.30
10	2.02	3.44	132.36	1.19	123.07	5.68	113.60	0.57	144.14	67.22
Sum					1566.41		782.20			2541.84

b = width of slice.
Δl = base length of slice.

9.3.3 Simplified Janbu method

In general, there is no reason for a slip surface to be circular. It is much preferred to use a generalized method of slices that allows non-circular slip surface be considered. This is especially true when there is an inclined thin soft soil layer in the slope where a straight slip surface is likely to pass through that soil layer. Janbu (1954, 1957, 1973) proposed such a method that allows a non-circular slip surface to be applied in the analysis. The method considered the force and moment equilibrium of a typical slice and force equilibrium of the sliding block as a whole. An iterative procedure was used to calculate the interslice forces and FS. The generalized Janbu method can be included in more recent and comprehensive procedures, thus only a simplified Janbu method is presented below. A later section is devoted to a discussion of the more recent and comprehensive slope stability analysis methods. As the simplified Bishop method, the interslice forces are practically ignored in the simplified Janbu method.

The simplified Janbu method involves the following procedure and simplifications:

1. All interslice shear forces ($Z \sin \theta$ in Figure 9.13) are ignored, as in the simplified Bishop method. The main difference for the simplified Janbu method is that the FS is determined considering overall horizontal force equilibrium, as described below.
2. Consider force equilibrium in the vertical direction for each slice (see Figure 9.13) while ignoring the interslice shear forces, then

$$N' = \frac{1}{m_a} \left[W(1 - k_v) - \frac{\bar{C} \sin \alpha}{FS} - U_a \cos \alpha + Q \cos \delta \right] \tag{9.42}$$

where

$$m_a = \cos \alpha \left[1 + \frac{\tan \alpha \, \tan \bar{\varnothing}}{FS} \right] \tag{9.43}$$

and FS is an assumed factor of safety.

3. Consider horizontal force equilibrium for the entire sliding block and rearrange the equation results in and expression for the FS of the slip surface:

$$FS = \frac{\sum_{i=1}^{n}(\bar{C} + N' \tan \bar{\varnothing}) \cos \alpha}{\sum_{i=1}^{n} A_7 + \sum_{i=1}^{n} N' \sin \alpha} \tag{9.44}$$

$$A_7 = U_\alpha \sin \alpha + W k_h + Q \sin \delta \tag{9.45}$$

4. For a sliding block with n slices, the procedure is determinate with $n+1$ equilibrium equations (n equations for $\sum$ forces in vertical direction for all slices and 1 equation for $\sum$ forces in horizontal direction for the entire sliding block) and $n+1$ unknowns (factor of safety and n normal force on the base of all slices).

The horizontal interslice forces from two successive slices have the same magnitude but opposite directions, thus are canceled in the overall horizontal force equilibrium computation, and practically ignored as in the simplified Bishop method.

The FS is determined using an iteration procedure. An FS is assumed first in Equations 9.42 and 9.43 of Step 2 for the determination of N'. FS is calculated again using Equations 9.44 and 9.45 in Step 3. The procedure is repeated until the difference between the assumed and calculated FS becomes small. By neglecting the interslice shear forces, the FS calculated using simplified Janbu method is likely to be too low. Janbu suggested that the calculated factor of safety from the above procedure, $FS_{calculated}$, be adjusted by multiplying a modification factor, f_o, to obtain the reported factor of safety, FS_{Janbu}, or

$$FS_{Janbu} = f_o \cdot FS_{calculated} \tag{9.46}$$

where

$$f_o = 1 + b_1 \left[\frac{d}{L} - 1.4 \left(\frac{d}{L} \right)^2 \right] \tag{9.47}$$

For soils with c only: $b_1 = 0.69$
For soils with $\varnothing$ only: $b_1 = 0.31$
For soils with c and $\varnothing$: $b_1 = 0.50$

The definition of L and d is given in Figure 9.16, where L is the chord length of the slip surface and d is the maximum depth of the slip surface, measured from the chord.

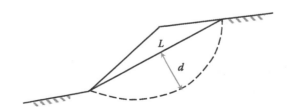

Figure 9.16 Definition of L and d.

EXAMPLE 9.6

Given

For the same slope geometry in Example 9.5 but consider a non-circular slip surface. The cross-section of a slope and its non-circular slip surface are shown in Figure E9.6. The soil parameters are

$\gamma = 19 \, \text{kN/m}^3$
$c' = 20 \, \text{kPa}$
$\varnothing' = 25°$

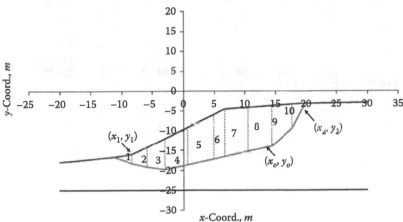

Figure E9.6 Profile of the slope, non-circular slip surface and the slices.

Required

Determine the factor of safety for the slip surface using the simplified Janbu method.

Solution

The sliding block is divided into 10 slices (numbered from left to right) as shown in Figure E9.6.

This example does not involve any surcharge load ($Q = 0$), pore water pressure ($U_\alpha = 0$), and there is no seismic load ($k_h = k_v = 0$). Under these conditions, Equation 9.42 becomes

$$N' = \frac{1}{m_a}\left[W - \frac{\bar{C}\sin\alpha}{\text{FS}}\right]$$

Equation 9.43 remains the same as

$$m_a = \cos\alpha\left[1 + \frac{\tan\alpha\,\tan\varnothing'}{\text{FS}}\right] \tag{9.43}$$

invoke Equation 9.44,

$$\text{FS} = \frac{\sum_{i=1}^{n}(\bar{C} + N'\tan\varnothing')\cos\alpha}{\sum_{i=1}^{n} A_7 + \sum_{i=1}^{n} N'\sin\alpha} \tag{9.44}$$

and $A_7 = 0$ for all slices (Equation 9.45).

The Excel spreadsheet program was used to set up the coordinates of the slope profile and compute the related values. A series of trial FS values was used and compared with the computed FS values. Table E9.6 summarizes the results of the computations when assumed FS = 1.83, after many iterations.

Table E9.6 Summary of the simplified Janbu method computations

Slice no.	W (kN)	α, radian	Δl (m)	$C = c'\Delta l$ (kN)	N' (kN)	m_a	$\dfrac{(\bar{C} + N'\tan\bar{\varnothing})}{\cos\alpha}$	N' sin α
1	61.70	−0.50	3.35	66.93	104.46	0.76	101.66	−49.80
2	157.14	−0.37	2.48	49.70	198.09	0.84	132.64	−70.95
3	349.08	−0.23	2.96	59.16	389.35	0.92	234.37	−88.82
4	591.06	0.33	3.86	77.25	561.30	1.03	321.22	179.36
5	838.57	0.33	4.58	91.69	799.38	1.03	440.10	255.43
6	402.01	0.33	1.94	38.77	384.10	1.03	206.46	122.73
7	816.72	0.33	3.94	78.73	780.35	1.03	419.41	249.35
8	807.19	0.33	4.20	83.96	770.19	1.03	419.88	246.11
9	502.15	0.95	5.53	110.61	572.12	0.79	220.96	463.81
10	117.93	1.24	6.24	124.83	94.49	0.57	54.74	89.39
Sum							2551.43	1396.61

Δl = base length of slice.

$$\sum_{i=1}^{n}(\bar{C} + N'\tan\bar{\varnothing}) = 2551.43\,\text{kN}$$

and

$$\sum_{i=1}^{n} N'\sin\alpha = 1396.61\,\text{kN}$$

$$FS = \frac{\sum_{i=1}^{n}(\bar{C} + N'\tan\bar{\varnothing})}{\sum_{i=1}^{n} N'\sin\alpha} = \frac{(2551.43)}{(1396.61)} = 1.83$$

The computed FS is essentially the same as the assumed value. The calculated factor of safety, $FS_{calculated}$, is corrected according to Equations 9.46 and 9.47, and

$$FS_{Janbu} = \left\{ 1 + b_1 \left[\frac{d}{L} - 1.4 \left(\frac{d}{L} \right)^2 \right] \right\} \cdot FS_{calculated}$$

For soils with c and $\varnothing$, $b_1 = 0.50$ (see Equation 9.47). Take the distance between (x_1, y_1) and (x_2, y_2) in Figure E9.6 as L. Consider the distance between (x_o, y_o) and the line that connects (x_1, y_1) and (x_2, y_2) in Figure E9.6 as d. Replacing $FS_{calculated}$, d, L, and b_1 into Equation 9.47,

$$FS_{Janbu} = \left\{ 1 + 0.5 \left[\frac{7.33}{30.69} - 1.4 \left(\frac{7.33}{30.69} \right)^2 \right] \right\} FS_{calculated} = (1.08)(1.83) = 1.97$$

9.3.4 General limit equilibrium (GLE) method

So far, three methods have been introduced for slope stability analysis under the category of method of slices. These are part of a much larger family of methods utilizing limit equilibrium method of slices that were proposed in a period of three decades since the 1950s.

Other noticeable methods in the family include those by Morgenstern and Price (1965) and Spencer (1967, 1973). The safety factor equations have been derived considering the summation of forces in two directions and/or the summation of moments about a point of rotation, the basic elements of statics. As described earlier in this chapter and in Table 9.2, these elements of statics, along with the failure criterion, are insufficient to make the slope stability analysis determinate except for very limited conditions (i.e., a planar slip surface). In order to render the analysis determinate, all methods of slices mentioned so far had made assumptions regarding the direction and/or magnitude of some of the forces involved.

Fredlund et al. (1981) indicated that the similarities and differences among the available limit equilibrium methods of slices were obscure, largely because of (1) the lack of uniformity in formulating the equations of equilibrium, (2) the ambiguity concerning interslice forces, and (3) the unknown limitations imposed by non-circular slip surfaces. Fredlund et al. (1981) proposed a common formulation of the equilibrium equations called the general limit equilibrium (GLE) method that can be used to encompass most of the above-mentioned methods. For the above reasons, this section will concentrate on the GLE method and follow the generalized formulations by Chugh (1986). It can be shown that most of the other methods of slices can be considered as special cases of the GLE method.

The GLE method starts by using a function to describe the distribution of the interslice force angles (i.e., θ_R and θ_L in Figure 9.13) as follows,

$$\theta_i = \lambda \cdot f(x_i) \tag{9.48}$$

where θ_i represents the interslice force angle on the right of slice i (θ_R). Note that θ_R of slice i equals to θ_L of the neighboring slice $i + 1$. $f(x_i)$ is user defined and can be a continuous function or a set of specified discrete values, as shown in Figure 9.17. The value of $f(x_i)$ is between 0 and 1 in keeping track of the characteristics of the interslice force angles. λ is treated as an unknown and serves as a scale factor to give the range of θ_i beyond 0 and 1. The function of $f(x_i)$ must be defined a priori in executing GLE method.

For simplicity, seismic effects, external load, and pore water pressure are not considered (i.e., $k_v = k_h = 0$, $Q = 0$ and $U_\alpha = 0$) in the following introduction. The remaining forces acting on a given slice are shown in Figure 9.18.

Consider force equilibrium for each slice tangential to the base of that slice,

$$S_m + Z_L \cos(\alpha - \theta_L) - Z_R \cos(\alpha - \theta_R) - W \sin \alpha = 0 \tag{9.49}$$

Invoking the Mohr–Coulomb failure criterion,

$$S_m = \frac{S_a}{\text{FS}} = \frac{\bar{C} + N' \tan \bar{\varnothing}}{\text{FS}} \tag{9.50}$$

Replace S_m into Equation 9.49,

$$N' \tan \bar{\varnothing} = \text{FS} \cdot \{Z_R \cos(\alpha - \theta_R) - Z_L \cos(\alpha - \theta_L) + W \sin \alpha\} - \bar{C} \tag{9.51}$$

Now consider force equilibrium for each slice normal to the base of that slice,

$$N' = Z_L \sin(\alpha - \theta_L) - Z_R \sin(\alpha - \theta_R) + W \cos \alpha \tag{9.52}$$

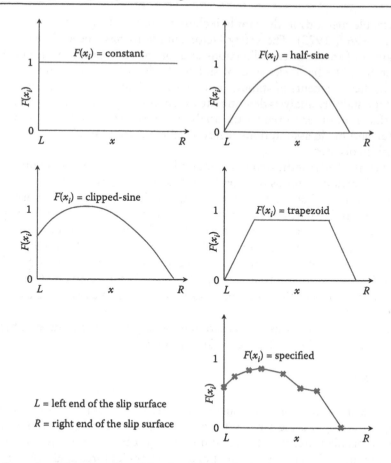

L = left end of the slip surface

R = right end of the slip surface

Figure 9.17 Examples of functions describing the variations in the direction of the interslice forces with respect to the x direction. (After Fredlund, D.G. et al., 1981. The relationship between limit equilibrium slope stability methods. *Proceedings of the 10th International Conference on Soil Mechanics and Foundation Engineering*, Stockholm, A.A. Balkema, Vol. 3, pp. 409–416.)

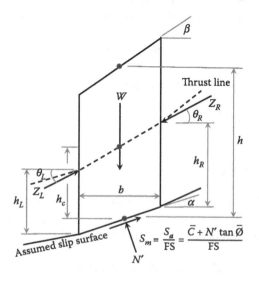

Figure 9.18 Key parameters involved in the GLE method.

Replace N' from Equation 9.52 into Equation 9.51,

$$Z_R = A_8 Z_L \cos(\theta_L - \alpha)\left[1 - \frac{1}{FS}\tan(\theta_L - \alpha)\tan\bar{\varnothing}\right]$$
$$+ A_8\left[\frac{\bar{C}b}{FS \cdot \cos\alpha} - W\sin\alpha + \frac{W\cos\alpha\tan\bar{\varnothing}}{FS}\right] \quad (9.53)$$

and

$$A_8 = \frac{1}{\cos(\theta_R - \alpha)\left[1 - \dfrac{1}{FS}\tan(\theta_R - \alpha)\tan\bar{\varnothing}\right]} \quad (9.54)$$

Consider moment equilibrium at the center of the base for each slice,

$$Z_L \cos\theta_L\left[h_L - \frac{b}{2}\tan\alpha\right] + Z_L\frac{b}{2}\sin\theta_L - Z_R\cos\theta_R\left[h_R + \frac{b}{2}\tan\alpha\right]$$
$$+ Z_R\frac{b}{2}\sin\theta_R = 0 \quad (9.55)$$

Rearrange Equation 9.55 so that,

$$h_R = \frac{Z_L \cos\theta_L}{Z_R \cos\theta_R}h_L + \frac{b}{2}\frac{1}{\cos\theta_R\cos\alpha}\left[\sin(\theta_R - \alpha) + \frac{Z_L}{Z_R}\sin(\theta_L - \alpha)\right] \quad (9.56)$$

The effective normal stress at the base of each slice, σ'_n, can be calculated based on the respective N' and base length, Δl, as follows:

$$\sigma'_n = \frac{N'}{\Delta l} = \frac{N'}{b\sec\alpha} \quad (9.57)$$

Replace N' from Equation 9.52 into Equation 9.57,

$$\sigma'_n = \frac{1}{b\sec\alpha}\{Z_L\sin(\alpha - \theta_L) - Z_R\sin(\alpha - \theta_R) + W\cos\alpha\} \quad (9.58)$$

The safety factor for a given slip surface can be computed using the following procedure:

1. Number the slices in the sliding block from left to right as described in Figure 9.12, where the toe is on the left and crest is on the right.
2. Determine θ_L, h_L, and Z_L for the first slice according to its boundary conditions on the left end of the sliding block. Specify the interslice force angle function, $f(x_i)$, such as those depicted in Figure 9.17.
3. Assume a λ, then use Equation 9.48 to determine the θ_R values for the rest of the slices. Note that θ_L and θ_R will be in radians and positive for interslice forces located counterclockwise to the normal of the side of the slice as shown in Figure 9.18. θ_R of slice i equals to θ_L of slice $i + 1$.
4. Assume an FS and compute Z_R using Equation 9.53 sequentially from the first to the last slice. Note that Z_R of slice i equals to Z_L of slice $i + 1$, and Z_L for the first slice is specified in Step 2.
5. Adjust FS and repeat Step 4 until Z_R for the last slice approaches the boundary force. This Z_R can be the hydrostatic force if there exists a crack in the crest and the crack is filled with water.

6. Using the θ_L/θ_R values from Step 3 and Z_L/Z_R values generated in Step 4, determine h_R sequentially from the first to the last slice following Equation 9.56. Note that h_L for the first slice is specified in Step 2.

7. The h_R for the last slice should be compatible with the boundary condition at the slope crest. If there is no crack or no water in the crack, $h_R = 0$, or $h_R = 1/3$ of the water depth for water-filled crack.

8. Adjust λ and repeat Steps 3 through 7 until a reasonable h_R for the last slice is obtained.

9. The computation involves iterations of λ and FS. Practice and experience are needed to obtain convergence. In certain cases, variations of the $f(x_i)$ may be necessary. There is no rigorous method to verify if the result is correct (e.g., comparison between the assumed and computed FS). Methods to evaluate the reasonableness of the computations can include plotting σ'_n for each slice according to Equation 9.58, position of the thrust line (a line that connects the location of all interslice forces, as shown in Figure 9.18), and interslice shear stresses (interslice shear force divided by the corresponding height of the slice wall). In most cases, the distribution of σ'_n should be compressive; it starts low and increases with x (horizontal distance to the toe) and then diminishes toward the crest. The thrust line should remain within the sliding block. The interslice shear stresses should be less than the respective shear strength of the material within the sliding block.

10. For a sliding block of n slices, the solution involves $3n$ equilibrium equations (n equations for $\sum$ forces in vertical direction, n equations for $\sum$ forces in horizontal direction, and n equations for $\sum$ moments about any selected point) and $3n$ unknowns (factor of safety, the scale factor λ, n normal forces on the base of slices, $n - 1$ interslice forces, and $n - 1$ locations of the interslice forces).

EXAMPLE 9.7

Given

For the same slope and non-circular slip surface in Example 9.6. The cross-section of the slope, its non-circular slip surface, and soil parameters are shown in Figure E9.7.

Required

Determine the factor of safety for the slip surface using the GLE method and use $f(x_i) = 1$.

Solution

The sliding block is divided into 10 slices as shown in Figure E9.7, and the slices are numbered from left to right. The computation follows the procedure stated above. The left side of slice 1 is a point and there is no external force applied to this side, therefore θ_L is irrelevant, Z_L and $h_L = 0$. The Excel spreadsheet program was used to set up the coordinates of the slope profile and make computations according to GLE procedure. A series of trial FS and λ values were used in the process. Table E9.7 summarize the results of the GLE computations when assumed FS $= 2.85$ and $\lambda = 0.4$.

The shear stress is computed according to the Z_R tangential to the right side wall ($= Z_R \sin \theta_R$) divided by height of the right side wall (h) of each slice.

The normal stress required in the computation of shear strength is determined according to the Z_R normal to the right side wall ($= Z_R \cos \theta_R$) divided by height of the right side wall of each slice.

The position of the thrust line is included in Figure E9.7. Distributions of σ'_n along the base of slip surface and shear stress ratios (ratio of the computed shear stress on the side wall over the available shear strength) on the vertical face of the slices versus x

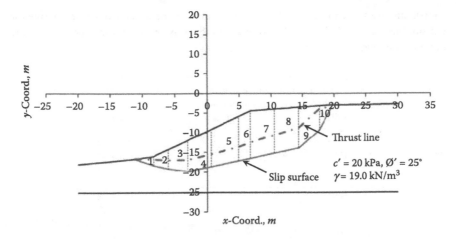

Figure E9.7 Profile of the slope, non-circular slip surface for the GLE method.

Table E9.7 Summary of GLE method computations

Slice no.	b (m)	h (m)	W (kN)	α, radian	Δl (m)	$C = c'\Delta l$ (kN)	θ_L radian	θ_R radian	A_B	Z_L (kN)	Z_R (kN)
1	2.94	1.10	61.70	−0.50	3.35	66.93	–	0.4	1.30	0.00	150.97
2	2.32	3.56	157.14	−0.37	2.48	49.70	0.4	0.4	1.23	150.97	264.40
3	2.88	6.38	349.08	−0.23	2.96	59.16	0.4	0.4	1.17	264.40	449.01
4	3.66	8.50	591.06	0.33	3.86	77.25	0.4	0.4	1.01	449.01	454.34
5	4.34	10.16	838.57	0.33	4.58	91.69	0.4	0.4	1.01	454.34	461.46
6	1.84	11.52	402.01	0.33	1.94	38.77	0.4	0.4	1.01	461.46	419.35
7	3.73	11.52	816.72	0.33	3.94	78.73	0.4	0.4	1.01	419.35	391.17
8	3.98	10.68	807.19	0.33	4.20	83.96	0.4	0.4	1.01	391.17	379.74
9	3.24	8.16	502.15	0.95	5.53	110.61	0.4	0.4	0.90	379.74	191.23
10	2.02	3.07	117.93	1.24	6.24	124.83	0.4	0.4	0.70	191.23	222.68

Slice no.	h_L (m)	h_R (m)	N' (kN)	Normal stress[a]		Thrust line[b]		Shear stress
				x (m)	σ'_n (kPa)	x (m)	y (m)	Shear strength
1	0.00	1.42	172.21	1.47	51.46	−11.2518	−16.60	0.33
2	1.42	2.28	225.38	4.10	90.70	−8.31	−16.77	0.11
3	2.28	2.85	448.66	6.70	151.68	−5.99	−16.80	0.06
4	2.85	3.12	560.47	9.97	145.11	−3.11	−16.91	0.05
5	3.12	3.44	795.14	13.97	173.45	0.55	−15.40	0.04
6	3.44	3.96	377.79	17.06	194.87	4.894	−13.61	0.03
7	3.96	4.57	771.80	19.85	196.07	6.731	−12.48	0.03
8	4.57	5.05	764.02	23.70	181.99	10.461	−10.61	0.04
9	5.05	5.39	391.78	27.31	70.84	14.439	−8.78	0.05
10	5.39	−0.07	14.79	29.94	2.37	17.677	−3.97	0.13

Δl = base length of slice.

Note

[a] x coordinate at the center of the base of each slice.

[b] x coordinate at the right side of each slice.

coordinate are shown in Figure E9.7a. The results fit the criteria described above for reasonableness check. The GLE offers significantly more information than the previous methods.

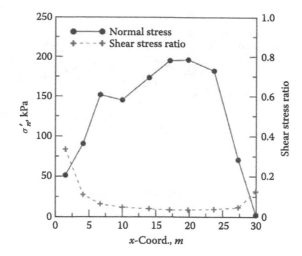

Figure E9.7a Distribution of normal stresses and shear stress ratio.

9.4 OTHER NUMERICAL METHODS

Other numerical methods that consider the slope as a continuous material, assembly of rigid or deformable discrete blocks/particles have also been developed (Sitar et al., 2005). The most noticeable continuous mechanics methods are those based on finite element or finite difference numerical schemes. The analysis is based on stress rather than force as in the case of limit equilibrium methods. The slope material is continuous, deformable, and can be layered with nonlinear stress–strain relationships. There is no need for a predefined slip surface. These methods are determinate and provide a unique solution. The finite element or finite difference methods are versatile and powerful. They have been used to perform seepage analysis and to determine the state of stress and strain within a slope under gravity and various loading conditions. However, when factor of safety is close to 1, the continuous mechanics-based methods tend to become numerically unstable. The finite element or finite difference methods are thus not well suited for searching the critical slip surface and determining the factor of safety of a slope.

The distinct or discrete element method (DEM) (Cundall and Strack, 1979) that considers the slope material as an assembly of rigid particles (spherical or irregular) has been used to simulate the development of a landslide. The DEM method allows the slope stability analysis without a pre-assumed slip surface. The discontinuous deformation analysis (DDA) (Shi and Goodman, 1985) considers the slope as a collection of deformable blocks. With the consideration of stress equilibrium, the DDA method is determinate and yields a unique solution for a given boundary and initial conditions (Huang and Ma, 1992). The results of DEM or DDA depend on the selection of particle or block contact properties, which are difficult to determine.

Details of limit analysis have been described in Chapter 2. Limit analysis solutions are rigorous in the sense that the stress field associated with a lower-bound solution is in equilibrium with imposed loads at every point in the soil mass, while the velocity field associated with an upper-bound solution is compatible with imposed displacements. The best (highest) lower bound to the true collapse load can be found by analyzing various trial statically admissible stress fields. The best (lowest) upper-bound to the true collapse load can be found by examining various kinematically admissible velocity fields. To obtain optimum solutions (high-lower bounds and low-upper bounds) to slope stability problems, analyses will have to be done with many trial statically admissible stress fields and trial kinematically admissible velocity fields—a rather tedious task that is not possible to tackle without the help of a numerical scheme. Because of the difficulties in constructing statically admissible stress field manually, the application of limit analysis had been limited to the upper-bound method (Chen, 1975; Chen and Liu, 1990). Extending the work by Sloan (1988), Yu and Sloan (1991), and Yu et al. (1998) demonstrated that, by discretization of the soil mass using three-noded triangular finite elements and linear programming, the optimized upper-bound and lower-bound solutions can be readily obtained. The method, called finite element limit analysis, shows great potential because with the help of a finite element mesh, various soil, boundary, and seepage conditions can be considered (Kim et al., 1999), and the solution is rigorous and unique.

While these numerical methods have demonstrated significant potential, they are still in various stages of research and are not routinely used in engineering practice.

9.5 CONDITIONS FOR SLOPE STABILITY ANALYSIS

So far, the introduction to slope stability analysis has been concentrating on the methodology itself; variations of soil parameters in the analysis have been kept to a minimum for simplicity. In practice, we generally divide the analysis into total and effective stress analysis as follows:

Total stress (undrained) analysis—Based on undrained strength parameters $(c, \emptyset)$ for the determination of shear resistance, in the case of $\emptyset = 0$ analysis $(c = s_u,$ the undrained shear strength), use total soil unit weight $(\gamma$ or $\gamma_{sat})$ for the computation of normal contact forces and driving forces, and pore pressure is ignored.

Effective stress (drained) analysis—Based on drained shear strength $(c', \emptyset')$ for the determination of shear resistance, use buoyant unit weight (γ') for the computation of normal contact forces and driving forces below groundwater table or phreatic surface, and pore pressure is computed according to the piezometric conditions of the slope.

The choice between total and effective stress analysis depends on the conditions of the slope to be considered and the types of soils involved.

9.5.1 Stability analysis for built embankments

Figure 9.19 shows the time history for a case of constructing an embankment on a saturated clay foundation. The average shear stress along a given slip surface (Figure 9.19b) increases with the height of fill. The shear stress reaches its maximum when the embankment reaches its "end of construction," and height of fill reaches its final maximum level. Pore pressure within the saturated clay increases with the fill height. The pore pressure reaches its maximum at the end of construction. Upon completion of the embankment, the pore pressure dissipates and eventually reaches an equilibrium level that is controlled by the long-term groundwater table, as shown in Figure 9.19c. Concurrent with pore pressure dissipation, the saturated clay consolidates and gains strength. The dissipation of pore pressure also causes the effective normal stress, σ'_n, along the slip surface to increase. The permeability of clay is usually too low for full pore pressure dissipation during the stage of construction ("rapid construction" in Figure 9.19d). Since it is difficult to accurately estimate the increase of pore pressure during rapid construction, we typically use

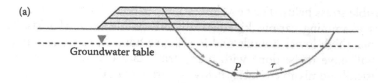

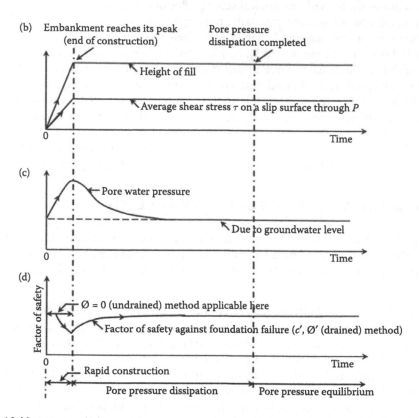

Figure 9.19 Variations of shear stress, pore pressure, and factor of safety with time for a built embankment on saturated clay. (a) Construction of an embankment on saturated clay. (b) Change of embankment height and average shear stress with time. (c) Variation of pore water pressure in clay with time. (d) Variation of FS against foundation failure with me. (Adapted from Bishop, A.W. and Bjerrum, L. 1960. The relevance of the triaxial test to the solution of stability problems. *Proceedings of the ASCE Research Conference on the Shear Strength of Cohesive Soils*, Boulder, CO, pp. 437–501.)

total stress analysis for this stage. The safety factor reaches its minimum value at the end of construction as the induced pore pressure and shear stress along the slip surface reach their respective maximum level. Because of these reasons, the "end of construction" is an important stability analysis that reflects the safety of an embankment at a critical stage. To determine the long-term stability of the embankment, effective stress analysis is conducted based on pore pressure values in equilibrium with the groundwater table.

In the case of an excavation in clay, as shown in Figure 9.20, the pore pressure drops initially due to unloading, as shown in Figure 9.20b. The pore pressure starts to increase after end of construction (i.e., excavation) and eventually reaches an equilibrium that is controlled by the long-term groundwater table. The safety factor decreases monotonically (see Figure 9.20c) in this case, and eventually reaches a stabilized value when the pore pressure stabilizes. In this case, the long-term, instead of end of construction, is the most critical stage.

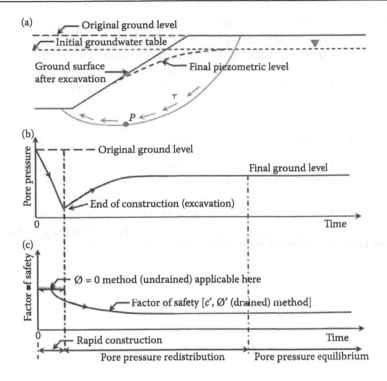

Figure 9.20 Variations of pore pressure and factor of safety with time for an excavation in saturated clay. (a) Excavation in clay. (b) Variation of pore water pressure in clay with time. (c) Variation of FS against foundation failure with time. (Adapted from Bishop, A.W. and Bjerrum, L. 1960. The relevance of the triaxial test to the solution of stability problems. *Proceedings of the ASCE Research Conference on the Shear Strength of Cohesive Soils*, Boulder, CO, pp. 437–501.)

If the materials involved are highly permeable, such as clean sand or gravel, then effective stress analysis can be used throughout. For intermediate materials such as silt or silty fine sand, the end of construction analysis is performed using both total stress and effective stress analysis. A range of safety factor values obtained is then considered.

In the case of an embankment dam, positive pore pressure can develop during construction. This is especially true as the zoned embankment dam core is usually made of clay compacted on the wet side of the optimum (water content higher than the optimum water content). The end of construction is again part of an important stability analysis. A few more relevant conditions that will have to be considered for an embankment dam are described below.

Long-term stability of an earth dam under steady state seepage—Upon completion of the earth dam and dissipation of the construction induced positive pore pressure, water is impounded in the reservoir and seepage develops through the embankment. The piezometric or phreatic surface (surface where the pore pressure is zero) associated with steady state seepage and various reservoir levels can affect the long term stability of the embankment dam. A series of drained stability analysis is performed to consider the various reservoir levels, their related seepage conditions, and positions of the phreatic surface.

Stability of an earth dam under rapid drawdown—a "rapid or sudden drawdown" can occur in a reservoir when the pool level drops quickly due to breakage of the dam or other forms of significant leakage. The drawdown occurs rapidly before the water within the slope has a chance to seep out of the slope and develop a phreatic surface that is in equilibrium with the lowered pool level. We take a conservative approach in the rapid drawdown stability analysis. The driving force is determined based on total soil unit weight. The shear strength is computed according to effective stress analysis using buoyant unit weight for soil located below the phreatic surface prior to rapid drawdown. As demonstrated in a simple case in Section 9.2, rapid drawdown can lower the safety factor significantly.

9.5.2 Stability analysis for natural slopes

End of construction is not an issue in this case. Effective stress analysis is usually applied for natural slopes. The groundwater table required in the analysis can be estimated, for example, on the basis of open-end piezometer observations (see Chapter 3 for details of open-end piezometers). For waterfront slopes, rapid drawdown analysis may also be required. Rainfall-induced landslide can be a major concern for natural slopes. The pore pressure induced by surface infiltration of rainfall water can be determined using seepage analysis. Rahardjo et al. (2007) reported such a series of seepage and slope stability analysis. Figure 9.21a

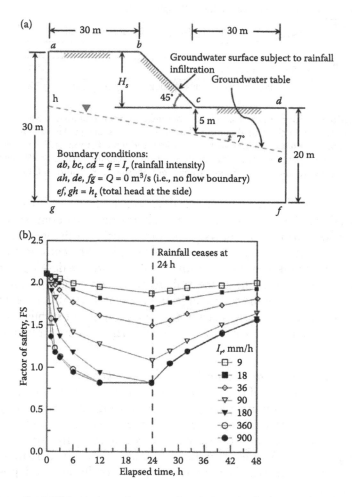

Figure 9.21 Effects of rainfall intensity and seepage duration on the factor of safety for a homogeneous fine grained sandy slope with saturated hydraulic conductivity $k = 10^{-2}$ cm/sec. (a) Cross section and initial groundwater conditions. (b) Variation of FS with time and rainfall intensity. (From Rahardjo, H. et al., 2007. *Journal of Geotechnical and Geoenvironmental Engineering*, 133, 12. Figure 5a, p. 1538. Reproduced with permission of the ASCE.)

shows the cross-section and initial groundwater conditions of a homogeneous fine-grained sandy slope in their studies. Soil immediately below the slope surface was unsaturated initially with negative pore pressure. The soil becomes saturated and develops positive pore pressure as rainwater infiltrates from the ground surface. This downward movement of seepage continues during the rainfall. A time history of the safety factor during and after the rainfall, considering different rainfall intensities (I_r), is shown in Figure 9.21b. The results show that the FS decreases during rainfall and slowly increases after the rainfall stops, and higher rainfall intensity causes lower FS.

EXAMPLE 9.8

Given

For the same slope and circular slip surface as shown in Example 9.4, but now with the phreatic surface above the slip surface. The cross-section of the slope and its slip surface with a radius of 20.1 m are shown in Figure E9.8.

$$\gamma = 19 \, kN/m^3$$
$$c' = 20 \, kPa$$
$$\emptyset' = 25°$$

The phreatic surface between the right end of slice 1 and left end of slice 10 is defined by the coordinates in Table E9.8.

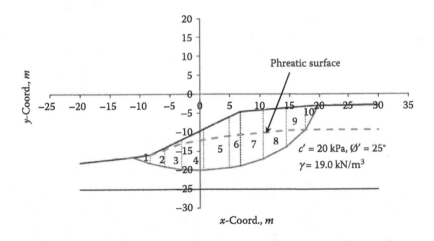

Figure E9.8 Profile of the slope with a phreatic surface and slices.

Table E9.8 Coordinates that define the phreatic surface

x, m	−11.25	−8.31	−5.99	−3.11	0.55	4.89	6.73	10.46	14.44	17.68
y, m	−16.53	−15.98	−14.80	−13.00	−11.80	−11.00	−10.50	−10.00	−9.50	−9.35

Required

Determine the safety factor for the slip surface using the simplified Bishop method.

Solution

The sliding block is divided into 10 slices as shown in Figure E9.8, and the slices are numbered from left to right. This example does not involve any surcharge load ($Q = 0$), and there is no seismic load ($k_h = k_v = 0$). Under these conditions, Equation 9.36 becomes

$$N' = \frac{1}{m_a}\left[W - \frac{\bar{C}\sin\alpha}{FS} - U_\alpha\cos\alpha\right]$$

Equation 9.37 remains the same:

$$m_a = \cos\alpha\left[1 + \frac{\tan\alpha\,\tan\bar{\varnothing}}{FS}\right]$$

Equation 9.38 becomes

$$FS = \frac{\sum_{i=1}^{n}(\bar{C} + N'\tan\bar{\varnothing})}{\sum_{i=1}^{n}A_4}$$

and

$$A_4 = W\sin\alpha$$

The Excel spreadsheet program was used to set up the coordinates of the slope profile and make the computations. A series of trial FS values were used and compared with the computed FS values. Table E9.8a summarizes the results of the computations when assumed FS = 1.57, after a few iterations. The average hydrostatic pressure at the base of each slice, u, is calculated by multiplying the average depth of water in that slice with the unit weight of water. The average depth of water is

Table E9.8a Simplified Bishop method computations with a phreatic surface

Slice no.	b (m)	h (m)	W (kN)	α, radian	W sin α (kN)	Δl (m)	C = c'Δl (kN)	m_a	u (kPa)	U_α (kN)	N' (kN)	N' tan Ø̄ (kN)
1	2.94	1.10	61.66	−0.51	−30.15	3.37	67.41	0.73	10.85	36.57	69.82	32.56
2	2.32	3.56	157.14	−0.37	−56.18	2.48	49.68	0.83	31.85	79.13	114.24	53.27
3	2.88	6.38	349.07	−0.23	−79.41	2.96	59.15	0.91	54.14	160.14	222.59	103.80
4	3.66	9.23	642.17	−0.06	−41.10	3.67	73.35	0.98	73.33	268.92	384.90	179.48
5	4.34	12.06	995.56	0.14	135.50	4.38	87.70	1.03	81.35	356.69	615.44	286.99
6	1.84	13.88	484.54	0.29	140.82	1.92	38.40	1.04	82.04	157.50	313.22	146.06
7	3.73	13.65	967.70	0.44	415.92	4.13	82.62	1.03	75.44	311.63	644.02	300.31
8	3.98	11.61	877.69	0.67	546.36	5.08	101.66	0.97	55.84	283.85	635.90	296.53
9	3.24	8.16	502.15	0.93	403.18	5.43	108.64	0.83	21.28	115.60	452.54	211.02
10	2.02	3.07	117.93	1.21	110.20	5.68	113.60	0.63	0.00	0.00	79.41	37.03
Sum					1545.14		782.20					1647.04

b = width of slice.
Δl = base length of slice.
u = average hydrostatic pressure at the base of each slice.
$U_\alpha = u\Delta l$ for each slice.

the vertical distance between the midpoint of the phreatic surface and that of the base of the slice.

$$\sum_{i=1}^{n} (\bar{C} + N' \tan \bar{\varnothing}) = (782.20) + (1647.04) = 2429.24 \, \text{kN}$$

and

$$\sum_{i=1}^{n} A_4 = \sum_{i=1}^{n} W \sin \alpha = 1545.14 \, \text{kN}$$

$$FS = \frac{\sum_{i=1}^{n} (\bar{C} + N' \tan \bar{\varnothing})}{\sum_{i=1}^{n} A_4} = \frac{(2429.24)}{(1545.14)} = 1.57$$

The computed FS after iterations is essentially the same as the assumed value.

9.5.3 Consideration of seismic effects

Seismic ground motions such as those induced by earthquakes can weaken the slope material, generate positive pore pressure (for certain types of soils below groundwater table), and impose inertial forces on the sliding block. These effects tend to lower the factor of safety against slope failure. Many methods have been developed to consider the effects of earthquake ground motions in the slope stability analysis. These methods include:

Pseudostatic method—The effects of cyclic motions are simulated by inclusion of a static horizontal and vertical force in the limit equilibrium analysis.

Newmark's displacement method (Newmark, 1965)—The method considers that permanent displacement of the sliding block occurs and accumulates when ground acceleration exceeds the yield (or threshold) acceleration for the material along the slip surface.

Finite element analysis—The cyclic stress–strain and pore pressure response of the slope material are computed in a coupled numerical model. The effects of soil cyclic strain softening and generation of positive excess pore pressure can be considered in this method (Finn, 1988).

Among the three methods mentioned above, pseudostatic method is the simplest and is often used in practice. The other two methods are more involved in numerical procedures and beyond the scope of this book. For these reasons, the following discussion will concentrate on the pseudostatic approach and its implementation in the method of slices only. In the pseudostatic method, the back-and-forth loading induced by an earthquake is simulated by static horizontal and vertical forces acting in directions that are least favorable to the stability of the slope. Consider a typical slice shown in Figure 9.13; these pseudostatic forces are $k_h W$ (pointing away from slope) and $k_v W$ (pointing upward), respectively, where W = weight of the slice, k_h = horizontal seismic coefficient, and k_v = vertical seismic coefficient. Both k_h and k_v are dimensionless. The selection of k_h and k_v considers intensity and other characteristics of the earthquakes such as frequency content and duration of the earthquake. This is an extremely conservative simplification. In most cases, we only consider k_h and usually ranges from 0.05 to 0.25.

EXAMPLE 9.9

Given

For the same slope and circular slip surface as shown in Example 9.8, but now add the effects of earthquake motions. The cross-section of the slope and its slip surface are shown in Figure E9.8. The circular slip surface is the same as that used in Example 9.1 where the radius of the slip surface $R = 20.1$ m.

$\gamma = 19.0 \, \text{kN/m}^3$
$c' = 20 \, \text{kPa}$
$\emptyset' = 25°$
$k_h = 0.1$

Required

Determine the safety factor for the slip surface using the simplified Bishop method.

Solution

The sliding block is divided into 10 slices as shown in Figure E9.8 and the slices are numbered from left to right.

This example does not involve any surcharge load $(Q = 0)$

$k_h = 0.1$ and $k_v = 0$. Under these conditions, Equation 9.36 becomes

$$N' = \frac{1}{m_a}\left[W - \frac{\bar{C}\sin\alpha}{FS} - U_a\cos\alpha\right]$$

Equation 9.37 remains the same,

$$m_a = \cos\alpha\left[1 + \frac{\tan\alpha\tan\emptyset}{FS}\right]$$

where FS is an assumed factor of safety. The factor of safety for the assumed slip surface is computed according to Equation 9.38:

$$FS = \frac{\sum_{i=1}^{n}(\bar{C} + N'\tan\emptyset)}{\sum_{i=1}^{n}A_4 + \sum_{i=1}^{n}A_6}$$

$$A_4 = W\sin\alpha$$

$$A_6 = k_h W\left(\cos\alpha - \frac{h_c}{R}\right)$$

and $A_5 = 0$.

The Excel spreadsheet program was used for the computations. A series of trial FS values were used and compared with the computed FS values. Table E9.9 summarizes the results of the computations when assumed FS = 1.21, after a few iterations.

Table E9.9 Simplified Bishop method computations with a phreatic surface and $k_h = 0.1$

Slice no.	W (kN)	α, radian	W sin α (kN)	h_c (m)	A_6	Δl (m)	C= $c'\Delta l$ (kN)	m_a	u (kPa)	U_α (kN)	N' (kN)	N' tan Ø (kN)
1	64.73	−0.51	−31.50	0.58	5.46	3.37	67.41	0.68	33.01	111.10	−7.21	−3.36
2	161.87	−0.36	−57.58	1.84	13.64	2.48	49.68	0.80	47.52	117.97	83.38	38.88
3	354.71	−0.23	−80.30	3.24	28.82	2.96	59.15	0.89	69.87	206.59	185.83	86.65
4	649.17	−0.06	−41.34	4.67	49.71	3.67	73.35	0.97	84.12	308.52	354.64	165.37
5	1003.95	0.14	135.96	6.08	69.08	4.38	87.70	1.04	88.72	388.98	583.55	272.11
6	488.20	0.29	141.18	6.99	29.73	1.92	38.40	1.07	87.97	168.81	297.01	138.50
7	975.62	0.44	417.24	6.88	54.68	4.13	82.62	1.07	81.44	336.04	601.61	280.54
8	887.57	0.67	549.76	5.87	43.54	5.08	101.66	1.02	60.29	305.48	583.07	271.89
9	513.10	0.93	409.92	4.17	19.94	5.43	108.64	0.91	23.74	127.80	402.90	187.87
10	132.36	1.19	123.07	1.72	3.58	5.68	113.60	0.72	0.00	0	62.32	29.06
Sum			1545.14			318.18	782.20					1467.51

b and h are the same as those in Example 9.8 and Table E9.8.

$$A_6 = k_h W\left(\cos\alpha - \frac{h_c}{R}\right)$$

h_c = height to centroid of slice (taken as half of the slice height)
R = radius of slip surface
Δl = base length of slice
u = average hydrostatic pressure at the base of each slice
$U_\alpha = u\Delta l$ for each slice

$$\sum_{i=1}^{n}(\bar{C} + N'\tan\bar{Ø}) = 782.20 + 1467.51 = 2249.72\,\text{kN}$$

and

$$\sum_{i=1}^{n} A_4 = \sum_{i-1}^{n} W\sin\alpha = 1545.14\,\text{kN}$$

$$\sum_{i=1}^{n} A_6 = \sum_{i=1}^{n} k_h W\left(\cos\alpha - \frac{h_c}{R}\right) = 318.18\,\text{kN}$$

$$FS = \frac{\sum_{i=1}^{n}(\bar{C} + N'\tan\bar{Ø})}{\sum_{i=1}^{n} A_4 + \sum_{i=1}^{n} A_6} = \frac{(2249.72)}{(1545.14 + 318.18)} = 1.21$$

The computed FS after iterations is essentially the same as the assumed value.

9.5.4 Consideration of progressive failure

The limit equilibrium analysis is based on the assumption that the safety factor is constant along the assumed slip surface, and the analysis does not consider material deformation. If the computed value of FS is greater than 1, it is implied that there is no failure anywhere along the slip surface. This approach is valid if all materials along the slip surface are strain hardening, or their strength increases monotonically with strain or deformation. Certain geotechnical materials such as overconsolidated clays or fractured rock mass can have strain softening or strain weakening behavior. The shear strength can decrease significantly after reaching a peak, and eventually stabilizes at a residual strength as the strain continues to increase. A phenomenon called progressive failure can develop if the materials along parts of the slip surface are strain softening. The strain softening parts of the slip surface progressively lose their shear resistance while the rest of the slip surface develops higher resistance as the slope undergoes deformation. The collapse of Carsington Dam in Derbyshire, England (Chen et al., 1992; Skempton and Vaughan, 1993) during construction in 1984 is a case of progressive failure. Extensive studies attributed the failure to softening of the natural, underlying foundation and the dam core materials as new fill was placed and the dam was raised to higher elevations. Traditional limit equilibrium analysis cannot be used alone to obtain a rational solution for progressive failure problems because the deformation of the structure must be taken into account in the analysis. The limit equilibrium analysis conducted for the Carsington Dam indicated that the factors of safety were over 1.4 using peak strength. Factors of safety were less than unity if residual strengths were used. The extent and degree of weakening along the potential slip surface were calculated using stress-based finite element analysis with strain-weakening models. The calculated shear strength was then used in the limit equilibrium analysis, and the safety factor was found to be close to the actual value of 1.0.

9.6 SLOPE STABILITY MITIGATION

For a slope that is deemed unstable from stability analysis and/or previous observations, to mitigate the situation, we often take the following measures into consideration:

- Monitor the conditions of the slope. The purpose of monitoring may include: (a) determining the reasons for the instability of the slope and/or its scope, (b) verifying the effectiveness of schemes to stabilize the slope, and (c) providing warning of an imminent slope failure.
- Undertake engineering measures to enhance the stability of the slope.

The following sections briefly describe some of the commonly used techniques in slope stability monitoring and stabilization.

9.6.1 Field slope monitoring

Many methods have been developed as means to provide warning against potential slope failure, or to determine the extent of a landslide. The following sections describe some of the available monitoring techniques.

9.6.2 Surface observation

The stability of a slope can be observed from ground surface. Many methods are available for this purpose, which can include simple visual observation of cracks in the ground, structural walls, or other topographic characteristics visible from the ground surface and their variations with time. Video cameras either installed on ground surface or carried by an unmanned aerial vehicle (UAV) have been used effectively to conduct remote observations of a slope area.

Many quantitative methods have been developed in the past few decades. Extensometers with reference posts installed at the opposite sides of a slip surface (Figure 9.22) have been used to monitor the slope movement. Movement of the slope causes the string attached between the two posts to be stretched. Measurement of the string extension reveals the amount of slope movement. Ground movement inevitably involves some rotation of the

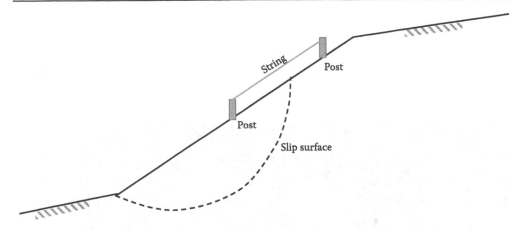

Figure 9.22 Slope monitoring with a surface extensometer.

surface material. Taking advantage of this phenomenon, electronic tilt sensors planted on the slope surface have been used as an option for slope stability monitoring. GPS (global positioning system), total stations, and other optical surveying tools have been used to keep track of slope surface movement at selected locations in a monitored area.

Remote-sensing techniques such as LIDAR (LIght Detection And Ranging), aerial, and satellite imaging have been used to identify the existence or extent of large-scale slope movement. With the help of digital filtering, a digital terrain model (DTM) that removes the effects of surface vegetation coverage (i.e., trees and grass) and reveals the topography of the bare ground surface can be established from LIDAR surveys. Figure 9.23 shows a SPOT-5 (SPOT is acronym of the French "satellite pour l'observation de la terre") image of the Daguangbao landslide region after the 2008 magnitude 7.9 Wenchuan earthquake in Sichuan province of western China. The Daguangbao landslide is considered one of the largest earthquake-triggered landslides in the world in the past century. Images obtained from Google map can often be valuable in mapping out the area affected by a potential landslide.

9.6.3 Ground displacement profile monitoring

The system described in Figure 9.24 that consists of a casing (inclinometer casing) installed in a near-vertical direction in the ground and a sensor probe (inclinometer probe, IP) that measures the inclination of the casing is probably the most widely used technique in the detection of lateral ground movements below ground surface. The IP, approximately 0.5 m in length, is basically a tilt sensor that measures the inclination of the casing (Mikkelsen, 1996) at different depths. The inclinometer casing, made of plastic or aluminum, with its outside diameter typically less than 100 mm, is inserted in a borehole. The space between the casing and drill-hole wall is backfilled with grout or pea gravel. The inclinometer casing has four grooves that separate in 90° as shown in Figure 9.24. The IP has spring-loaded guide wheels to occupy two of the four casing grooves and to assure its center position inside the casing as the IP is moved up and down the inclinometer casing. An electric cable is used to raise and lower the IP sensor unit in the casing and transmits electric signals to the ground surface. The IP system monitors the casing response to the lateral ground movement. The sensor unit measures its inclination angle ($\delta\theta$) with respect to the true verticality. Distance between successive readings is usually kept at a distance, L. Assuming a fixed casing bottom, the accumulated lateral displacement, δ_L, is then

$$\delta_L = \sum L \sin \delta\theta \qquad (9.59)$$

Figure 9.23 Post-Wenchuan earthquake SPOT-5 image of the Daguangbao landslide. (Courtesy of Prof. Jia-Jyun Dong, National Central University, Jhongli, Taiwan.)

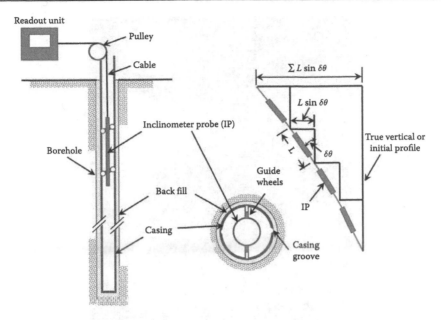

Figure 9.24 The inclinometer-sensing system. (Adapted from Mikkelsen, P.E., 1996. Chapter 11, Field instrumentation, in landslides, investigation and mitigation. In: Turner, A.K. and Schuster, R.L., eds., *TRB Special Report 247*, Transportation Research Board, National Research Council, Washington, DC, pp. 278–316.)

With the help of handheld electronic devices (e.g., tablet computer or smart phone) as shown in Figure 9.25, the modern IP control/readout unit is highly portable and easy to operate in the field. In-place inclinometers (IPI) have been developed for long-term automated ground displacement monitoring. In this case, multiple units of IPIs are connected to a string and inserted into the inclinometer casing on a long-term basis. Readings can be taken and transmitted to the office automatically once the IPI string is inserted in the ground.

9.6.4 Rainfall monitoring

Rainfall water on slope surface can infiltrate into the ground, saturate the surface soils, and cause the pore water pressure to increase. These factors have negative impact on slope stability; therefore rainfall measurement can be an important part of slope stability monitoring. Figure 9.26 shows a tipping bucket rain gauge that is commonly used for rainfall monitoring. The rain gauge consists of a funnel that collects and channels the precipitation into a small container. After a preset amount of precipitation falls, the lever tips, dumping the collected water and sending an electrical signal. The rain gauge can be used to keep track of the total accumulated rainfall as well as rainfall intensity (i.e., the amount of rainfall per unit time period).

9.6.5 Groundwater/pore pressure monitoring

Pore pressure affects the effective stress in soil mass, which in turn controls the strength and thus stability of the slope. Under hydrostatic conditions, the pore pressure at a given location can be calculated from its depth below the groundwater table. The Casagrande type open-end standpipe piezometer (Figure 3.20) described in Chapter 3 can be used for

Figure 9.25 Taking IP readings in the field. (Courtesy of DECL, Taipei, Taiwan.)

Figure 9.26 Tipping bucket rain gauge. (Courtesy of DECL, Taipei, Taiwan.)

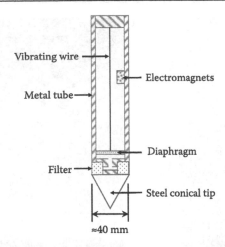

Figure 9.27 Vibrating wire piezometer.

long-term groundwater monitoring. In this setup, groundwater enters the standpipe through a porous element (the open end). A tape dropped into the piezometer is used to measure the water level in the pipe. A time lag may exist between the water level within the pipe and the surrounding groundwater.

In transient seepage conditions, such as surface water infiltration during a heavy rainfall, the pore pressure distribution may deviate significantly from hydrostatic and vary with time. In this case, a diaphragm type of piezometer such as the one shown in Figure 9.27 is preferred to measure the pore water pressure. The amount of diaphragm deflection induced by water pressure is sensed by a strain gauge (e.g., vibrating wire strain gauge). In this design the pore pressure readings can be easily recorded automatically, and time lag is minimal.

Huang et al. (2012) introduced an optical fiber Bragg grating (FBG) sensored piezometer. This is also a diaphragm piezometer (Figure 9.28a). Taking advantage of the partially distributive nature of FBG, multiple FBG piezometers can be cascaded into an array (Figure 9.28b) and installed in a single borehole for pore water pressure profile monitoring.

9.6.6 Automation in slope monitoring

Essentially all electronic and optical fiber sensors mentioned above can be connected to an automated data logging/transmission system. A fully automated monitoring system can consist of

- Electrical/fiber optic sensors—in-place inclinometers, rain gauge, and piezometers
- Power system (battery + solar panel + uninterruptable power supply)
- Automated data analysis system (ADAS)
  - Multiplexer—controls switching among the installed sensors
  - Analog/digital conversion—converts analog signal into digital signal if necessary
- Data transmission system—transmits data via communication system (e.g., internet, GSM, WiFi, etc.)
- Data storage—memory device to store digital data

Figure 9.29 shows a typical setup of automated field slope monitoring system.

(a)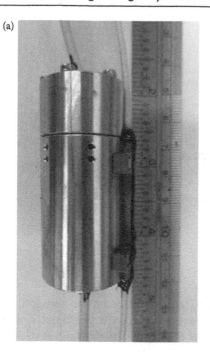
(b)

Figure 9.28 Use of fiber optic piezometers for pore pressure profile monitoring. (a) The fiber optic piezometer. (b) The piezometer array. (Courtesy of An-Bin Huang, Hsinchu, Taiwan.)

9.6.7 Warning of slope failure

The result of a slope failure can be catastrophic and may involve economic as well as human losses. The analysis methods described in this chapter can help us identify areas that may be affected by slope failures if they do occur. An ultimate solution to mitigate

Figure 9.29 Solar-powered automated field data logging. (Courtesy of DECL, Taipei, Taiwan.)

the situation would be to avoid human activities in the hazardous areas altogether, before the slope failure actually occurs. This can include relocating villages and rerouting infrastructures. These options are often not practical, especially in densely populated and/or highly developed areas. A warning system that can offer ample time for reaction (e.g., temporarily relocation of local residents and blockage of highways) in a case of an imminent slope failure would be highly desirable, especially if economically feasible. Many warning methods have been developed against potential failure of natural slopes, a few of which are described in the following sections.

9.6.8 Warning based on ground displacement

This warning method is based on the premises that ground displacement accumulates and goes through initial, steady, and tertiary stages, as shown in Figure 9.30, before the failure occurs. The amount of ground displacement in the tertiary stage increases rapidly just prior to a slope failure. The quantitative ground displacement monitoring techniques described above can all be considered in this warning method as a source of displacement measurement. To carry out this method, periodic displacement readings in reference to an initial state are taken. The results are plotted against time in order to evaluate the displacement trend with time. A predetermined threshold displacement value is used as a basis for the warning of a slope failure.

As time rate may be more indicative than the absolute displacement value of an imminent slope failure, methods have also been proposed to issue slope failure warning based on rate of ground displacement. In this case, the derivatives of displacement measurements with time are plotted against time. A predetermined threshold displacement rate is used as a basis for the warning of a slope failure.

The displacement-based warning method is useful in dealing with a slowly developing slope failure where long-term readings are collected and evaluated. Unless readings in short-time intervals can be taken remotely and automatically, this method is not practical in the case of a rainfall-induced slope failure. Depending on the displacement measurement methods involved, the cost of this warning method can range from low to very high.

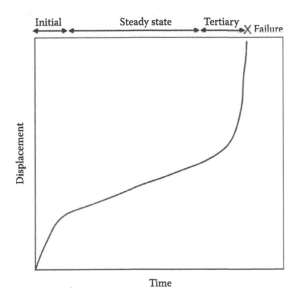

Figure 9.30 Warning of slope failure based on ground displacement. (After Liao, J.T. et al., 2013. *Sino-Geotechnics*, 136, 59–70 [in Chinese].)

9.6.9 Warning based on rainfall duration and intensity

Keefer et al. (1987) reported a warning method based on combinations of rainfall duration and intensity. A threshold curve of rainfall duration versus rainfall intensity is empirically developed based on previous rainfall and landslide records for the target area. Figure 9.31 shows such a threshold curve in concept. If the characteristics of the rainfall are such that the plot of duration versus intensity falls to the upper right side of the threshold, a slope failure is then deemed imminent.

The rain gauge is the key element involved in this method for collecting data in the development of the threshold curve and providing real-time readings in a rainfall event. A rain gauge and its automation are relatively low-cost in comparison with other monitoring methods. Warning can be issued based on weather reports and prediction of a rainfall event. In this case, the time for undertaking precautionary measures can be sufficient. Alternatively, a warning can be issued based on real-time rain gauge readings during a rainfall event. Because of its versatility and relatively low cost, the rainfall-based scheme is probably the most widely applied method in the world against rainfall-induced slope failure. The rainfall-based method, however, lacks the consideration of antecedent soil moisture conditions. The same rainfall conditions can impose different effects on the slope stability due to differences in the pre-existing hydrogeological conditions (e.g., groundwater level) before the rainfall. Unless the potential drawbacks are properly considered, the rainfall-based method can be misleading.

9.6.10 Warning based on pore pressure measurements

Pore-water pressure is probably the most indicative of slope instability in its early stage, among the viable physical quantities that can be monitored in the field. The groundwater table in a slope is usually significantly below the slope surface. The soil near the ground surface is likely to be unsaturated, with negative pore pressure. Monitoring of positive and negative pore pressures using piezometers and tensiometers, respectively, has been reported for slopes in different parts of the world (Johnson and Sitar, 1990; Fannin and Jaakkola, 1999; Ng et al., 2008). Fannin and Jaakkola (1999) reported from their experience that the field pore pressure measurements rarely showed a linear distribution with

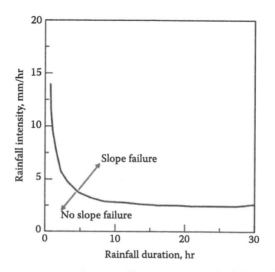

Figure 9.31 Conceptual description of a threshold curve for rainfall-induced slope failure.

depth. This is consistent with the transient seepage analysis shown in Figure 9.11. As rainfall infiltrates from the slope surface downward, the soil near the ground surface becomes saturated first, and pore pressure changes from negative to zero. As shown in Figure 9.11, this surficial near-zero pore pressure gradually merges with the original negative pore pressure with depth in the unsaturated zone until the groundwater table is reached where the soil then becomes saturated and pore pressure positive.

Assume the slope is uniform and infinitely long with a slope angle of β and the slope fails along a planar slip surface. Considering the equilibrium of gravity, available soil resistance, and pore pressure imposing on the slice, a relationship among the critical depth for infinite slope failure (d_{cr}), soil strength parameters (with strength parameters c' and $\emptyset'$), and pore-water pressure (u) can be established according to Equation 9.30 as

$$d_{cr} = \frac{c' - u \cdot \tan \emptyset'}{\gamma \cos^2 \beta (\tan \beta - \tan \emptyset')} \tag{9.30}$$

where γ is the total unit weight of the slope soil. The $d_{cr} - u$ correlation can be used as a reference for a stress-based warning system for rainfall-induced slope failure. This equation ignores the effects of negative pore pressure and therefore is valid only in cases where slip surface occurs in saturated soil with positive pore pressure.

Huang et al. (2012) reported their deployment of optical fiber-based piezometers for pore pressure profile monitoring at a landslide research site in southern Taiwan. Ten piezometers at 5 m intervals were installed in a single, 60 m deep borehole with automated data logging. The pore pressure readings along with the rain gauge readings during Typhoon Morakot in August, 2009 are shown in Figure 9.32. The monitored slope had an average slope angle (β) of 22° and $\gamma \approx 18$ kN/m³. Within a period of 4 days, the monitored area received an accumulated rainfall of nearly 2500 mm (Figure 9.32b). Pore pressure readings peaked at 11:05 on August 9, as shown in Figure 9.32a, with the maximum pore pressure reached 430 kPa. According to this set of pore pressure measurements and Equation 9.30, the monitored slope (consisted of fracture rock with $c' = 0$) would fail if its $\emptyset'$ value is less than 40°.

The pore pressure- or stress-based warning method has a rigorous theoretical basis. The need for empiricism or judgement in the interpretation of available data is minimized.

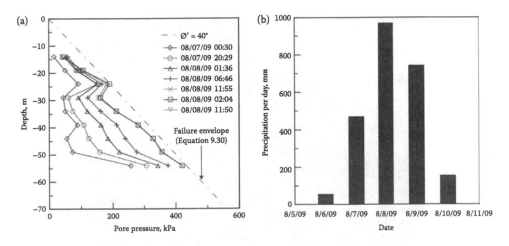

Figure 9.32 Pore pressure profile and rain gauge readings during Typhoon Morakot. (a) Variation of pore pressure profiles. (b) Rainfall records. (Adapted from Huang, A.B. et al., 2012. *Soils and Foundations*, 52(4), 737–747.)

The pore pressure-based method is thus expected to be more reliable than the other two methods described above. The cost of installing the piezometer array and automated data logging system, however, is relatively high.

9.6.11 Improving the stability of a slope

If a slope that has a marginal safety factor and a collapse can cause serious consequences, it may be necessary to improve the stability of the slope by engineering means. A number of methods have been developed and used for improving the slope stability. The methods can generally be divided into two categories: (1) increase the resisting force against slope failure, or (2) reduce the driving force that tends to destabilize the slope. In many cases, methods involved in both categories are applied to a given slope.

Widely used methods of slope stabilization include (a) slope drainage (surface and/or subsurface), (b) modification of the slope geometry by cut-and-fill operations, (c) use of restraining structures (such as retaining walls, anchors, and piles), and (d) use of geosynthetically reinforced earth structures. The choice of method depends on the type of slope, local geology, potential failure mechanisms, local expertise, and the performance required after improvement. Figure 9.33 shows the stabilization for a slope at 85 km (from the north end of the expressway) on Taiwan National Expressway No. 3. Reinforced concrete frame coupled with earth anchors was applied to the slope surface. The original toe of the slope was cut to make room for the expressway, thus making the slope less stable. A series of tangent piles (piles touching each other) were installed at the toe of the slope and tied together with a pile cap to increase stability. Earth anchors were also applied to the tangent piles. Fiber optic sensors were deployed to monitor lateral displacement of the slope and piles, and pore pressure distribution within the slope.

9.7 REMARKS

We have discussed the use of limit equilibrium method in bearing capacity, lateral earth pressure, and slope stability analysis, some of the most important elements in foundation engineering analysis. The limit equilibrium method is well suited for limit state analysis

Figure 9.33 Stabilization of a slope at 85 km on Taiwan National Expressway No. 3. (Courtesy of An-Bin Huang, Hsinchu, Taiwan.)

where our major concern is the factor of safety against the ultimate failure conditions, such as the slope stability analysis described in this chapter. Unfortunately, when applied to slope stability analysis, the system becomes indeterminate, except for very restricted conditions. The various methods of slices basically circle around the issue of how to derive a unique solution by matching the number of unknowns and equations. With the help of commercially available software, the methods of slices are widely used in engineering practice. Because of the differences in the simplification involved in the methods of slices, the selection of method can result in variations in the factor of safety. How the critical slip surfaces were searched can also affect the obtained minimum factor of safety. These are some of the inherent disadvantages of limit equilibrium methods that the reader should be aware of when applying the methods of slices.

HOMEWORK

9.1. For the slope shown in Figure H9.1. The slope surface can be defined by the coordinates shown in the table below. The center of the circular arc is at coordinate (0, 0) with a radius of 20 m. The soil unit weight, $\gamma = 19$ kN/m^3. Determine the arc length of the circular slip surface, L_a, weight of the sliding block, W, and distance between the center of the circular slip surface and the line of action of W.

x, m	y, m
−20.0	−16.0
−12.0	−16.0
19.8	−3.0
30.0	−3.0

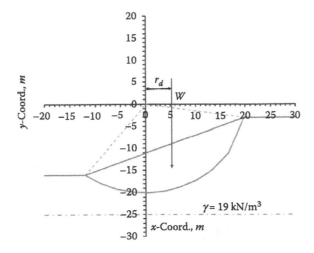

Figure H9.1 Slope of cohesive soil with a circular slip surface.

9.2. Consider 1 m thick sliding block in the direction perpendicular to the paper. For the circular slip surface shown in Figure H9.1, $s_u = 60$ kPa ($\varnothing = 0$). Using the necessary parameters derived in Problem 9.1, determine the FS of the circular slip surface using the circular arc method.

9.3. For the same slope and circular slip surface as shown in Problem 9.1. Consider 1 m thick sliding block in the direction perpendicular to the paper. For the circular slip surface shown in Figure H9.1, the material above the slip surface has $\gamma = 19 \, kN/m^3$, $c = 20 \, kPa$, $\emptyset = 25°$. Using the necessary parameters derived in Problem 9.1, determine the FS using the friction circle method.

9.4. For the same slip surface and soil conditions in Problem 9.3. Repeat Problem 9.3 using the ordinary method of slices. For the analysis, divide the sliding block into 10 even slices as shown in Figure H9.4.

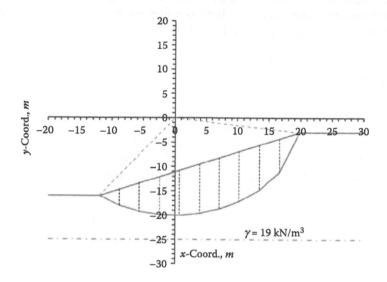

Figure H9.4 The circular sliding block and slices for the analysis.

9.5. For the same slope and circular slip surface as shown in Problem 9.4. $\gamma = 19 \, kN/m^3$, $c = 20 \, kPa$ and $\emptyset' = 25°$. Determine the safety factor for the slip surface using the simplified Bishop method.

9.6. For the same slope in Problem 9.4 but consider a non-circular slip surface. The cross-section of the slope and its non-circular slip surface are shown in Figure H9.6. $\gamma = 19 \, kN/m^3$, $c' = 20 \, kPa$, and $\emptyset' = 25°$. Coordinates at the top and bottom of each slice wall are given below.

No.	Upper end		Lower end	
	x, m	y, m	x, m	y, m
1	−12.00	−16.0	−12.00	−16.0
2	−8.82	−14.7	−8.82	−17.5
3	−5.65	−13.4	−5.65	−18
4	−2.47	−12.1	−2.47	−18
5	0.71	−10.8	0.71	−17.5
6	3.89	−9.5	3.89	−16.8
7	7.06	−8.2	7.06	−15.6
8	10.24	−6.9	10.24	−14.5
9	13.42	−5.6	13.42	−13
10	16.60	−4.3	16.60	−10.5
11	19.77	−3.0	19.77	−3.0

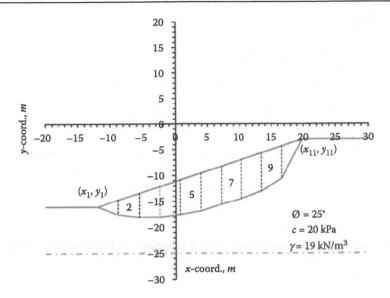

Figure H9.6 Profile of the slope, non-circular slip surface, and the slices.

9.7. For the same slope and circular slip surface as shown in Example 9.4, but now with the phreatic surface above the slip surface. The cross-section of the slope and its phreatic surface are shown in Figure H9.7. $\gamma = 19\,\text{kN/m}^3$, $c' = 20\,\text{kPa}$, and $\emptyset' = 25°$. Determine the factor of safety for the slip surface using the simplified Bishop method. Estimate the coordinates of the phreatic surface.

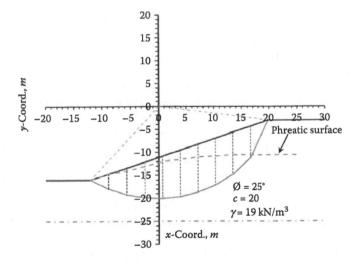

Figure H9.7 The slope profile and its phreatic surface.

9.8. For the same slope and circular slip surface as shown in Problem 9.7, but now add the effects of earthquake motions. The cross-section of the slope and its slip surface are shown in Figure H9.7. Use $k_h = 0.1$. Determine the safety factor for the slip surface using the simplified Bishop method. Estimate the coordinates of the phreatic surface.

REFERENCES

Abramson, L.W., Lee, T.S., Sharma, S., and Boyce, G.M. 1996. *Slope Stability and Stabilization Methods*. Wiley, New York, 629p.

Bishop, A.W. 1955. The use of the slip circle in the stability analysis of slopes. *Geotechnique*, 5, 7–17.

Bishop, A.W. and Bjerrum, L. 1960. The relevance of the triaxial test to the solution of stability problems. *Proceedings of the ASCE Research Conference on the Shear Strength of Cohesive Soils*, Boulder, CO, pp. 437–501.

Chen, W.F. 1975. *Limit Analysis and Soil Plasticity*. Elsevier, Amsterdam, 638pp.

Chen, R.H. and Chameau, J.L. 1982. Three-dimensional slope stability analysis. *Proceedings 4th International Conference on Numerical Methods in Geomechanics*, Edmonton, Vol. 2, pp. 671–677.

Chen, W.F. and Liu, X.L. 1990. *Limit Analysis in Soil Mechanics*. Elsevier Science Publishers, BV, Amsterdam, The Netherlands.

Chen, Z., Morgenstern, N.R., and Chan, D.H. 1992. Progressive failure of the Carsington Dam: A numerical study. *Canadian Geotechnical Journal*, 29(6), 971–988.

Chugh, A.K. 1986. Variable interslice force inclination in slope stability analysis. *Soils and Foundations*, 26(1), 115–121.

Collins, B.D. and Znidarcic, D. 2004. Stability analyses of rainfall induced landslides. *Journal of Geotechnical and Geoenvironmental Engineering*, 130(4), 362–372.

Cruden, D.M. and Varnes, D.J. 1996. Landslide types and processes. In: *Landslides, Investigation and Mitigation*. Special Report 247. Transportation Research Board, National Research Council. National Academy Press, Washington, DC.

Cundall, P.A. and Strack, O.D.L. 1979. A discrete numerical model for granular assemblies. *Géotechnique*. 29(1), 47–65. doi: 10.1680/geot.1979.29.1.47.

Duncan, J.M. 1992. State-of-the-art: Static stability and deformation analysis. Invited lecture. *Stability and Performance of Slopes and Embankments II*. Geotechnical Special Publication No. 31, ASCE, Vol. 1, pp. 222–266.

Fannin, R.J. and Jaakkola, J. 1999. Hydrological response of hillslope soils above a debris-slide headscarp. *Canadian Geotechnical Journal*, 36, 1111–1122.

Fellenius, W. 1927. *Erdstatische Berechnungen Mit Reibung and Kohaesion*. Ernst, Berlin.

Fellenius, W. 1936. Calculation of stability of earth dams. *Transactions, 2nd Congress Large Dams*, Vol. 4, 445.

Finn, W.D.L. 1988. Dynamic analysis in geotechnical engineering. In: Von Thun, J.L., ed., *Proceedings of Earthquake Engineering and Soil Dynamics II—Recent Advances in Ground Motion Evaluation*, ASCE, Park City, UT. Geotechnical Special Publication No. 20.

Fredlund, D.G., Krahn, J., and Pufahl, D.E. 1981. The relationship between limit equilibrium slope stability methods. *Proceedings of the 10th International Conference on Soil Mechanics and Foundation Engineering*, Stockholm, A.A. Balkema, Vol. 3, pp. 409–416.

GEO-SLOPE International Ltd. 1994. Computer program SEEP/W-for finite element seepage analysis. User's guide, Version 3, Calgary, Alta, Canada.

Huang, A.B., Lee, J.T., Ho, Y.T., Chiu, Y.F., and Cheng, S.Y. 2012. Stability monitoring of rainfall induced deep landslides through pore pressure profile measurements. *Soils and Foundations*, 52(4), 737–747.

Huang, A.B. and Ma, M.Y. 1992. Discontinuous deformation slope stability analysis. *Proceedings, Conference on Stability and Performance of Slopes and Embankments-II*, Berkeley, CA.

Janbu, N. 1954. Application of composite slip surfaces for stability analysis. *European Conference on Stability of Earth Slopes*, Stockholm, Sweden.

Janbu, N. 1957. Earth pressure and bearing capacity calculations by generalized procedure of slices. *Proceedings, 4th International Conference on Soil Mechanical Foundation Engineering*. Vol. 2, pp. 207–212.

Janbu, N. 1973. Soil stability computations. In: Hirschfeld, R.C. and Poulos, S.J., eds., *Embankment Dam Engineering, Casagrande Volume*. Wiley, New York, 47–87.

Johnson, K.A. and Sitar, N. 1990. Hydrologic conditions leading to debris-flow initiation. *Canadian Geotechnical Journal*, 27, 789–801.

Keefer, D.K., Wilson, R.C., Mark, R.K., Brabb, E.E., Brown, W.M., Ellen, S.D., Harp, E.L., Wieczorek, G.F., Alger, C.S., and Zatkih, R.S. 1987. Real-time landslide warning during heavy rainfall. *Science*, 238, 921–925.

Kim, J.M., Salgado, R., and Yu, H.S. 1999. Limit analysis of soil slopes subjected to pore-water pressures. *Journal of Geotechnical and Geoenvironmental Engineering*, 125(1), 49–58.

Liao, J.T., Chen, C.W., Chi, C.C., and Lin, H.H. 2013. Study of slope displacement for management from landslide monitoring cases in Taiwan. *Sino-Geotechnics*, 136, 59–70. [in Chinese]

Mikkelsen, P.E. 1996. Chapter 11, Field instrumentation, in landslides, investigation and mitigation. In: Turner, A.K. and Schuster, R.L., eds. *TRB Special Report 247*, Transportation Research Board, National Research Council, Washington, DC, pp. 278–316.

Morgenstern, N.R. and Price, V.E. 1965. The analysis of the stability of general slip surfaces. *Geotechnique*, 15(1), 79–93.

Newmark, N.M. 1965. Effects of earthquakes on dams and embankments. *Geotechnique*, 15(2), 129–160.

Ng, C.W.W., Springman, S.M., and Alonso, E.E. 2008. Monitoring the performance of unsaturated soil slopes. *Geotechnical and Geological Engineering*, 26(6), 799–816.

Rahardjo, H., Ong, T.H., Rezaur, R.B., and Leong, E.C. 2007. Factors controlling instability of homogeneous soil slopes under rainfall. *Journal of Geotechnical and Geoenvironmental Engineering*, 133(12), 1532–1543.

Schuster, R.L., Salcedo, D.A., and Valenzuela, L. 2002. Overview of catastrophic landslides of South America in the twentieth century. In: Evans, S.G. and DeGraff, J.V., eds., *Catastrophic Landslides: Effects, Occurrence and Mechanisms, Reviews in Engineering Geology XV*, The Geological Society of America, Boulder, Colorado, USA, pp. 1–34.

Shi, G.H. and Goodman, R.E. 1985. Two dimensional discontinuous deformation analysis. *International Journal for Numerical and Analytical Methods in Geomechanics*, 9(6), 541–556.

Sitar, N., MacLaughlin, M.M., and Doolin, D.M. 2005. Influence of kinematics on landslide mobility and failure mode. *Journal of Geotechnical and Geoenvironmental Engineering*, 131 (6), 716–728.

Skempton, A.W. and Vaughan, P.R. 1993. Failure of Carsington dam. *Geotechnique*, 43(1), 151–173. Discussion in Vol. 45, No. 4, 719–739.

Sloan, S.W. 1988. Lower bound limit analysis using finite elements and linear programming. *Internal Journal of Numerical and Analytical Methods in Geomechanics*, 12, 61–77.

Spencer, E. 1967. A method of analysis of the stability of embankments assuming parallel interslice forces. *Geotechnique*, 17, 11–26.

Spencer, E. 1973. The thrust line criterion in embankment stability analysis. *Geotechnique*, 23, 85–101.

Taylor, D.W. 1937. Stability of earth slopes. *Journal of the Boston Society of Civil Engineering*, 24, 197.

Taylor, D.W. 1948. *Fundamentals of Soil Mechanics*. Wiley, New York, 700pp.

Varnes, D.J. 1978. Slope movement types and processes. Chapter 2, Landslides: Analysis and control. *Special Report 176*, TRB, National Academy of Sciences, Washington, DC, 234p.

Yu, H.S. and Sloan, S.W. 1991. Lower bound limit analysis of plane problems in soil mechanics. *Proceedings, Internal Conference of Non-linear Engineering Computations*, Swansea, Wales, pp. 329–338.

Yu, H.S., Salgado, R., Sloan, S.W., and Kim, J.M. 1998. Limit analysis versus limit equilibrium for slope stability. *Journal of Geotechnical and Geoenvironmental Engineering*, 124(1), 1–11.

Index